Introduction to Ordinary Differential Equations

Introduction to **Ordinary Differential Equations**

SECOND EDITION

Shepley L. Ross UNIVERSITY OF NEW HAMPSHIRE

XEROX

Xerox College Publishing

Lexington, Massachusetts Toronto

ISB Number: 0-536-00829-9
Library of Congress Catalog Card Number: 73-79298
Printed in the United States of America.

Preface

This second edition, like the first, is an introduction to the basic concepts, theory, methods, and applications of ordinary differential equations. The text is essentially the same as the first nine chapters of the second edition of the author's longer book, entitled *Differential Equations* (Xerox College Publishing, Lexington, Mass., 1974). A knowledge of elementary calculus is presupposed.

Designed for a one-semester introductory course in ordinary differential equations, the book covers and emphasizes the most fundamental methods of the subject and also contains traditional applications and brief introductions to fundamental theory. An examination of the table of contents will reveal just what topics are presented.

The detailed style of presentation which characterized the first edition has been retained. Indeed, many sections have been taken verbatim from the previous edition, while others have been rewritten or rearranged somewhat with the sole intention of making them clearer and smoother. As in the first edition, the text contains many thoroughly worked-out examples. Further, over one hundred new exercises have been added.

The new text material added for this edition deals with Bessel's equation and Bessel functions (Section 6.3) and with additional theory and methods of systems of linear differential equations. Chapter 7, "Systems of Linear Differential Equations," has been rearranged and greatly expanded; and new textual material on the solution of linear systems using the Laplace transform has been added (Section 9.4).

I believe that Chapter 7 now provides considerable flexibility for the study of linear systems. Several possible presentations of this chapter are listed here:

1. Sections 7.1 and 7.2 (methods and applications)
2. Sections 7.1A, 7.3, and 7.4 (theory and methods for the case $n = 2$, but no proofs)
3. Sections 7.1A, 7.3, 7.4, 7.5, 7.6, and 7.7 (same as presentation 2, plus the following: theory and methods for the general case, with almost all proofs)
4. Sections 7.1A, 7.5, 7.6, and 7.7 judiciously combined with 7.4 (an alternative approach to presentation 3)

It should be pointed out that Section 7.5 is a very elementary introduction to the surprisingly few concepts of vectors and matrices which are needed in Sections 7.6 and

7.7. It may therefore be omitted or very rapidly reviewed if the class has already studied elementary matrix algebra.

I am very pleased to record my great appreciation and sincere thanks to Professor Elmer Haskins of the State University of New York, Potsdam, N.Y., and Major Francis W. Farrell of the United States Military Academy, West Point, N.Y. Both Professor Haskins and Major Farrell carefully read the entire manuscript and made many valuable comments and suggestions which resulted in a variety of improvements —major and minor. I also express my appreciation to the mathematics department staff of West Point for their careful work in detecting and correcting errors in the first edition. Special thanks are also given to Professor Stanley M. Lukawecki of Clemson University, Clemson, S.C., who carefully read Chapter 7 and made a number of constructive suggestions which helped me considerably in reaching a final decision on the material and arrangement of this chapter.

Further thanks are given to Professor F. A. Ficken, New York University, New York, N.Y.; Professor Arnold Seiken, Union College, Schenectady, N.Y.; and my colleague, Professor William Bonnice, University of New Hampshire, Durham, N.H., for their worthwhile comments and suggestions.

I am very grateful to Solange Abbott for her excellent work in typing the revised portions of the manuscript. I am pleased to record my appreciation to Arthur Evans, Marret McCorkle, and others of the staff of Xerox College Publishing for their constant helpfulness and cooperation.

Special thanks go to my wife for her great encouragement, understanding and patience, as well as for her considerable assistance in the many different tasks required in the writing and revising of a text. Thank you, Gin!

SLR

Contents

4

Explicit Methods of Solving Higher-Order Linear Differential Equations 92

5

Applications of Second-Order Linear Differential Equations with Constant Coefficients 155

6

Series Solutions of Linear Differential Equations 199

7

Systems of Linear Differential Equations 243

Differential Equations and Their Solutions

The subject of differential equations constitutes a large and very important branch of modern mathematics. From the early days of the calculus the subject has been an area of great theoretical research and practical applications, and it continues to be so in our day. This much stated, several questions naturally arise. Just what is a differential equation and what does it signify? Where and how do differential equations originate and of what use are they? Confronted with a differential equation, what does one do with it, how does one do it, and what are the results of such activity? These questions indicate three major aspects of the subject: theory, method, and application. The purpose of this chapter is to introduce the reader to the basic aspects of the subject and at the same time give a brief survey of the three aspects just mentioned. In the course of the chapter, we shall find answers to the general questions raised above, answers which will become more and more meaningful as we proceed with the study of differential equations in the following chapters.

1.1 Classification of Differential Equations; Their Origin and Application

A. *Differential Equations and Their Classification*

DEFINITION

An equation involving derivatives of one or more dependent variables with respect to one or more independent variables is called a differential equation.*

▶ Example 1.1. For examples of differential equations we list the following:

$$\frac{d^2y}{dx^2} + xy\left(\frac{dy}{dx}\right)^2 = 0, \tag{1.1}$$

* In connection with this basic definition, we do *not* include in the class of differential equations those equations which are actually derivative identities. For example, we exclude such expressions as

$$\frac{d}{dx}(e^{ax}) = ae^{ax}, \qquad \frac{d}{dx}(uv) = u\frac{dv}{dx} + v\frac{du}{dx}, \qquad \text{and so forth.}$$

$$\frac{d^4x}{dt^4} + 5\frac{d^2x}{dt^2} + 3x = \sin t, \tag{1.2}$$

$$\frac{\partial v}{\partial s} + \frac{\partial v}{\partial t} = v, \tag{1.3}$$

$$\frac{\partial^2 u}{\partial x^2} + \frac{\partial^2 u}{\partial y^2} + \frac{\partial^2 u}{\partial z^2} = 0. \tag{1.4}$$

From the brief list of differential equations in Example 1.1 it is clear that the various variables and derivatives involved in a differential equation can occur in a variety of ways. Clearly some kind of classification must be made. To begin with, we classify differential equations according to whether there is one or more than one independent variable involved.

DEFINITION

A differential equation involving ordinary derivatives of one or more dependent variables with respect to a single independent variable is called an ordinary differential equation.

▶ Example 1.2. Equations (1.1) and (1.2) are ordinary differential equations. In Equation (1.1) the variable x is the single independent variable, and y is a dependent variable. In Equation (1.2) the independent variable is t, whereas x is dependent.

DEFINITION

A differential equation involving partial derivatives of one or more dependent variables with respect to more than one independent variable is called a partial differential equation.

▶ Example 1.3. Equations (1.3) and (1.4) are partial differential equations. In Equation (1.3) the variables s and t are independent variables and v is a dependent variable. In Equation (1.4) there are three independent variables, x, y, and z; in this equation u is dependent.

We further classify differential equations, both ordinary and partial, according to the order of the highest derivative appearing in the equation. For this purpose we give the following definition.

DEFINITION

The order of the highest ordered derivative involved in a differential equation is called the order *of the differential equation.*

▶ Example 1.4. The ordinary differential equation (1.1) is of the second order, since the highest derivative involved is a second derivative. Equation (1.2) is an ordinary differential equation of the fourth order. The partial differential equations (1.3) and (1.4) are of the first and second orders, respectively.

Proceeding with our study of ordinary differential equations, we now introduce the important concept of *linearity* applied to such equations. This concept will enable us to classify these equations still further.

DEFINITION

A linear ordinary differential equation *of order n, in the dependent variable y and the independent variable x, is an equation which is in, or can be expressed in, the form*

$$a_0(x)\frac{d^n y}{dx^n} + a_1(x)\frac{d^{n-1}y}{dx^{n-1}} + \cdots + a_{n-1}(x)\frac{dy}{dx} + a_n(x)y = b(x),$$

where a_0 is not identically zero.

Observe (1) that the dependent variable y and its various derivatives occur to the first degree only, (2) that no products of y and/or any of its derivatives are present, and (3) that no transcendental functions of y and/or its derivatives occur.

▶ Example 1.5. The following ordinary differential equations are both linear. In each case y is the dependent variable. Observe that y and its various derivatives occur to the first degree only and that no products of y and/or any of its derivatives are present.

$$\frac{d^2y}{dx^2} + 5\frac{dy}{dx} + 6y = 0, \tag{1.5}$$

$$\frac{d^4y}{dx^4} + x^2\frac{d^3y}{dx^3} + x^3\frac{dy}{dx} = xe^x. \tag{1.6}$$

DEFINITION

A nonlinear ordinary differential equation *is an ordinary differential equation which is not linear.*

▶ Example 1.6. The following ordinary differential equations are all nonlinear:

$$\frac{d^2y}{dx^2} + 5\frac{dy}{dx} + 6y^2 = 0, \tag{1.7}$$

$$\frac{d^2y}{dx^2} + 5\left(\frac{dy}{dx}\right)^3 + 6y = 0, \tag{1.8}$$

$$\frac{d^2y}{dx^2} + 5y\frac{dy}{dx} + 6y = 0. \tag{1.9}$$

Equation (1.7) is nonlinear because the dependent variable y appears to the second degree in the term $6y^2$. Equation (1.8) owes its nonlinearity to the presence of the term $5(dy/dx)^3$, which involves the third power of the first derivative. Finally, Equation (1.9) is nonlinear because of the term $5y(dy/dx)$, which involves the product of the dependent variable and its first derivative.

Linear ordinary differential equations are further classified according to the nature of the coefficients of the dependent variables and their derivatives. For example, Equation (1.5) is said to be linear with *constant coefficients*, while Equation (1.6) is linear with *variable coefficients*.

B. Origin and Application of Differential Equations

Having classified differential equations in various ways, let us now consider briefly where, and how, such equations actually originate. In this way we shall obtain some indication of the great variety of subjects to which the theory and methods of differential equations may be applied.

Differential equations occur in connection with numerous problems which are encountered in the various branches of science and engineering. We indicate a few such problems in the following list, which could easily be extended to fill many pages.

1. The problem of determining the motion of a projectile, rocket, satellite, or planet.
2. The problem of determining the charge or current in an electric circuit.
3. The problem of the conduction of heat in a rod or in a slab.
4. The problem of determining the vibrations of a wire or a membrane.
5. The study of the rate of decomposition of a radioactive substance or the rate of growth of a population.
6. The study of the reactions of chemicals.
7. The problem of the determination of curves which have certain geometrical properties.

The mathematical formulation of such problems gives rise to differential equations. But just how does this occur? In the situations under consideration in each of the above problems the objects involved obey certain scientific laws. These laws involve various rates of change of one or more quantities with respect to other quantities. Let us recall that such rates of change are expressed mathematically by derivatives. In the mathematical formulation of each of the above situations, the various rates of change are thus expressed by various derivatives and the scientific laws themselves become mathematical equations involving derivatives, that is, differential equations.

In this process of mathematical formulation, certain simplifying assumptions generally have to be made in order that the resulting differential equations be tractable. For example, if the actual situation in a certain aspect of the problem is of a relatively complicated nature, we are often forced to modify this by assuming instead an approximate situation which is of a comparatively simple nature. Indeed, certain relatively unimportant aspects of the problem must often be entirely eliminated. The result of such changes from the actual nature of things means that the resulting differential equation is actually that of an idealized situation. Nonetheless, the information obtained from such an equation is of the greatest value to the scientist.

A natural question now is the following: How does one obtain useful information from a differential equation? The answer is essentially that if it is possible to do so, one solves the differential equation to obtain a solution; if this is not possible, one uses the theory of differential equations to obtain information *about* the solution. To understand the meaning of this answer, we must discuss what is meant by a solution of a differential equation; this is done in the next section.

Exercises

Classify each of the following differential equations as ordinary or partial differential equations; state the order of each equation; and determine whether the equation under consideration is linear or nonlinear.

1. $\dfrac{dy}{dx} + x^2 y = xe^x.$

2. $\dfrac{d^3y}{dx^3} + 4\dfrac{d^2y}{dx^2} - 5\dfrac{dy}{dx} + 3y = \sin x.$

3. $\dfrac{\partial^2 u}{\partial x^2} + \dfrac{\partial^2 u}{\partial y^2} = 0.$

4. $x^2\,dy + y^2\,dx = 0.$

5. $\dfrac{d^4y}{dx^4} + 3\left(\dfrac{d^2y}{dx^2}\right)^5 + 5y = 0.$

6. $\dfrac{\partial^4 u}{\partial x^2\,\partial y^2} + \dfrac{\partial^2 u}{\partial x^2} + \dfrac{\partial^2 u}{\partial y^2} + u = 0.$

7. $\dfrac{d^2y}{dx^2} + y \sin x = 0.$

8. $\dfrac{d^2y}{dx^2} + x \sin y = 0.$

9. $\dfrac{d^6x}{dt^6} + \left(\dfrac{d^4x}{dt^4}\right)\left(\dfrac{d^3x}{dt^3}\right) + x = t.$

10. $\left(\dfrac{dr}{ds}\right)^3 = \sqrt{\dfrac{d^2r}{ds^2} + 1}.$

1.2 Solutions

A. Nature of Solutions

We now consider the concept of a solution of the nth-order ordinary differential equation.

DEFINITION

Consider the nth-order ordinary differential equation

$$F\left[x, y, \frac{dy}{dx}, \ldots, \frac{d^n y}{dx^n}\right] = 0, \tag{1.10}$$

where F is a real function of its $(n + 2)$ *arguments* $x, y, \dfrac{dy}{dx}, \ldots, \dfrac{d^n y}{dx^n}$.

1. Let f be a real function defined for all x in a real interval I and having an nth derivative (and hence also all lower ordered derivatives) for all $x \in I$. *The function f is called an* explicit solution *of the differential equation (1.10) on I if it fulfills the following two requirements:*

$$F[x, f(x), f'(x), \ldots, f^{(n)}(x)] \tag{A}$$

is defined for all $x \in I$, and

$$F[x, f(x), f'(x), \ldots, f^{(n)}(x)] = 0 \tag{B}$$

for all $x \in I$. That is, the substitution of $f(x)$ and its various derivatives for y and its corresponding derivatives, respectively, in (1.10) reduces (1.10) to an identity on I.

2. *A relation $g(x, y) = 0$ is called an* implicit solution *of (1.10) if this relation defines at least one real function f of the variable x on an interval I such that this function is an explicit solution of (1.10) on this interval.*

3. *Both explicit solutions and implicit solutions will usually be called simply* solutions.

Roughly speaking, then, we may say that a solution of the differential equation (1.10) is a relation, explicit or implicit, between x and y, not containing derivatives, which identically satisfies (1.10).

▶ Example 1.7. The function f defined for all real x by

$$f(x) = 2 \sin x + 3 \cos x \tag{1.11}$$

is an explicit solution of the differential equation

$$\frac{d^2y}{dx^2} + y = 0 \tag{1.12}$$

for all real x. First note that f is defined and has a second derivative for all real x. Next observe that

$$f'(x) = 2 \cos x - 3 \sin x,$$

$$f''(x) = -2 \sin x - 3 \cos x.$$

Upon substituting $f''(x)$ for $\frac{d^2y}{dx^2}$ and $f(x)$ for y in the differential equation (1.12), it reduces to the identity

$$(-2 \sin x - 3 \cos x) + (2 \sin x + 3 \cos x) = 0,$$

which holds for all real x. Thus the function f defined by (1.11) is an explicit solution of the differential equation (1.12) for all real x.

▶ Example 1.8. The relation

$$x^2 + y^2 - 25 = 0 \tag{1.13}$$

is an implicit solution of the differential equation

$$x + y\frac{dy}{dx} = 0 \tag{1.14}$$

on the interval I defined by $-5 < x < 5$. For the relation (1.13) defines the two real functions f_1 and f_2 given by

$$f_1(x) = \sqrt{25 - x^2}$$

and

$$f_2(x) = -\sqrt{25 - x^2},$$

respectively, for all real $x \in I$, and both of these functions are explicit solutions of the differential equations (1.14) on I.

Let us illustrate this for the function f_1. Since

$$f_1(x) = \sqrt{25 - x^2},$$

we see that

$$f_1'(x) = \frac{-x}{\sqrt{25 - x^2}}$$

for all real $x \in I$. Substituting $f_1(x)$ for y and $f_1'(x)$ for dy/dx in (1.14), we obtain the identity

$$x + (\sqrt{25 - x^2})\left(\frac{-x}{\sqrt{25 - x^2}}\right) = 0 \quad \text{or} \quad x - x = 0,$$

which holds for all real $x \in I$. Thus the function f_1 is an explicit solution of (1.14) on the interval I.

Now consider the relation

$$x^2 + y^2 + 25 = 0. \tag{1.15}$$

Is this also an implicit solution of Equation (1.14)? Let us differentiate the relation (1.15) implicitly with respect to x. We obtain

$$2x + 2y\frac{dy}{dx} = 0 \quad \text{or} \quad \frac{dy}{dx} = -\frac{x}{y}.$$

Substituting this into the differential equation (1.14), we obtain the *formal* identity

$$x + y\left(-\frac{x}{y}\right) = 0.$$

Thus the relation (1.15) *formally* satisfies the differential equation (1.14). Can we conclude from this alone that (1.15) is an implicit solution of (1.14)? The answer to this question is "no," for we have no assurance from this that the relation (1.15) defines any function which is an explicit solution of (1.14) on any real interval I. All that we have shown is that (1.15) is a relation between x and y which, upon implicit differentiation and substitution, *formally* reduces the differential equation (1.14) to a *formal* identity. It is called a *formal* solution; it has the *appearance* of a solution; but that is all that we know about it at this stage of our investigation.

Let us investigate a little further. Solving (1.15) for y, we find that

$$y = \pm\sqrt{-25 - x^2}.$$

Since this expression yields nonreal values of y for all real values of x, we conclude that the relation (1.15) does not define any real function on any interval. Thus the relation (1.15) is not truly an implicit solution but merely a *formal solution* of the differential equation (1.14).

In applying the methods of the following chapters we shall often obtain relations which we can readily verify are at least formal solutions. Our main objective will be to gain familiarity with the methods themselves and we shall often be content to refer to the relations so obtained as "solutions," although we have no assurance that these relations are actually true implicit solutions. If a critical examination of the situation is required, one must undertake to determine whether or not these formal solutions so obtained are actually true implicit solutions which define explicit solutions.

In order to gain further insight into the significance of differential equations and their solutions, we now examine the simple equation of the following example.

▶ Example 1.9. Consider the first-order differential equation

$$\frac{dy}{dx} = 2x. \tag{1.16}$$

The function f_0 defined for all real x by $f_0(x) = x^2$ is a solution of this equation. So also are the functions f_1, f_2, and f_3 defined for all real x by $f_1(x) = x^2 + 1, f_2(x) = x^2 + 2$, and $f_3(x) = x^2 + 3$, respectively. In fact, for each real number c, the function f_c defined for all real x by

$$f_c(x) = x^2 + c \tag{1.17}$$

is a solution of the differential equation (1.16). In other words, the formula (1.17) defines an infinite family of functions, one for each real constant c, and every function of this family is a solution of (1.16). We call the constant c in (1.17) an *arbitrary constant* or *parameter* and refer to the family of functions defined by (1.17) as a *one-parameter family of solutions* of the differential equation (1.16). We write this one-parameter family of solutions as

$$y = x^2 + c. \tag{1.18}$$

Although it is clear that every function of the family defined by (1.18) is a solution of (1.16), we have not shown that the family of functions defined by (1.18) includes *all* of the solutions of (1.16). However, we point out (without proof) that this is indeed the case here; that is, every solution of (1.16) is actually of the form (1.18) for some appropriate real number c.

Note. We must not conclude from the last sentence of Example 1.9 that *every* first-order ordinary differential equation has a so-called one-parameter family of solutions which contains *all* solutions of the differential equation, for this is by no means the case. Indeed, some first-order differential equations have no solution at all (see Exercise 7(a) at the end of this section), while others have a one-parameter family of solutions plus one or more "extra" solutions which appear to be "different" from all those of the family (see Exercise 7(b) at the end of this section).

The differential equation of Example 1.9 enables us to obtain a better understanding of the analytic significance of differential equations. Briefly stated, the differential equation of that example *defines functions*, namely, its solutions. We shall see that this is the case with many other differential equations of both first and higher order. Thus we may say that a differential equation is merely an expression involving derivatives which may serve as a means of defining a certain set of functions: its solutions. Indeed, many of the now familiar functions originally appeared in the form of differential equations which define them.

We now consider the geometric significance of differential equations and their solutions. We first recall that a real function F may be represented geometrically by a curve $y = F(x)$ in the xy plane and that the value of the derivative of F at x, $F'(x)$, may be interpreted as the slope of the curve $y = F(x)$ at x. Thus the general first-order differential equation

$$\frac{dy}{dx} = f(x, y), \tag{1.19}$$

where f is a real function, may be interpreted geometrically as defining a slope $f(x, y)$ at every point (x, y) at which the function f is defined. Now assume that the differential equation (1.19) has a so-called one-parameter family of solutions which can be written in the form

$$y = F(x, c), \tag{1.20}$$

where c is the arbitrary constant or parameter of the family. The one-parameter family of functions defined by (1.20) is represented geometrically by a so-called *one-parameter family of curves* in the xy plane, the slopes of which are given by the differential equation (1.19). These curves, the graphs of the solutions of the differential equation (1.19), are called the *integral curves* of the differential equation (1.19).

▶ **Example 1.10.** Consider again the first-order differential equation

$$\frac{dy}{dx} = 2x \tag{1.16}$$

of Example 1.9. This differential equation may be interpreted as defining the slope $2x$ at the point with coordinates (x, y) for every real x. Now, we observed in Example 1.9 that the differential equation (1.16) has a one-parameter family of solutions of the form

$$y = x^2 + c, \tag{1.18}$$

where c is the arbitrary constant or parameter of the family. The one-parameter family of functions defined by (1.18) is represented geometrically by a one-parameter family of curves in the xy plane, namely, the family of *parabolas* with Equation (1.18). The slope of each of these parabolas is given by the differential equation (1.16) of the family. Thus we see that the family of parabolas (1.18) defined by differential equation (1.16) is that family of parabolas, each of which has slope $2x$ at the point (x, y) for every real x, and all of which have the y-axis as axis. These parabolas are the integral curves of the differential equation (1.16). See Figure 1.1.

B. *Methods of Solution*

When we say that we shall solve a differential equation we mean that we shall find one or more of its solutions. How is this done and what does it really mean? The greater

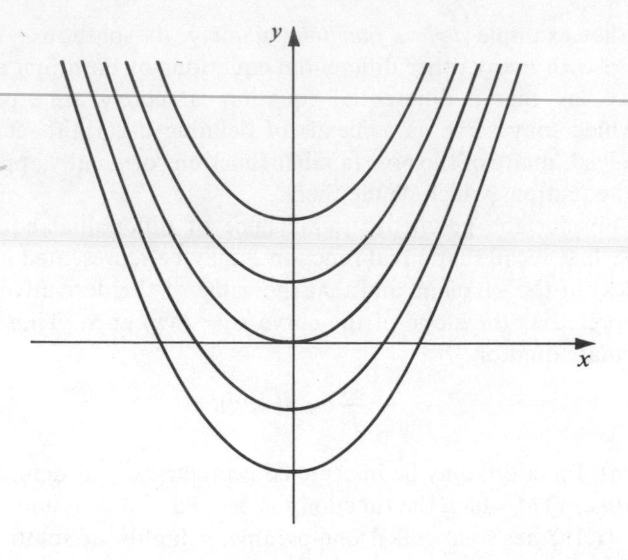

FIGURE 1.1

part of this text is concerned with various methods of solving differential equations. The method to be employed depends upon the type of differential equation under consideration, and we shall not enter into the details of specific methods here.

But suppose we solve a differential equation, using one or another of the various methods. Does this necessarily mean that we have found an explicit solution f expressed in the so-called closed form of a finite sum of known elementary functions? That is, roughly speaking, when we have solved a differential equation, does this necessarily mean that we have found a "formula" for the solution? The answer is "no." Comparatively few differential equations have solutions so expressible; in fact, a closed-form solution is really a luxury in differential equations. In Chapters 2 and 4 we shall consider certain types of differential equations which do have such closed-form solutions and study the exact methods available for finding these desirable solutions. But, as we have just noted, such equations are actually in the minority and we must consider what it means to "solve" equations for which exact methods are unavailable. Such equations are solved approximately by various methods, some of which are considered in Chapters 6 and 8. Among such methods are series methods, numerical methods, and graphical methods. What do such approximate methods actually yield? The answer to this depends upon the method under consideration.

Series methods yield solutions in the form of infinite series; numerical methods give approximate values of the solution functions corresponding to selected values of the independent variables; and graphical methods produce approximately the graphs of solutions (the integral curves). These methods are not so desirable as exact methods because of the amount of work involved in them and because the results obtained from them are only approximate; but if exact methods are not applicable, one has no choice but to turn to approximate methods. Modern science and engineering problems continue to give rise to differential equations to which exact methods do not apply, and approximate methods are becoming increasingly more important.

Exercises

1. Show that each of the functions defined in Column I is a solution of the corresponding differential equation in Column II on every interval $a < x < b$ of the x axis.

	I	II
(a)	$f(x) = x + 3e^{-x}$	$\dfrac{dy}{dx} + y = x + 1$
(b)	$f(x) = 2e^{3x} - 5e^{4x}$	$\dfrac{d^2y}{dx^2} - 7\dfrac{dy}{dx} + 12y = 0$
(c)	$f(x) = e^x + 2x^2 + 6x + 7$	$\dfrac{d^2y}{dx^2} - 3\dfrac{dy}{dx} + 2y = 4x^2$
(d)	$f(x) = \dfrac{1}{1 + x^2}$	$(1 + x^2)\dfrac{d^2y}{dx^2} + 4x\dfrac{dy}{dx} + 2y = 0$

2. (a) Show that $x^3 + 3xy^2 = 1$ is an implicit solution of the differential equation $2xy\dfrac{dy}{dx} + x^2 + y^2 = 0$ on the interval $0 < x < 1$.

   (b) Show that $5x^2y^2 - 2x^3y^2 = 1$ is an implicit solution of the differential equation $x\dfrac{dy}{dx} + y = x^3y^3$ on the interval $0 < x < \frac{5}{2}$.

3. (a) Show that every function f defined by

$$f(x) = (x^3 + c)e^{-3x},$$

   where c is an arbitrary constant, is a solution of the differential equation

$$\frac{dy}{dx} + 3y = 3x^2e^{-3x}.$$

   (b) Show that every function f defined by

$$f(x) = 2 + ce^{-2x^2},$$

   where c is an arbitrary constant, is a solution of the differential equation

$$\frac{dy}{dx} + 4xy = 8x.$$

4. (a) Show that every function f defined by $f(x) = c_1e^{4x} + c_2e^{-2x}$, where c_1 and c_2 are arbitrary constants, is a solution of the differential equation

$$\frac{d^2y}{dx^2} - 2\frac{dy}{dx} - 8y = 0.$$

   (b) Show that every function g defined by $g(x) = c_1e^{2x} + c_2xe^{2x} + c_3e^{-2x}$, where c_1, c_2, and c_3 are arbitrary constants, is a solution of the differential equation

$$\frac{d^3y}{dx^3} - 2\frac{d^2y}{dx^2} - 4\frac{dy}{dx} + 8y = 0.$$

5. (a) For certain values of the constant m the function f defined by $f(x) = e^{mx}$ is a solution of the differential equation

$$\frac{d^3y}{dx^3} - 3\frac{d^2y}{dx^2} - 4\frac{dy}{dx} + 12y = 0.$$

Determine all such values of m.

(b) For certain values of the constant n the function g defined by $g(x) = x^n$ is a solution of the differential equation

$$x^3\frac{d^3y}{dx^3} + 2x^2\frac{d^2y}{dx^2} - 10x\frac{dy}{dx} - 8y = 0.$$

Determine all such values of n.

6. (a) Show that the function f defined by $f(x) = (2x^2 + 2e^{3x} + 3)e^{-2x}$ satisfies the differential equation

$$\frac{dy}{dx} + 2y = 6e^x + 4xe^{-2x}$$

and also the condition $f(0) = 5$.

(b) Show that the function f defined by $f(x) = 3e^{2x} - 2xe^{2x} - \cos 2x$ satisfies the differential equation

$$\frac{d^2y}{dx^2} - 4\frac{dy}{dx} + 4y = -8\sin 2x$$

and also the conditions that $f(0) = 2$ and $f'(0) = 4$.

7. (a) Show that the first-order differential equation

$$\left|\frac{dy}{dx}\right| + |y| + 1 = 0$$

has *no* (real) solutions.

(b) Show that the first-order differential equation

$$\left(\frac{dy}{dx}\right)^2 - 4y = 0$$

has a one-parameter family of solutions of the form $f(x) = (x + c)^2$, where c is an arbitrary constant, plus the "extra" solution $g(x) = 0$ which is *not* a member of this family $f(x) = (x + c)^2$ for *any* choice of the constant c.

1.3 Initial-Value Problems, Boundary-Value Problems, and Existence of Solutions

A. *Initial-Value Problems and Boundary-Value Problems*

We shall begin this section by considering the rather simple problem of the following example.

▶ Example 1.11

Problem. Find a solution f of the differential equation

$$\frac{dy}{dx} = 2x \tag{1.21}$$

such that at $x = 1$ this solution f has the value 4.

Explanation. First let us be certain that we thoroughly understand this problem. We seek a real function f which fulfills the two following requirements:

1. The function f must satisfy the differential equation (1.21). That is, the function f must be such that $f'(x) = 2x$ for all real x in a real interval I.
2. The function f must have the value 4 at $x = 1$. That is, the function f must be such that $f(1) = 4$.

Notation. The stated problem may be expressed in the following somewhat abbreviated notation:

$$\frac{dy}{dx} = 2x,$$

$$y(1) = 4.$$

In this notation we may regard y as representing the desired solution. Then the differential equation itself obviously represents requirement 1, and the statement $y(1) = 4$ stands for requirement 2. More specifically, the notation $y(1) = 4$ states that the desired solution y must have the value 4 at $x = 1$; that is, $y = 4$ at $x = 1$.

Solution. We observed in Example 1.9 that the differential equation (1.21) has a one-parameter family of solutions which we write as

$$y = x^2 + c, \tag{1.22}$$

where c is an arbitrary constant, and that each of these solutions satisfies requirement 1. Let us now attempt to determine the constant c so that (1.22) satisfies requirement 2, that is, $y = 4$ at $x = 1$. Substituting $x = 1$, $y = 4$ into (1.22), we obtain $4 = 1 + c$, and hence $c = 3$. Now substituting the value $c = 3$ thus determined back into (1.22), we obtain

$$y = x^2 + 3,$$

which is indeed a solution of the differential equation (1.21), which has the value 4 at $x = 1$. In other words, the function f defined by

$$f(x) = x^2 + 3,$$

satisfies both of the requirements set forth in the problem.

Comment on Requirement 2 and Its Notation. In a problem of this type, requirement 2 is regarded as a *supplementary condition* which the solution of the differential equation must also satisfy. The abbreviated notation $y(1) = 4$ which we used to express this condition is in some way undesirable, but it has the advantages of being both customary and convenient.

In the application of both first- and higher-order differential equations the problems most frequently encountered are similar to the above introductory problem in that they involve *both* a differential equation *and* one or more supplementary conditions which the solution of the given differential equation must satisfy. If all of the associated supplementary conditions relate to *one* x value, the problem is called an *initial-value problem* (or one-point boundary-value problem). If the conditions relate to *two* different x values, the problem is called a *two-point boundary-value problem* (or simply a boundary-value problem). We shall illustrate these concepts with examples and then consider one such type of problem in detail. Concerning notation, we generally employ abbreviated notations for the supplementary conditions which are similar to the abbreviated notation introduced in Example 1.11.

▶ Example 1.12

$$\frac{d^2y}{dx^2} + y = 0,$$
$$y(1) = 3,$$
$$y'(1) = -4.$$

This problem consists in finding a solution of the differential equation

$$\frac{d^2y}{dx^2} + y = 0,$$

which assumes the value 3 at $x = 1$ and whose first derivative assumes the value -4 at $x = 1$. Both of these conditions relate to one x value, namely, $x = 1$. Thus this is an initial-value problem. We shall see later that this problem has a unique solution.

▶ Example 1.13

$$\frac{d^2y}{dx^2} + y = 0,$$
$$y(0) = 1,$$
$$y\left(\frac{\pi}{2}\right) = 5.$$

In this problem we again seek a solution of the same differential equation, but this time the solution must assume the value 1 at $x = 0$ and the value 5 at $x = \pi/2$. That is, the conditions relate to the *two* different x values, 0 and $\pi/2$. This is a (two-point) boundary-value problem. This problem also has a unique solution; but the boundary-value problem

$$\frac{d^2y}{dx^2} + y = 0,$$

$$y(0) = 1,$$

$$y(\pi) = 5,$$

has no solution at all! This simple fact may lead one to the correct conclusion that boundary-value problems are not to be taken lightly!

We now turn to a more detailed consideration of the initial-value problem for a first-order differential equation.

DEFINITION

Consider the first-order differential equation

$$\frac{dy}{dx} = f(x, y), \tag{1.23}$$

where f is a continuous function of x and y in some domain D of the xy plane; and let (x_0, y_0) be a point of D. The* initial-value problem *associated with (1.23) is to find a solution ϕ of the differential equation (1.23), defined on some real interval containing x_0, and satisfying the* initial condition

$$\phi(x_0) = y_0.$$

In the customary abbreviated notation, this initial-value problem may be written

$$\frac{dy}{dx} = f(x, y),$$

$$y(x_0) = y_0.$$

To solve this problem, we must find a function ϕ which not only satisfies the differential equation (1.23) but which also satisfies the initial condition that it has the value y_0 when x has the value x_0. The geometric interpretation of the initial condition, and hence of the entire initial-value problem, is easily understood. The graph of the desired solution ϕ must pass through the point with coordinates (x_0, y_0). That is, interpreted geometrically, the initial-value problem is to find an integral curve of the differential equation (1.23) which passes through the point (x_0, y_0).

The method of actually finding the desired solution ϕ depends upon the nature of the differential equation of the problem, that is, upon the form of $f(x, y)$. Certain special types of differential equations have a one-parameter family of solutions whose equation may be found exactly by following definite procedures (see Chapter 2). If

---

* A *domain* is an open, connected set. For those unfamiliar with such concepts, D may be regarded as the interior of some simple closed curve in the plane.

the differential equation of the problem is of some such special type, one first obtains the equation of its one-parameter family of solutions and then applies the initial condition to this equation in an attempt to obtain a "particular" solution ϕ which satisfies the entire initial-value problem. We shall explain this situation more precisely in the next paragraph. Before doing so, however, we point out that in general one cannot find the equation of a one-parameter family of solutions of the differential equation; approximate methods must then be used (see Chapter 8).

Now suppose one can determine the equation

$$g(x, y, c) = 0 \tag{1.24}$$

of a one-parameter family of solutions of the differential equation of the problem. Then, since the initial condition requires that $y = y_0$ at $x = x_0$, we let $x = x_0$ and $y = y_0$ in (1.24) and thereby obtain

$$g(x_0, y_0, c) = 0.$$

Solving this for c, in general we obtain a particular value of c which we denote here by c_0. We now replace the arbitrary constant c by the particular constant c_0 in (1.24), thus obtaining the particular solution

$$g(x, y, c_0) = 0.$$

The particular explicit solution satisfying the two conditions (differential equation and initial condition) of the problem is then determined from this, if possible.

We have already solved one initial-value problem in Example 1.11. We now give another example in order to illustrate the concepts and procedures more thoroughly.

▶ **Example 1.14.** Solve the initial-value problem

$$\frac{dy}{dx} = -\frac{x}{y}, \tag{1.25}$$

$$y(3) = 4, \tag{1.26}$$

given that the differential equation (1.25) has a one-parameter family of solutions which may be written in the form

$$x^2 + y^2 = c^2. \tag{1.27}$$

The condition (1.26) means that we seek the solution of (1.25) such that $y = 4$ at $x = 3$. Thus the pair of values (3, 4) must satisfy the relation (1.27). Substituting $x = 3$ and $y = 4$ into (1.27), we find

$$9 + 16 = c^2 \quad \text{or} \quad c^2 = 25.$$

Now substituting this value of c^2 into (1.27), we have

$$x^2 + y^2 = 25.$$

Solving this for y, we obtain

$$y = \pm\sqrt{25 - x^2}.$$

Obviously the positive sign must be chosen to give y the value $+4$ at $x = 3$. Thus the function f defined by

$$f(x) = \sqrt{25 - x^2}, \qquad -5 < x < 5,$$

is the solution of the problem. In the usual abbreviated notation, we write this solution as $y = \sqrt{25 - x^2}$.

B. Existence of Solutions

In Example 1.14 we were able to find a solution of the initial-value problem under consideration. But do all initial-value and boundary-value problems have solutions? We have already answered this question in the negative, for we have pointed out that the boundary-value problem

$$\frac{d^2y}{dx^2} + y = 0,$$

$$y(0) = 1,$$

$$y(\pi) = 5,$$

mentioned at the end of Example 1.13, has no solution! Thus arises the question of *existence* of solutions: given an initial-value or boundary-value problem, does it actually have a solution? Let us consider the question for the initial-value problem defined on page 15. Here we can give a definite answer. Every initial-value problem that satisfies the definition on page 15 has *at least one* solution.

But now another question is suggested, the question of *uniqueness*. Does such a problem ever have *more than one* solution? Let us consider the initial-value problem

$$\frac{dy}{dx} = y^{1/3},$$

$$y(0) = 0.$$

One may verify that the functions f_1 and f_2 defined, respectively, by

$$f_1(x) = 0 \quad \text{for all real } x;$$

and

$$f_2(x) = (\tfrac{2}{3}x)^{3/2}, \quad x \geq 0; \qquad f_2(x) = 0, \quad x \leq 0;$$

are *both* solutions of this initial-value problem! In fact, this problem has infinitely many solutions! The answer to the uniqueness question is clear: the initial-value problem, as stated, need not have a *unique* solution. In order to ensure uniqueness, some additional requirement must certainly be imposed. We shall see what this is in Theorem 1.1, which we shall now state.

THEOREM 1.1. Basic Existence and Uniqueness Theorem

Hypothesis. *Consider the differential equation*

$$\frac{dy}{dx} = f(x, y), \tag{1.28}$$

where

1. The function f is a continuous function of x and y in some domain D of the xy plane, and

2. The partial derivative $\frac{\partial f}{\partial y}$ is also a continuous function of x and y in D;

and let (x_0, y_0) be a point in D.

Conclusion. *There exists a unique solution ϕ of the differential equation (1.28), defined on some interval $|x - x_0| \leq h$, where h is sufficiently small, which satisfies the condition*

$$\phi(x_0) = y_0. \tag{1.29}$$

Explanatory Remarks. This basic theorem is the first theorem from the theory of differential equations which we have encountered. We shall therefore attempt to explain its meaning in detail.

1. It is an *existence and uniqueness theorem.* This means that it is a theorem which tells us that under certain conditions (stated in the hypothesis) something *exists* (the solution described in the conclusion) and is *unique* (there is *only one* such solution). It gives no hint whatsoever concerning *how* to find this solution but merely tells us that the problem *has* a solution.

2. The *hypothesis* tells us what conditions are required of the quantities involved. It deals with two objects: the differential equation (1.28) and the point (x_0, y_0). As far as the differential equation (1.28) is concerned, the hypothesis requires that *both* the function f *and* the function $\frac{\partial f}{\partial y}$ (obtained by differentiating $f(x, y)$ partially with respect to y) must be continuous in some domain D of the xy plane. As far as the point (x_0, y_0) is concerned, it must be a point in this same domain D, where f and $\frac{\partial f}{\partial y}$ are so well behaved (that is, continuous).

3. The *conclusion* tells us of what we can be assured when the stated hypothesis is satisfied. It tells us that we are assured that there exists one and only one solution ϕ of the differential equation, which is defined on some interval $|x - x_0| \leq h$ centered about x_0 and which assumes the value y_0 when x takes on the value x_0. That is, it tells us that, under the given hypothesis on $f(x, y)$, the *initial-value problem*

$$\frac{dy}{dx} = f(x, y),$$

$$y(x_0) = y_0,$$

has a *unique solution* which is valid in some interval about the initial point x_0.

4. The *proof* of this theorem is omitted. It is proved under somewhat less restrictive hypotheses in Chapter 10 of the author's *Differential Equations.*

5. The *value* of an existence theorem may be worth a bit of attention. What good is it, one might ask, if it does not tell us how to obtain the solution? The answer to this question is quite simple: an existence theorem will assure us that there *is* a solution to look for! It would be rather pointless to spend time, energy, and even money in trying to find a solution when there was actually no solution to be found! As for the value of the uniqueness, it would be equally pointless to waste time and energy finding one particular solution only to learn later that there were others and that the one found was not the one wanted!

We have included this rather lengthy discussion in the hope that the student, who has probably never before encountered a theorem of this type, will obtain a clearer idea of what this important theorem really means. We further hope that this discussion will help him to analyze theorems which he will encounter in the future, both in this book and elsewhere. We now consider two simple examples which illustrate Theorem 1.1.

▶ **Example 1.15.** Consider the initial-value problem

$$\frac{dy}{dx} = x^2 + y^2,$$

$$y(1) = 3.$$

Let us apply Theorem 1.1. We first check the hypothesis. Here $f(x, y) = x^2 + y^2$ and $\dfrac{\partial f(x, y)}{\partial y} = 2y$. Both of the functions f and $\dfrac{\partial f}{\partial y}$ are continuous in every domain D of the xy plane. The initial condition $y(1) = 3$ means that $x_0 = 1$ and $y_0 = 3$, and the point (1, 3) certainly lies in some such domain D. Thus all hypotheses are satisfied and the conclusion holds. That is, there is a unique solution ϕ of the differential equation $dy/dx = x^2 + y^2$, defined on some interval $|x - 1| \leq h$ about $x_0 = 1$, which satisfies the initial condition, that is, which is such that $\phi(1) = 3$.

▶ **Example 1.16.** Consider the two problems:

1. $$\frac{dy}{dx} = \frac{y}{\sqrt{x}},$$

$$y(1) = 2,$$

2. $$\frac{dy}{dx} = \frac{y}{\sqrt{x}},$$

$$y(0) = 2.$$

Here

$$f(x, y) = \frac{y}{x^{1/2}} \quad \text{and} \quad \frac{\partial f(x, y)}{\partial y} = \frac{1}{x^{1/2}}.$$

These functions are both continuous *except* for $x = 0$ (that is, along the y axis). In problem 1, $x_0 = 1$, $y_0 = 2$. The square of side 1 centered about (1, 2) does *not* contain the y axis, and so both f and $\frac{\partial f}{\partial y}$ satisfy the required hypotheses in this square. Its interior may thus be taken to be the domain D of Theorem 1.1; and (1, 2) certainly lies within it. Thus the conclusion of Theorem 1.1 applies to problem 1 and we know the problem has a unique solution defined in some sufficiently small interval about $x_0 = 1$.

Now let us turn to problem 2. Here $x_0 = 0$, $y_0 = 2$. At this point neither f nor $\frac{\partial f}{\partial y}$ are continuous. In other words, the point (0, 2) cannot be included in a domain D where the required hypotheses are satisfied. Thus we can *not* conclude from Theorem 1.1 that problem 2 has a solution. We are *not* saying that it does *not* have one. Theorem 1.1 simply gives no information one way or the other.

Exercises

1. Show that

$$y = 4e^{2x} + 2e^{-3x}$$

is a solution of the initial-value problem

$$\frac{d^2y}{dx^2} + \frac{dy}{dx} - 6y = 0,$$

$$y(0) = 6,$$

$$y'(0) = 2.$$

Is $y = 2e^{2x} + 4e^{-3x}$ also a solution of this problem? Explain why or why not.

2. Given that every solution of

$$\frac{dy}{dx} + y = 2xe^{-x}$$

may be written in the form $y = (x^2 + c)e^{-x}$, for some choice of the arbitrary constant c, solve the following initial-value problems:

(a) $\frac{dy}{dx} + y = 2xe^{-x}$,

$y(0) = 2.$

(b) $\frac{dy}{dx} + y = 2xe^{-x}$,

$y(-1) = e + 3.$

3. Given that every solution of

$$\frac{d^2y}{dx^2} - \frac{dy}{dx} - 12y = 0$$

may be written in the form

$$y = c_1e^{4x} + c_2e^{-3x},$$

for some choice of the arbitrary constants c_1 and c_2, solve the following initial-value problems:

(a) $\frac{d^2y}{dx^2} - \frac{dy}{dx} - 12y = 0,$

$y(0) = 5,$

$y'(0) = 6.$

(b) $\frac{d^2y}{dx^2} - \frac{dy}{dx} - 12y = 0,$

$y(0) = -2,$

$y'(0) = 6.$

4. Every solution of the differential equation

$$\frac{d^2y}{dx^2} + y = 0$$

may be written in the form $y = c_1 \sin x + c_2 \cos x$, for some choice of the arbitrary constants c_1 and c_2. Using this information, show that boundary problems (a) and (b) possess solutions but that (c) does not.

(a) $\frac{d^2y}{dx^2} + y = 0,$

$y(0) = 0,$

$y(\pi/2) = 1.$

(b) $\frac{d^2y}{dx^2} + y = 0,$

$y(0) = 1,$

$y'(\pi/2) = -1.$

(c) $\frac{d^2y}{dx^2} + y = 0,$

$y(0) = 0,$

$y(\pi) = 1.$

5. Given that every solution of

$$x^3 \frac{d^3y}{dx^3} - 3x^2 \frac{d^2y}{dx^2} + 6x \frac{dy}{dx} - 6y = 0$$

may be written in the form $y = c_1x + c_2x^2 + c_3x^3$ for some choice of the arbitrary constants c_1, c_2, and c_3, solve the initial-value problem consisting of the above differential equation plus the three conditions

$$y(2) = 0, \quad y'(2) = 2, \quad y''(2) = 6.$$

6. Apply Theorem 1.1 to show that each of the following initial-value problems has a unique solution defined on some sufficiently small interval $|x - 1| \leq h$ about $x_0 = 1$:

(a) $\frac{dy}{dx} = x^2 \sin y,$

$y(1) = -2.$

(b) $\frac{dy}{dx} = \frac{y^2}{x - 2},$

$y(1) = 0.$

7. Consider the initial-value problem

$$\frac{dy}{dx} = P(x)y^2 + Q(x)y,$$

$$y(2) = 5,$$

where $P(x)$ and $Q(x)$ are both third-degree polynomials in x. Has this problem a unique solution on some interval $|x - 2| \leq h$ about $x_0 = 2$? Explain why or why not.

8. On page 17 we stated that the initial-value problem

$$\frac{dy}{dx} = y^{1/3},$$

$$y(0) = 0,$$

has infinitely many solutions.

(a) Verify that this is indeed the case by showing that

$$y = \begin{cases} 0, & x \leq c, \\ [\frac{2}{3}(x - c)]^{3/2}, & x \geq c, \end{cases}$$

is a solution of the stated problem for every real number $c \geq 0$.

(b) Carefully graph the solution for which $c = 0$. Then, using this particular graph, also graph the solutions for which $c = 1$, $c = 2$, and $c = 3$.

Suggested Reading

AGNEW, R. P., *Differential Equations*, 2nd ed. (McGraw-Hill, New York, 1960).

BRAUER, F., and J. NOHEL, *Ordinary Differential Equations: A First Course* (Benjamin, New York, 1967).

CODDINGTON, E., *An Introduction to Ordinary Differential Equations* (Prentice-Hall, Englewood Cliffs, N.J., 1961).

FORD, L., *Differential Equations*, 2nd ed. (McGraw-Hill, New York, 1955).

KAPLAN, W., *Ordinary Differential Equations* (Addison-Wesley, Reading, Mass., 1958).

LEIGHTON, W., *Ordinary Differential Equations*, 3rd ed. (Wadsworth, Belmont, Cal., 1970).

First-Order Equations for Which Exact Solutions are Obtainable

In this chapter we consider certain basic types of first-order equations for which exact solutions may be obtained by definite procedures. The purpose of this chapter is to gain ability to recognize these various types and to apply the corresponding methods of solutions. Of the types considered here, the so-called exact equations considered in Section 2.1 are in a sense the most basic, while the separable equations of Section 2.2 are in a sense the "easiest." The most important, from the point of view of applications, are the separable equations of Section 2.2 and the linear equations of Section 2.3. The remaining types are of various very special forms, and the corresponding methods of solution involve various devices. In short, we might describe this chapter as a collection of special "methods," "devices," "tricks," or "recipes," in descending order of kindness!

2.1 Exact Differential Equations and Integrating Factors

A. Standard Forms of First-Order Differential Equations

The first-order differential equations to be studied in this chapter may be expressed in either the derivative form

$$\frac{dy}{dx} = f(x, y) \tag{2.1}$$

or the differential form

$$M(x, y)\,dx + N(x, y)\,dy = 0. \tag{2.2}$$

An equation in one of these forms may readily be written in the other form. For example, the equation

$$\frac{dy}{dx} = \frac{x^2 + y^2}{x - y}$$

is of the form (2.1). It may be written

$$(x^2 + y^2)\,dx + (y - x)\,dy = 0,$$

which is of the form (2.2). The equation

$$(\sin x + y)\, dx + (x + 3y)\, dy = 0,$$

which is of the form (2.2), may be written in the form (2.1) as

$$\frac{dy}{dx} = -\frac{\sin x + y}{x + 3y}.$$

In the form (2.1) it is clear from the notation itself that y is regarded as the dependent variable and x as the independent one; but in the form (2.2) we may actually regard either variable as the dependent one and the other as the independent. However, in this text, in all differential equations of the form (2.2) in x and y, we shall regard y as dependent and x as independent, unless the contrary is specifically stated.

B. *Exact Differential Equations*

DEFINITION

Let F be a function of two real variables such that F has continuous first partial derivatives in a domain D. The total differential *dF of the function F is defined by the formula*

$$dF(x, y) = \frac{\partial F(x, y)}{\partial x}\, dx + \frac{\partial F(x, y)}{\partial y}\, dy$$

for all $(x, y) \in D$.

▶ Example 2.1. Let F be the function of two real variables defined by

$$F(x, y) = xy^2 + 2x^3y$$

for all real (x, y). Then

$$\frac{\partial F(x, y)}{\partial x} = y^2 + 6x^2y, \qquad \frac{\partial F(x, y)}{\partial y} = 2xy + 2x^3,$$

and the total differential dF is defined by

$$dF(x, y) = (y^2 + 6x^2y)\, dx + (2xy + 2x^3)\, dy$$

for all real (x, y).

DEFINITION

The expression

$$M(x, y)\, dx + N(x, y)\, dy \tag{2.3}$$

is called an exact differential *in a domain D if there exists a function F of two real variables such that this expression equals the total differential $dF(x, y)$ for all $(x, y) \in D$. That is, expression (2.3) is an exact differential in D if there exists a function F such that*

$$\frac{\partial F(x, y)}{\partial x} = M(x, y) \quad \text{and} \quad \frac{\partial F(x, y)}{\partial y} = N(x, y)$$

for all $(x, y) \in D$.

If $M(x, y)\,dx + N(x, y)\,dy$ *is an exact differential, then the differential equation*

$$M(x, y)\,dx + N(x, y)\,dy = 0$$

is called an exact differential equation.

▶ Example 2.2. The differential equation

$$y^2\,dx + 2xy\,dy = 0 \tag{2.4}$$

is an exact differential equation, since the expression $y^2\,dx + 2xy\,dy$ is an exact differential. Indeed, it is the total differential of the function F defined for all (x, y) by $F(x, y) = xy^2$, since the coefficient of dx is $\frac{\partial F(x, y)}{\partial x} = y^2$ and that of dy is $\frac{\partial F(x, y)}{\partial y} = 2xy$. On the other hand, the more simple appearing equation

$$y\,dx + 2x\,dy = 0, \tag{2.5}$$

obtained from (2.4) by dividing through by y, is *not* exact.

In Example 2.2 we stated without hesitation that the differential equation (2.4) is exact but the differential equation (2.5) is not. In the case of Equation (2.4), we verified our assertion by actually exhibiting the function F of which the expression $y^2\,dx + 2xy\,dy$ is the total differential. But in the case of Equation (2.5), we did not back up our statement by showing that there is no function F such that $y\,dx + 2x\,dy$ is its total differential. It is clear that we need a simple test to determine whether or not a given differential equation is exact. This is given by the following theorem.

THEOREM 2.1

Consider the differential equation

$$M(x, y)\,dx + N(x, y)\,dy = 0, \tag{2.6}$$

where M and N have continuous first partial derivatives at all points (x, y) in a rectangular domain D.

1. If *the differential equation* (2.6) *is exact in D*, then

$$\frac{\partial M(x, y)}{\partial y} = \frac{\partial N(x, y)}{\partial x} \tag{2.7}$$

for all $(x, y) \in D$.

2. Conversely, if

$$\frac{\partial M(x, y)}{\partial y} = \frac{\partial N(x, y)}{\partial x}$$

for all $(x, y) \in D$, then *the differential equation* (2.6) *is exact in D.*

Proof. Part 1. If the differential equation (2.6) is exact in D, then $M\,dx + N\,dy$ is an exact differential in D. By definition of an exact differential, there exists a function F such that

$$\frac{\partial F(x, y)}{\partial x} = M(x, y) \quad \text{and} \quad \frac{\partial F(x, y)}{\partial y} = N(x, y)$$

for all $(x, y) \in D$. Then

$$\frac{\partial^2 F(x, y)}{\partial y\, \partial x} = \frac{\partial M(x, y)}{\partial y} \quad \text{and} \quad \frac{\partial^2 F(x, y)}{\partial x\, \partial y} = \frac{\partial N(x, y)}{\partial x}$$

for all $(x, y) \in D$. But, using the continuity of the first partial derivatives of M and N, we have

$$\frac{\partial^2 F(x, y)}{\partial y\, \partial x} = \frac{\partial^2 F(x, y)}{\partial x\, \partial y}$$

and therefore

$$\frac{\partial M(x, y)}{\partial y} = \frac{\partial N(x, y)}{\partial x}$$

for all $(x, y) \in D$.

Part 2. This being the converse of Part 1, we start with the hypothesis that

$$\frac{\partial M(x, y)}{\partial y} = \frac{\partial N(x, y)}{\partial x}$$

for all $(x, y) \in D$, and set out to show that $M\,dx + N\,dy = 0$ is exact in D. This means that we must prove that there exists a function F such that

$$\frac{\partial F(x, y)}{\partial x} = M(x, y) \tag{2.8}$$

and

$$\frac{\partial F(x, y)}{\partial y} = N(x, y) \tag{2.9}$$

for all $(x, y) \in D$. We can certainly find some $F(x, y)$ satisfying either (2.8) or (2.9), but what about both? Let us assume that F satisfies (2.8) and proceed. Then

$$F(x, y) = \int M(x, y)\, \partial x + \phi(y), \tag{2.10}$$

where $\int M(x, y)\, \partial x$ indicates a partial integration with respect to x, holding y constant, and ϕ is an arbitrary function of y only. This $\phi(y)$ is needed in (2.10) so that $F(x, y)$ given by (2.10) will represent *all* solutions of (2.8). It corresponds to a constant of integration in the "one-variable" case. Differentiating (2.10) partially with respect to y, we obtain

$$\frac{\partial F(x, y)}{\partial y} = \frac{\partial}{\partial y} \int M(x, y)\, \partial x + \frac{d\phi(y)}{dy}.$$

Now if (2.9) is to be satisfied, we must have

$$N(x, y) = \frac{\partial}{\partial y} \int M(x, y)\, \partial x + \frac{d\phi(y)}{dy} \tag{2.11}$$

and hence

$$\frac{d\phi(y)}{dy} = N(x, y) - \frac{\partial}{\partial y} \int M(x, y)\, \partial x.$$

Since ϕ is a function of y only, the derivative $d\phi/dy$ must also be independent of x. That is, in order for (2.11) to hold,

$$N(x, y) - \frac{\partial}{\partial y}\int M(x, y)\,\partial x \tag{2.12}$$

must be independent of x.

We shall show that

$$\frac{\partial}{\partial x}\left[N(x, y) - \frac{\partial}{\partial y}\int M(x, y)\,\partial x\right] = 0.$$

We at once have

$$\frac{\partial}{\partial x}\left[N(x, y) - \frac{\partial}{\partial y}\int M(x, y)\,\partial x\right] = \frac{\partial N(x, y)}{\partial x} - \frac{\partial^2}{\partial x\,\partial y}\int M(x, y)\,\partial x.$$

If (2.8) and (2.9) are to be satisfied, then using the hypothesis (2.7), we must have

$$\frac{\partial^2}{\partial x\,\partial y}\int M(x, y)\,\partial x = \frac{\partial^2 F(x, y)}{\partial x\,\partial y} = \frac{\partial^2 F(x, y)}{\partial y\,\partial x} = \frac{\partial^2}{\partial y\,\partial x}\int M(x, y)\,\partial x.$$

Thus we obtain

$$\frac{\partial}{\partial x}\left[N(x, y) - \frac{\partial}{\partial y}\int M(x, y)\,\partial x\right] = \frac{\partial N(x, y)}{\partial x} - \frac{\partial^2}{\partial y\,\partial x}\int M(x, y)\,\partial x$$

and hence

$$\frac{\partial}{\partial x}\left[N(x, y) - \frac{\partial}{\partial y}\int M(x, y)\,\partial x\right] = \frac{\partial N(x, y)}{\partial x} - \frac{\partial M(x, y)}{\partial y}.$$

But by hypothesis (2.7),

$$\frac{\partial M(x, y)}{\partial y} = \frac{\partial N(x, y)}{\partial x}$$

for all $(x, y) \in D$. Thus

$$\frac{\partial}{\partial x}\left[N(x, y) - \frac{\partial}{\partial y}\int M(x, y)\,\partial x\right] = 0$$

for all $(x, y) \in D$, and so (2.12) *is* independent of x. Thus we may write

$$\phi(y) = \int\left[N(x, y) - \int\frac{\partial M(x, y)}{\partial y}\,\partial x\right] dy.$$

Substituting this into Equation (2.10), we have

$$F(x, y) = \int M(x, y)\,\partial x + \int\left[N(x, y) - \int\frac{\partial M(x, y)}{\partial y}\,\partial x\right] dy. \tag{2.13}$$

This $F(x, y)$ thus satisfies both (2.8) and (2.9) for all $(x, y) \in D$, and so $M\,dx + N\,dy = 0$ is exact in D.

Q.E.D.

Students well versed in the terminology of higher mathematics will recognize that Theorem 2.1 may be stated in the following words: A necessary and sufficient condition that Equation (2.6) be exact in D is that condition (2.7) hold for all $(x, y) \in D$.

For students not so well versed, let us emphasize that condition (2.7),

$$\frac{\partial M(x, y)}{\partial y} = \frac{\partial N(x, y)}{\partial x},$$

is the criterion for exactness. If (2.7) holds, then (2.6) is exact; if (2.7) does *not* hold, then (2.6) is *not* exact.

▶ **Example 2.3.** We apply the exactness criterion (2.7) to Equations (2.4) and (2.5), introduced in Example 2.2. For the equation

$$y^2\,dx + 2xy\,dy = 0 \tag{2.4}$$

we have

$$M(x, y) = y^2, \qquad N(x, y) = 2xy,$$

$$\frac{\partial M(x, y)}{\partial y} = 2y = \frac{\partial N(x, y)}{\partial x}$$

for all (x, y). Thus Equation (2.4) is exact in every rectangular domain D. On the other hand, for the equation

$$y\,dx + 2x\,dy = 0, \tag{2.5}$$

we have

$$M(x, y) = y, \qquad N(x, y) = 2x,$$

$$\frac{\partial M(x, y)}{\partial y} = 1 \neq 2 = \frac{\partial N(x, y)}{\partial x}$$

for all (x, y). Thus Equation (2.5) is not exact in any rectangular domain D.

▶ **Example 2.4.** Consider the differential equation

$$(2x \sin y + y^3 e^x)\,dx + (x^2 \cos y + 3y^2 e^x)\,dy = 0.$$

Here

$$M(x, y) = 2x \sin y + y^3 e^x,$$

$$N(x, y) = x^2 \cos y + 3y^2 e^x,$$

$$\frac{\partial M(x, y)}{\partial y} = 2x \cos y + 3y^2 e^x = \frac{\partial N(x, y)}{\partial x}$$

in every rectangular domain D. Thus this differential equation is exact in every such domain.

These examples illustrate the use of the test given by (2.7) for determining whether or not an equation of the form $M(x, y)\,dx + N(x, y)\,dy = 0$ is exact. It should be observed that the equation *must* be in the standard form $M(x, y)\,dx + N(x, y)\,dy = 0$ in order to use the exactness test (2.7). Note this carefully: an equation may be encountered in the *non*standard form $M(x, y)\,dx = N(x, y)\,dy$, and in this form the test (2.7) does *not* apply.

C. *The Solution of Exact Differential Equations*

Now that we have a test with which to determine exactness, let us proceed to solve exact differential equations. If the equation $M(x, y)\,dx + N(x, y)\,dy = 0$ is exact in a rectangular domain D, then there exists a function F such that

$$\frac{\partial F(x, y)}{\partial x} = M(x, y) \quad \text{and} \quad \frac{\partial F(x, y)}{\partial y} = N(x, y) \quad \text{for all } (x, y) \in D.$$

Then the equation may be written

$$\frac{\partial F(x, y)}{\partial x}\,dx + \frac{\partial F(x, y)}{\partial y}\,dy = 0 \quad \text{or simply} \quad dF(x, y) = 0.$$

The relation $F(x, y) = c$ is obviously a solution of this, where c is an arbitrary constant. We summarize this observation in the following theorem.

THEOREM 2.2

Suppose the differential equation $M(x, y)\,dx + N(x, y)\,dy = 0$ satisfies the differentiability requirements of Theorem 2.1 and is exact in a rectangular domain D. Then a one-parameter family of solutions of this differential equation is given by $F(x, y) = c$, where F is a function such that

$$\frac{\partial F(x, y)}{\partial x} = M(x, y) \quad \textit{and} \quad \frac{\partial F(x, y)}{\partial y} = N(x, y) \quad \textit{for all } (x, y) \in D$$

and c is an arbitrary constant.

Referring to Theorem 2.1, we observe that $F(x, y)$ is given by formula (2.13). However, in solving exact differential equations it is neither necessary nor desirable to use this formula. Instead one obtains $F(x, y)$ either by proceeding as in the proof of Theorem 2.1, Part 2, or by the so-called "method of grouping," which will be explained in the following examples.

▶ Example 2.5. Solve the equation

$$(3x^2 + 4xy)\,dx + (2x^2 + 2y)\,dy = 0.$$

Our first duty is to determine whether or not the equation is exact. Here

$$M(x, y) = 3x^2 + 4xy, \qquad N(x, y) = 2x^2 + 2y,$$

$$\frac{\partial M(x, y)}{\partial y} = 4x, \qquad \frac{\partial N(x, y)}{\partial x} = 4x,$$

for all real (x, y), and so the equation is exact in every rectangular domain D. Thus we must find F such that

$$\frac{\partial F(x, y)}{\partial x} = M(x, y) = 3x^2 + 4xy \quad \text{and} \quad \frac{\partial F(x, y)}{\partial y} = N(x, y) = 2x^2 + 2y.$$

From the first of these,

$$F(x, y) = \int M(x, y)\,\partial x + \phi(y) = \int (3x^2 + 4xy)\,\partial x + \phi(y)$$
$$= x^3 + 2x^2y + \phi(y).$$

Then

$$\frac{\partial F(x, y)}{\partial y} = 2x^2 + \frac{d\phi(y)}{dy}.$$

But we must have

$$\frac{\partial F(x, y)}{\partial y} = N(x, y) = 2x^2 + 2y.$$

Thus

$$2x^2 + 2y = 2x^2 + \frac{d\phi(y)}{dy}$$

or

$$\frac{d\phi(y)}{dy} = 2y.$$

Thus $\phi(y) = y^2 + c_0$, where c_0 is an arbitrary constant, and so

$$F(x, y) = x^3 + 2x^2y + y^2 + c_0.$$

Hence a one-parameter family of solutions is $F(x, y) = c_1$, or

$$x^3 + 2x^2y + y^2 + c_0 = c_1.$$

Combining the constants c_0 and c_1 we may write this solution as

$$x^3 + 2x^2y + y^2 = c,$$

where $c = c_1 - c_0$ is an arbitrary constant. The student will observe that there is no loss in generality by taking $c_0 = 0$ and writing $\phi(y) = y^2$. We now consider an alternative procedure.

Method of Grouping. We shall now solve the differential equation of this example by grouping the terms in such a way that its left member appears as the sum of certain exact differentials. We write the differential equation

$$(3x^2 + 4xy)\,dx + (2x^2 + 2y)\,dy = 0$$

in the form

$$3x^2\,dx + (4xy\,dx + 2x^2\,dy) + 2y\,dy = 0.$$

We now recognize this as

$$d(x^3) + d(2x^2y) + d(y^2) = d(c),$$

where c is an arbitrary constant, or

$$d(x^3 + 2x^2y + y^2) = d(c).$$

From this we have at once

$$x^3 + 2x^2y + y^2 = c.$$

Clearly this procedure is much quicker, but it requires a good "working knowledge" of differentials and a certain amount of ingenuity to determine just how the terms should be grouped. The standard method may require more "work" and take longer, but it is perfectly straightforward. It is recommended for those who like to follow a pattern and for those who have a tendency to jump at conclusions.

Just to make certain that we have both procedures well in hand, we shall consider an initial-value problem involving an exact differential equation.

▶ Example 2.6. Solve the initial-value problem

$$(2x \cos y + 3x^2y)\, dx + (x^3 - x^2 \sin y - y)\, dy = 0,$$

$$y(0) = 2.$$

We first observe that the equation is exact in every rectangular domain D, since

$$\frac{\partial M(x, y)}{\partial y} = -2x \sin y + 3x^2 = \frac{\partial N(x, y)}{\partial x}$$

for all real (x, y).

Standard Method. We must find F such that

$$\frac{\partial F(x, y)}{\partial x} = M(x, y) = 2x \cos y + 3x^2y$$

and

$$\frac{\partial F(x, y)}{\partial y} = N(x, y) = x^3 - x^2 \sin y - y.$$

Then

$$F(x, y) = \int M(x, y)\, \partial x + \phi(y)$$

$$= \int (2x \cos y + 3x^2y)\, \partial x + \phi(y)$$

$$= x^2 \cos y + x^3y + \phi(y),$$

$$\frac{\partial F(x, y)}{\partial y} = -x^2 \sin y + x^3 + \frac{d\phi(y)}{dy}.$$

But also

$$\frac{\partial F(x, y)}{\partial y} = N(x, y) = x^3 - x^2 \sin y - y$$

and so

$$\frac{d\phi(y)}{dy} = -y$$

and hence

$$\phi(y) = -\frac{y^2}{2} + c_0.$$

Thus

$$F(x, y) = x^2 \cos y + x^3 y - \frac{y^2}{2} + c_0.$$

Hence a one-parameter family of solutions is $F(x, y) = c_1$, which may be expressed as

$$x^2 \cos y + x^3 y - \frac{y^2}{2} = c.$$

Applying the initial condition $y = 2$ when $x = 0$, we find $c = -2$. Thus the solution of the given initial-value problem is

$$x^2 \cos y + x^3 y - \frac{y^2}{2} = -2.$$

Method of Grouping. We group the terms as follows:

$$(2x \cos y \, dx - x^2 \sin y \, dy) + (3x^2 y \, dx + x^3 \, dy) - y \, dy = 0.$$

Thus we have

$$d(x^2 \cos y) + d(x^3 y) - d\left(\frac{y^2}{2}\right) = d(c);$$

and so

$$x^2 \cos y + x^3 y - \frac{y^2}{2} = c$$

is a one-parameter family of solutions of the differential equation. Of course the initial condition $y(0) = 2$ again yields the particular solution already obtained.

D. Integrating Factors

Given the differential equation

$$M(x, y)\, dx + N(x, y)\, dy = 0,$$

if

$$\frac{\partial M(x, y)}{\partial y} = \frac{\partial N(x, y)}{\partial x},$$

then the equation is exact and we can obtain a one-parameter family of solutions by one of the procedures explained above. But if

$$\frac{\partial M(x, y)}{\partial y} \neq \frac{\partial N(x, y)}{\partial x},$$

then the equation is *not* exact and the above procedures do not apply. What shall we do in such a case? Perhaps we can multiply the nonexact equation by some expression which will transform it into an essentially equivalent exact equation. If so, we can proceed to solve the resulting exact equation by one of the above procedures. Let us consider again the equation

$$y\,dx + 2x\,dy - 0, \tag{2.5}$$

which was introduced in Example 2.2. In that example we observed that this equation is *not* exact. However, if we multiply Equation (2.5) by y, it is transformed into the essentially equivalent equation

$$y^2\,dx + 2xy\,dy = 0, \tag{2.4}$$

which is exact (see Example 2.2). Since this resulting exact equation (2.4) is integrable, we call y an *integrating factor* of Equation (2.5). In general, we have the following definition:

DEFINITION

If the differential equation

$$M(x, y)\,dx + N(x, y)\,dy = 0 \tag{2.14}$$

is not *exact in a domain D but the differential equation*

$$\mu(x, y)M(x, y)\,dx + \mu(x, y)N(x, y)\,dy = 0 \tag{2.15}$$

is exact *in D, then $\mu(x, y)$ is called an* integrating factor *of the differential equation (2.14).*

▶ **Example 2.7.** Consider the differential equation

$$(3y + 4xy^2)\,dx + (2x + 3x^2y)\,dy = 0. \tag{2.16}$$

This equation is of the form (2.14), where

$$M(x, y) = 3y + 4xy^2, \qquad N(x, y) = 2x + 3x^2y,$$

$$\frac{\partial M(x, y)}{\partial y} = 3 + 8xy, \qquad \frac{\partial N(x, y)}{\partial x} = 2 + 6xy.$$

Since

$$\frac{\partial M(x, y)}{\partial y} \neq \frac{\partial N(x, y)}{\partial x}$$

except for (x, y) such that $2xy + 1 = 0$, Equation (2.16) is *not* exact in any rectangular domain D.

Let $\mu(x, y) = x^2y$. Then the corresponding differential equation of the form (2.15) is

$$(3x^2y^2 + 4x^3y^3)\,dx + (2x^3y + 3x^4y^2)\,dy = 0.$$

This equation is exact in every rectangular domain D, since

$$\frac{\partial[\mu(x, y)M(x, y)]}{\partial y} = 6x^2y + 12x^3y^2 = \frac{\partial[\mu(x, y)N(x, y)]}{\partial x}$$

for all real (x, y). Hence $\mu(x, y) = x^2y$ is an integrating factor of Equation (2.16).

Multiplication of a nonexact differential equation by an integrating factor thus transforms the nonexact equation into an exact one. We have referred to this resulting exact equation as "essentially equivalent" to the original. This so-called essentially equivalent exact equation has the same one-parameter family of solutions as the nonexact original. However, the multiplication of the original equation by the integrating factor may result in either (1) the loss of (one or more) solutions of the original, or (2) the gain of (one or more) functions which are solutions of the "new" equation but *not* of the original, or (3) both of these phenomena. Hence, whenever we transform a nonexact equation into an exact one by multiplication by an integrating factor, we should check carefully to determine whether any solutions may have been lost or gained. We shall illustrate an important special case of these phenomena when we consider separable equations in Section 2.2. See also Exercise 22 at the end of this section.

The question now arises: How is an integrating factor found? We shall not attempt to answer this question at this time. Instead we shall proceed to a study of the important classes of separable equations in Section 2.2 and linear equations in Section 2.3. We shall see that separable equations always possess integrating factors which are perfectly obvious, while linear equations always have integrating factors of a certain special form. We shall return to the question raised above in Section 2.4. Our object here has been merely to introduce the concept of an integrating factor.

Exercises

In Exercises 1 through 10 determine whether or not each of the given equations is exact; solve those which are exact.

1. $(3x + 2y)\,dx + (2x + y)\,dy = 0.$
2. $(y^2 + 3)\,dx + (2xy - 4)\,dy = 0.$
3. $(2xy + 1)\,dx + (x^2 + 4y)\,dy = 0.$
4. $(3x^2y + 2)\,dx - (x^3 + y)\,dy = 0.$
5. $(6xy + 2y^2 - 5)\,dx + (3x^2 + 4xy - 6)\,dy = 0.$
6. $(\theta^2 + 1)\cos r\,dr + 2\theta \sin r\,d\theta = 0.$
7. $(y \sec^2 x + \sec x \tan x)\,dx + (\tan x + 2y)\,dy = 0.$
8. $\left(\dfrac{x}{y^2} + x\right) dx + \left(\dfrac{x^2}{y^3} + y\right) dy = 0.$
9. $\left(\dfrac{2s - 1}{t}\right) ds + \left(\dfrac{s - s^2}{t^2}\right) dt = 0.$
10. $\dfrac{2y^{3/2} + 1}{x^{1/2}}\,dx + (3x^{1/2}y^{1/2} - 1)\,dy = 0.$

Solve the initial-value problems in Exercises 11 through 16.

11. $(2xy - 3)\,dx + (x^2 + 4y)\,dy = 0, \quad y(1) = 2.$
12. $(3x^2y^2 - y^3 + 2x)\,dx + (2x^3y - 3xy^2 + 1)\,dy = 0, \quad y(-2) = 1.$
13. $(2y \sin x \cos x + y^2 \sin x)\,dx + (\sin^2 x - 2y \cos x)\,dy = 0, \quad y(0) = 3.$

14. $(ye^x + 2e^x + y^2)\,dx + (e^x + 2xy)\,dy = 0, \qquad y(0) = 6.$

15. $\left(\dfrac{3 - y}{x^2}\right) dx + \left(\dfrac{y^2 - 2x}{xy^2}\right) dy = 0, \qquad y(-1) = 2.$

16. $\dfrac{1 - y^{2/3}}{x^{2/3}y^{1/3}}\,dx + \dfrac{2x^{4/3}y^{2/3} - x^{1/3}}{y^{4/3}}\,dy = 0, \qquad y(1) = 8.$

17. In each of the following equations determine the constant A such that the equation is exact, and solve the resulting exact equation:

(a) $(x^2 + 3xy)\,dx + (Ax^2 + 4y)\,dy = 0.$

(b) $\left(\dfrac{1}{x^2} + \dfrac{1}{y^2}\right) dx + \left(\dfrac{Ax + 1}{y^3}\right) dy = 0.$

18. In each of the following equations determine the constant A such that the equation is exact, and solve the resulting exact equation:

(a) $(Ax^2y + 2y^2)\,dx + (x^3 + 4xy)\,dy = 0.$

(b) $\left(\dfrac{Ay}{x^3} + \dfrac{y}{x^2}\right) dx + \left(\dfrac{1}{x^2} - \dfrac{1}{x}\right) dy = 0.$

19. In each of the following equations determine the most general function $N(x, y)$ such that the equation is exact:

(a) $(x^3 + xy^2)\,dx + N(x, y)\,dy = 0.$

(b) $(x^{-2}y^{-2} + xy^{-3})\,dx + N(x, y)\,dy = 0.$

20. In each of the following equations determine the most general function $M(x, y)$ such that the equation is exact:

(a) $M(x, y)\,dx + (2x^2y^3 + x^4y)\,dy = 0.$

(b) $M(x, y)\,dx + (2ye^x + y^2e^{3x})\,dy = 0.$

21. Consider the differential equation

$$(4x + 3y^2)\,dx + 2xy\,dy = 0.$$

(a) Show that this equation is not exact.

(b) Find an integrating factor of the form x^n, where n is a positive integer.

(c) Multiply the given equation through by the integrating factor found in (b) and solve the resulting exact equation.

22. Consider the differential equation

$$(y^2 + 2xy)\,dx - x^2\,dy = 0.$$

(a) Show that this equation is not exact.

(b) Multiply the given equation through by y^n, where n is an integer, and then determine n so that y^n is an integrating factor of the given equation.

(c) Multiply the given equation through by the integrating factor found in (b) and solve the resulting exact equation.

(d) Show that $y = 0$ is a solution of the original nonexact equation but is not a solution of the essentially equivalent exact equation found in step (c).

(e) Graph several integral curves of the original equation, including all those whose equations are (or can be written) in some "special" form.

23. Consider a differential equation of the form

$$[y + xf(x^2 + y^2)]\,dx + [yf(x^2 + y^2) - x]\,dy = 0.$$

(a) Show that an equation of this form is not exact.

(b) Show that $1/(x^2 + y^2)$ is an integrating factor of an equation of this form.

24. Use the result of Exercise 23(b) to solve the equation

$$[y + x(x^2 + y^2)^2]\,dx + [y(x^2 + y^2)^2 - x]\,dy = 0.$$

2.2 Separable Equations and Equations Reducible to This Form

A. Separable Equations

DEFINITION

An equation of the form

$$F(x)G(y)\,dx + f(x)g(y)\,dy = 0 \tag{2.17}$$

is called an equation with variables separable *or simply a* separable equation.

For example, the equation $(x - 4)y^4\,dx - x^3(y^2 - 3)\,dy = 0$ is a separable equation.

In general the separable equation (2.17) is not exact, but it possesses an obvious integrating factor, namely $1/f(x)G(y)$. For if we multiply Equation (2.17) by this expression, we separate the variables, reducing (2.17) to the essentially equivalent equation

$$\frac{F(x)}{f(x)}\,dx + \frac{g(y)}{G(y)}\,dy = 0. \tag{2.18}$$

This equation is exact, since

$$\frac{\partial}{\partial y}\left[\frac{F(x)}{f(x)}\right] = 0 = \frac{\partial}{\partial x}\left[\frac{g(y)}{G(y)}\right].$$

Denoting $F(x)/f(x)$ by $M(x)$ and $g(y)/G(y)$ by $N(y)$, Equation (2.18) takes the form $M(x)\,dx + N(y)\,dy = 0$. Since M is a function of x only and N is a function of y only, we see at once that a one-parameter family of solutions is

$$\int M(x)\,dx + \int N(y)\,dy = c, \tag{2.19}$$

where c is the arbitrary constant. Thus the problem of finding such a family of solutions of the separable equation (2.17) has reduced to that of performing the integrations indicated in Equation (2.19). It is in this sense that separable equations are the simplest first-order differential equations.

Since we obtained the separated exact equation (2.18) from the nonexact equation (2.17) by multiplying (2.17) by the integrating factor $1/f(x)G(y)$, solutions may have been lost or gained in this process. We now consider this more carefully. In formally multiplying by the integrating factor $1/f(x)G(y)$, we actually divided by $f(x)G(y)$. We did this under the tacit assumption that neither $f(x)$ nor $G(y)$ is zero; and, under this assumption, we proceeded to obtain the one-parameter family of solutions given by (2.19). Now, we should investigate the possible loss or gain of solutions which may have occurred in this formal process. In particular, regarding y as the dependent variable as usual, we consider the situation that occurs if $G(y)$ is zero. Writing the original differential equation (2.17) in the derivative form

$$f(x)g(y)\frac{dy}{dx} + F(x)G(y) = 0,$$

we immediately note the following: If y_0 is any real number such that $G(y_0) = 0$, then $y = y_0$ is a (constant) solution of the original differential equation; and this solution may (or may not) have been lost in the formal separation process.

In finding a one-parameter family of solutions of a separable equation, we shall always make the assumption that any factors by which we divide in the formal separation process are not zero. Then we must find the solutions $y = y_0$ of the equation $G(y) = 0$ and determine whether any of these are solutions of the original equation which were lost in the formal separation process.

▶ Example 2.8. Solve the equation

$$(x - 4)y^4\,dx - x^3(y^2 - 3)\,dy = 0.$$

The equation is separable; separating the variables by dividing by x^3y^4, we obtain

$$\frac{(x - 4)\,dx}{x^3} - \frac{(y^2 - 3)\,dy}{y^4} = 0$$

or

$$(x^{-2} - 4x^{-3})\,dx - (y^{-2} - 3y^{-4})\,dy = 0.$$

Integrating, we have the one-parameter family of solutions

$$-\frac{1}{x} + \frac{2}{x^2} + \frac{1}{y} - \frac{1}{y^3} = c,$$

where c is the arbitrary constant.

In dividing by x^3y^4 in the separation process, we assumed that $x^3 \neq 0$ and $y^4 \neq 0$. We now consider the solution $y = 0$ of $y^4 = 0$. It is not a member of the one-parameter family of solutions which we obtained. However, writing the original differential equation of the problem in the derivative form

$$\frac{dy}{dx} = \frac{(x - 4)y^4}{x^3(y^2 - 3)},$$

it is obvious that $y = 0$ *is* a solution of the original equation. We conclude that it is a solution which was lost in the separation process.

▶ **Example 2.9.** Solve the initial-value problem that consists of the differential equation

$$x \sin y \, dx + (x^2 + 1) \cos y \, dy = 0 \tag{2.20}$$

and the initial condition

$$y(1) = \frac{\pi}{2}. \tag{2.21}$$

We first obtain a one-parameter family of solutions of the differential equation (2.20). Separating the variables by dividing by $(x^2 + 1) \sin y$, we obtain

$$\frac{x}{x^2 + 1} dx + \frac{\cos y}{\sin y} dy = 0.$$

Thus

$$\int \frac{x \, dx}{x^2 + 1} + \int \frac{\cos y}{\sin y} dy = c_0,$$

where c_0 is an arbitrary constant. Recall that

$$\int \frac{du}{u} = \ln |u| + C \quad \text{and} \quad |u| = \begin{cases} u & \text{if } u \geq 0, \\ -u & \text{if } u \leq 0. \end{cases}$$

Then, carrying out the integrations, we find

$$\tfrac{1}{2} \ln (x^2 + 1) + \ln |\sin y| = c_0. \tag{2.22}$$

We could leave the family of solutions in this form, but we can put it in a neater form in the following way. Since each term of the left member of this equation involves the logarithm of a function, it would seem reasonable that something might be accomplished by writing the arbitrary constant c_0 in the form $\ln |c_1|$. This we do, obtaining

$$\tfrac{1}{2} \ln (x^2 + 1) + \ln |\sin y| = \ln |c_1|.$$

Multiplying by 2, we have

$$\ln (x^2 + 1) + 2 \ln |\sin y| = 2 \ln |c_1|.$$

Since

$$2 \ln |\sin y| = \ln (\sin y)^2,$$

and

$$2 \ln |c_1| = \ln c_1^2 = \ln c,$$

where

$$c = c_1^2 \geq 0,$$

we now have

$$\ln (x^2 + 1) + \ln \sin^2 y = \ln c.$$

Since $\ln A + \ln B = \ln AB$, this equation may be written

$$\ln (x^2 + 1) \sin^2 y = \ln c.$$

From this we have at once

$$(x^2 + 1) \sin^2 y = c. \tag{2.23}$$

Clearly (2.23) is of a neater form than (2.22).

In dividing by $(x^2 + 1) \sin y$ in the separation process, we assumed that $\sin y \neq 0$. Now consider the solutions of $\sin y = 0$. These are given by $y = n\pi$ $(n = 0, \pm 1, \pm 2, \ldots)$. Writing the original differential equation (2.20) in the derivative form, it is clear that each of these solutions $y = n\pi$ $(n = 0, \pm 1, \pm 2, \ldots)$, of $\sin y = 0$ is a constant solution of the original differential equation. Now, each of these constant solutions $y = n\pi$ is a member of the one-parameter family (2.23) of solutions of (2.20) for $c = 0$. Thus none of these solutions was lost in the separation process.

We now apply the initial condition (2.21) to the family of solutions (2.23). We have

$$(1^2 + 1) \sin^2 \frac{\pi}{2} = c$$

and so $c = 2$. Therefore the solution of the initial-value problem under consideration is

$$(x^2 + 1) \sin^2 y = 2.$$

B. *Homogeneous Equations*

We now consider a class of differential equations which can be reduced to separable equations by a change of variables.

DEFINITION

The first-order differential equation $M(x, y)\,dx + N(x, y)\,dy = 0$ *is said to be* homogeneous *if, when written in the derivative form* $\dfrac{dy}{dx} = f(x, y)$, *there exists a function* g *such that* $f(x, y)$ *can be expressed in the form* $g(y/x)$.

▶ **Example 2.10.** The differential equation $(x^2 - 3y^2)\,dx + 2xy\,dy = 0$ is homogeneous. To see this, we first write this equation in the derivative form

$$\frac{dy}{dx} = \frac{3y^2 - x^2}{2xy}.$$

Now observing that

$$\frac{3y^2 - x^2}{2xy} = \frac{3y}{2x} - \frac{x}{2y} = \frac{3}{2}\left(\frac{y}{x}\right) - \frac{1}{2}\left(\frac{1}{y/x}\right),$$

we see that the differential equation under consideration may be written as

$$\frac{dy}{dx} = \frac{3}{2}\left(\frac{y}{x}\right) - \frac{1}{2}\left(\frac{1}{y/x}\right),$$

in which the right member is of the form $g(y/x)$ for a certain function g.

▶ **Example 2.11.** The equation

$$(y + \sqrt{x^2 + y^2})\,dx - x\,dy = 0$$

is homogeneous. When written in the form

$$\frac{dy}{dx} = \frac{y + \sqrt{x^2 + y^2}}{x},$$

the right member may be expressed as

$$\frac{y}{x} \pm \frac{\sqrt{x^2 + y^2}}{\sqrt{x^2}}$$

or

$$\frac{y}{x} \pm \sqrt{1 + \left(\frac{y}{x}\right)^2},$$

depending on the sign of x. This is obviously of the form $g(y/x)$.

Before proceeding to the actual solution of homogeneous equations we shall consider a slightly different procedure for recognizing such equations. A function F is called *homogeneous of degree n* if $F(tx, ty) = t^nF(x, y)$. This means that if tx and ty are substituted for x and y, respectively, in $F(x, y)$, and if t^n is then factored out, the other factor which remains is the original expression $F(x, y)$ itself. For example, the function F given by $F(x, y) = x^2 + y^2$ is homogeneous of degree 2, since

$$F(tx, ty) = (tx)^2 + (ty)^2 = t^2(x^2 + y^2) = t^2F(x, y).$$

Now suppose the functions M and N in the differential equation $M(x, y)\,dx + N(x, y)\,dy = 0$ are both homogeneous of the *same* degree n. Then since $M(tx, ty) = t^nM(x, y)$, if we let $t = 1/x$, we have

$$M\left(\frac{1}{x} \cdot x, \frac{1}{x} \cdot y\right) = \left(\frac{1}{x}\right)^n M(x, y).$$

Clearly this may be written more simply as

$$M\left(1, \frac{y}{x}\right) = \left(\frac{1}{x}\right)^n M(x, y);$$

and from this we at once obtain

$$M(x, y) = \left(\frac{1}{x}\right)^{-n} M\left(1, \frac{y}{x}\right).$$

Likewise, we find

$$N(x, y) = \left(\frac{1}{x}\right)^{-n} N\left(1, \frac{y}{x}\right).$$

Now writing the differential equation $M(x, y)\,dx + N(x, y)\,dy = 0$ in the form

$$\frac{dy}{dx} = -\frac{M(x, y)}{N(x, y)},$$

we find

$$\frac{dy}{dx} = -\frac{\left(\frac{1}{x}\right)^{-n} M\left(1, \frac{y}{x}\right)}{\left(\frac{1}{x}\right)^{-n} N\left(1, \frac{y}{x}\right)} = -\frac{M\left(1, \frac{y}{x}\right)}{N\left(1, \frac{y}{x}\right)}.$$

Clearly the expression on the right is of the form $g(y/x)$, and so the equation $M(x, y)\,dx + N(x, y)\,dy = 0$ is homogeneous in the sense of the original definition of homogeneity. Thus we conclude that if M and N in $M(x, y)\,dx + N(x, y)\,dy = 0$ are both homogeneous functions of the same degree n, then the differential equation is a homogeneous differential equation.

Let us now look back at Examples 2.10 and 2.11 in this light. In Example 2.10, $M(x, y) = x^2 - 3y^2$ and $N(x, y) = 2xy$. Both M and N are homogeneous of degree 2. Thus we know at once that the equation $(x^2 - 3y^2)\,dx + 2xy\,dy = 0$ is a homogeneous equation. In Example 2.11, $M(x, y) = y + \sqrt{x^2 + y^2}$ and $N(x, y) = -x$. Clearly N is homogeneous of degree 1. Since

$$M(tx, ty) = ty + \sqrt{(tx)^2 + (ty)^2} = t(y + \sqrt{x^2 + y^2}) = t^1 M(x, y),$$

we see that M is also homogeneous of degree 1. Thus we conclude that the equation

$$(y + \sqrt{x^2 + y^2})\,dx - x\,dy = 0$$

is indeed homogeneous.

We now show that every homogeneous equation can be reduced to a separable equation by proving the following theorem.

THEOREM 2.3

If

$$M(x, y)\,dx + N(x, y)\,dy = 0 \tag{2.24}$$

is a homogeneous equation, then the change of variables $y = vx$ transforms (2.24) into a separable equation in the variables v and x.

Proof. Since $M(x, y)\,dx + N(x, y)\,dy = 0$ is homogeneous, it may be written in the form

$$\frac{dy}{dx} = g\left(\frac{y}{x}\right).$$

Let $y = vx$. Then

$$\frac{dy}{dx} = v + x\frac{dv}{dx}$$

and (2.24) becomes

$$v + x\frac{dv}{dx} = g(v)$$

or

$$[v - g(v)]\,dx + x\,dv = 0.$$

This equation is separable. Separating the variables we obtain

$$\frac{dv}{v - g(v)} + \frac{dx}{x} = 0. \tag{2.25}$$

Q.E.D.

Thus to solve a homogeneous differential equation of the form (2.24), we let $y = vx$ and transform the homogeneous equation into a separable equation of the form (2.25). From this, we have

$$\int \frac{dv}{v - g(v)} + \int \frac{dx}{x} = c,$$

where c is an arbitrary constant. Letting $F(v)$ denote

$$\int \frac{dv}{v - g(v)}$$

and returning to the original dependent variable y, the solution takes the form

$$F\left(\frac{y}{x}\right) + \ln |x| = c.$$

▶ Example 2.12. Solve the equation

$$(x^2 - 3y^2)\,dx + 2xy\,dy = 0.$$

We have already observed that this equation is homogeneous. Writing it in the form

$$\frac{dy}{dx} = -\frac{x}{2y} + \frac{3y}{2x}$$

and letting $y = vx$, we obtain

$$v + x\frac{dv}{dx} = -\frac{1}{2v} + \frac{3v}{2},$$

or

$$x\frac{dv}{dx} = -\frac{1}{2v} + \frac{v}{2},$$

or, finally,

$$x\frac{dv}{dx} = \frac{v^2 - 1}{2v}.$$

This equation is separable. Separating the variables, we obtain

$$\frac{2v\,dv}{v^2 - 1} = \frac{dx}{x}.$$

Integrating, we find

$$\ln |v^2 - 1| = \ln |x| + \ln |c|,$$

and hence

$$|v^2 - 1| = |cx|,$$

where c is an arbitrary constant. The reader should observe that no solutions were lost in the separation process. Now, replacing v by y/x we obtain the solutions in the form

$$\left|\frac{y^2}{x^2} - 1\right| = |cx|$$

or

$$|y^2 - x^2| = |cx|x^2.$$

If $y \geq x \geq 0$, then this may be expressed somewhat more simply as

$$y^2 - x^2 = cx^3.$$

▶ Example 2.13. Solve the initial-value problem

$$(y + \sqrt{x^2 + y^2})\,dx - x\,dy = 0,$$

$$y(1) = 0.$$

We have seen that the differential equation is homogeneous. As before, we write it in the form

$$\frac{dy}{dx} = \frac{y + \sqrt{x^2 + y^2}}{x}.$$

Since the initial x value is 1, we consider $x > 0$ and take $x = \sqrt{x^2}$ and obtain

$$\frac{dy}{dx} = \frac{y}{x} + \sqrt{1 + \left(\frac{y}{x}\right)^2}.$$

We let $y = vx$ and obtain

$$v + x\frac{dv}{dx} = v + \sqrt{1 + v^2}$$

or

$$x\frac{dv}{dx} = \sqrt{1 + v^2}.$$

Separating variables, we find

$$\frac{dv}{\sqrt{v^2 + 1}} = \frac{dx}{x}.$$

Using tables, we perform the required integrations to obtain

$$\ln|v + \sqrt{v^2 + 1}| = \ln|x| + \ln|c|,$$

or

$$v + \sqrt{v^2 + 1} = cx.$$

Now replacing v by y/x, we obtain the general solution of the differential equation in the form

$$\frac{y}{x} + \sqrt{\frac{y^2}{x^2} + 1} = cx$$

or

$$y + \sqrt{x^2 + y^2} = cx^2.$$

The initial condition requires that $y = 0$ when $x = 1$. This gives $c = 1$ and hence

$$y + \sqrt{x^2 + y^2} = x^2,$$

from which it follows that

$$y = \tfrac{1}{2}(x^2 - 1).$$

Exercises

Solve each of the differential equations in Exercises 1 through 14.

1. $4xy\,dx + (x^2 + 1)\,dy = 0.$
2. $(xy + 2x + y + 2)\,dx + (x^2 + 2x)\,dy = 0.$
3. $2r(s^2 + 1)\,dr + (r^4 + 1)\,ds = 0.$
4. $\csc y\,dx + \sec x\,dy = 0.$
5. $\tan\theta\,dr + 2r\,d\theta = 0.$
6. $(e^v + 1)\cos u\,du + e^v(\sin u + 1)\,dv = 0.$
7. $(x + 4)(y^2 + 1)\,dx + y(x^2 + 3x + 2)\,dy = 0.$
8. $(x + y)\,dx - x\,dy = 0.$
9. $(2xy + 3y^2)\,dx - (2xy + x^2)\,dy = 0.$
10. $v^3\,du + (u^3 - uv^2)\,dv = 0.$
11. $\left(x\tan\dfrac{y}{x} + y\right)dx - x\,dy = 0.$
12. $(2s^2 + 2st + t^2)\,ds + (s^2 + 2st - t^2)\,dt = 0.$
13. $(x^3 + y^2\sqrt{x^2 + y^2})\,dx - xy\sqrt{x^2 + y^2}\,dy = 0.$
14. $(\sqrt{x + y} + \sqrt{x - y})\,dx + (\sqrt{x - y} - \sqrt{x + y})\,dy = 0.$

Solve the initial-value problems in Exercises 15 through 20.

15. $(y + 2)\,dx + y(x + 4)\,dy = 0, \quad y(-3) = -1.$
16. $8\cos^2 y\,dx + \csc^2 x\,dy = 0, \quad y\left(\dfrac{\pi}{12}\right) = \dfrac{\pi}{4}.$
17. $(3x + 8)(y^2 + 4)\,dx - 4y(x^2 + 5x + 6)\,dy = 0, \quad y(1) = 2.$
18. $(x^2 + 3y^2)\,dx - 2xy\,dy = 0, \quad y(2) = 6.$
19. $(2x - 5y)\,dx + (4x - y)\,dy = 0, \quad y(1) = 4.$
20. $(3x^2 + 9xy + 5y^2)\,dx - (6x^2 + 4xy)\,dy = 0, \quad y(2) = -6.$
21. (a) Show that the homogeneous equation

$$(Ax + By)\,dx + (Cx + Dy)\,dy = 0$$

is exact if and only if $B = C$.

(b) Show that the homogeneous equation

$$(Ax^2 + Bxy + Cy^2)\,dx + (Dx^2 + Exy + Fy^2)\,dy = 0$$

is exact if and only if $B = 2D$ and $E = 2C$.

22. Solve each of the following by two methods (see Exercise 21(a)):

(a) $(x + 2y)\,dx + (2x - y)\,dy = 0$.

(b) $(3x - y)\,dx - (x + y)\,dy = 0$.

23. Solve each of the following by two methods (see Exercise 21(b)):

(a) $(x^2 + 2y^2)\,dx + (4xy - y^2)\,dy = 0$.

(b) $(2x^2 + 2xy + y^2)\,dx + (x^2 + 2xy)\,dy = 0$.

24. (a) Prove that if $M\,dx + N\,dy = 0$ is a homogeneous equation, then the change of variables $x = uy$ transforms this equation into a separable equation in the variables u and x.

(b) Use the result of (a) to solve the equation of Example 2.12 of the text.

(c) Use the result of (a) to solve the equation of Example 2.13 of the text.

25. Suppose the equation $M\,dx + N\,dy = 0$ is homogeneous. Show that the transformation $x = r\cos\theta$, $y = r\sin\theta$ reduces this equation to a separable equation in the variables r and θ.

26. (a) Use the method of Exercise 25 to solve Exercise 8.

(b) Use the method of Exercise 25 to solve Exercise 9.

27. Suppose the equation

$$M\,dx + N\,dy = 0 \tag{A}$$

is homogeneous.

(a) Show that Equation (A) is invariant under the transformation

$$x = k\xi, \qquad y = k\eta, \tag{B}$$

where k is a constant.

(b) Show that the general solution of Equation (A) can be written in the form

$$x = c\phi\left(\frac{y}{x}\right), \tag{C}$$

where c is an arbitrary constant.

(c) Use the result of (b) to show that the solution (C) is also invariant under the transformation (B).

(d) Interpret geometrically the results proved in (a) and (c).

2.3 Linear Equations and Bernoulli Equations

A. *Linear Equations*

In Chapter 1 we gave the definition of the linear ordinary differential equation of order n; we now consider the linear ordinary differential equation of the first order.

DEFINITION

A first-order ordinary differential equation is linear *in the dependent variable y and the independent variable x if it is, or can be, written in the form*

$$\frac{dy}{dx} + P(x)y = Q(x). \tag{2.26}$$

For example, the equation

$$x\frac{dy}{dx} + (x + 1)y = x^3$$

is a first-order linear differential equation, for it can be written as

$$\frac{dy}{dx} + \left(1 + \frac{1}{x}\right)y = x^2,$$

which is of the form (2.26) with $P(x) = 1 + (1/x)$ and $Q(x) = x^2$.

Let us write Equation (2.26) in the form

$$[P(x)y - Q(x)]\,dx + dy = 0. \tag{2.27}$$

Equation (2.27) is of the form

$$M(x, y)\,dx + N(x, y)\,dy = 0,$$

where

$$M(x, y) = P(x)y - Q(x) \quad \text{and} \quad N(x, y) = 1.$$

Since

$$\frac{\partial M(x, y)}{\partial y} = P(x) \quad \text{and} \quad \frac{\partial N(x, y)}{\partial x} = 0,$$

Equation (2.27) is *not* exact unless $P(x) = 0$, in which case Equation (2.26) degenerates into a simple separable equation. However, Equation (2.27) possesses an integrating factor which depends on x only and may easily be found. Let us proceed to find it. Let us multiply Equation (2.27) by $\mu(x)$, obtaining

$$[\mu(x)P(x)y - \mu(x)Q(x)]\,dx + \mu(x)\,dy = 0. \tag{2.28}$$

By definition, $\mu(x)$ is an integrating factor of Equation (2.28) if and only if Equation (2.28) is exact; that is, if and only if

$$\frac{\partial}{\partial y}[\mu(x)P(x)y - \mu(x)Q(x)] = \frac{\partial}{\partial x}[\mu(x)].$$

This condition reduces to

$$\mu(x)P(x) = \frac{d}{dx}[\mu(x)]. \tag{2.29}$$

In (2.29), P is a known function of the independent variable x, but μ is an unknown function of x which we are trying to determine. Thus we write (2.29) as the differential equation

$$\mu P(x) = \frac{d\mu}{dx},$$

in the dependent variable μ and the independent variable x, where P is a known function of x. This differential equation is separable; separating the variables, we have

$$\frac{d\mu}{\mu} = P(x)\,dx.$$

Integrating, we obtain the particular solution

$$\ln|\mu| = \int P(x)\,dx$$

or

$$\mu = e^{\int P(x)\,dx}, \tag{2.30}$$

where it is clear that $\mu > 0$. Thus the linear equation (2.26) possesses an integrating factor of the form (2.30). Multiplying (2.26) by (2.30) gives

$$e^{\int P(x)\,dx}\frac{dy}{dx} + e^{\int P(x)\,dx}P(x)y = Q(x)e^{\int P(x)\,dx},$$

which is precisely

$$\frac{d}{dx}\left[e^{\int P(x)\,dx}y\right] = Q(x)e^{\int P(x)\,dx}.$$

Integrating this we obtain the solution of Equation (2.26) in the form

$$e^{\int P(x)\,dx}y = \int e^{\int P(x)\,dx}Q(x)\,dx + c,$$

where c is an arbitrary constant.

Summarizing this discussion, we have the following theorem:

THEOREM 2.4

The linear differential equation

$$\frac{dy}{dx} + P(x)y = Q(x) \tag{2.26}$$

has an integrating factor of the form

$$e^{\int P(x)\,dx}.$$

A one-parameter family of solutions of this equation is

$$ye^{\int P(x)\,dx} = \int e^{\int P(x)\,dx}Q(x)\,dx + c;$$

that is,

$$y = e^{-\int P(x)\,dx}\left[\int e^{\int P(x)\,dx}Q(x)\,dx + c\right].$$

Furthermore, it can be shown that this one-parameter family of solutions of the linear equation (2.26) includes all *solutions of (2.26).*

We consider several examples.

▶ Example 2.14

$$\frac{dy}{dx} + \left(\frac{2x + 1}{x}\right) y = e^{-2x}. \tag{2.31}$$

Here

$$P(x) = \frac{2x + 1}{x}$$

and hence an integrating factor is

$$\exp\left[\int P(x)\,dx\right] = \exp\left[\int\left(\frac{2x + 1}{x}\right) dx\right] = \exp\,(2x + \ln|x|)$$

$$= \exp\,(2x)\exp\,(\ln|x|) = x\exp\,(2x).*$$

Multiplying Equation (2.31) through by this integrating factor, we obtain

$$xe^{2x}\frac{dy}{dx} + e^{2x}(2x + 1)y = x$$

or

$$\frac{d}{dx}(xe^{2x}y) = x.$$

Integrating, we obtain the solutions

$$xe^{2x}y = \frac{x^2}{2} + c$$

or

$$y = \tfrac{1}{2}xe^{-2x} + \frac{c}{x}e^{-2x},$$

where c is an arbitrary constant.

▶ Example 2.15. Solve the initial-value problem that consists of the differential equation

$$(x^2 + 1)\frac{dy}{dx} + 4xy = x \tag{2.32}$$

and the initial condition

$$y(2) = 1. \tag{2.33}$$

The differential equation (2.32) is not in the form (2.26). We therefore divide by $x^2 + 1$ to obtain

$$\frac{dy}{dx} + \frac{4x}{x^2 + 1}y = \frac{x}{x^2 + 1}. \tag{2.34}$$

Equation (2.34) is in the standard form (2.26), where

$$P(x) = \frac{4x}{x^2 + 1}.$$

---

* The expressions e^x and $\exp x$ are identical.

An integrating factor is

$$\exp\left[\int P(x)\,dx\right] = \exp\left(\int \frac{4x\,dx}{x^2+1}\right) = \exp\left[\ln (x^2+1)^2\right] = (x^2+1)^2.$$

Multiplying Equation (2.34) through by this integrating factor, we have

$$(x^2+1)^2\frac{dy}{dx} + 4x(x^2+1)y = x(x^2+1)$$

or

$$\frac{d}{dx}[(x^2+1)^2 y] = x^3 + x.$$

We now integrate to obtain a one-parameter family of solutions of Equation (2.32) in the form

$$(x^2+1)^2 y = \frac{x^4}{4} + \frac{x^2}{2} + c.$$

Applying the initial condition (2.33), we have

$$25 = 6 + c.$$

Thus $c = 19$ and the solution of the initial-value problem under consideration is

$$(x^2+1)^2 y = \frac{x^4}{4} + \frac{x^2}{2} + 19.$$

▶ Example 2.16. Consider the differential equation

$$y^2\,dx + (3xy - 1)\,dy = 0. \tag{2.35}$$

Solving for dy/dx, this becomes

$$\frac{dy}{dx} = \frac{y^2}{1 - 3xy},$$

which is clearly *not* linear in y. Also, Equation (2.35) is *not* exact, separable, or homogeneous. It appears to be of a type which we have not yet encountered; but let us look a little closer. In Section 2.1, we pointed out that in the differential form of a first-order differential equation the roles of x and y are interchangeable, in the sense that either variable may be regarded as the dependent variable and the other as the independent variable. Considering differential equation (2.35) with this in mind, let us now regard x as the dependent variable and y as the independent variable. With this interpretation, we now write (2.35) in the derivative form

$$\frac{dx}{dy} = \frac{1 - 3xy}{y^2}$$

or

$$\frac{dx}{dy} + \frac{3}{y}x = \frac{1}{y^2}. \tag{2.36}$$

Now observe that Equation (2.36) is of the form

$$\frac{dx}{dy} + P(y)x = Q(y)$$

and so is *linear in x*. Thus the theory developed in this section may be applied to Equation (2.36) merely by interchanging the roles played by x and y. Thus an integrating factor is

$$\exp\left[\int P(y)\, dy\right] = \exp\left(\int \frac{3}{y}\, dy\right) = \exp\,(\ln |y|^3) = y^3.$$

Multiplying (2.36) by y^3 we obtain

$$y^3 \frac{dx}{dy} + 3y^2 x = y$$

or

$$\frac{d}{dy}[y^3 x] = y.$$

Integrating, we find the solutions in the form

$$y^3 x = \frac{y^2}{2} + c$$

or

$$x = \frac{1}{2y} + \frac{c}{y^3},$$

where c is an arbitrary constant.

B. Bernoulli Equations

We now consider a rather special type of equation which can be reduced to a linear equation by an appropriate transformation. This is the so-called Bernoulli equation.

DEFINITION

An equation of the form

$$\frac{dy}{dx} + P(x)y = Q(x)y^n \tag{2.37}$$

is called a Bernoulli differential equation.

We observe that if $n = 0$ or 1, then the Bernoulli equation (2.37) is actually a linear equation and is therefore readily solvable as such. However, in the general case in which $n \neq 0$ or 1, this simple situation does not hold and we must proceed in a different manner. We now state and prove Theorem 2.5, which gives a method of solution in the general case.

THEOREM 2.5

Suppose $n \neq 0$ or 1. Then the transformation $v = y^{1-n}$ reduces the Bernoulli equation

$$\frac{dy}{dx} + P(x)y = Q(x)y^n \tag{2.37}$$

to a linear equation in v.

Proof. We first multiply Equation (2.37) by y^{-n}, thereby expressing it in the equivalent form

$$y^{-n}\frac{dy}{dx} + P(x)y^{1-n} = Q(x). \tag{2.38}$$

If we let $v = y^{1-n}$, then

$$\frac{dv}{dx} = (1 - n)y^{-n}\frac{dy}{dx}$$

and Equation (2.38) transforms into

$$\frac{1}{1-n}\frac{dv}{dx} + P(x)v = Q(x)$$

or, equivalently,

$$\frac{dv}{dx} + (1 - n)P(x)v = (1 - n)Q(x).$$

Letting

$$P_1(x) = (1 - n)P(x)$$

and

$$Q_1(x) = (1 - n)Q(x),$$

this may be written

$$\frac{dv}{dx} + P_1(x)v = Q_1(x),$$

which is linear in v. *Q.E.D.*

▶ Example 2.17

$$\frac{dy}{dx} + y = xy^3. \tag{2.39}$$

This is a Bernoulli differential equation, where $n = 3$. We first multiply the equation through by y^{-3}, thereby expressing it in the equivalent form

$$y^{-3}\frac{dy}{dx} + y^{-2} = x.$$

If we let $v = y^{1-n} = y^{-2}$, then $dv/dx = -2y^{-3}(dy/dx)$ and the preceding differential equation transforms into the linear equation

$$-\frac{1}{2}\frac{dv}{dx} + v = x.$$

Writing this linear equation in the standard form

$$\frac{dv}{dx} - 2v = -2x, \tag{2.40}$$

we see that an integrating factor for this equation is

$$e^{\int P(x)\,dx} = e^{-\int 2\,dx} = e^{-2x}.$$

Multiplying (2.40) by e^{-2x}, we find

$$e^{-2x}\frac{dv}{dx} - 2e^{-2x}v = -2xe^{-2x}$$

or

$$\frac{d}{dx}(e^{-2x}v) = -2xe^{-2x}.$$

Integrating, we find

$$e^{-2x}v = \tfrac{1}{2}e^{-2x}(2x + 1) + c,$$

$$v = x + \tfrac{1}{2} + ce^{2x},$$

where c is an arbitrary constant. But

$$v = \frac{1}{y^2}.$$

Thus we obtain the solutions of (2.39) in the form

$$\frac{1}{y^2} = x + \tfrac{1}{2} + ce^{2x}.$$

Note. Consider the equation

$$\frac{df(y)}{dy}\frac{dy}{dx} + P(x)f(y) = Q(x), \tag{2.41}$$

where f is a known function of y. Letting $v = f(y)$, we have

$$\frac{dv}{dx} = \frac{dv}{dy}\frac{dy}{dx} = \frac{df(y)}{dy}\frac{dy}{dx},$$

and Equation (2.41) becomes

$$\frac{dv}{dx} + P(x)v = Q(x),$$

which is linear in v. We now observe that the Bernoulli differential equation (2.37) is a special case of Equation (2.41). Writing (2.37) in the form

$$y^{-n}\frac{dy}{dx} + P(x)y^{1-n} = Q(x)$$

and then multiplying through by $(1 - n)$, we have

$$(1 - n)y^{-n}\frac{dy}{dx} + P_1(x)y^{1-n} = Q_1(x),$$

where $P_1(x) = (1 - n)P(x)$ and $Q_1(x) = (1 - n)Q(x)$. This is of the form (2.41), where $f(y) = y^{1-n}$; letting $v = y^{1-n}$, it becomes

$$\frac{dv}{dx} + P_1(x)v = Q_1(x),$$

which is linear in v. For other special cases of (2.41), see Exercise 37.

Exercises

Solve the given differential equations in Exercises 1 through 18.

1. $\dfrac{dy}{dx} + \dfrac{3y}{x} = 6x^2.$

2. $x^4\dfrac{dy}{dx} + 2x^3y = 1.$

3. $\dfrac{dy}{dx} + 3y = 3x^2e^{-3x}.$

4. $\dfrac{dy}{dx} + 4xy = 8x.$

5. $\dfrac{dx}{dt} + \dfrac{x}{t^2} = \dfrac{1}{t^2}.$

6. $(u^2 + 1)\dfrac{dv}{du} + 4uv = 3u.$

7. $x\dfrac{dy}{dx} + \dfrac{2x + 1}{x + 1}y = x - 1.$

8. $(x^2 + x - 2)\dfrac{dy}{dx} + 3(x + 1)y = x - 1.$

9. $x\,dy + (xy + y - 1)\,dx = 0.$

10. $y\,dx + (xy^2 + x - y)\,dy = 0.$

11. $\dfrac{dr}{d\theta} + r\tan\theta = \cos\theta.$

12. $\cos\theta\,dr + (r\sin\theta - \cos^4\theta)\,d\theta = 0.$

13. $(\cos^2 x - y\cos x)\,dx - (1 + \sin x)\,dy = 0.$

14. $(y\sin 2x - \cos x)\,dx + (1 + \sin^2 x)\,dy = 0.$

15. $\dfrac{dy}{dx} - \dfrac{y}{x} = -\dfrac{y^2}{x}.$

16. $x\dfrac{dy}{dx} + y = -2x^6y^4.$

17. $dy + (4y - 8y^{-3})x\,dx = 0.$

18. $\dfrac{dx}{dt} + \dfrac{t + 1}{2t}x = \dfrac{t + 1}{xt}.$

Solve the initial-value problems in Exercises 19 through 30.

19. $x\dfrac{dy}{dx} - 2y = 2x^4, \quad y(2) = 8.$

20. $\dfrac{dy}{dx} + 3x^2y = x^2, \quad y(0) = 2.$

21. $e^x[y - 3(e^x + 1)^2]\,dx + (e^x + 1)\,dy = 0, \quad y(0) = 4.$

22. $2x(y + 1)\,dx - (x^2 + 1)\,dy = 0, \quad y(1) = -5.$

23. $\dfrac{dr}{d\theta} + r\tan\theta = \cos^2\theta, \quad r\left(\dfrac{\pi}{4}\right) = 1.$

24. $\dfrac{dx}{dt} - x = \sin 2t, \quad x(0) = 0.$

25. $\dfrac{dy}{dx} + \dfrac{y}{2x} = \dfrac{x}{y^3}, \quad y(1) = 2.$

26. $x\dfrac{dy}{dx} + y = (xy)^{3/2}, \quad y(1) = 4.$

27. $\dfrac{dy}{dx} + y = f(x)$, where $f(x) = \begin{cases} 2, & 0 \le x < 1, \\ 0, & x \ge 1, \end{cases}$

    $y(0) = 0.$

28. $\dfrac{dy}{dx} + y = f(x)$, where $f(x) = \begin{cases} 5, & 0 \le x < 10, \\ 1, & x \ge 10, \end{cases}$

    $y(0) = 6.$

29. $\dfrac{dy}{dx} + y = f(x)$, where $f(x) = \begin{cases} e^{-x}, & 0 \le x < 2, \\ e^{-2}, & x \ge 2, \end{cases}$

    $y(0) = 1.$

30. $(x + 2)\dfrac{dy}{dx} + y = f(x)$, where $f(x) = \begin{cases} 2x, & 0 \le x < 2, \\ 4, & x \ge 2, \end{cases}$

    $y(0) = 4.$

31. Consider the equation $a(dy/dx) + by = ke^{-\lambda x}$, where a, b, and k are positive constants and λ is a nonnegative constant.

    (a) Solve this equation.

    (b) Show that if $\lambda = 0$ every solution approaches k/b as $x \to \infty$ but if $\lambda > 0$ every solution approaches 0 as $x \to \infty$.

32. Consider the differential equation

$$\frac{dy}{dx} + P(x)y = 0.$$

(a) Show that if f and g are two solutions of this equation and c_1 and c_2 are arbitrary constants, then $c_1 f + c_2 g$ is also a solution of this equation.

(b) Extending the result of (a), show that if $f_1, f_2, \ldots, f_n$ are n solutions of this equation and $c_1, c_2, \ldots, c_n$ are n arbitrary constants, then

$$\sum_{k=1}^{n} c_k f_k$$

is also a solution of this equation.

33. Consider the differential equation

$$\frac{dy}{dx} + P(x)y = 0, \tag{A}$$

where P is continuous on a real interval I.

(a) Show that the function f such that $f(x) = 0$ for all $x \in I$ is a solution of this equation.

(b) Show that if f is a solution of (A) such that $f(x_0) = 0$ for some $x_0 \in I$, then $f(x) = 0$ for all $x \in I$.

(c) Show that if f and g are two solutions of (A) such that $f(x_0) = g(x_0)$ for some $x_0 \in I$, then $f(x) = g(x)$ for all $x \in I$.

34. (a) Prove that if f and g are two different solutions of

$$\frac{dy}{dx} + P(x)y = Q(x), \tag{A}$$

then $f - g$ is a solution of the equation

$$\frac{dy}{dx} + P(x)y = 0.$$

(b) Thus show that if f and g are two different solutions of Equation (A) and c is an arbitrary constant, then

$$c(f - g) + f$$

is a one-parameter family of solutions of (A).

35. (a) Let f_1 be a solution of

$$\frac{dy}{dx} + P(x)y = Q_1(x)$$

and f_2 be a solution of

$$\frac{dy}{dx} + P(x)y = Q_2(x),$$

where P, Q_1, and Q_2 are all defined on the same real interval I. Prove that

$f_1 + f_2$ is a solution of

$$\frac{dy}{dx} + P(x)y = Q_1(x) + Q_2(x)$$

on I.

(b) Use the result of (a) to solve the equation

$$\frac{dy}{dx} + y = 2 \sin x + 5 \sin 2x.$$

36. (a) Extend the result of Exercise 35(a) to cover the case of the equation

$$\frac{dy}{dx} + P(x)y = \sum_{k=1}^{n} Q_k(x),$$

where P, Q_k $(k = 1, 2, \ldots, n)$ are all defined on the same real interval I.

(b) Use the result obtained in (a) to solve the equation

$$\frac{dy}{dx} + y = \sum_{k=1}^{5} \sin kx.$$

37. Solve each of the following equations of the form (2.41):

(a) $\cos y \dfrac{dy}{dx} + \dfrac{1}{x} \sin y = 1.$

(b) $(y + 1) \dfrac{dy}{dx} + x(y^2 + 2y) = x.$

38. The equation

$$\frac{dy}{dx} = A(x)y^2 + B(x)y + C(x) \qquad \text{(A)}$$

is called *Riccati's equation*.

(a) Show that if $A(x) = 0$ for all x, then Equation (A) is a linear equation, whereas if $C(x) = 0$ for all x, then Equation (A) is a Bernoulli equation.

(b) Show that if f is any solution of Equation (A), then the transformation

$$y = f + \frac{1}{v}$$

reduces (A) to a linear equation in v.

In each of Exercises 39–41, use the result of Exercise 38(b) and the given solution to find a one-parameter family of solutions of the given Riccati equation:

39. $\dfrac{dy}{dx} = (1 - x)y^2 + (2x - 1)y - x$; given solution $f(x) = 1$.

40. $\dfrac{dy}{dx} = -y^2 + xy + 1$; given solution $f(x) = x$.

41. $\dfrac{dy}{dx} = -8xy^2 + 4x(4x + 1)y - (8x^3 + 4x^2 - 1)$; given solution $f(x) = x$.

2.4 Special Integrating Factors and Transformations

We have thus far encountered five distinct types of first-order equations for which solutions may be obtained by exact methods, namely, exact, separable, homogeneous, linear, and Bernoulli equations. In the case of exact equations, we follow a definite procedure to directly obtain solutions. For the other four types definite procedures for solution are also available, but in these cases the procedures are actually not quite so direct. In the cases of both separable and linear equations we actually multiply by appropriate integrating factors which reduce the given equations to equations which are of the more basic exact type. For both homogeneous and Bernoulli equations we make appropriate transformations which reduce such equations to equations which are of the more basic separable and linear types, respectively.

This suggests two general plans of attack to be used in solving a differential equation which is *not* of one of the five types mentioned. Either (1) we might multiply the given equation by an appropriate integrating factor and directly reduce it to an exact equation, or (2) we might make an appropriate transformation which will reduce the given equation to an equation of some more basic type (say, one of the five types already studied). Unfortunately no general directions can be given for finding an appropriate integrating factor or transformation in all cases. However, there is a variety of special types of equations which either possess special types of integrating factors or to which special transformations may be applied. We shall consider a few of these in this section. Since these types are relatively unimportant, in most cases we shall simply state the relevant theorem and leave the proof to the exercises.

A. *Finding Integrating Factors*

The so-called separable equations considered in Section 2.2 always possess integrating factors which may be determined by immediate inspection. While it is true that some nonseparable equations also possess integrating factors which may be determined "by inspection," such equations are rarely encountered except in differential equations texts on pages devoted to an exposition of this dubious "method." Even then a considerable amount of knowledge and skill are often required.

Let us attempt to attack the problem more systematically. Suppose the equation

$$M(x, y)\,dx + N(x, y)\,dy = 0 \tag{2.42}$$

is *not* exact and that $\mu(x, y)$ is an integrating factor of it. Then the equation

$$\mu(x, y)M(x, y)\,dx + \mu(x, y)N(x, y)\,dy = 0 \tag{2.43}$$

is exact. Now using the criterion (2.7) for exactness, Equation (2.43) is exact if and only if

$$\frac{\partial}{\partial y}[\mu(x, y)M(x, y)] = \frac{\partial}{\partial x}[\mu(x, y)N(x, y)].$$

This condition reduces to

$$N(x, y)\frac{\partial \mu(x, y)}{\partial x} - M(x, y)\frac{\partial \mu(x, y)}{\partial y} = \left[\frac{\partial M(x, y)}{\partial y} - \frac{\partial N(x, y)}{\partial x}\right]\mu(x, y),$$

Here M and N are known functions of x and y, but μ is an unknown function of x and y which we are trying to determine. Thus we write the preceding condition in the form

$$N(x, y)\frac{\partial \mu}{\partial x} - M(x, y)\frac{\partial \mu}{\partial y} = \left[\frac{\partial M(x, y)}{\partial y} - \frac{\partial N(x, y)}{\partial x}\right]\mu. \tag{2.44}$$

Hence μ is an integrating factor of the differential equation (2.42) if and only if it is a solution of the differential equation (2.44). Equation (2.44) is a partial differential equation for the general integrating factor μ, and we are in no position to attempt to solve such an equation. Let us instead attempt to determine integrating factors of certain special types. But what special types might we consider? Let us recall that the linear differential equation

$$\frac{dy}{dx} + P(x)y = Q(x)$$

always possesses the integrating factor $e^{\int P(x)\,dx}$, which depends only upon x. Perhaps other equations also have integrating factors which depend only upon x. We therefore multiply Equation (2.42) by $\mu(x)$, where μ depends upon x alone. We obtain

$$\mu(x)M(x, y)\,dx + \mu(x)N(x, y)\,dy = 0.$$

This is exact if and only if

$$\frac{\partial}{\partial y}[\mu(x)M(x, y)] = \frac{\partial}{\partial x}[\mu(x)N(x, y)].$$

Now M and N are known functions of both x and y, but here the integrating factor μ depends only upon x. Thus the above condition reduces to

$$\mu(x)\frac{\partial M(x, y)}{\partial y} = \mu(x)\frac{\partial N(x, y)}{\partial x} + N(x, y)\frac{d\mu(x)}{dx}$$

or

$$\frac{d\mu(x)}{\mu(x)} = \frac{1}{N(x, y)}\left[\frac{\partial M(x, y)}{\partial y} - \frac{\partial N(x, y)}{\partial x}\right]dx. \tag{2.45}$$

If

$$\frac{1}{N(x, y)}\left[\frac{\partial M(x, y)}{\partial y} - \frac{\partial N(x, y)}{\partial x}\right]$$

involves the variable y, this equation then involves two dependent variables and we again have difficulties. However, if

$$\frac{1}{N(x, y)}\left[\frac{\partial M(x, y)}{\partial y} - \frac{\partial N(x, y)}{\partial x}\right]$$

depends upon x only, Equation (2.45) is a separated ordinary equation in the single independent variable x and the single dependent variable μ. In this case we may integrate to obtain the integrating factor

$$\mu(x) = \exp\left\{\int \frac{1}{N(x, y)}\left[\frac{\partial M(x, y)}{\partial y} - \frac{\partial N(x, y)}{\partial x}\right]dx\right\}.$$

In like manner, if

$$\frac{1}{M(x, y)}\left[\frac{\partial N(x, y)}{\partial x} - \frac{\partial M(x, y)}{\partial y}\right]$$

depends upon y only, then we may obtain an integrating factor which depends only on y.

We summarize these observations in the following theorem.

THEOREM 2.6

Consider the differential equation

$$M(x, y)\,dx + N(x, y)\,dy = 0. \tag{2.42}$$

If

$$\frac{1}{N(x, y)}\left[\frac{\partial M(x, y)}{\partial y} - \frac{\partial N(x, y)}{\partial x}\right] \tag{2.46}$$

depends upon x only, then

$$\exp\left\{\int \frac{1}{N(x, y)}\left[\frac{\partial M(x, y)}{\partial y} - \frac{\partial N(x, y)}{\partial x}\right] dx\right\} \tag{2.47}$$

is an integrating factor of Equation (2.42). If

$$\frac{1}{M(x, y)}\left[\frac{\partial N(x, y)}{\partial x} - \frac{\partial M(x, y)}{\partial y}\right] \tag{2.48}$$

depends upon y only, then

$$\exp\left\{\int \frac{1}{M(x, y)}\left[\frac{\partial N(x, y)}{\partial x} - \frac{\partial M(x, y)}{\partial y}\right] dy\right\} \tag{2.49}$$

is an integrating factor of Equation (2.42).

We emphasize that, given a differential equation, we have no assurance in general that either of these procedures will apply. It may well turn out that (2.46) involves y and (2.48) involves x for the differential equation under consideration. Then we must seek other procedures. However, since the calculation of the expressions (2.46) and (2.48) is generally quite simple, it is often worthwhile to calculate them before trying something more complicated.

▶ Example 2.18. Consider the differential equation

$$(2x^2 + y)\,dx + (x^2y - x)\,dy = 0. \tag{2.50}$$

Let us first observe that this equation is *not* exact, separable, homogeneous, linear, or Bernoulli. Let us then see if Theorem 2.6 applies. Here $M(x, y) = 2x^2 + y$, and $N(x, y) = x^2y - x$, and the expression (2.46) becomes

$$\frac{1}{x^2y - x}\,[1 - (2xy - 1)] = \frac{2(1 - xy)}{x(xy - 1)} = -\frac{2}{x}.$$

This depends upon x only, and so

$$\exp\left(-\int \frac{2}{x}\,dx\right) = \exp(-2\ln|x|) = \frac{1}{x^2}$$

is an integrating factor of Equation (2.50). Multiplying (2.50) by this integrating factor, we obtain the equation

$$\left(2 + \frac{y}{x^2}\right) dx + \left(y - \frac{1}{x}\right) dy = 0. \tag{2.51}$$

The student may readily verify that Equation (2.51) is indeed exact and that the solution is

$$2x + \frac{y^2}{2} - \frac{y}{x} = c.$$

More and more specialized results concerning particular types of integrating factors corresponding to particular types of equations are known. However, instead of going into such special cases we shall now proceed to investigate certain useful transformations.

B. A Special Transformation

We have already made use of transformations in reducing both homogeneous and Bernoulli equations to more tractable types. Another type of equation which can be reduced to a more basic type by means of a suitable transformation is an equation of the form

$$(a_1x + b_1y + c_1)\,dx + (a_2x + b_2y + c_2)\,dy = 0.$$

We state the following theorem concerning this equation.

THEOREM 2.7

Consider the equation

$$(a_1x + b_1y + c_1)\,dx + (a_2x + b_2y + c_2)\,dy = 0, \tag{2.52}$$

where a_1, b_1, c_1, a_2, b_2, and c_2 are constants.

Case 1. *If $a_2/a_1 \neq b_2/b_1$, then the transformation*

$$x = X + h,$$

$$y = Y + k,$$

where (h, k) is the solution of the system

$$a_1h + b_1k + c_1 = 0,$$

$$a_2h + b_2k + c_2 = 0,$$

reduces Equation (2.52) to the homogeneous equation

$$(a_1X + b_1Y)\,dX + (a_2X + b_2Y)\,dY = 0$$

in the variables X and Y.

Case 2. If $a_2/a_1 = b_2/b_1 = k$, then the transformation $z = a_1 x + b_1 y$ reduces the equation (2.52) to a separable equation in the variables x and z.

Examples 2.19 and 2.20 illustrate the two cases of this theorem.

▶ Example 2.19

$$(x - 2y + 1)\,dx + (4x - 3y - 6)\,dy = 0. \tag{2.53}$$

Here $a_1 = 1$, $b_1 = -2$, $a_2 = 4$, $b_2 = -3$, and so

$$\frac{a_2}{a_1} = 4 \quad \text{but} \quad \frac{b_2}{b_1} = \frac{3}{2} \neq \frac{a_2}{a_1}.$$

Therefore this is Case 1 of Theorem 2.7. We make the transformation

$$x = X + h,$$
$$y = Y + k,$$

where (h, k) is the solution of the system

$$h - 2k + 1 = 0,$$
$$4h - 3k - 6 = 0.$$

The solution of this system is $h = 3$, $k = 2$, and so the transformation is

$$x = X + 3,$$
$$y = Y + 2.$$

This reduces Equation (2.53) to the homogeneous equation

$$(X - 2Y)\,dX + (4X - 3Y)\,dY = 0. \tag{2.54}$$

Now following the procedure in Section 2.2 we first put this homogeneous equation in the form

$$\frac{dY}{dX} = \frac{1 - 2(Y/X)}{3(Y/X) - 4}$$

and let $Y = vX$ to obtain

$$v + X\frac{dv}{dX} = \frac{1 - 2v}{3v - 4}.$$

This reduces to

$$\frac{(3v - 4)\,dv}{3v^2 - 2v - 1} = -\frac{dX}{X}. \tag{2.55}$$

Integrating (we recommend the use of tables here), we obtain

$$\tfrac{1}{2}\ln|3v^2 - 2v - 1| - \tfrac{3}{4}\ln\left|\frac{3v - 3}{3v + 1}\right| = -\ln|X| + \ln|c_1|,$$

or

$$\ln(3v^2 - 2v - 1)^2 - \ln\left|\frac{3v - 3}{3v + 1}\right|^3 = \ln\left(\frac{c_1^4}{X^4}\right),$$

or

$$\ln\left|\frac{(3v+1)^5}{v-1}\right| = \ln\left(\frac{c_1^4}{X^4}\right),$$

or, finally,

$$X^4|(3v+1)^5| = c|v-1|,$$

where $c = c_1^4$. These are the solutions of the separable equation (2.55). Now replacing v by Y/X, we obtain the solutions of the homogeneous equation (2.54) in the form

$$|3Y+X|^5 = c|Y-X|.$$

Finally, replacing X by $x-3$ and Y by $y-2$ from the original transformation, we obtain the solutions of the differential equation (2.53) in the form

$$|3(y-2)+(x-3)|^5 = c|y-2-x+3|$$

or

$$|x+3y-9|^5 = c|y-x+1|.$$

▶ **Example 2.20**

$$(x+2y+3)\,dx + (2x+4y-1)\,dy = 0. \tag{2.56}$$

Here $a_1 = 1$, $b_1 = 2$, $a_2 = 2$, $b_2 = 4$, and $a_2/a_1 = b_2/b_1 = 2$. Therefore this is Case 2 of Theorem 2.7. We therefore let

$$z = x + 2y,$$

and Equation (2.56) transforms into

$$(z+3)\,dx + (2z-1)\left(\frac{dz-dx}{2}\right) = 0$$

or

$$7\,dx + (2z-1)\,dz = 0,$$

which is separable. Integrating, we have

$$7x + z^2 - z = c.$$

Replacing z by $x+2y$ we obtain the solution of Equation (2.56) in the form

$$7x + (x+2y)^2 - (x+2y) = c$$

or

$$x^2 + 4xy + 4y^2 + 6x - 2y = c.$$

C. *Other Special Types and Methods; An Important Reference*

Many other special types of first-order equations exist for which corresponding special methods of solution are known. We shall not go into such highly specialized types in this book. Instead we refer the reader to the book *Differentialgleichungen: Losungsmethoden und Losungen*, by E. Kamke (Chelsea, New York, 1948). This remarkable volume contains discussions of a large number of special types of equations and their solutions. We strongly suggest that the reader consult this book whenever he encounters an unfamiliar type of equation. Of course one may encounter an equation for which no exact method of solution is known. In such a case one must resort to various methods of approximation. We shall consider some of these general methods in Chapter 8.

Exercises

Solve each differential equation in Exercises 1 through 4 by first finding an integrating factor.

1. $(5xy + 4y^2 + 1)\,dx + (x^2 + 2xy)\,dy = 0$.
2. $(2x + \tan y)\,dx + (x - x^2 \tan y)\,dy = 0$.
3. $[y^2(x + 1) + y]\,dx + (2xy + 1)\,dy = 0$.
4. $(2xy^2 + y)\,dx + (2y^3 - x)\,dy = 0$.

In each of Exercises 5 and 6 find an integrating factor of the form $x^p y^q$ and solve.

5. $(4xy^2 + 6y)\,dx + (5x^2y + 8x)\,dy = 0$.
6. $(8x^2y^3 - 2y^4)\,dx + (5x^3y^2 - 8xy^3)\,dy = 0$.

Solve each differential equation in Exercises 7 through 10 by making a suitable transformation.

7. $(5x + 2y + 1)\,dx + (2x + y + 1)\,dy = 0$.
8. $(3x - y + 1)\,dx - (6x - 2y - 3)\,dy = 0$.
9. $(x - 2y - 3)\,dx + (2x + y - 1)\,dy = 0$.
10. $(10x - 4y + 12)\,dx - (x + 5y + 3)\,dy = 0$.

Solve the initial-value problems in Exercises 11 through 14.

11. $(6x + 4y + 1)\,dx + (4x + 2y + 2)\,dy = 0, \quad y(\tfrac{1}{2}) = 3$.
12. $(3x - y - 6)\,dx + (x + y + 2)\,dy = 0, \quad y(2) = -2$.
13. $(2x + 3y + 1)\,dx + (4x + 6y + 1)\,dy = 0, \quad y(-2) = 2$.
14. $(4x + 3y + 1)\,dx + (x + y + 1)\,dy = 0, \quad y(3) = -4$.
15. Prove Theorem 2.6.
16. Prove Theorem 2.7.
17. Show that if $\mu(x, y)$ and $\nu(x, y)$ are integrating factors of

$$M(x, y)\,dx + N(x, y)\,dy = 0 \tag{A}$$

such that $\mu(x, y)/\nu(x, y)$ is not constant, then

$$\mu(x, y) = c\nu(x, y)$$

is a solution of Equation (A) for every constant c.

18. Show that if the equation

$$M(x, y)\,dx + N(x, y)\,dy = 0 \tag{A}$$

is homogeneous and $M(x, y)x + N(x, y)y \neq 0$, then $1/[M(x, y)x + N(x, y)y]$ is an integrating factor of (A).

19. Show that if the equation $M(x, y)\,dx + N(x, y)\,dy = 0$ is both homogeneous and exact and if $M(x, y)x + N(x, y)y$ is not a constant, then the solution of this equation is $M(x, y)x + N(x, y)y = c$, where c is an arbitrary constant.

20. An equation which is of the form

$$y = px + f(p), \tag{A}$$

where $p \equiv dy/dx$ and f is a given function, is called a *Clairaut equation*. Given such an equation, proceed as follows:

1. Differentiate (A) with respect to x and simplify to obtain

$$[x + f'(p)]\frac{dp}{dx} = 0. \tag{B}$$

Observe that (B) is a first-order differential equation in x and p.

2. Assume $x + f'(p) \neq 0$, divide through by this factor, and solve the resulting equation to obtain

$$p = c, \tag{C}$$

where c is an arbitrary constant.

3. Eliminate p between (A) and (C) to obtain

$$y = cx + f(c). \tag{D}$$

Note that (D) is a one-parameter family of solutions of (A) and compare the *form* of differential equation (A) with the *form* of the family of solutions (D).

4. *Remark*. Assuming $x + f'(p) = 0$ and then eliminating p between (A) and $x + f'(p) = 0$ may lead to an "extra" solution which is *not* a member of the one-parameter family of solutions of the form (D). Such an extra solution is usually called a *singular solution*. For a specific example, see Exercise 21.

21. Consider the *Clairaut equation*

$$y = px + p^2, \quad \text{where } p \equiv \frac{dy}{dx}.$$

(a) Find a one-parameter family of solutions of this equation.

(b) Proceed as in the Remark of Exercise 20 and find an "extra" solution which is not a member of the one-parameter family found in part (a).

(c) Graph the integral curves corresponding to several members of the one-parameter family of part (a); graph the integral curve corresponding to the "extra" solution of part (b); and describe the geometric relationship between the graphs of the members of the one-parameter family and the graph of the "extra" solution.

Suggested Reading

AGNEW, R., *Differential Equations*, 2nd ed. (McGraw-Hill, New York, 1960).

BOYCE, W., and R. DIPRIMA, *Elementary Differential Equations*, 2nd ed. (Wiley, New York, 1969).

BRAUER, F., and J. NOHEL, *Ordinary Differential Equations: A First Course* (Benjamin, New York, 1967).

FORD, L., *Differential Equations*, 2nd ed. (McGraw-Hill, New York, 1955).

KAPLAN, W., *Ordinary Differential Equations* (Addison-Wesley, Reading, Mass., 1958).

KREIDER, D., R. KULLER, and D. OSTBERG, *Elementary Differential Equations* (Addison-Wesley, Reading, Mass., 1968).

RAINVILLE, E., and P. BEDIENT, *Elementary Differential Equations*, 4th ed. (Macmillan, New York, 1969).

RITGER, P., and N. ROSE, *Differential Equations with Applications* (McGraw-Hill, New York, 1968).

Applications of First-Order Equations

In Chapter 1 we pointed out that differential equations originate from the mathematical formulation of a great variety of problems in science and engineering. In this chapter we consider problems that give rise to some of the types of first-order ordinary differential equations studied in Chapter 2. First, we formulate the problem mathematically, thereby obtaining a differential equation. Then we solve the equation and attempt to interpret the solution in terms of the quantities involved in the original problem.

3.1 Orthogonal and Oblique Trajectories

A. *Orthogonal Trajectories*

DEFINITION

Let

$$F(x, y, c) = 0 \tag{3.1}$$

be a given one-parameter family of curves in the xy plane. A curve which intersects the curves of the family (3.1) at right angles is called an orthogonal trajectory *of the given family.*

▶ Example 3.1. Consider the family of circles

$$x^2 + y^2 = c^2 \tag{3.2}$$

with center at the origin and radius c. Each straight line through the origin,

$$y = kx, \tag{3.3}$$

is an orthogonal trajectory of the family of circles (3.2). Conversely, each circle of the family (3.2) is an orthogonal trajectory of the family of straight lines (3.3). The families (3.2) and (3.3) are orthogonal trajectories of each other. In Figure 3.1 several members of the family of circles (3.2), drawn solidly, and several members of the family of straight lines (3.3), drawn with dashes, are shown.

The problem of finding the orthogonal trajectories of a given family of curves arises in many physical situations. For example, in a two-dimensional electric field the lines

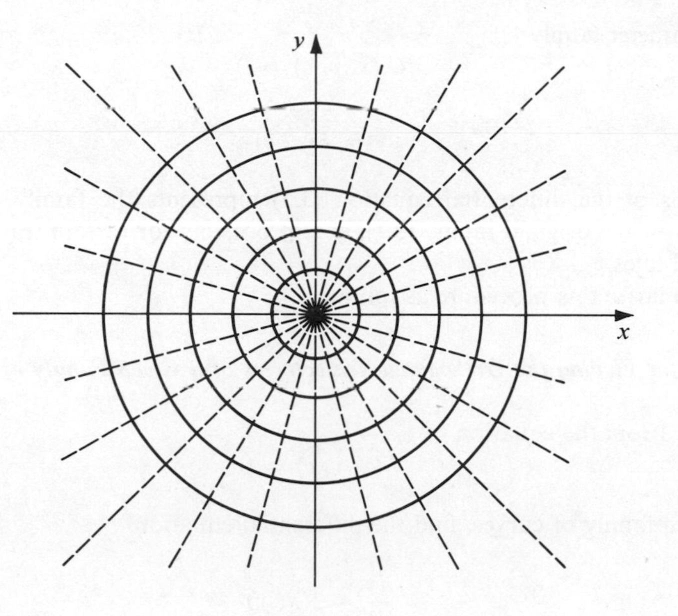

FIGURE 3.1

of force (flux lines) and the equipotential curves are orthogonal trajectories of each other.

We now proceed to find the orthogonal trajectories of a family of curves

$$F(x, y, c) = 0. \tag{3.1}$$

We obtain the differential equation of the family (3.1) by first differentiating Equation (3.1) implicitly with respect to x and then eliminating the parameter c between the derived equation so obtained and the given equation (3.1) itself. We assume that the resulting differential equation of the family (3.1) can be expressed in the form

$$\frac{dy}{dx} = f(x, y). \tag{3.4}$$

Thus the curve C of the given family (3.1) which passes through the point (x, y) has the slope $f(x, y)$ there. Since an orthogonal trajectory of the given family intersects each curve of the family at right angles, the slope of the orthogonal trajectory to C at (x, y) is

$$-\frac{1}{f(x, y)}.$$

Thus the differential equation of the family of orthogonal trajectories is

$$\frac{dy}{dx} = -\frac{1}{f(x, y)}. \tag{3.5}$$

A one-parameter family

$$G(x, y, c) = 0$$

or

$$y = F(x, c)$$

of solutions of the differential equation (3.5) represents the family of orthogonal trajectories of the original family (3.1), except possibly for certain trajectories which are vertical lines.

We summarize this procedure as follows:

Procedure for Finding the Orthogonal Trajectories of a Given Family of Curves

Step 1. From the equation

$$F(x, y, c) = 0 \tag{3.1}$$

of the given family of curves, find the differential equation

$$\frac{dy}{dx} = f(x, y) \tag{3.4}$$

of this family.

Step 2. In the differential equation $dy/dx = f(x, y)$ so found in Step 1, replace $f(x, y)$ by its negative reciprocal $-1/f(x, y)$. This gives the differential equation

$$\frac{dy}{dx} = -\frac{1}{f(x, y)} \tag{3.5}$$

of the orthogonal trajectories.

Step 3. Obtain a one-parameter family

$$G(x, y, c) = 0 \quad \text{or} \quad y = F(x, c)$$

of solutions of the differential equation (3.5), thus obtaining the desired family of orthogonal trajectories (except possibly for certain trajectories which are vertical lines and which must be determined separately).

Caution. In Step 1, in finding the differential equation (3.4) of the given family, be sure to eliminate the parameter c during the process.

▶ Example 3.2. In Example 3.1 we stated that the set of orthogonal trajectories of the family of circles

$$x^2 + y^2 = c^2 \tag{3.2}$$

is the family of straight lines

$$y = kx. \tag{3.3}$$

Let us verify this using the procedure outlined above.

Step 1. Differentiating the equation

$$x^2 + y^2 = c^2 \tag{3.2}$$

of the given family, we obtain

$$x + y\frac{dy}{dx} = 0.$$

From this we obtain the differential equation

$$\frac{dy}{dx} = -\frac{x}{y} \tag{3.6}$$

of the given family (3.2). (Note that the parameter c was automatically eliminated in this case.)

Step 2. We replace $-x/y$ by its negative reciprocal y/x in the differential equation (3.6) to obtain the differential equation

$$\frac{dy}{dx} = \frac{y}{x} \tag{3.7}$$

of the orthogonal trajectories.

Step 3. We now solve the differential equation (3.7). Separating variables, we have

$$\frac{dy}{y} = \frac{dx}{x};$$

integrating, we obtain

$$y = kx. \tag{3.3}$$

This is a one-parameter family of solutions of the differential equation (3.7) and thus represents the family of orthogonal trajectories of the given family of circles (3.2) (except for the single trajectory which is the vertical line $x = 0$ and which may be determined by inspection).

▶ **Example 3.3.** Find the orthogonal trajectories of the family of parabolas $y = cx^2$.

Step 1. We first find the differential equation of the given family

$$y = cx^2. \tag{3.8}$$

Differentiating, we obtain

$$\frac{dy}{dx} = 2cx. \tag{3.9}$$

Eliminating the parameter c between Equations (3.8) and (3.9), we obtain the differential equation of the family (3.8) in the form

$$\frac{dy}{dx} = \frac{2y}{x}. \tag{3.10}$$

Step 2. We now find the differential equation of the orthogonal trajectories by replacing $2y/x$ in (3.10) by its negative reciprocal, obtaining

$$\frac{dy}{dx} = -\frac{x}{2y}. \tag{3.11}$$

Step 3. We now solve the differential equation (3.11). Separating variables, we have

$$2y\,dy = -x\,dx.$$

Integrating, we obtain the one-parameter family of solutions of (3.11) in the form

$$x^2 + 2y^2 = k^2,$$

where k is an arbitrary constant. This is the family of orthogonal trajectories of (3.8); it is clearly a family of ellipses with centers at the origin and major axes along the x-axis. Some members of the original family of parabolas and some of the orthogonal trajectories (the ellipses) are shown in Figure 3.2.

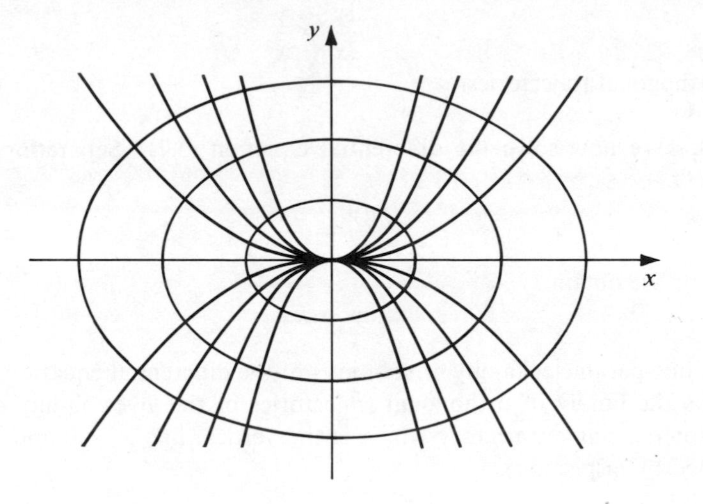

FIGURE 3.2

B. Oblique Trajectories

DEFINITION

Let

$$F(x, y, c) = 0 \tag{3.12}$$

be a one-parameter family of curves. A curve which intersects the curves of the family (3.12) at a constant angle $\alpha \neq 90°$ *is called an* oblique trajectory *of the given family.*

Suppose the differential equation of a family is

$$\frac{dy}{dx} = f(x, y). \tag{3.13}$$

Then the curve of the family (3.13) through the point (x, y) has slope $f(x, y)$ at (x, y) and hence its tangent line has angle of inclination $\tan^{-1}[f(x, y)]$ there. The tangent line of an oblique trajectory which intersects this curve at the angle α will thus have angle of inclination

$$\tan^{-1}[f(x, y)] + \alpha$$

at the point (x, y). Hence the slope of this oblique trajectory is given by

$$\tan\{\tan^{-1}[f(x, y)] + \alpha\} = \frac{f(x, y) + \tan\alpha}{1 - f(x, y)\tan\alpha}.$$

Thus the differential equation of such a family of oblique trajectories is given by

$$\frac{dy}{dx} = \frac{f(x, y) + \tan\alpha}{1 - f(x, y)\tan\alpha}.$$

Thus to obtain a family of oblique trajectories intersecting a given family of curves at the constant angle $\alpha \neq 90°$, we may follow the three steps in the above procedure (page 68) for finding the orthogonal trajectories, except that we replace Step 2 by the following step:

Step 2′. In the differential equation $dy/dx = f(x, y)$ of the given family, replace $f(x, y)$ by the expression

$$\frac{f(x, y) + \tan\alpha}{1 - f(x, y)\tan\alpha}. \tag{3.14}$$

▶ **Example 3.4.** Find a family of oblique trajectories that intersect the family of straight lines $y = cx$ at angle 45°.

Step 1. From $y = cx$, we find $dy/dx = c$. Eliminating c, we obtain the differential equation

$$\frac{dy}{dx} = \frac{y}{x} \tag{3.15}$$

of the given family of straight lines.

Step 2′. We replace $f(x, y) = y/x$ in Equation (3.15) by

$$\frac{f(x, y) + \tan\alpha}{1 - f(x, y)\tan\alpha} = \frac{y/x + 1}{1 - y/x} = \frac{x + y}{x - y}$$

($\tan\alpha = \tan 45° = 1$ here). Thus the differential equation of the desired oblique trajectories is

$$\frac{dy}{dx} = \frac{x + y}{x - y}. \tag{3.16}$$

Step 3. We now solve the differential equation (3.16). Observing that it is a homogeneous differential equation, we let $y = vx$ to obtain

$$v + x\frac{dv}{dx} = \frac{1 + v}{1 - v}.$$

After simplifications this becomes

$$\frac{(v - 1)\,dv}{v^2 + 1} = -\frac{dx}{x}.$$

Integrating we obtain

$$\tfrac{1}{2}\ln(v^2 + 1) - \arctan v = -\ln|x| - \ln|c|$$

or

$$\ln c^2x^2(v^2 + 1) - 2\arctan v = 0.$$

Replacing v by y/x, we obtain the family of oblique trajectories in the form

$$\ln c^2(x^2 + y^2) - 2\arctan\frac{y}{x} = 0.$$

Exercises

In Exercises 1–9 find the orthogonal trajectories of each given family of curves. In each case sketch several members of the family and several of the orthogonal trajectories on the same set of axes.

1. $y = cx^3$.
2. $y^2 = cx$.
3. $cx^2 + y^2 = 1$.
4. $y = e^{cx}$.
5. $y = x - 1 + ce^{-x}$.
6. $x - y = cx^2$.
7. $x^2 + y^2 = cx^3$.
8. $x^2 = 2y - 1 + ce^{-2y}$.
9. $x = \dfrac{y^2}{4} + \dfrac{c}{y^2}$.
10. Find the orthogonal trajectories of the family of ellipses having center at the origin, a focus at the point $(c, 0)$, and semimajor axis of length $2c$.
11. Find the orthogonal trajectories of the family of circles which are tangent to the y axis at the origin.
12. Find the value of K such that the parabolas $y = c_1x^2 + K$ are the orthogonal trajectories of the family of ellipses $x^2 + 2y^2 - y = c_2$.
13. Find the value of n such that the curves $x^n + y^n = c_1$ are the orthogonal trajectories of the family
$$y = \frac{x}{1 - c_2x}.$$
14. A given family of curves is said to be *self-orthogonal* if its family of orthogonal trajectories is the same as the given family. Show that the family of parabolas $y^2 = 2cx + c^2$ is self-orthogonal.
15. Find a family of oblique trajectories that intersect the family of circles $x^2 + y^2 = c^2$ at angle 45°.

16. Find a family of oblique trajectories which intersect the family of parabolas $y^2 = cx$ at angle 60°.

17. Find a family of oblique trajectories which intersect the family of curves $x + y = cx^2$ at angle α such that $\tan \alpha = 2$.

3.2 Problems in Mechanics

A. Introduction

Before we apply our knowledge of differential equations to certain problems in mechanics, let us briefly recall certain principles of that subject. The *momentum* of a body is defined to be the product mv of its mass m and its velocity v. The velocity v and hence the momentum are vector quantities. We now state the following basic law of mechanics:

Newton's Second Law. The time rate of change of momentum of a body is proportional to the resultant force acting on the body and is in the direction of this resultant force.

In mathematical language, this law states that

$$\frac{d}{dt}(mv) = KF,$$

where m is the mass of the body, v is its velocity, F is the resultant force acting upon it, and K is a constant of proportionality. If the mass m is considered constant, this reduces to

$$m\frac{dv}{dt} = KF,$$

or

$$a = K\frac{F}{m}, \tag{3.17}$$

or

$$F = kma, \tag{3.18}$$

where $k = 1/K$ and $a = dv/dt$ is the acceleration of the body. The form (3.17) is a direct mathematical statement of the manner in which Newton's second law is usually expressed in words, the mass being considered constant. However, we shall make use of the equivalent form (3.18). The magnitude of the constant of proportionality k depends upon the units employed for force, mass, and acceleration. Obviously the simplest systems of units are those for which $k = 1$. When such a system is used (3.18) reduces to

$$F = ma. \tag{3.19}$$

It is in this form that we shall use Newton's second law. Observe that Equation (3.19) is a vector equation.

Several systems of units for which $k = 1$ are in use. In this text we shall use only two: the centimeter-gram-second system (cgs) and the British gravitational system (British). We summarize the various units of these two systems in Table 3.1.

TABLE 3.1

	British system	*cgs system*
force	pound	dyne
mass	slug	gram
distance	foot	centimeter
time	second	second
acceleration	ft/sec^2	cm/sec^2

Recall that the force of gravitational attraction which the earth exerts on a body is called the weight of the body. The weight, being a force, is expressed in force units. Thus in the British system the weight is measured in pounds, and in the cgs system in dynes.

Let us now apply Newton's second law to a freely falling body (a body falling toward the earth in the absence of air resistance). Let the mass of the body be m and let w denote its weight. The only force acting on the body is its weight and so this is the resultant force. The acceleration is that due to gravity, denoted by g, which is approximately 32 ft/sec^2 in the British system or 980 cm/sec^2 in the cgs system (for points near the earth's surface). Newton's second law $F = ma$ thus reduces to $w = mg$. Thus

$$m = \frac{w}{g}, \tag{3.20}$$

a relation which we shall frequently employ.

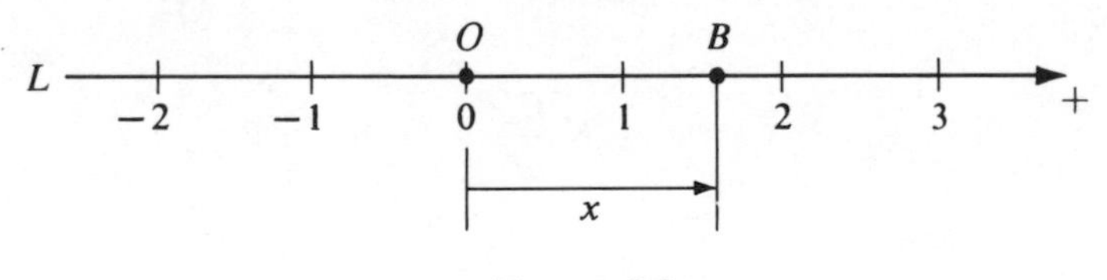

FIGURE 3.3

Let us now consider a body B in rectilinear motion, that is, in motion along a straight line L. On L we choose a fixed reference point as origin O, a fixed direction as positive, and a unit of distance. Then the coordinate x of the position of B from the origin O tells us the distance or displacement of B. (See Figure 3.3.) The *instantaneous velocity* of B is the time rate of change of x:

$$v = \frac{dx}{dt};$$

and the *instantaneous acceleration* of B is the time rate of change of v:

$$a = \frac{dv}{dt} = \frac{d^2x}{dt^2}.$$

Note that x, v, and a are vector quantities. All forces, displacements, velocities, and accelerations in the positive direction on L are positive quantities; while those in the negative direction are negative quantities.

If we now apply Newton's second law $F = ma$ to the motion of B along L, noting that

$$\frac{dv}{dt} = \frac{dv}{dx}\frac{dx}{dt} = v\frac{dv}{dx},$$

we may express the law in any of the following three forms:

$$m\frac{dv}{dt} = F, \tag{3.21}$$

$$m\frac{d^2x}{dt^2} = F, \tag{3.22}$$

$$mv\frac{dv}{dx} = F, \tag{3.23}$$

where F is the resultant force acting on the body. The form to use depends upon the way in which F is expressed. For example, if F is a function of time t only and we desire to obtain the velocity v as a function of t, we would use (3.21); whereas if F is expressed as a function of the displacement x and we wish to find v as a function of x, we would employ (3.23).

B. *Falling Body Problems*

We shall now consider some examples of a body falling through air toward the earth. In such a circumstance the body encounters air resistance as it falls. The amount of air resistance depends upon the velocity of the body, but no general law exactly expressing this dependence is known. In some instances the law $R = kv$ appears to be quite satisfactory, while in others $R = kv^2$ appears to be more exact. In any case, the constant of proportionality k in turn depends on several circumstances. In the examples which follow we shall assume certain reasonable resistance laws in each case. Thus we shall actually be dealing with idealized problems in which the true resistance law is approximated and in which certain comparatively negligible factors are disregarded.

▶ Example 3.5. A body weighing 8 lb falls from rest toward the earth from a great height. As it falls, air resistance acts upon it, and we shall assume that this resistance (in pounds) is numerically equal to $2v$, where v is the velocity (in feet per second). Find the velocity and distance fallen at time t seconds.

Formulation. We choose the positive x axis vertically downward along the path of the body B and the origin at the point from which the body fell. The forces acting on the body are:

1. F_1, its weight, 8 lb, which acts downward and hence is positive.
2. F_2, the air resistance, numerically equal to $2v$, which acts upward and hence is the negative quantity $-2v$.

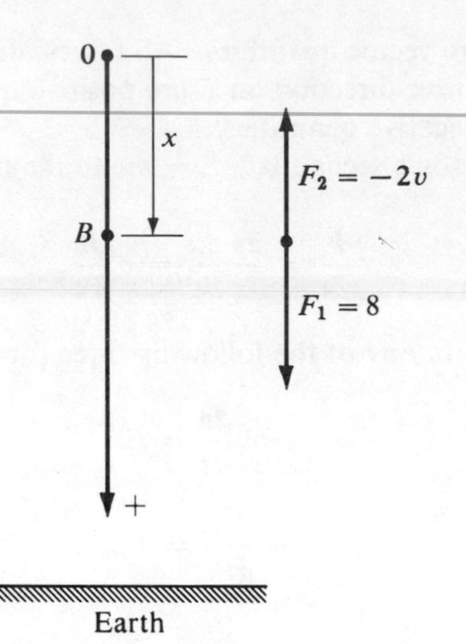

FIGURE 3.4

See Figure 3.4, where these forces are indicated.

Newton's second law, $F = ma$, becomes

$$m \frac{dv}{dt} = F_1 + F_2$$

or, taking $g = 32$ and using $m = w/g = \frac{8}{32} = \frac{1}{4}$,

$$\frac{1}{4}\frac{dv}{dt} = 8 - 2v. \tag{3.24}$$

Since the body was initially at rest, we have the initial condition

$$v(0) = 0. \tag{3.25}$$

Solution. Equation (3.24) is separable. Separating variables, we have

$$\frac{dv}{8 - 2v} = 4\,dt.$$

Integrating we find

$$-\tfrac{1}{2} \ln |8 - 2v| = 4t + c_0,$$

which reduces to

$$8 - 2v = c_1 e^{-8t}.$$

Applying the condition (3.25) we find $c_1 = 8$. Thus the velocity at time t is given by

$$v = 4(1 - e^{-8t}). \tag{3.26}$$

Now to determine the distance fallen at time t, we write (3.26) in the form

$$\frac{dx}{dt} = 4(1 - e^{-8t})$$

and note that $x(0) = 0$. Integrating the above equation, we obtain

$$x = 4(t + \tfrac{1}{8}e^{-8t}) + c_2.$$

Since $x = 0$ when $t = 0$, we find $c_2 = -\frac{1}{2}$ and hence the distance fallen is given by

$$x = 4(t + \tfrac{1}{8}e^{-8t} - \tfrac{1}{8}). \tag{3.27}$$

Interpretation of Results. Equation (3.26) shows us that as $t \to \infty$, the velocity v approaches the *limiting velocity* 4 (ft/sec). We also observe that this limiting velocity is approximately attained in a very short time. Equation (3.27) states that as $t \to \infty$, x also $\to \infty$. Does this imply that the body will plow through the earth and continue forever? Of course not; for when the body reaches the earth's surface its motion will certainly cease. How then do we reconcile this obvious end to the motion with the statement of Equation (3.27)? It is simple; when the body reaches the earth's surface, the differential equation (3.24) and hence Equation (3.27) no longer apply!

▶ Example 3.6. A skydiver equipped with parachute and other essential equipment falls from rest toward the earth. The total weight of the man plus the equipment is 160 lb. Before the parachute opens, the air resistance (in pounds) is numerically equal to $\frac{1}{2}v$, where v is the velocity (in feet per second). The parachute opens 5 sec after the fall begins; after it opens, the air resistance (in pounds) is numerically equal to $\frac{5}{8}v^2$, where v is the velocity (in feet per second). Find the velocity of the skydiver (A) before the parachute opens, and (B) after the parachute opens.

Formulation. We again choose the positive x axis vertically downward with the origin at the point where the fall began. The statement of the problem suggests that we break it into two parts: (A) *before* the parachute opens; (B) *after* it opens.

We first consider problem (A). Before the parachute opens, the forces acting upon the skydiver are:

1. F_1, the weight, 160 lb, which acts downward and hence is positive.
2. F_2, the air resistance, numerically equal to $\frac{1}{2}v$, which acts upward and hence is the negative quantity $-\frac{1}{2}v$.

We use Newton's second law $F = ma$, where $F = F_1 + F_2$, let $m = w/g$, and take $g = 32$. We obtain

$$5\frac{dv}{dt} = 160 - \tfrac{1}{2}v.$$

Since the skydiver was initially at rest, $v = 0$ when $t = 0$. Thus, problem (A), concerned with the time *before* the parachute opens, is formulated as follows:

$$5\frac{dv}{dt} = 160 - \tfrac{1}{2}v. \tag{3.28}$$

$$v(0) = 0. \tag{3.29}$$

We now turn to the formulation of problem (B). Reasoning as before, we see that after the parachute opens, the forces acting upon the skydiver are:

1. $F_1 = 160$, exactly as before.
2. $F_2 = -\frac{5}{8}v^2$ (instead of $-\frac{1}{2}v$).

Thus, proceeding as above, we obtain the differential equation

$$5\frac{dv}{dt} = 160 - \tfrac{5}{8}v^2.$$

Since the parachute opens 5 sec after the fall begins, we have $v = v_1$ when $t = 5$, where v_1 is the velocity attained when the parachute opened. Thus, problem (B), concerned with the time *after* the parachute opens, is formulated as follows:

$$5\frac{dv}{dt} = 160 - \tfrac{5}{8}v^2, \tag{3.30}$$

$$v(5) = v_1. \tag{3.31}$$

Solution. We shall first consider problem (A). We find a one-parameter family of solutions of

$$5\frac{dv}{dt} = 160 - \tfrac{1}{2}v. \tag{3.28}$$

Separating variables, we obtain

$$\frac{dv}{v - 320} = -\tfrac{1}{10}\,dt.$$

Integration yields

$$\ln (v - 320) = -\tfrac{1}{10}t + c_0,$$

which readily simplifies to the form

$$v = 320 + ce^{-t/10}.$$

Applying the initial condition (3.29) that $v = 0$ at $t = 0$, we find that $c = -320$. Hence the solution to problem (A) is

$$v = 320(1 - e^{-t/10}), \tag{3.32}$$

which is valid for $0 \le t \le 5$. In particular, where $t = 5$, we obtain

$$v_1 = 320(1 - e^{-1/2}) \approx 126, \tag{3.33}$$

which is the velocity when the parachute opens.

Now let us consider problem (B). We first find a one-parameter family of solutions of the differential equation

$$5\frac{dv}{dt} = 160 - \tfrac{5}{8}v^2. \tag{3.30}$$

Simplifying and separating variables, we obtain

$$\frac{dv}{v^2 - 256} = -\frac{dt}{8}.$$

Integration yields

$$\frac{1}{32}\ln\frac{v - 16}{v + 16} = -\frac{t}{8} + c_2$$

or

$$\ln\frac{v - 16}{v + 16} = -4t + c_1.$$

This readily simplifies to the form

$$\frac{v - 16}{v + 16} = ce^{-4t}, \tag{3.34}$$

and solving this for v we obtain

$$v = \frac{16(ce^{-4t} + 1)}{1 - ce^{-4t}}. \tag{3.35}$$

Applying the initial condition (3.31) that $v = v_1$ at $t = 5$, where v_1 is given by (3.33) and is approximately 126, to (3.34), we obtain

$$c = \tfrac{110}{142}e^{20}.$$

Substituting this into (3.35) we obtain

$$v = \frac{16(\frac{110}{142}e^{20-4t} + 1)}{1 - \frac{110}{142}e^{20-4t}}, \tag{3.36}$$

which is valid for $t \geq 5$.

Interpretation of Results. Let us first consider the solution of problem (A), given by Equation (3.32). According to this, as $t \to \infty$, v approaches the limiting velocity 320 ft/sec. Thus if the parachute never opened, the velocity would have been approximately 320 ft/sec at the time when the unfortunate skydiver would have struck the earth! But, according to the statement of the problem, the parachute *does* open 5 sec after the fall begins (we tacitly and thoughtfully assume $5 \ll T$, where T is the time when the earth *is* reached!). Then, referring to the solution of problem (B), Equation (3.36), we see that as $t \to \infty$, v approaches the limiting velocity 16 ft/sec. Thus, assuming that the parachute opens at a considerable distance above the earth, the velocity is approximately 16 ft/sec when the earth is finally reached. We thus obtain the well-known fact that the velocity of impact with the open parachute is a small fraction of the impact velocity which would have occurred if the parachute had not opened. The calculations in this problem are somewhat complicated, but the moral is clear: Make certain that the parachute opens!

C. Frictional Forces

If a body moves on a rough surface, it will encounter not only air resistance but also another resistance force due to the roughness of the surface. This additional force is called *friction*. It is shown in physics that the friction is given by μN, where

1. μ is a constant of proportionality called the *coefficient of friction*, which depends upon the roughness of the given surface; and
2. N is the normal (that is, perpendicular) force which the surface exerts on the body.

We now apply Newton's second law to a problem in which friction is involved.

▶ Example 3.7. An object weighing 48 lb is released from rest at the top of a plane metal slide which is inclined 30° to the horizontal. Air resistance (in pounds) is numerically equal to one-half the velocity (in feet per second), and the coefficient of friction is one-quarter.

A. What is the velocity of the object 2 sec after it is released?
B. If the slide is 24 ft long, what is the velocity when the object reaches the bottom?

Formulation. The line of motion is along the slide. We choose the origin at the top and the positive x direction down the slide. If we temporarily neglect the friction and air resistance, the forces acting upon the object A are:

1. Its weight, 48 lb, which acts vertically downward; and
2. The normal force, N, exerted by the slide which acts in an upward direction perpendicular to the slide. (See Figure 3.5.)

The components of the weight parallel and perpendicular to the slide have magnitude

$$48 \sin 30° = 24$$

and

$$48 \cos 30° = 24\sqrt{3},$$

respectively. The components perpendicular to the slide are in equilibrium and hence the normal force N has magnitude $24\sqrt{3}$.

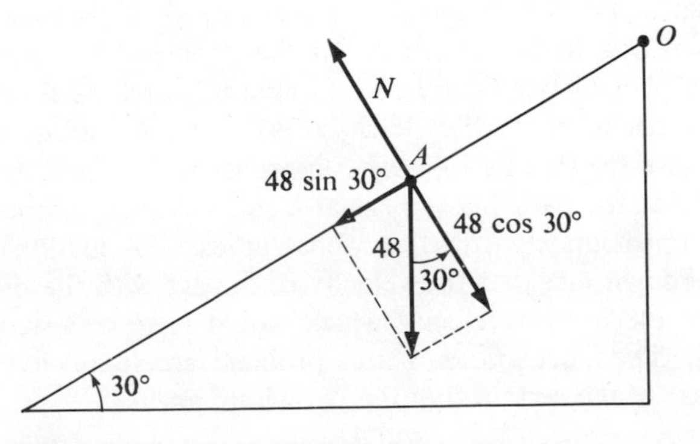

FIGURE 3.5

Now, taking into consideration the friction and air resistance, we see that the forces acting on the object as it moves along the slide are the following:

1. F_1, the component on the weight parallel to the plane, having numerical value 24. Since this force acts in the positive (downward) direction along the slide, we have

$$F_1 = 24.$$

2. F_2, the frictional force, having numerical value $\mu N = \frac{1}{4}(24\sqrt{3})$. Since this acts in the negative (upward) direction along the slide, we have

$$F_2 = -6\sqrt{3}.$$

3. F_3, the air resistance, having numerical value $\frac{1}{2}v$. Since $v > 0$ and this also acts in the negative direction, we have

$$F_3 = -\tfrac{1}{2}v.$$

We apply Newton's second law $F = ma$. Here $F = F_1 + F_2 + F_3 = 24 - 6\sqrt{3} - \frac{1}{2}v$ and $m = w/g = \frac{48}{32} = \frac{3}{2}$. Thus we have the differential equation

$$\frac{3}{2}\frac{dv}{dt} = 24 - 6\sqrt{3} - \tfrac{1}{2}v. \tag{3.37}$$

Since the object is released from rest, the initial condition is

$$v(0) = 0. \tag{3.38}$$

Solution. Equation (3.37) is separable; separating variables we have

$$\frac{dv}{48 - 12\sqrt{3} - v} = \frac{dt}{3}.$$

Integrating and simplifying, we find

$$v = 48 - 12\sqrt{3} - c_1 e^{-t/3}.$$

The condition (3.38) gives $c_1 = 48 - 12\sqrt{3}$. Thus we obtain

$$v = (48 - 12\sqrt{3})(1 - e^{-t/3}). \tag{3.39}$$

Question A is thus answered by letting $t = 2$ in Equation (3.39). We find

$$v(2) = (48 - 12\sqrt{3})(1 - e^{-2/3}) \approx 10.2 \text{ (ft/sec)}.$$

In order to answer question B, we integrate (3.39) to obtain

$$x = (48 - 12\sqrt{3})(t + 3e^{-t/3}) + c_2.$$

Since $x(0) = 0$, $c_2 = -(48 - 12\sqrt{3})(3)$. Thus the distance covered at time t is given by

$$x = (48 - 12\sqrt{3})(t + 3e^{-t/3} - 3).$$

Since the slide is 24 ft long, the object reaches the bottom at the time T determined from the transcendental equation

$$24 = (48 - 12\sqrt{3})(T + 3e^{-T/3} - 3),$$

which may be written as

$$3e^{-T/3} = \frac{47 + 2\sqrt{3}}{13} - T.$$

The value of T which satisfies this equation is approximately 2.6. Thus from Equation (3.39) the velocity of the object when it reaches the bottom is given approximately by

$$(48 - 12\sqrt{3})(1 - e^{-0.9}) \approx 12.3 \text{ (ft/sec)}.$$

Exercises

1. A stone weighing 4 lb falls from rest toward the earth from a great height. As it falls it is acted upon by air resistance which is numerically equal to $\frac{1}{2}v$ (in pounds), where v is the velocity (in feet per second).
   (a) Find the velocity and distance fallen at time t sec.
   (b) Find the velocity and distance fallen at the end of 5 sec.

2. A ball weighing 6 lb is thrown vertically downward toward the earth from a height of 1000 ft with an initial velocity of 6 ft/sec. As it falls it is acted upon by air resistance which is numerically equal to $\frac{2}{3}v$ (in pounds), where v is the velocity (in feet per second).
   (a) What is the velocity and distance fallen at the end of one minute?
   (b) With what velocity does the ball strike the earth?

3. A ball weighing $\frac{3}{4}$ lb is thrown vertically upward from a point 6 ft above the surface of the earth with an initial velocity of 20 ft/sec. As it rises it is acted upon by air resistance which is numerically equal to $\frac{1}{64}v$ (in pounds), where v is the velocity (in feet per second). How high will the ball rise?

4. A ship which weighs 32,000 tons starts from rest under the force of a constant propeller thrust of 100,000 lb. The resistance in pounds is numerically equal to $8000v$, where v is in feet per second.
   (a) Find the velocity of the ship as a function of the time.
   (b) Find the limiting velocity (that is, the limit of v as $t \to +\infty$).
   (c) Find how long it takes the ship to attain a velocity of 80% of the limiting velocity.

5. Two men are riding in a motorboat and the combined weight of men, motor, boat, and equipment is 640 lb. The motor exerts a constant force of 20 lb on the boat in the direction of motion, while the resistance (in pounds) is numerically equal to one and one-half times the velocity (in feet per second). If the boat started from rest, find the velocity of the boat after (a) 20 sec, (b) 1 min.

6. A boat weighing 150 lb with a single rider weighing 170 lb is being towed in a certain direction at the rate of 20 mph. At time $t = 0$ the tow rope is suddenly cast off and the rider begins to row in the same direction, exerting a force equivalent to a constant force of 12 lb in this direction. The resistance (in pounds) is numerically equal to twice the velocity (in feet per second).
   (a) Find the velocity of the boat 15 sec after the tow rope was cast off.
   (b) How many seconds after the tow rope is cast off will the velocity be one-half that at which the boat was being towed?
7. A bullet weighing 2 oz is fired vertically downward from a stationary helicopter with a muzzle velocity of 1200 ft/sec. The air resistance (in pounds) is numerically equal to $10^{-5}v^2$, where v is the velocity (in feet per second). Find the velocity of the bullet as a function of the time.
8. A shell weighing 1 lb is fired vertically upward from the earth's surface with a muzzle velocity of 1000 ft/sec. The air resistance (in pounds) is numerically equal to $10^{-4}v^2$, where v is the velocity (in feet per second).
   (a) Find the velocity of the rising shell as a function of the time.
   (b) How long will the shell rise?
9. An object weighing 16 lb is dropped from rest on the surface of a calm lake and thereafter starts to sink. While its weight tends to force it downward, the buoyancy of the object tends to force it back upward. If this buoyancy force is one of 6 lb and the resistance of the water (in pounds) is numerically equal to twice the square of the velocity (in feet per second), find the formula for the velocity of the sinking object as a function of the time.
10. An object weighing 12 lb is placed beneath the surface of a calm lake. The buoyancy of the object is 30 lb; because of this the object begins to rise. If the resistance of the water (in pounds) is numerically equal to the square of the velocity (in feet per second) and the object surfaces in 5 sec, find the velocity of the object at the instant when it reaches the surface.
11. A man is pushing a loaded sled across a level field of ice at the constant speed of 10 ft/sec. When the man is halfway across the ice field, he stops pushing and lets the loaded sled continue on. The combined weight of the sled and its load is 80 lb; the air resistance (in pounds) is numerically equal to $\frac{3}{4}v$, where v is the velocity of the sled (in feet per second); and the coefficient of friction of the runners on the ice is 0.04. How far will the sled continue to move after the man stops pushing?
12. A boy on his sled has just slid down a hill onto a level field of ice and is starting to slow down. At the instant when their speed is 5 ft/sec, the boy's father runs up and begins to push the sled forward, exerting a constant force of 15 lb in the direction of motion. The combined weight of the boy and the sled is 96 lb, the air resistance (in pounds) is numerically equal to one-half the velocity (in feet per second), and the coefficient of friction of the runners on the ice is 0.05. How fast is the sled moving 10 sec after the father begins pushing?

13. A case of canned milk weighing 24 lb is released from rest at the top of a plane metal slide which is 30 ft long and inclined 45° to the horizontal. Air resistance (in pounds) is numerically equal to one-third the velocity (in feet per second) and the coefficient of friction is 0.4.

    (a) What is the velocity of the moving case 1 sec after it is released?

    (b) What is the velocity when the case reaches the bottom of the slide?

14. A boy goes sledding down a long 30° slope. The combined weight of the boy and his sled is 72 lb and the air resistance (in pounds) is numerically equal to twice their velocity (in feet per second). If they started from rest and their velocity at the end of 5 sec is 10 ft/sec, what is the coefficient of friction of the sled runners on the snow?

15. An object weighing 32 lb is released from rest 50 ft above the surface of a calm lake. Before the object reaches the surface of the lake, the air resistance (in pounds) is given by $2v$, where v is the velocity (in feet per second). After the object passes beneath the surface, the water resistance (in pounds) is given by $6v$. Further, the object is then buoyed up by a buoyancy force of 8 lb. Find the velocity of the object 2 sec after it passes beneath the surface of the lake.

3.3 Rate Problems

In certain problems the rate at which a quantity changes is a known function of the amount present and/or the time, and it is desired to find the quantity itself. If x denotes the amount of the quantity present at time t, then dx/dt denotes the rate at which the quantity changes and we are at once led to a differential equation. In this section we consider certain problems of this type.

A. Rate of Growth and Decay

▶ Example 3.8. The rate at which radioactive nuclei decay is proportional to the number of such nuclei that are present in a given sample. Half of the original number of radioactive nuclei have undergone disintegration in a period of 1500 years.

1. What percentage of the original radioactive nuclei will remain after 4500 years?
2. In how many years will only one-tenth of the original number remain?

Mathematical Formulation. Let x be the amount of radioactive nuclei present after t years. Then dx/dt represents the rate at which the nuclei decay. Since the nuclei decay at a rate proportional to the amount present, we have

$$\frac{dx}{dt} = Kx, \tag{3.40}$$

where K is a constant of proportionality. The amount x is clearly positive; further, since x is decreasing, $dx/dt < 0$. Thus, from Equation (3.40), we must have $K < 0$. In order to emphasize that x is decreasing, we prefer to replace K by a positive constant preceded by a minus sign. Thus we let $k = -K > 0$ and write the differential equation (3.40) in the form

$$\frac{dx}{dt} = -kx. \tag{3.41}$$

Letting x_0 denote the amount initially present, we also have the initial condition

$$x(0) = x_0. \tag{3.42}$$

We know that we shall need such a condition in order to determine the arbitrary constant which will appear in a one-parameter family of solutions of the differential equation (3.41). However, we shall apparently need something else, for Equation (3.41) contains an unknown constant of proportionality k. This "something else" appears in the statement of the problem, for we are told that half of the original number disintegrate in 1500 years. Thus half also remain at that time, and this at once gives the condition

$$x(1500) = \tfrac{1}{2}x_0. \tag{3.43}$$

Solution. The differential equation (3.41) is clearly separable; separating variables, integrating, and simplifying, we have at once

$$x = ce^{-kt}.$$

Applying the initial condition (3.42), $x = x_0$ when $t = 0$, we find that $c = x_0$ and hence we obtain

$$x = x_0e^{-kt}. \tag{3.44}$$

We have not yet determined k. Thus we now apply condition (3.43), $x = \frac{1}{2}x_0$ when $t = 1500$, to Equation (3.44). We find

$$\tfrac{1}{2}x_0 = x_0e^{-1500k},$$

or

$$(e^{-k})^{1500} = \tfrac{1}{2},$$

or finally

$$e^{-k} = (\tfrac{1}{2})^{1/1500}. \tag{3.45}$$

From this equation we could determine k explicitly and substitute the result into Equation (3.44). However, we see from Equation (3.44) that we actually do not need k itself but rather only e^{-k}, which we have just obtained in Equation (3.45). Thus we substitute e^{-k} from (3.45) into (3.44) to obtain

$$x = x_0(e^{-k})^t = x_0[(\tfrac{1}{2})^{1/1500}]^t$$

or

$$x = x_0(\tfrac{1}{2})^{t/1500}. \tag{3.46}$$

Equation (3.46) gives the number x of radioactive nuclei which are present at time t. Question 1 asks us what percentage of the original number will remain after 4500 years. We thus let $t = 4500$ in Equation (3.46) and find

$$x = x_0(\tfrac{1}{2})^3 = \tfrac{1}{8}x_0.$$

Thus, one-eighth or 12.5% of the original number remain after 4500 years. Question 2 asks us when only one-tenth will remain. Thus we let $x = \frac{1}{10}x_0$ in Equation (3.46) and solve for t. We have

$$\tfrac{1}{10} = (\tfrac{1}{2})^{t/1500}.$$

Using logarithms, we then obtain

$$\ln (\tfrac{1}{10}) = \ln (\tfrac{1}{2})^{t/1500} = \frac{t}{1500} \ln (\tfrac{1}{2}).$$

From this it follows at once that

$$\frac{t}{1500} = \frac{\ln \frac{1}{10}}{\ln \frac{1}{2}}$$

or

$$t = \frac{1500 \ln 10}{\ln 2} \approx 4985 \text{ (years)}.$$

B. *Mixture Problems*

We now consider rate problems involving mixtures. A substance S is allowed to flow into a certain mixture in a container at a certain rate, and the mixture is kept uniform by stirring. Further, in one such situation, this uniform mixture simultaneously flows out of the container at another (generally different) rate; in another situation this may not be the case. In either case we seek to determine the quantity of the substance S present in the mixture at time t.

Letting x denote the amount of S present at time t, the derivative dx/dt denotes the rate of change of x with respect to t. If IN denotes the rate at which S enters the mixture and OUT the rate at which it leaves, we have at once the basic equation

$$\frac{dx}{dt} = \text{IN} - \text{OUT} \tag{3.47}$$

from which to determine the amount x of S at time t. We now consider examples.

▶ Example 3.9. A tank initially contains 50 gal of pure water. Starting at time $t = 0$ a brine containing 2 lb of dissolved salt per gallon flows into the tank at the rate of 3 gal/min. The mixture is kept uniform by stirring and the well-stirred mixture simultaneously flows out of the tank at the same rate.

1. How much salt is in the tank at any time $t > 0$?
2. How much salt is present at the end of 25 min?
3. How much salt is present after a long time?

Mathematical Formulation. Let x denote the amount of salt in the tank at time t. We apply the basic equation (3.47),

$$\frac{dx}{dt} = \text{IN} - \text{OUT}.$$

The brine flows in at the rate of 3 gal/min, and each gallon contains 2 lb of salt. Thus

$$\text{IN} = (2 \text{ lb/gal})(3 \text{ gal/min}) = 6 \text{ lb/min}.$$

Since the rate of outflow equals the rate of inflow, the tank contains 50 gal of the mixture at any time t. This 50 gal contains x lb of salt at time t, and so the concentration of salt at time t is $\frac{1}{50}x$ lb/gal. Thus, since the mixture flows out at the rate of 3 gal/min, we have

$$\text{OUT} = \left(\frac{x}{50} \text{ lb/gal}\right)(3 \text{ gal/min}) = \frac{3x}{50} \text{ lb/min}.$$

Thus the differential equation for x as a function of t is

$$\frac{dx}{dt} = 6 - \frac{3x}{50}. \tag{3.48}$$

Since initially there was no salt in the tank, we also have the initial condition

$$x(0) = 0. \tag{3.49}$$

Solution. Equation (3.48) is both linear and separable. Separating variables, we have

$$\frac{dx}{100 - x} = \frac{3}{50}\,dt.$$

Integrating and simplifying, we obtain

$$x = 100 + ce^{-3t/50}.$$

Applying the condition (3.49), $x = 0$ at $t = 0$, we find that $c = -100$. Thus we have

$$x = 100(1 - e^{-3t/50}). \tag{3.50}$$

This is the answer to question 1. As for question 2, at the end of 25 min, $t = 25$, and Equation (3.50) gives

$$x(25) = 100(1 - e^{-1.5}) \approx 78(\text{lb}).$$

Question 3 essentially asks us how much salt is present as $t \to \infty$. To answer this we let $t \to \infty$ in Equation (3.50) and observe that $x \to 100$.

▶ Example 3.10. A large tank initially contains 50 gal of brine in which there is dissolved 10 lb of salt. Brine containing 2 lb of dissolved salt per gallon flows into the tank at the rate of 5 gal/min. The mixture is kept uniform by stirring, and the stirred mixture simultaneously flows out at the slower rate of 3 gal/min. How much salt is in the tank at any time $t > 0$?

Mathematical Formulation. Let $x =$ the amount of salt at time t. Again we shall use Equation (3.47):

$$\frac{dx}{dt} = \text{IN} - \text{OUT}.$$

Proceeding as in Example 3.9,

$$\text{IN} = (2 \text{ lb/gal})(5 \text{ gal/min}) = 10 \text{ lb/min};$$

also, once again

$$\text{OUT} = (C \text{ lb/gal})(3 \text{ gal/min}),$$

where C lb/gal denotes the concentration. But here, since the rate of outflow is different from that of inflow, the concentration is not quite so simple. At time $t = 0$, the tank contains 50 gal of brine. Since brine flows in at the rate of 5 gal/min but flows out at the slower rate of 3 gal/min, there is a net gain of $5 - 3 = 2$ gal/min of brine in the tank. Thus at the end of t minutes the amount of brine in the tank is

$$50 + 2t \text{ gal}.$$

Hence the concentration at time t minutes is

$$\frac{x}{50 + 2t} \text{ lb/gal},$$

and so

$$\text{OUT} = \frac{3x}{50 + 2t} \text{ lb/min}.$$

Thus the differential equation becomes

$$\frac{dx}{dt} = 10 - \frac{3x}{50 + 2t}. \tag{3.51}$$

Since there was initially 10 lb of salt in the tank, we have the initial condition

$$x(0) = 10. \tag{3.52}$$

Solution. The differential equation (3.51) is *not* separable but it *is* linear. Putting it in standard form,

$$\frac{dx}{dt} + \frac{3}{2t + 50} x = 10,$$

we find the integrating factor

$$\exp\left(\int \frac{3}{2t + 50} \, dt\right) = (2t + 50)^{3/2}.$$

Multiplying through by this, we have

$$(2t + 50)^{3/2} \frac{dx}{dt} + 3(2t + 50)^{1/2} x = 10(2t + 50)^{3/2}$$

or

$$\frac{d}{dt}[(2t + 50)^{3/2} x] = 10(2t + 50)^{3/2}.$$

Thus

$$(2t + 50)^{3/2} x = 2(2t + 50)^{5/2} + c$$

or

$$x = 4(t + 25) + \frac{c}{(2t + 50)^{3/2}}.$$

Applying condition (3.52), $x = 10$ at $t = 0$, we find

$$10 = 100 + \frac{c}{(50)^{3/2}}$$

or

$$c = -(90)(50)^{3/2} = -22{,}500\sqrt{2}.$$

Thus the amount of salt at any time $t > 0$ is given by

$$x = 4t + 100 - \frac{22{,}500\sqrt{2}}{(2t + 50)^{3/2}}.$$

Exercises

1. Assume that the rate at which radioactive nuclei decay is proportional to the number of such nuclei that are present in a given sample. In a certain sample 10% of the original number of radioactive nuclei have undergone disintegration in a period of 200 years.
   (a) What percentage of the original radioactive nuclei will remain after 1000 years?
   (b) In how many years will only one-fourth of the original number remain?

2. A certain chemical is converted into another chemical by a chemical reaction. The rate at which the first chemical is converted is proportional to the amount of this chemical present at any instant. Ten percent of the original amount of the first chemical has been converted in 5 min.
   (a) What percent of the first chemical will have been converted in 20 min?
   (b) In how many minutes will 60% of the first chemical have been converted?

3. A chemical reaction converts a certain chemical into another chemical, and the rate at which the first chemical is converted is proportional to the amount of this chemical present at any time. At the end of one hour, 50 gm of the first chemical remain; while at the end of three hours, only 25 gm remain.
   (a) How many grams of the first chemical were present initially?
   (b) How many grams of the first chemical will remain at the end of five hours?
   (c) In how many hours will only 2 gm of the first chemical remain?

4. Assume that the population of a certain city increases at a rate proportional to the number of inhabitants at any time. If the population doubles in 40 years, in how many years will it triple?

5. The population of the city of Bingville increases at a rate proportional to the number of its inhabitants present at any time t. If the population of Bingville was 30,000 in 1960 and 35,000 in 1970, what will be the population of Bingville in 1980?

6. In a certain bacteria culture the rate of increase in the number of bacteria is proportional to the number present.
   (a) If the number triples in 5 hr, how many will be present in 10 hr?

(b) When will the number present be 10 times the number initially present?

7. An amount of invested money is said to draw interest *compounded continuously* if the amount of money increases at a rate proportional to the amount present. Suppose \$1000 is invested and draws interest compounded continuously, where the annual interest rate is 6%.

   (a) How much money will be present 10 years after the original amount was invested?

   (b) How long will it take the original amount of money to double?

8. Suppose a certain amount of money is invested and draws interest compounded continuously.

   (a) If the original amount doubles in two years, then what is the annual interest rate?

   (b) If the original amount increases 50% in six months, then how long will it take the original amount to double?

9. A tank initially contains 100 gal of brine in which there is dissolved 20 lb of salt. Starting at time $t = 0$, brine containing 3 lb of dissolved salt per gallon flows into the tank at the rate of 4 gal/min. The mixture is kept uniform by stirring and the well-stirred mixture simultaneously flows out of the tank at the same rate.

   (a) How much salt is in the tank at the end of 10 min?

   (b) When is there 160 lb of salt in the tank?

10. A large tank initially contains 100 gal of brine in which 10 lb of salt is dissolved. Starting at $t = 0$, pure water flows into the tank at the rate of 5 gal/min. The mixture is kept uniform by stirring and the well-stirred mixture simultaneously flows out at the slower rate of 2 gal/min.

    (a) How much salt is in the tank at the end of 15 min and what is the concentration at that time?

    (b) If the capacity of the tank is 250 gal, what is the concentration at the instant the tank overflows?

11. A tank initially contains 100 gal of pure water. Starting at $t = 0$, a brine containing 4 lb of salt per gallon flows into the tank at the rate of 5 gal/min. The mixture is kept uniform by stirring and the well-stirred mixture flows out at the slower rate of 3 gal/min.

    (a) How much salt is in the tank at the end of 20 min?

    (b) When is there 50 lb of salt in the tank?

12. A large tank initially contains 200 gal of brine in which 15 lb of salt is dissolved. Starting at $t = 0$, brine containing 4 lb of salt per gallon flows into the tank at the rate of 3.5 gal/min. The mixture is kept uniform by stirring and the well-stirred mixture leaves the tank at the rate of 4 gal/min.

    (a) How much salt is in the tank at the end of one hour?

    (b) How much salt is in the tank when the tank contains only 50 gal of brine?

13. The air in a room whose volume is 10,000 cu ft tests 0.15% carbon dioxide. Starting at $t = 0$, outside air testing 0.05% carbon dioxide is admitted at the rate of 5000 cu ft/min.

    (a) What is the percentage of carbon dioxide in the air in the room after 3 min?

    (b) When does the air in the room test 0.1% carbon dioxide?

14. The air in a room 50 ft by 20 ft by 8 ft tests 0.2% carbon dioxide. Starting at $t = 0$, outside air testing 0.05% carbon dioxide is admitted to the room. How many cubic feet of this outside air must be admitted per minute in order that the air in the room test 0.1% at the end of 30 min?

15. Newton's law of cooling states that the rate at which a body cools is proportional to the difference between the temperature of the body and that of the medium in which it is situated. A body of temperature 80 °F is placed at time $t = 0$ in a medium the temperature of which is maintained at 50 °F. At the end of 5 min, the body has cooled to a temperature of 70 °F.

    (a) What is the temperature of the body at the end of 10 min?

    (b) When will the temperature of the body be 60 °F?

16. A body cools from 60 °C to 50 °C in 15 min in air which is maintained at 30 °C. How long will it take this body to cool from 100 °C to 80 °C in air which is maintained at 50 °C? Assume Newton's law of cooling (Exercise 15).

17. The rate at which a certain substance dissolves in water is proportional to the product of the amount undissolved and the difference $c_1 - c_2$, where c_1 is the concentration in the saturated solution and c_2 is the concentration in the actual solution. If saturated, 50 gm of water would dissolve 20 gm of the substance. If 10 gm of the substance is placed in 50 gm of water and half of the substance is then dissolved in 90 min, how much will be dissolved in 3 hr?

18. Under natural circumstances the population of mice on a certain island would increase at a rate proportional to the number of mice present at any time, provided the island had no cats. There were no cats on the island from the beginning of 1960 to the beginning of 1970, and during this time the mouse population doubled, reaching an all-time high of 100,000 at the beginning of 1970. At this time the people of the island, alarmed by the increasing number of mice, imported a number of cats to kill the mice. If the indicated natural rate of increase of mice was thereafter offset by the work of the cats, who killed 1000 mice a month, how many mice remained at the beginning of 1971?

Suggested Reading

AGNEW, R. P., *Differential Equations*, 2nd ed. (McGraw-Hill, New York, 1960).

BOYCE, W., and R. DIPRIMA, *Elementary Differential Equations*, 2nd ed. (Wiley, New York, 1969).

KAPLAN, W., *Ordinary Differential Equations* (Addison-Wesley, Reading, Mass., 1958).

RAINVILLE, E., and P. BEDIENT, *Elementary Differential Equations*, 4th ed. (Macmillan, New York, 1969).

RITGER, P., and N. ROSE, *Differential Equations with Applications* (McGraw-Hill, New York, 1968).

SPIEGEL, M. R., *Applied Differential Equations*, 2nd ed. (Prentice-Hall, Englewood Cliffs, N.J., 1967).

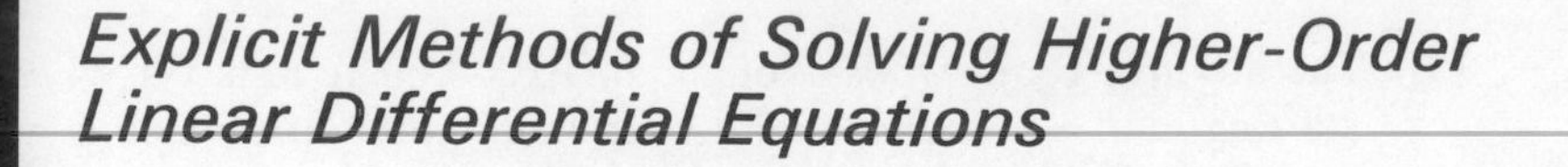

4 Explicit Methods of Solving Higher-Order Linear Differential Equations

The subject of ordinary linear differential equations is one of great theoretical and practical importance. Theoretically, the subject is one of simplicity and elegance. Practically, linear differential equations originate in a variety of applications to science and engineering. Fortunately many of the linear differential equations which thus occur are of a special type, linear with constant coefficients, for which explicit methods of solution are available. The main purpose of this chapter is to study certain of these methods. First, however, we need to consider certain basic theorems which will be used throughout the chapter. These theorems are stated and illustrated in Section 4.1, but proofs are omitted in this introductory section. In the final section of the chapter we return to this fundamental theory and present theorems *and* proofs in the most important special case. Proofs in the general case are given in Chapter 11 of the author's *Differential Equations.*

4.1 Basic Theory of Linear Differential Equations

A. Definition and Basic Existence Theorem

DEFINITION

A linear ordinary differential equation of order *n in the dependent variable y and the independent variable x is an equation which is in, or can be expressed in, the form*

$$a_0(x)\frac{d^ny}{dx^n} + a_1(x)\frac{d^{n-1}y}{dx^{n-1}} + \cdots + a_{n-1}(x)\frac{dy}{dx} + a_n(x)y = F(x), \tag{4.1}$$

where a_0 is not identically zero. We shall assume that $a_0, a_1, \ldots, a_n$ and F are continuous real functions on a real interval $a \le x \le b$ and that $a_0(x) \ne 0$ for any x on $a \le x \le b$. The right-hand member F(x) is called the nonhomogeneous term. *If F is identically zero Equation (4.1) reduces to*

$$a_0(x)\frac{d^ny}{dx^n} + a_1(x)\frac{d^{n-1}y}{dx^{n-1}} + \cdots + a_{n-1}(x)\frac{dy}{dx} + a_n(x)y = 0 \tag{4.2}$$

and is then called homogeneous.

▶ Example 4.1. The equation

$$\frac{d^2y}{dx^2} + 3x\frac{dy}{dx} + x^3y = e^x$$

is a linear ordinary differential equation of the second order.

▶ Example 4.2. The equation

$$\frac{d^3y}{dx^3} + x\frac{d^2y}{dx^2} + 3x^2\frac{dy}{dx} - 5y = \sin x$$

is a linear ordinary differential equation of the third order.

We now state the basic existence theorem for initial-value problems associated with an nth-order linear ordinary differential equation:

THEOREM 4.1

Hypothesis

1. Consider the nth-order linear differential equation

$$a_0(x)\frac{d^ny}{dx^n} + a_1(x)\frac{d^{n-1}y}{dx^{n-1}} + \cdots + a_{n-1}(x)\frac{dy}{dx} + a_n(x)y = F(x), \qquad (4.1)$$

where $a_0, a_1, \ldots, a_n$ and F are continuous real functions on a real interval $a \le x \le b$ and $a_0(x) \ne 0$ for any x on $a \le x \le b$.

2. Let x_0 be any point of the interval $a \le x \le b$, and let $c_0, c_1, \ldots, c_{n-1}$ be n arbitrary real constants.

Conclusion. *There exists a unique solution f of (4.1) such that*

$$f(x_0) = c_0,\ f'(x_0) = c_1, \ldots,\ f^{(n-1)}(x_0) = c_{n-1},$$

and this solution is defined over the entire interval $a \le x \le b$.

Suppose that we are considering an nth-order linear differential equation (4.1), the coefficients and nonhomogeneous term of which all possess the continuity requirements set forth in Hypothesis 1 of Theorem 4.1 on a certain interval of the x axis. Then, given *any* point x_0 of this interval and *any* n real numbers $c_0, c_1, \ldots, c_{n-1}$, the theorem assures us that there is *precisely one* solution of the differential equation which assumes the value c_0 at $x = x_0$ and whose kth derivative assumes the value c_k for each $k = 1, 2, \ldots, n - 1$ at $x = x_0$. Further, the theorem asserts that this unique solution is defined for *all* x in the above-mentioned interval.

▶ Example 4.3. Consider the initial-value problem

$$\frac{d^2y}{dx^2} + 3x\frac{dy}{dx} + x^3y = e^x,$$

$$y(1) = 2,$$

$$y'(1) = -5.$$

The coefficients 1, $3x$, and x^3, as well as the nonhomogeneous term e^x, in this second-order differential equation are all continuous for all values of x, $-\infty < x < \infty$. The point x_0 here is the point 1, which certainly belongs to this interval; and the real numbers c_0 and c_1 are 2 and -5, respectively. Thus Theorem 4.1 assures us that a solution of the given problem exists, is unique, and is defined for all x, $-\infty < x < \infty$.

▶ Example 4.4. Consider the initial-value problem

$$2\frac{d^3y}{dx^3} + x\frac{d^2y}{dx^2} + 3x^2\frac{dy}{dx} - 5y = \sin x,$$

$$y(4) = 3,$$

$$y'(4) = 5,$$

$$y''(4) = -\tfrac{7}{2}.$$

Here we have a third-order problem. The coefficients 2, x, $3x^2$, and -5, as well as the nonhomogeneous term $\sin x$, are all continuous for all x, $-\infty < x < \infty$. The point $x_0 = 4$ certainly belongs to this interval; the real numbers c_0, c_1, and c_2 in this problem are 3, 5, and $-\frac{7}{2}$, respectively. Theorem 4.1 assures us that this problem also has a unique solution which is defined for all x, $-\infty < x < \infty$.

A useful corollary to Theorem 4.1 is the following:

COROLLARY

Hypothesis. *Let f be a solution of the nth-order* homogeneous *linear differential equation*

$$a_0(x)\frac{d^ny}{dx^n} + a_1(x)\frac{d^{n-1}y}{dx^{n-1}} + \cdots + a_{n-1}(x)\frac{dy}{dx} + a_n(x)y = 0 \tag{4.2}$$

such that

$$f(x_0) = 0,\ f'(x_0) = 0, \ldots,\ f^{(n-1)}(x_0) = 0,$$

where x_0 is a point of the interval $a \le x \le b$ in which the coefficients $a_0, a_1, \ldots, a_n$ are all continuous and $a_0(x) \neq 0$.

Conclusion. *Then $f(x) = 0$ for all x on $a \le x \le b$.*

Let us suppose that we are considering a homogeneous equation of the form (4.2), all the coefficients of which are continuous on a certain interval of the x axis. Suppose further that we have a solution f of this equation which is such that f and its first $n - 1$ derivatives all equal zero at a point x_0 of this interval. Then this corollary states that this solution is the "trivial" solution f such that $f(x) = 0$ for *all* x on the above-mentioned interval.

▶ Example 4.5. The unique solution f of the third-order homogeneous equation

$$\frac{d^3y}{dx^3} + 2\frac{d^2y}{dx^2} + 4x\frac{dy}{dx} + x^2y = 0,$$

which is such that

$$f(2) = f'(2) = f''(2) = 0,$$

is the trivial solution f such that $f(x) = 0$ for *all* x.

B. *The Homogeneous Equation*

We now consider the fundamental results concerning the homogeneous equation (4.2). We first state the following basic theorem:

THEOREM 4.2 Basic Theorem on Linear Homogeneous Differential Equations

Hypothesis. *Let $f_1, f_2, \ldots, f_m$ be any m solutions of the homogeneous linear differential equation (4.2).*

Conclusion. *Then $c_1f_1 + c_2f_2 + \cdots + c_mf_m$ is also a solution of (4.2), where $c_1, c_2, \ldots, c_m$ are m arbitrary constants.*

Theorem 4.2 states that if m known solutions of (4.2) are each multiplied by an arbitrary constant and the resulting products are then added together, the resulting sum is also a solution of (4.2). We may put this theorem in a very simple form by means of the concept of linear combination, which we now introduce.

DEFINITION

If $f_1, f_2, \ldots, f_m$ are m given functions, and $c_1, c_2, \ldots, c_m$ are m constants, then the expression

$$c_1f_1 + c_2f_2 + \cdots + c_mf_m$$

is called a linear combination *of $f_1, f_2, \ldots, f_m$.*

In terms of this concept, Theorem 4.2 may be stated as follows:

THEOREM 4.2 (Restated)

Any linear combination of solutions of the homogeneous linear differential equation (4.2) is also a solution of (4.2).

▶ Example 4.6. The student will readily verify that $\sin x$ and $\cos x$ are solutions of

$$\frac{d^2y}{dx^2} + y = 0.$$

Theorem 4.2 states that the linear combination $c_1 \sin x + c_2 \cos x$ is also a solution for any constants c_1 and c_2. For example, the particular linear combination

$$5 \sin x + 6 \cos x$$

is a solution.

▶ Example 4.7. The student may verify that e^x, e^{-x}, and e^{2x} are solutions of

$$\frac{d^3y}{dx^3} - 2\frac{d^2y}{dx^2} - \frac{dy}{dx} + 2y = 0.$$

Theorem 4.2 states that the linear combination $c_1e^x + c_2e^{-x} + c_3e^{2x}$ is also a solution for any constants c_1, c_2, and c_3. For example, the particular linear combination

$$2e^x - 3e^{-x} + \tfrac{2}{3}e^{2x}$$

is a solution.

We now consider what constitutes the so-called general solution of (4.2). To understand this we first introduce the concepts of *linear dependence* and *linear independence.*

DEFINITION

The n functions $f_1, f_2, \ldots, f_n$ are called linearly dependent *on $a \leq x \leq b$ if there exist constants $c_1, c_2, \ldots, c_n$,* not all zero, *such that*

$$c_1f_1(x) + c_2f_2(x) + \cdots + c_nf_n(x) = 0$$

for all x such that $a \leq x \leq b$.

In particular, two functions f_1 and f_2 are linearly dependent *on $a \leq x \leq b$ if there exist constants c_1, c_2,* not both zero, *such that*

$$c_1f_1(x) + c_2f_2(x) = 0$$

for all x such that $a \leq x \leq b$.

▶ Example 4.8. We observe that x and $2x$ are linearly dependent on the interval $0 \leq x \leq 1$. For there exist constants c_1 and c_2, *not both zero,* such that

$$c_1x + c_2(2x) = 0$$

for all x on the interval $0 \leq x \leq 1$. For example, let $c_1 = 2$, $c_2 = -1$.

▶ Example 4.9. We observe that $\sin x$, $3\sin x$, and $-\sin x$ are linearly dependent on the interval $-1 \leq x \leq 2$. For there exist constants c_1, c_2, c_3, *not all zero,* such that

$$c_1 \sin x + c_2(3\sin x) + c_3(-\sin x) = 0$$

for all x on the interval $-1 \leq x \leq 2$. For example, let $c_1 = 1$, $c_2 = 1$, $c_3 = 4$.

DEFINITION

The n functions $f_1, f_2, \ldots, f_n$ are called linearly independent *on the interval $a \leq x \leq b$ if they are* not *linearly dependent there. That is, the functions $f_1, f_2, \ldots, f_n$ are* linearly independent *on $a \leq x \leq b$* if the relation

$$c_1f_1(x) + c_2f_2(x) + \cdots + c_nf_n(x) = 0$$

for all x such that $a \leq x \leq b$ implies that

$$c_1 = c_2 = \cdots = c_n = 0.$$

In other words, the only linear combination of $f_1, f_2, \ldots, f_n$ which is identically zero on $a \leq x \leq b$ is the trivial linear combination

$$0 \cdot f_1 + 0 \cdot f_2 + \cdots + 0 \cdot f_n.$$

▶ Example 4.10. We observe that x and x^2 are linearly independent on $0 \leq x \leq 1$, since $c_1 x + c_2 x^2 = 0$ for *all* x on $0 \leq x \leq 1$ implies that both $c_1 = 0$ and $c_2 = 0$. (Why?)

The next theorem is concerned with the existence of sets of linearly independent solutions of an nth-order homogeneous linear differential equation and with the significance of such linearly independent sets.

THEOREM 4.3

The nth-order homogeneous linear differential equation (4.2) always possesses n solutions which are linearly independent. Further, if $f_1, f_2, \ldots, f_n$ are n linearly independent solutions of (4.2), then every solution f of (4.2) can be expressed as a linear combination

$$c_1 f_1 + c_2 f_2 + \cdots + c_n f_n$$

of these n linearly independent solutions by proper choice of the constants $c_1, c_2, \ldots, c_n$.

Given an nth-order homogeneous linear differential equation, this theorem assures us first that a set of n linearly independent solutions actually exists. The existence of such a linearly independent set assured, the theorem goes on to tell us that *any solution whatsoever* of (4.2) can be written as a linear combination of such a linearly independent set of n solutions by suitable choice of the constants $c_1, c_2, \ldots, c_n$.

Now let $f_1, f_2, \ldots, f_n$ be a set of n linearly independent solutions of (4.2). Then by Theorem 4.2 we know that the linear combination

$$c_1 f_1 + c_2 f_2 + \cdots + c_n f_n, \tag{4.3}$$

where $c_1, c_2, \ldots, c_n$ are n *arbitrary* constants, is also a solution of (4.2). On the other hand, by Theorem 4.3 we know that if f is *any* solution of (4.2), then it can be expressed as a linear combination (4.3) of the n linearly independent solutions $f_1, f_2, \ldots, f_n$, by a suitable choice of the constants $c_1, c_2, \ldots, c_n$. Thus a linear combination (4.3) of the n linearly independent solutions $f_1, f_2, \ldots, f_n$ in which $c_1, c_2, \ldots, c_n$ are *arbitrary* constants must include *all* solutions of (4.2). For this reason, we refer to a set of n linearly independent solutions of (4.2) as a "fundamental set" of (4.2) and call a "general" linear combination of n linearly independent solutions a "general solution" of (4.2), in accordance with the following definition:

DEFINITION

If $f_1, f_2, \ldots, f_n$ are n linearly independent solutions of the nth-order homogeneous linear differential equation (4.2) on $a \leq x \leq b$, then the set $f_1, f_2, \ldots, f_n$ is called a fundamental set *of solutions of (4.2) and the function f defined by*

$$f(x) = c_1 f_1(x) + c_2 f_2(x) + \cdots + c_n f_n(x), \qquad a \leq x \leq b,$$

where $c_1, c_2, \ldots, c_n$ are arbitrary constants, is called a general solution *of (4.2) on $a \leq x \leq b$.*

Therefore, if we can find n linearly independent solutions of (4.2), we can at once write the general solution of (4.2) as a general linear combination of these n solutions.

▶ Example 4.11. We have observed that $\sin x$ and $\cos x$ are solutions of

$$\frac{d^2y}{dx^2} + y = 0$$

for all x, $-\infty < x < \infty$. Further, one can show that these two solutions are linearly independent. Thus, they constitute a fundamental set of solutions of the given differential equation, and its general solution may be expressed as the linear combination

$$c_1 \sin x + c_2 \cos x,$$

where c_1 and c_2 are arbitrary constants. We write this as $y = c_1 \sin x + c_2 \cos x$.

▶ Example 4.12. The solutions e^x, e^{-x}, and e^{2x} of

$$\frac{d^3y}{dx^3} - 2\frac{d^2y}{dx^2} - \frac{dy}{dx} + 2y = 0$$

may be shown to be linearly independent for all x, $-\infty < x < \infty$. Thus, e^x, e^{-x}, and e^{2x} constitute a fundamental set of the given differential equation, and its general solution may be expressed as the linear combination

$$c_1e^x + c_2e^{-x} + c_3e^{2x},$$

where c_1, c_2, and c_3 are arbitrary constants. We write this as

$$y = c_1e^x + c_2e^{-x} + c_3e^{2x}.$$

The next theorem gives a simple criterion for determining whether or not n solutions of (4.2) are linearly independent. We first introduce another concept.

DEFINITION

Let $f_1, f_2, \ldots, f_n$ be n real functions each of which has an $(n - 1)$st derivative on a real interval $a \le x \le b$. The determinant

$$W(f_1, f_2, \ldots, f_n) = \begin{vmatrix} f_1 & f_2 & \cdots & f_n \\ f_1' & f_2' & \cdots & f_n' \\ \vdots & & & \vdots \\ f_1^{(n-1)} & f_2^{(n-1)} & \cdots & f_n^{(n-1)} \end{vmatrix},$$

in which primes denote derivatives, is called the Wronskian *of these n functions. We observe that $W(f_1, f_2, \ldots, f_n)$ is itself a real function defined on $a \le x \le b$. Its value at x is denoted by $W(f_1, f_2, \ldots, f_n)(x)$ or by $W[f_1(x), f_2(x), \ldots, f_n(x)]$.*

THEOREM 4.4

The n solutions $f_1, f_2, \ldots, f_n$ of the nth-order homogeneous linear differential equation (4.2) are linearly independent on $a \le x \le b$ if and only if the Wronskian of $f_1, f_2, \ldots, f_n$ is different from zero for some x on the interval $a \le x \le b$.

We have further:

THEOREM 4.5

The Wronskian of n solutions $f_1, f_2, \ldots, f_n$ of (4.2) is either identically zero on $a \le x \le b$ or else is never zero on $a \le x \le b$.

Thus if we can find n solutions of (4.2), we can apply the Theorems 4.4 and 4.5 to determine whether or not they are linearly independent. If they are linearly independent, then we can form the general solution as a linear combination of these n linearly independent solutions.

In the case of the general *second*-order homogeneous linear differential equation

$$a_0(x)\frac{d^2y}{dx^2} + a_1(x)\frac{dy}{dx} + a_2(x)y = 0,$$

the Wronskian of two solutions f_1 and f_2 is the second-order determinant

$$\begin{vmatrix} f_1 & f_2 \\ f_1' & f_2' \end{vmatrix} = f_1 f_2' - f_1' f_2.$$

▶ **Example 4.13.** We apply Theorem 4.4 to show that the solutions $\sin x$ and $\cos x$ of

$$\frac{d^2y}{dx^2} + y = 0$$

are linearly independent. We find that

$$W(\sin x, \cos x) = \begin{vmatrix} \sin x & \cos x \\ \cos x & -\sin x \end{vmatrix} = -\sin^2 x - \cos^2 x = -1 \ne 0$$

for all real x. Thus, since $W(\sin x, \cos x) \ne 0$ for all real x, we conclude that $\sin x$ and $\cos x$ are indeed linearly independent solutions of the given differential equation on every real interval.

▶ **Example 4.14.** The solutions e^x, e^{-x}, and e^{2x} of

$$\frac{d^3y}{dx^3} - 2\frac{d^2y}{dx^2} - \frac{dy}{dx} + 2y = 0$$

are linearly independent on every real interval, for

$$W(e^x, e^{-x}, e^{2x}) = \begin{vmatrix} e^x & e^{-x} & e^{2x} \\ e^x & -e^{-x} & 2e^{2x} \\ e^x & e^{-x} & 4e^{2x} \end{vmatrix} = e^{2x}\begin{vmatrix} 1 & 1 & 1 \\ 1 & -1 & 2 \\ 1 & 1 & 4 \end{vmatrix} = -6e^{2x} \ne 0$$

for all real x.

C. *Reduction of Order*

In Section 4.2 we shall begin to study methods for obtaining explicit solutions of higher-order linear differential equations. There and in later sections we shall find that the following theorem on reduction of order is often quite useful.

THEOREM 4.6

Hypothesis. *Let f be a nontrivial solution of the nth-order homogeneous linear differential equation*

$$a_0(x)\frac{d^ny}{dx^n} + a_1(x)\frac{d^{n-1}y}{dx^{n-1}} + \cdots + a_{n-1}(x)\frac{dy}{dx} + a_n(x)y = 0. \tag{4.2}$$

Conclusion. *The transformation $y = f(x)v$ reduces Equation (4.2) to an $(n - 1)$st-order homogeneous linear differential equation in the dependent variable $w = dv/dx$.*

This theorem states that if one nonzero solution of the nth-order homogeneous linear differential equation (4.2) is known, then by making the appropriate transformation we may reduce the given equation to another homogeneous linear equation which is one order lower than the original. Since this theorem will be most useful for us in connection with second-order homogeneous linear equations (the case where $n = 2$), we shall now investigate the second-order case in detail. Suppose f is a *known* nontrivial solution of the second-order homogeneous linear equation

$$a_0(x)\frac{d^2y}{dx^2} + a_1(x)\frac{dy}{dx} + a_2(x)y = 0. \tag{4.4}$$

Let us make the transformation

$$y = f(x)v, \tag{4.5}$$

where f is the *known* solution of (4.4) and v is a function of x which will be determined. Then, differentiating, we obtain

$$\frac{dy}{dx} = f(x)\frac{dv}{dx} + f'(x)v, \tag{4.6}$$

$$\frac{d^2y}{dx^2} = f(x)\frac{d^2v}{dx^2} + 2f'(x)\frac{dv}{dx} + f''(x)v. \tag{4.7}$$

Substituting (4.5), (4.6), and (4.7) into (4.4), we obtain

$$a_0(x)\left[f(x)\frac{d^2v}{dx^2} + 2f'(x)\frac{dv}{dx} + f''(x)v\right] + a_1(x)\left[f(x)\frac{dv}{dx} + f'(x)v\right] + a_2(x)f(x)v = 0$$

or

$$a_0(x)f(x)\frac{d^2v}{dx^2} + [2a_0(x)f'(x) + a_1(x)f(x)]\frac{dv}{dx} + [a_0(x)f''(x) + a_1(x)f'(x) + a_2(x)f(x)]v = 0.$$

Since f is a solution of (4.4), the coefficient of v is zero, and so the last equation reduces to

$$a_0(x)f(x)\frac{d^2v}{dx^2} + [2a_0(x)f'(x) + a_1(x)f(x)]\frac{dv}{dx} = 0.$$

Letting $w = dv/dx$, this becomes

$$a_0(x)f(x)\frac{dw}{dx} + [2a_0(x)f'(x) + a_1(x)f(x)]w = 0. \tag{4.8}$$

This is a *first*-order homogeneous linear differential equation in the dependent variable w. The equation is separable; thus assuming $f(x) \neq 0$ and $a_0(x) \neq 0$, we may write

$$\frac{dw}{w} = -\left[2\frac{f'(x)}{f(x)} + \frac{a_1(x)}{a_0(x)}\right] dx.$$

Thus integrating, we obtain

$$\ln |w| = -\ln [f(x)]^2 - \int \frac{a_1(x)}{a_0(x)}\, dx + \ln |c|$$

or

$$w = \frac{c \exp\left[-\int \frac{a_1(x)}{a_0(x)}\, dx\right]}{[f(x)]^2}.$$

This is the general solution of Equation (4.8); choosing the particular solution for which $c = 1$, recalling that $dv/dx = w$, and integrating again, we now obtain

$$v = \int \frac{\exp\left[-\int \frac{a_1(x)}{a_0(x)}\, dx\right]}{[f(x)]^2}\, dx.$$

Finally, from (4.5), we obtain

$$y = f(x)\int \frac{\exp\left[-\int \frac{a_1(x)}{a_0(x)}\, dx\right]}{[f(x)]^2}\, dx. \tag{4.9}$$

The function defined in the right member of (4.9), which we shall henceforth denote by g, is actually a solution of the original second-order equation (4.4). Furthermore, this new solution g and the original known solution f are linearly independent, since

$$W(f, g)(x) = \begin{vmatrix} f(x) & g(x) \\ f'(x) & g'(x) \end{vmatrix} = \begin{vmatrix} f(x) & f(x)v \\ f'(x) & f(x)v' + f'(x)v \end{vmatrix}$$

$$= [f(x)]^2 v' = \exp\left[-\int \frac{a_1(x)}{a_0(x)}\, dx\right] \neq 0.$$

Thus the linear combination

$$c_1 f + c_2 g$$

is the general solution of Equation (4.4). We now summarize this discussion in the following theorem.

THEOREM 4.7

Hypothesis. *Let f be a nontrivial solution of the second-order homogeneous linear differential equation*

$$a_0(x)\frac{d^2y}{dx^2} + a_1(x)\frac{dy}{dx} + a_2(x)y = 0. \tag{4.4}$$

Conclusion 1. *The transformation $y = f(x)v$ reduces Equation (4.4) to the first-order homogeneous linear differential equation*

$$a_0(x)f(x)\frac{dw}{dx} + [2a_0(x)f'(x) + a_1(x)f(x)]w = 0 \tag{4.8}$$

in the dependent variable w, where $w = \dfrac{dv}{dx}$.

Conclusion 2. *The particular solution*

$$w = \frac{\exp\left[-\int \frac{a_1(x)}{a_0(x)}\,dx\right]}{[f(x)]^2}$$

of Equation (4.8) gives rise to the function v, where

$$v(x) = \int \frac{\exp\left[-\int \frac{a_1(x)}{a_0(x)}\,dx\right]}{[f(x)]^2}\,dx.$$

The function g defined by $g(x) = f(x)v(x)$ is then a solution of the second-order equation (4.4).

Conclusion 3. *The original known solution f and the "new" solution g are linearly independent solutions of (4.4), and hence the general solution of (4.4) may be expressed as the linear combination*

$$c_1 f + c_2 g.$$

Let us emphasize the utility of this theorem and at the same time clearly recognize its limitations. Certainly its utility is by now obvious. It tells us that if one solution of the second-order equation (4.4) is known, then we can reduce the order to obtain a linearly independent solution and thereby obtain the general solution of (4.4). But the limitations of the theorem are equally obvious. One solution of Equation (4.4) must already be known to us in order to apply the theorem. How does one "already know" a solution? In general one does not. In some cases the form of the equation itself or related physical considerations suggest that there may be a solution of a certain special form: for example, an exponential solution or a linear solution. However, such cases are not too common and if no solution at all can be so ascertained, then the theorem will not aid us.

We now illustrate the method of reduction of order by means of the following example.

▶ Example 4.15. Given that $y = x$ is a solution of

$$(x^2 + 1)\frac{d^2y}{dx^2} - 2x\frac{dy}{dx} + 2y = 0, \qquad (4.10)$$

find a linearly independent solution by reducing the order.

Solution. First observe that $y = x$ *does* satisfy Equation (4.10). Then let

$$y = xv.$$

Then

$$\frac{dy}{dx} = x\frac{dv}{dx} + v \quad \text{and} \quad \frac{d^2y}{dx^2} = x\frac{d^2v}{dx^2} + 2\frac{dv}{dx}.$$

Substituting the expressions for y, $\dfrac{dy}{dx}$, and $\dfrac{d^2y}{dx^2}$ into Equation (4.10), we obtain

$$(x^2 + 1)\left(x\frac{d^2v}{dx^2} + 2\frac{dv}{dx}\right) - 2x\left(x\frac{dv}{dx} + v\right) + 2xv = 0$$

or

$$x(x^2 + 1)\frac{d^2v}{dx^2} + 2\frac{dv}{dx} = 0.$$

Letting $w = dv/dx$ we obtain the *first*-order homogeneous linear equation

$$x(x^2 + 1)\frac{dw}{dx} + 2w = 0.$$

Treating this as a separable equation, we obtain

$$\frac{dw}{w} = -\frac{2\,dx}{x(x^2 + 1)}$$

or

$$\frac{dw}{w} = \left(-\frac{2}{x} + \frac{2x}{x^2 + 1}\right)dx.$$

Integrating, we obtain the general solution

$$w = \frac{c(x^2 + 1)}{x^2}.$$

Choosing $c = 1$, we recall that $dv/dx = w$ and integrate to obtain the function v given by

$$v(x) = x - \frac{1}{x}.$$

Now forming $g = fv$, where $f(x)$ denotes the *known* solution x, we obtain the function g defined by

$$g(x) = x\left(x - \frac{1}{x}\right) = x^2 - 1.$$

By Theorem 4.7 we know that this is the desired linearly independent solution. The general solution of Equation (4.10) may thus be expressed as the linear combination $c_1x + c_2(x^2 - 1)$ of the linearly independent solutions f and g. We thus write the general solution of Equation (4.10) as

$$y = c_1x + c_2(x^2 - 1).$$

D. The Nonhomogeneous Equation

We now return briefly to the nonhomogeneous equation

$$a_0(x)\frac{d^ny}{dx^n} + a_1(x)\frac{d^{n-1}y}{dx^{n-1}} + \cdots + a_{n-1}(x)\frac{dy}{dx} + a_n(x)y = F(x). \tag{4.1}$$

The basic theorem dealing with this equation is the following.

THEOREM 4.8

Hypothesis. *(1) Let v be any solution of the given (nonhomogeneous) nth-order linear differential equation (4.1). (2) Let u be any solution of the corresponding homogeneous equation*

$$a_0(x)\frac{d^ny}{dx^n} + a_1(x)\frac{d^{n-1}y}{dx^{n-1}} + \cdots + a_{n-1}(x)\frac{dy}{dx} + a_n(x)y = 0. \tag{4.2}$$

Conclusion. *Then $u + v$ is also a solution of the given (nonhomogeneous) equation (4.1).*

▶ Example 4.16. Observe that $y = x$ is a solution of the nonhomogeneous equation

$$\frac{d^2y}{dx^2} + y = x.$$

and that $y = \sin x$ is a solution of the corresponding homogeneous equation

$$\frac{d^2y}{dx^2} + y = 0.$$

Then by Theorem 4.8 the sum

$$\sin x + x$$

is also a solution of the given nonhomogeneous equation

$$\frac{d^2y}{dx^2} + y = x.$$

The student should check that this is indeed true.

Now suppose u is the *general solution* of the homogeneous equation (4.2); that is, it is a "general" linear combination of n linearly independent solutions of (4.2). Let v be any particular solution of the nonhomogeneous equation (4.1), where v contains no arbitrary constants. Then applying Theorem 4.8, we see that $u + v$ is a solution of the nth-order nonhomogeneous equation and that this solution involves n arbitrary constants. We call such a solution a *general solution* of Equation (4.1), in accordance with the following definition:

DEFINITION

Consider the nth-order (nonhomogeneous) linear differential equation

$$a_0(x)\frac{d^ny}{dx^n} + a_1(x)\frac{d^{n-1}y}{dx^{n-1}} + \cdots + a_{n-1}(x)\frac{dy}{dx} + a_n(x)y = F(x) \qquad (4.1)$$

and the corresponding homogeneous equation

$$a_0(x)\frac{d^ny}{dx^n} + a_1(x)\frac{d^{n-1}y}{dx^{n-1}} + \cdots + a_{n-1}(x)\frac{dy}{dx} + a_n(x)y = 0. \qquad (4.2)$$

1. *The general solution of (4.2) is called the* complementary function *of Equation (4.1). We shall denote this by y_c.*
2. *Any particular solution of (4.1) involving no arbitrary constants is called a* particular integral *of (4.1). We shall denote this by y_p.*
3. *The solution $y_c + y_p$ of (4.1), where y_c is the complementary function and y_p is a particular integral of (4.1), is called the* general solution *of (4.1).*

Thus to find the general solution of (4.1), we need merely find:

1. *The complementary function*, that is, a "general" linear combination of n linearly independent solutions of the corresponding homogeneous equation (4.2); and
2. a *particular integral*, that is, any particular solution of (4.1) involving no arbitrary constants.

▶ **Example 4.17.** Consider the differential equation

$$\frac{d^2y}{dx^2} + y = x.$$

The complementary function is the general solution

$$y_c = c_1 \sin x + c_2 \cos x$$

of the corresponding homogeneous equation

$$\frac{d^2y}{dx^2} + y = 0.$$

A particular integral is given by

$$y_p = x.$$

Thus the general solution of the given equation may be written

$$y = y_c + y_p = c_1 \sin x + c_2 \cos x + x.$$

In the remaining sections of this chapter we shall proceed to study methods of obtaining the two constituent parts of the general solution.

We point out that if the nonhomogeneous member $F(x)$ of the linear differential equation (4.1) is expressed as a linear combination of two or more functions, then the following theorem may often be used to advantage in finding a particular integral.

THEOREM 4.9

Hypothesis

1. Let f_1 be a particular integral of

$$a_0(x)\frac{d^n y}{dx^n} + a_1(x)\frac{d^{n-1}y}{dx^{n-1}} + \cdots + a_{n-1}(x)\frac{dy}{dx} + a_n(x)y = F_1(x). \tag{4.11}$$

2. Let f_2 be a particular integral of

$$a_0(x)\frac{d^n y}{dx^n} + a_1(x)\frac{d^{n-1}y}{dx^{n-1}} + \cdots + a_{n-1}(x)\frac{dy}{dx} + a_n(x)y = F_2(x). \tag{4.12}$$

Conclusion. *Then $k_1 f_1 + k_2 f_2$ is a particular integral of*

$$a_0(x)\frac{d^n y}{dx^n} + a_1(x)\frac{d^{n-1}y}{dx^{n-1}} + \cdots + a_{n-1}(x)\frac{dy}{dx} + a_n(x)y = k_1F_1(x) + k_2F_2(x), \tag{4.13}$$

where k_1 and k_2 are constants.

▶ Example 4.18. Suppose we seek a particular integral of

$$\frac{d^2y}{dx^2} + y = 3x + 5\tan x. \tag{4.14}$$

We may then consider the two equations

$$\frac{d^2y}{dx^2} + y = x \tag{4.15}$$

and

$$\frac{d^2y}{dx^2} + y = \tan x. \tag{4.16}$$

We have already noted in Example 4.17 that a particular integral of Equation (4.15) is given by

$$y = x.$$

Further, we can verify (by direct substitution) that a particular integral of Equation (4.16) is given by

$$y = -(\cos x)\ln|\sec x + \tan x|.$$

Therefore, applying Theorem 4.9, a particular integral of Equation (4.16) is

$$y = 3x - 5(\cos x)\ln|\sec x + \tan x|.$$

This example makes the utility of Theorem 4.9 apparent. The particular integral $y = x$ of (4.15) can be quickly determined by the method of Section 4.3 (or by direct inspection!), whereas the particular integral

$$y = -(\cos x)\ln|\sec x + \tan x|$$

of (4.16) must be determined by the method of Section 4.4, and this requires considerably greater computation.

Exercises

1. Theorem 4.1 applies to one of the following problems but not to the other. Determine to which of the problems the theorem applies and state precisely the conclusion which can be drawn in this case. Explain why the theorem does not apply to the remaining problem.

(a) $\dfrac{d^2y}{dx^2} + 5\dfrac{dy}{dx} + 6y = e^x, \qquad y(0) = 5, \qquad y'(0) = 7.$

(b) $\dfrac{d^2y}{dx^2} + 5\dfrac{dy}{dx} + 6y = e^x, \qquad y(0) = 5, \qquad y'(1) = 7.$

2. Answer orally: What is the solution of the following initial-value problem? Why?

$$\frac{d^2y}{dx^2} + x\frac{dy}{dx} + x^2y = 0, \qquad y(1) = 0, \qquad y'(1) = 0.$$

3. Prove Theorem 4.2 for the case $m = n = 2$. That is, prove that if $f_1(x)$ and $f_2(x)$ are two solutions of

$$a_0(x)\frac{d^2y}{dx^2} + a_1(x)\frac{dy}{dx} + a_2(x)y = 0,$$

then $c_1f_1(x) + c_2f_2(x)$ is also a solution of this equation, where c_1 and c_2 are arbitrary constants.

4. Consider the differential equation

$$\frac{d^2y}{dx^2} - 4\frac{dy}{dx} + 3y = 0. \qquad \text{(A)}$$

(a) Show that each of the functions e^x and e^{3x} is a solution of differential equation (A) on the interval $a \le x \le b$, where a and b are arbitrary real numbers such that $a < b$.

(b) What theorem enables us to conclude at once that each of the functions

$$5e^x + 2e^{3x}, \quad 6e^x - 4e^{3x}, \quad \text{and} \quad -7e^x + 5e^{3x}$$

is also a solution of differential equation (A) on $a \le x \le b$?

(c) Each of the functions

$$3e^x, \quad -4e^x, \quad 5e^x, \quad \text{and} \quad 6e^x$$

is also a solution of differential equation (A) on $a \le x \le b$. Why?

5. Again consider the differential equation (A) of Exercise 4.

(a) Use the definition of linear dependence to show that the four functions of part (c) of Exercise 4 are linearly dependent on $a \le x \le b$.

(b) Use Theorem 4.4 to show that the four solutions of differential equation (A) listed in part (c) of Exercise 4 are linearly dependent on $a \le x \le b$.

6. Again consider the differential equation (A) of Exercise 4.
   (a) Use the definition of linear independence to show that the two functions e^x and e^{3x} are linearly independent on $a \le x \le b$.
   (b) Use Theorem 4.4 to show that the two solutions e^x and e^{3x} of differential equation (A) are linearly independent on $a \le x \le b$.
7. Consider the differential equation

$$\frac{d^2y}{dx^2} - 5\frac{dy}{dx} + 6y = 0.$$

   (a) Show that e^{2x} and e^{3x} are linearly independent solutions of this equation on the interval $-\infty < x < \infty$.
   (b) Write the general solution of the given equation.
   (c) Find the solution which satisfies the conditions $y(0) = 2$, $y'(0) = 3$. Explain why this solution is unique. Over what interval is it defined?
8. Consider the differential equation

$$\frac{d^2y}{dx^2} - 2\frac{dy}{dx} + y = 0.$$

   (a) Show that e^x and xe^x are linearly independent solutions of this equation on the interval $-\infty < x < \infty$.
   (b) Write the general solution of the given equation.
   (c) Find the solution which satisfies the conditions $y(0) = 1$, $y'(0) = 4$. Explain why this solution is unique. Over what interval is it defined?
9. Consider the differential equation

$$x^2\frac{d^2y}{dx^2} - 2x\frac{dy}{dx} + 2y = 0.$$

   (a) Show that x and x^2 are linearly independent solutions of this equation on the interval $0 < x < \infty$.
   (b) Write the general solution of the given equation.
   (c) Find the solution which satisfies the conditions $y(1) = 3$, $y'(1) = 2$. Explain why this solution is unique. Over what interval is this solution defined?
10. Consider the differential equation

$$x^2\frac{d^2y}{dx^2} + x\frac{dy}{dx} - 4y = 0.$$

   (a) Show that x^2 and $1/x^2$ are linearly independent solutions of this equation on the interval $0 < x < \infty$.
   (b) Write the general solution of the given equation.
   (c) Find the solution which satisfies the conditions $y(2) = 3$, $y'(2) = -1$.

Explain why this solution is unique. Over what interval is this solution defined?

11. Consider the differential equation

$$\frac{d^2y}{dx^2} - 5\frac{dy}{dx} + 4y = 0.$$

(a) Show that each of the functions e^x, e^{4x}, and $2e^x - 3e^{4x}$ is a solution of this equation on the interval $-\infty < x < \infty$.

(b) Show that the solutions e^x and e^{4x} are linearly independent on $-\infty < x < \infty$.

(c) Show that the solutions e^x and $2e^x - 3e^{4x}$ are also linearly independent on $-\infty < x < \infty$.

(d) Are the solutions e^{4x} and $2e^x - 3e^{4x}$ still another pair of linearly independent solutions on $-\infty < x < \infty$? Justify your answer.

12. Given that e^{-x}, e^{3x}, and e^{4x} are all solutions of

$$\frac{d^3y}{dx^3} - 6\frac{d^2y}{dx^2} + 5\frac{dy}{dx} + 12y = 0,$$

show that they are linearly independent on the interval $-\infty < x < \infty$ and write the general solution.

13. Given that x, x^2, and x^4 are all solutions of

$$x^3\frac{d^3y}{dx^3} - 4x^2\frac{d^2y}{dx^2} + 8x\frac{dy}{dx} - 8y = 0,$$

show that they are linearly independent on the interval $0 < x < \infty$ and write the general solution.

14. Verify the truth of Theorem 4.6 for the equation

$$(x^2 - 1)\frac{d^2y}{dx^2} - 2x\frac{dy}{dx} + 2y = 0,$$

given that $f(x) = x$ is a solution.

15. Given that $y = e^{2x}$ is a solution of

$$(2x + 1)\frac{d^2y}{dx^2} - 4(x + 1)\frac{dy}{dx} + 4y = 0,$$

find a linearly independent solution by reducing the order. Write the general solution.

16. Given that $y = x^2$ is a solution of

$$(x^3 - x^2)\frac{d^2y}{dx^2} - (x^3 + 2x^2 - 2x)\frac{dy}{dx} + (2x^2 + 2x - 2)y = 0,$$

find a linearly independent solution by reducing the order. Write the general solution.

17. Prove Theorem 4.8 for the case $n = 2$. That is, prove that if u is any solution of

$$a_0(x)\frac{d^2y}{dx^2} + a_1(x)\frac{dy}{dx} + a_2(x)y = 0$$

and v is any solution of

$$a_0(x)\frac{d^2y}{dx^2} + a_1(x)\frac{dy}{dx} + a_2(x)y = F(x),$$

then $u + v$ is also a solution of this latter nonhomogeneous equation.

18. Consider the nonhomogeneous differential equation

$$\frac{d^2y}{dx^2} - 3\frac{dy}{dx} + 2y = 4x^2.$$

(a) Show that e^x and e^{2x} are linearly independent solutions of the corresponding homogeneous equation

$$\frac{d^2y}{dx^2} - 3\frac{dy}{dx} + 2y = 0.$$

(b) What is the complementary function of the given nonhomogeneous equation?

(c) Show that $2x^2 + 6x + 7$ is a particular integral of the given equation.

(d) What is the general solution of the given equation?

19. Given that a particular integral of

$$\frac{d^2y}{dx^2} - 5\frac{dy}{dx} + 6y = 1 \quad \text{is} \quad y = \frac{1}{6},$$

a particular integral of

$$\frac{d^2y}{dx^2} - 5\frac{dy}{dx} + 6y = x \quad \text{is} \quad y = \frac{x}{6} + \frac{5}{36},$$

and a particular integral of

$$\frac{d^2y}{dx^2} - 5\frac{dy}{dx} + 6y = e^x \quad \text{is} \quad y = \frac{e^x}{2},$$

use Theorem 4.9 to find a particular integral of

$$\frac{d^2y}{dx^2} - 5\frac{dy}{dx} + 6y = 2 - 12x + 6e^x.$$

4.2 The Homogeneous Linear Equation with Constant Coefficients

A. *Introduction*

In this section we consider the special case of the nth-order homogeneous linear differential equation in which all of the coefficients are real constants. That is, we shall be concerned with the equation

$$a_0\frac{d^ny}{dx^n} + a_1\frac{d^{n-1}y}{dx^{n-1}} + \cdots + a_{n-1}\frac{dy}{dx} + a_ny = 0 \tag{4.17}$$

where $a_0, a_1, \ldots, a_{n-1}, a_n$ are real constants. We shall show that the general solution of this equation can be found explicitly.

In an attempt to find solutions of a differential equation we would naturally inquire whether or not any familiar type of function might possibly have the properties which would enable it to be a solution. The differential equation (4.17) requires a function f having the property such that if it and its various derivatives are each multiplied by certain constants, the a_i, and the resulting products, $a_i f^{(n-i)}$, are then added, the result will equal zero for all values of x for which this result is defined. For this to be the case we need a function such that its derivatives are constant multiples of itself. Do we know of functions f having this property that

$$\frac{d^k}{dx^k}[f(x)] = cf(x)$$

for all x? The answer is "yes," for the exponential function f such that $f(x) = e^{mx}$, where m is a constant, is such that

$$\frac{d^k}{dx^k}(e^{mx}) = m^k e^{mx}.$$

Thus we shall seek solutions of (4.17) of the form $y = e^{mx}$, where the constant m will be chosen such that e^{mx} *does* satisfy the equation. Assuming then that $y = e^{mx}$ is a solution for certain m, we have:

$$\frac{dy}{dx} = me^{mx},$$

$$\frac{d^2y}{dx^2} = m^2 e^{mx},$$

$$\vdots$$

$$\frac{d^ny}{dx^n} = m^n e^{mx}.$$

Substituting in (4.17), we obtain

$$a_0 m^n e^{mx} + a_1 m^{n-1} e^{mx} + \cdots + a_{n-1} m e^{mx} + a_n e^{mx} = 0$$

or

$$e^{mx}(a_0 m^n + a_1 m^{n-1} + \cdots + a_{n-1} m + a_n) = 0.$$

Since $e^{mx} \neq 0$, we obtain the polynomial equation in the unknown m:

$$a_0 m^n + a_1 m^{n-1} + \cdots + a_{n-1} m + a_n = 0. \tag{4.18}$$

This equation is called the *auxiliary equation* or the *characteristic equation* of the given differential equation (4.17). If $y = e^{mx}$ is a solution of (4.17) then we see that the constant m must satisfy (4.18). Hence, to solve (4.17), we write the auxiliary equation (4.18) and solve it for m. Observe that (4.18) is formally obtained from (4.17) by merely replacing the kth derivative in (4.17) by m^k ($k = 0, 1, 2, \ldots, n$). Three cases arise, according as the roots of (4.18) are real and distinct, real and repeated, or complex.

B *Case 1. Distinct Real Roots*

Suppose the roots of (4.18) are the n distinct real numbers

$$m_1, m_2, \ldots, m_n.$$

Then

$$e^{m_1 x}, e^{m_2 x}, \ldots, e^{m_n x}$$

are n distinct solutions of (3.17). Further, using the Wronskian determinant one may show that these n solutions are linearly independent. Thus we have the following result.

THEOREM 4.10

Consider the nth-order homogeneous linear differential equation (4.17) with constant coefficients. If the auxiliary equation (4.18) has the n distinct real roots $m_1, m_2, \ldots, m_n$, then the general solution of (4.17) is

$$y = c_1 e^{m_1 x} + c_2 e^{m_2 x} + \cdots + c_n e^{m_n x},$$

where $c_1, c_2, \ldots, c_n$ are arbitrary constants.

▶ **Example 4.19.** Consider the differential equation

$$\frac{d^2y}{dx^2} - 3\frac{dy}{dx} + 2y = 0.$$

The auxiliary equation is

$$m^2 - 3m + 2 = 0.$$

Hence

$$(m - 1)(m - 2) = 0, \qquad m_1 = 1, \qquad m_2 = 2.$$

The roots are real and distinct. Thus e^x and e^{2x} are solutions and the general solution may be written

$$y = c_1 e^x + c_2 e^{2x}.$$

We verify that e^x and e^{2x} are indeed linearly independent. Their Wronskian is

$$W(e^x, e^{2x}) = \begin{vmatrix} e^x & e^{2x} \\ e^x & 2e^{2x} \end{vmatrix} = e^{3x} \neq 0.$$

Thus by Theorem 4.4 we are assured of their linear independence.

▶ **Example 4.20.** Consider the differential equation

$$\frac{d^3y}{dx^3} - 4\frac{d^2y}{dx^2} + \frac{dy}{dx} + 6y = 0.$$

The auxiliary equation is

$$m^3 - 4m^2 + m + 6 = 0.$$

We observe that $m = -1$ is a root of this equation. By synthetic division we obtain the factorization

$$(m + 1)(m^2 - 5m + 6) = 0$$

or

$$(m + 1)(m - 2)(m - 3) = 0.$$

Thus the roots are the distinct real numbers

$$m_1 = -1, \qquad m_2 = 2, \qquad m_3 = 3,$$

and the general solution is

$$y = c_1 e^{-x} + c_2 e^{2x} + c_3 e^{3x}.$$

C. Case 2. Repeated Real Roots

We shall begin our study of this case by considering a simple example.

▶ **Example 4.21: Introductory Example.** Consider the differential equation

$$\frac{d^2y}{dx^2} - 6\frac{dy}{dx} + 9y = 0. \tag{4.19}$$

The auxiliary equation is

$$m^2 - 6m + 9 = 0$$

or

$$(m - 3)^2 = 0.$$

The roots of this equation are

$$m_1 = 3, \qquad m_2 = 3$$

(real but *not* distinct).

Corresponding to the root m_1 we have the solution e^{3x}, and corresponding to m_2 we have the *same* solution e^{3x}. The linear combination $c_1e^{3x} + c_2e^{3x}$ of these "two" solutions is clearly *not* the general solution of the differential equation (4.19), for it is *not* a linear combination of *two linearly independent* solutions. Indeed we may write the combination $c_1e^{3x} + c_2e^{3x}$ as simply c_0e^{3x}, where $c_0 = c_1 + c_2$; and clearly $y = c_0e^{3x}$, involving *one* arbitrary constant, is not the general solution of the given *second*-order equation.

We must find a linearly independent solution; but how shall we proceed to do so? Since we already know the one solution e^{3x}, we may apply Theorem 4.7 and reduce the order. We let

$$y = e^{3x}v,$$

where v is to be determined. Then

$$\frac{dy}{dx} = e^{3x}\frac{dv}{dx} + 3e^{3x}v,$$

$$\frac{d^2y}{dx^2} = e^{3x}\frac{d^2v}{dx^2} + 6e^{3x}\frac{dv}{dx} + 9e^{3x}v.$$

Substituting into Equation (4.19) we have

$$\left(e^{3x}\frac{d^2v}{dx^2} + 6e^{3x}\frac{dv}{dx} + 9e^{3x}v\right) - 6\left(e^{3x}\frac{dv}{dx} + 3e^{3x}v\right) + 9e^{3x}v = 0$$

or

$$e^{3x}\frac{d^2v}{dx^2} = 0.$$

Letting $w = dv/dx$, we have the first-order equation

$$e^{3x}\frac{dw}{dx} = 0$$

or simply

$$\frac{dw}{dx} = 0.$$

The solutions of this first-order equation are simply $w = c$, where c is an arbitrary constant. Choosing the particular solution $w = 1$ and recalling that $dv/dx = w$, we find

$$v(x) = x + c_0,$$

where c_0 is an arbitrary constant. By Theorem 4.7 we know that for any choice of the constant c_0, $v(x)e^{3x} = (x + c_0)e^{3x}$ is a solution of the given second-order equation (4.19). Further, by Theorem 4.7, we know that this solution and the previously known solution e^{3x} are linearly independent. Choosing $c_0 = 0$ we obtain the solution

$$y = xe^{3x},$$

and thus corresponding to the *double* root 3 we find the linearly independent solutions

$$e^{3x} \quad \text{and} \quad xe^{3x}$$

of Equation (4.19).

Thus the general solution of Equation (4.19) may be written

$$y = c_1e^{3x} + c_2xe^{3x} \tag{4.20}$$

or

$$y = (c_1 + c_2x)e^{3x}. \tag{4.21}$$

With this example as a guide, let us return to the general nth-order equation (4.17). If the auxiliary equation (4.18) has the *double* real root m, we would surely expect that e^{mx} and xe^{mx} would be the corresponding linearly independent solutions. This is indeed the case. Specifically, suppose the roots of (4.18) are the double real root m and the $(n - 2)$ distinct real roots

$$m_1, m_2, \ldots, m_{n-2}.$$

Then linearly independent solutions of (4.17) are

$$e^{mx}, xe^{mx}, e^{m_1x}, e^{m_2x}, \ldots, e^{m_{n-2}x},$$

and the general solution may be written

$$y = c_1e^{mx} + c_2xe^{mx} + c_3e^{m_1x} + c_4e^{m_2x} + \cdots + c_ne^{m_{n-2}x}$$

or

$$y = (c_1 + c_2x)e^{mx} + c_3e^{m_1x} + c_4e^{m_2x} + \cdots + c_ne^{m_{n-2}x}.$$

In like manner, if the auxiliary equation (4.18) has the triple real root m, corresponding linearly independent solutions are

$$e^{mx}, \quad xe^{mx}, \quad \text{and} \quad x^2e^{mx}.$$

The corresponding part of the general solution may be written

$$(c_1 + c_2x + c_3x^2)e^{mx}.$$

Proceeding further in like manner, we summarize Case 2 in the following theorem:

THEOREM 4.11

1. Consider the nth-order homogeneous linear differential equation (4.17) with constant coefficients. If the auxiliary equation (4.18) has the real root m occurring k times, then the part of the general solution of (4.17) corresponding to this k-fold repeated root is

$$(c_1 + c_2x + c_3x^2 + \cdots + c_kx^{k-1})e^{mx}.$$

2. If, further, the remaining roots of the auxiliary equation (4.18) are the distinct real numbers $m_{k+1}, \ldots, m_n$, then the general solution of (4.17) is

$$y = (c_1 + c_2x + c_3x^2 + \cdots + c_kx^{k-1})e^{mx} + c_{k+1}e^{m_{k+1}x} + \cdots + c_ne^{m_nx}.$$

3. If, however, any of the remaining roots are also repeated, then the parts of the general solution of (4.17) corresponding to each of these other repeated roots are expressions similar to that corresponding to m in part 1.

We now consider several examples.

▶ **Example 4.22.** Find the general solution of

$$\frac{d^3y}{dx^3} - 4\frac{d^2y}{dx^2} - 3\frac{dy}{dx} + 18y = 0.$$

The auxiliary equation

$$m^3 - 4m^2 - 3m + 18 = 0$$

has the roots, 3, 3, -2. The general solution is

$$y = c_1e^{3x} + c_2xe^{3x} + c_3e^{-2x}$$

or

$$y = (c_1 + c_2x)e^{3x} + c_3e^{-2x}.$$

▶ **Example 4.23.** Find the general solution of

$$\frac{d^4y}{dx^4} - 5\frac{d^3y}{dx^3} + 6\frac{d^2y}{dx^2} + 4\frac{dy}{dx} - 8y = 0.$$

The auxiliary equation is

$$m^4 - 5m^3 + 6m^2 + 4m - 8 = 0,$$

with roots 2, 2, 2, -1. The part of the general solution corresponding to the threefold root 2 is

$$y_1 = (c_1 + c_2x + c_3x^2)e^{2x}$$

and that corresponding to the simple root -1 is simply

$$y_2 = c_4e^{-x}.$$

Thus the general solution is $y = y_1 + y_2$, that is,

$$y = (c_1 + c_2x + c_3x^2)e^{2x} + c_4e^{-x}.$$

D. *Case 3. Conjugate Complex Roots*

Now suppose that the auxiliary equation has the complex number $a + bi$ (a, b real, $i^2 = -1$, $b \neq 0$) as a nonrepeated root. Then, since the coefficients are real, the conjugate complex number $a - bi$ is also a nonrepeated root. The corresponding part of the general solution is

$$k_1 e^{(a+bi)x} + k_2 e^{(a-bi)x},$$

where k_1 and k_2 are arbitrary constants. The solutions defined by $e^{(a+bi)x}$ and $e^{(a-bi)x}$ are complex functions of the real variable x. It is desirable to replace these by two *real* linearly independent solutions. This can be accomplished by using Euler's formula,

$$e^{i\theta} = \cos\theta + i\sin\theta,^*$$

which holds for all real θ. Using this we have:

$$\begin{aligned} k_1 e^{(a+bi)x} + k_2 e^{(a-bi)x} &= k_1 e^{ax} e^{bix} + k_2 e^{ax} e^{-bix} \\ &= e^{ax}[k_1 e^{ibx} + k_2 e^{-ibx}] \\ &= e^{ax}[k_1(\cos bx + i\sin bx) + k_2(\cos bx - i\sin bx)] \\ &= e^{ax}[(k_1 + k_2)\cos bx + i(k_1 - k_2)\sin bx] \\ &= e^{ax}[c_1 \sin bx + c_2 \cos bx], \end{aligned}$$

where $c_1 = i(k_1 - k_2)$, $c_2 = k_1 + k_2$ are two new arbitrary constants. Thus the part of the general solution corresponding to the nonrepeated conjugate complex roots $a \pm bi$ is

$$e^{ax}[c_1 \sin bx + c_2 \cos bx].$$

Combining this with the results of Case 2, we have the following theorem covering Case 3.

THEOREM 4.12

1. Consider the nth-order homogeneous linear differential equation (4.17) with constant coefficients. If the auxiliary equation (4.18) has the conjugate complex roots $a + bi$ and $a - bi$, neither repeated, then the corresponding part of the general solution of (4.17) may be written

$$y = e^{ax}(c_1 \sin bx + c_2 \cos bx).$$

2. If, however, $a + bi$ and $a - bi$ are each k-fold roots of the auxiliary equation (4.18), then the corresponding part of the general solution of (4.17) may be written

$$y = e^{ax}[(c_1 + c_2 x + c_3 x^2 + \cdots + c_k x^{k-1})\sin bx + (c_{k+1} + c_{k+2}x + c_{k+3}x^2 + \cdots + c_{2k}x^{k-1})\cos bx].$$

We now give several examples.

---

* We borrow this basic identity from complex variable theory, as well as the fact that $e^{ax+bix} = e^{ax}e^{ibx}$ holds for complex exponents.

▶ **Example 4.24.** Find the general solution of

$$\frac{d^2y}{dx^2} + y = 0.$$

We have already used this equation to illustrate the theorems of Section 4.1. Let us now obtain its solution using Theorem 4.12. The auxiliary equation $m^2 + 1 = 0$ has the roots $m = \pm i$. These are the pure imaginary complex numbers $a \pm bi$, where $a = 0$, $b = 1$. The general solution is thus

$$y = e^{0x}(c_1 \sin 1 \cdot x + c_2 \cos 1 \cdot x),$$

which is simply

$$y = c_1 \sin x + c_2 \cos x.$$

▶ **Example 4.25.** Find the general solution of

$$\frac{d^2y}{dx^2} - 6\frac{dy}{dx} + 25y = 0.$$

The auxiliary equation is $m^2 - 6m + 25 = 0$. Solving it, we find

$$m = \frac{6 \pm \sqrt{36 - 100}}{2} = \frac{6 \pm 8i}{2} = 3 \pm 4i.$$

Here the roots are the conjugate complex numbers $a \pm bi$, where $a = 3$, $b = 4$. The general solution may be written

$$y = e^{3x}(c_1 \sin 4x + c_2 \cos 4x).$$

▶ **Example 4.26.** Find the general solution of

$$\frac{d^4y}{dx^4} - 4\frac{d^3y}{dx^3} + 14\frac{d^2y}{dx^2} - 20\frac{dy}{dx} + 25y = 0.$$

The auxiliary equation is

$$m^4 - 4m^3 + 14m^2 - 20m + 25 = 0.$$

The solution of this equation presents some ingenuity and labor. Since our purpose in this example is not to display our mastery of the solution of algebraic equations but rather to illustrate the above principles of determining the general solution of differential equations, we unblushingly list the roots without further apologies.

They are

$$1 + 2i, \quad 1 - 2i, \quad 1 + 2i, \quad 1 - 2i.$$

Since each pair of conjugate complex roots is double, the general solution is

$$y = e^x[(c_1 + c_2x) \sin 2x + (c_3 + c_4x) \cos 2x]$$

or

$$y = c_1e^x \sin 2x + c_2xe^x \sin 2x + c_3e^x \cos 2x + c_4xe^x \cos 2x.$$

E. An Initial-Value Problem

We now apply the results concerning the general solution of a homogeneous linear equation with constant coefficients to an initial-value problem involving such an equation.

▶ Example 4.27. Solve the initial-value problem

$$\frac{d^2y}{dx^2} - 6\frac{dy}{dx} + 25y = 0, \tag{4.22}$$

$$y(0) = -3, \tag{4.23}$$

$$y'(0) = -1. \tag{4.24}$$

First let us note that by Theorem 4.1 this problem has a unique solution defined for all x, $-\infty < x < \infty$. We now proceed to find this solution; that is, we seek the particular solution of the differential equation (4.22) which satisfies the two initial conditions (4.23) and (4.24). We have already found the general solution of the differential equation (4.22) in Example 4.25. It is

$$y = e^{3x}(c_1 \sin 4x + c_2 \cos 4x). \tag{4.25}$$

From this, we find

$$\frac{dy}{dx} = e^{3x}[(3c_1 - 4c_2) \sin 4x + (4c_1 + 3c_2) \cos 4x]. \tag{4.26}$$

We now apply the initial conditions. Applying condition (4.23), $y(0) = -3$, to Equation (4.25), we find

$$-3 = e^0(c_1 \sin 0 + c_2 \cos 0),$$

which reduces at once to

$$c_2 = -3. \tag{4.27}$$

Applying condition (4.24), $y'(0) = -1$, to Equation (4.26), we obtain

$$-1 = e^0[(3c_1 - 4c_2) \sin 0 + (4c_1 + 3c_2) \cos 0],$$

which reduces to

$$4c_1 + 3c_2 = -1. \tag{4.28}$$

Solving Equations (4.27) and (4.28) for the unknowns c_1 and c_2, we find

$$c_1 = 2, \qquad c_2 = -3.$$

Replacing c_1 and c_2 in Equation (4.25) by these values, we obtain the unique solution of the given initial-value problem in the form

$$y = e^{3x}(2 \sin 4x - 3 \cos 4x).$$

Recall from trigonometry that a linear combination of a sine term and a cosine term having a common argument cx may be expressed as an appropriate constant multiple of the sine of the sum of this common argument cx and an appropriate constant angle ϕ. Thus the preceding solution can be reexpressed in an alternative form involving the factor $\sin(4x + \phi)$ for some suitable ϕ. To do this we first multiply and divide by $\sqrt{(2^2) + (-3)^2} = \sqrt{13}$, thereby obtaining

$$y = \sqrt{13}\, e^{3x} \left[\frac{2}{\sqrt{13}} \sin 4x - \frac{3}{\sqrt{13}} \cos 4x \right].$$

From this we may express the solution in the alternative form

$$y = \sqrt{13}\, e^{3x} \sin(4x + \phi),$$

where the angle ϕ is defined by the equations

$$\sin \phi = -\frac{3}{\sqrt{13}}, \qquad \cos \phi = \frac{2}{\sqrt{13}}.$$

Exercises

Find the general solution of each of the differential equations in Exercises 1 through 24.

1. $\dfrac{d^2y}{dx^2} - 5\dfrac{dy}{dx} + 6y = 0.$

2. $\dfrac{d^2y}{dx^2} - 2\dfrac{dy}{dx} - 3y = 0.$

3. $4\dfrac{d^2y}{dx^2} - 12\dfrac{dy}{dx} + 5y = 0.$

4. $3\dfrac{d^2y}{dx^2} - 14\dfrac{dy}{dx} - 5y = 0.$

5. $\dfrac{d^3y}{dx^3} - 3\dfrac{d^2y}{dx^2} - \dfrac{dy}{dx} + 3y = 0.$

6. $\dfrac{d^3y}{dx^3} - 6\dfrac{d^2y}{dx^2} + 5\dfrac{dy}{dx} + 12y = 0.$

7. $\dfrac{d^2y}{dx^2} - 8\dfrac{dy}{dx} + 16y = 0.$

8. $4\dfrac{d^2y}{dx^2} + 4\dfrac{dy}{dx} + y = 0.$

9. $\dfrac{d^2y}{dx^2} - 4\dfrac{dy}{dx} + 13y = 0.$

10. $\dfrac{d^2y}{dx^2} + 6\dfrac{dy}{dx} + 25y = 0.$

11. $\dfrac{d^2y}{dx^2} + 9y = 0.$

12. $4\dfrac{d^2y}{dx^2} + y = 0.$

13. $\dfrac{d^3y}{dx^3} - 5\dfrac{d^2y}{dx^2} + 7\dfrac{dy}{dx} - 3y = 0.$

14. $4\dfrac{d^3y}{dx^3} + 4\dfrac{d^2y}{dx^2} - 7\dfrac{dy}{dx} + 2y = 0.$

15. $\dfrac{d^3y}{dx^3} - 6\dfrac{d^2y}{dx^2} + 12\dfrac{dy}{dx} - 8y = 0.$

16. $\dfrac{d^3y}{dx^3} + 4\dfrac{d^2y}{dx^2} + 5\dfrac{dy}{dx} + 6y = 0.$

17. $\dfrac{d^3y}{dx^3} - \dfrac{d^2y}{dx^2} + \dfrac{dy}{dx} - y = 0.$

18. $\dfrac{d^4y}{dx^4} + 8\dfrac{d^2y}{dx^2} + 16y = 0.$

19. $\dfrac{d^5y}{dx^5} - 2\dfrac{d^4y}{dx^4} + \dfrac{d^3y}{dx^3} = 0.$

20. $\dfrac{d^4y}{dx^4} - \dfrac{d^3y}{dx^3} - 3\dfrac{d^2y}{dx^2} + \dfrac{dy}{dx} + 2y = 0.$

21. $\dfrac{d^4y}{dx^4} - 3\dfrac{d^3y}{dx^3} - 2\dfrac{d^2y}{dx^2} + 2\dfrac{dy}{dx} + 12y = 0.$

22. $\dfrac{d^4y}{dx^4} + 6\dfrac{d^3y}{dx^3} + 15\dfrac{d^2y}{dx^2} + 20\dfrac{dy}{dx} + 12y = 0.$

23. $\dfrac{d^4y}{dx^4} + y = 0.$

24. $\dfrac{d^5y}{dx^5} = 0.$

Solve the initial-value problems in Exercises 25 through 36.

25. $\dfrac{d^2y}{dx^2} - \dfrac{dy}{dx} - 12y = 0, \quad y(0) = 3, \quad y'(0) = 5.$

26. $\dfrac{d^2y}{dx^2} + 7\dfrac{dy}{dx} + 10y = 0, \quad y(0) = -4, \quad y'(0) = 2.$

27. $\dfrac{d^2y}{dx^2} + 4\dfrac{dy}{dx} + 4y = 0, \quad y(0) = 3, \quad y'(0) = 7.$

28. $9\dfrac{d^2y}{dx^2} - 6\dfrac{dy}{dx} + y = 0, \quad y(0) = 3, \quad y'(0) = -1.$

29. $\dfrac{d^2y}{dx^2} - 4\dfrac{dy}{dx} + 29y = 0, \quad y(0) = 0, \quad y'(0) = 5.$

30. $\dfrac{d^2y}{dx^2} + 6\dfrac{dy}{dx} + 58y = 0, \quad y(0) = -1, \quad y'(0) = 5.$

31. $9\dfrac{d^2y}{dx^2} + 6\dfrac{dy}{dx} + 5y = 0, \quad y(0) = 6, \quad y'(0) = 0.$

32. $4\dfrac{d^2y}{dx^2} + 4\dfrac{dy}{dx} + 37y = 0, \quad y(0) = 2, \quad y'(0) = -4.$

33. $\dfrac{d^3y}{dx^3} - 6\dfrac{d^2y}{dx^2} + 11\dfrac{dy}{dx} - 6y = 0, \quad y(0) = 0, \quad y'(0) = 0, \quad y''(0) = 2.$

34. $\dfrac{d^3y}{dx^3} - 2\dfrac{d^2y}{dx^2} + 4\dfrac{dy}{dx} - 8y = 0, \quad y(0) = 2, \quad y'(0) = 0, \quad y''(0) = 0.$

35. $\dfrac{d^3y}{dx^3} - 3\dfrac{d^2y}{dx^2} + 4y = 0, \quad y(0) = 1, \quad y'(0) = -8, \quad y''(0) = -4.$

36. $\dfrac{d^3y}{dx^3} - 5\dfrac{d^2y}{dx^2} + 9\dfrac{dy}{dx} - 5y = 0, \quad y(0) = 0, \quad y'(0) = 1, \quad y''(0) = 6.$

37. The roots of the auxiliary equation, corresponding to a certain 10th-order homogeneous linear differential equation with constant coefficients, are

$$4, \quad 4, \quad 4, \quad 4, \quad 2 + 3i, \quad 2 - 3i, \quad 2 + 3i, \quad 2 - 3i, \quad 2 + 3i, \quad 2 - 3i,$$

Write the general solution.

38. The roots of the auxiliary equation, corresponding to a certain 12th-order homogeneous linear differential equation with constant coefficients, are

$$2, \quad 2, \quad 2, \quad 2, \quad 2, \quad 2, \quad 3 + 4i, \quad 3 - 4i, \quad 3 + 4i, \quad 3 - 4i, \quad 3 + 4i, \quad 3 - 4i.$$

Write the general solution.

39. Given that $\sin x$ is a solution of

$$\frac{d^4y}{dx^4} + 2\frac{d^3y}{dx^3} + 6\frac{d^2y}{dx^2} + 2\frac{dy}{dx} + 5y = 0,$$

find the general solution.

40. Given that $e^x \sin 2x$ is a solution of

$$\frac{d^4y}{dx^4} + 3\frac{d^3y}{dx^3} + \frac{d^2y}{dx^2} + 13\frac{dy}{dx} + 30y = 0,$$

find the general solution.

4.3 The Method of Undetermined Coefficients

A. *The Method*

We now consider the (nonhomogeneous) differential equation

$$a_0\frac{d^ny}{dx^n} + a_1\frac{d^{n-1}y}{dx^{n-1}} + \cdots + a_{n-1}\frac{dy}{dx} + a_ny = F(x), \tag{4.29}$$

where the coefficients $a_0, a_1, \ldots, a_n$ are constants but where the nonhomogeneous term F is (in general) a nonconstant function of x. Recall that the general solution of (4.29) may be written

$$y = y_c + y_p,$$

where y_c is the *complementary function*, that is, the general solution of the corresponding homogeneous equation (Equation (4.29) with F replaced by 0), and y_p is a *particular integral*, that is, any solution of (4.29) containing no arbitrary constants. In Section 4.2 we learned how to find the complementary function; now we consider methods of determining a particular integral.

We consider first the method of *undetermined coefficients*. Mathematically speaking, the class of functions F to which this method applies is actually quite restricted; but this mathematically narrow class includes functions of frequent occurrence and considerable importance in various physical applications. And this method has one distinct advantage—when it *does apply*, it is relatively simple!

We first introduce certain preliminary definitions.

DEFINITION

We shall call a function a UC function *if it is* either *(1) a function defined by one of the following:*

(*i*) x^n, *where n is a positive integer or zero.*
(*ii*) e^{ax}, *where a is a constant* $\neq 0$.
(*iii*) $\sin(bx + c)$, *where b and c are constants,* $b \neq 0$.
(*iv*) $\cos(bx + c)$, *where b and c are constants,* $b \neq 0$.

or (*2*) *a function defined as a finite product of two or more functions of these four types.*

The method of undetermined coefficients applies when the nonhomogeneous function F in the differential equation is a finite linear combination of UC functions. Observe that given a UC function f, each successive derivative of f is either itself a constant multiple of a UC function or else a linear combination of UC functions.

DEFINITION

Consider a UC function f. The set of functions consisting of f itself and all linearly independent UC functions of which the successive derivatives of f are either constant multiples or linear combinations will be called the UC set *of f.*

▶ Example 4.28. The function f defined for all real x by $f(x) = x^3$ is a UC function. Computing derivatives of f, we find

$$f'(x) = 3x^2, \quad f''(x) = 6x, \quad f'''(x) = 6 = 6 \cdot 1, \quad f^{(n)}(x) = 0 \text{ for } n > 3.$$

The linearly independent UC functions of which the successive derivatives of f are either constant multiples or linear combinations are those given by

$$x^2, \quad x, \quad 1.$$

Thus the *UC set* of x^3 is the set $S = \{x^3, x^2, x, 1\}$.

▶ Example 4.29. The function f defined for all real x by $f(x) = \sin 2x$ is a UC function. Computing derivatives of f, we find

$$f'(x) = 2\cos 2x, \quad f''(x) = -4\sin 2x, \quad \ldots.$$

The only linearly independent UC function of which the successive derivatives of f are constant multiples or linear combinations is that given by $\cos 2x$. Thus the *UC set* of $\sin 2x$ is the set $S = \{\sin 2x, \cos 2x\}$.

▶ Example 4.30. The function f defined for all real x by $f(x) = x^2 \sin x$ is the product of the two UC functions defined by x^2 and $\sin x$. Hence f is itself a UC function. Computing derivatives of f, we find

$$f'(x) = 2x\sin x + x^2\cos x,$$

$$f''(x) = 2\sin x + 4x\cos x - x^2\sin x,$$

$$f'''(x) = 6\cos x - 6x\sin x - x^2\cos x,$$

$$\ldots.$$

No "new" types of functions will occur from further differentiation. Each derivative of f is a linear combination of certain of the six UC functions given by $x^2 \sin x$, $x^2 \cos x$, $x \sin x$, $x \cos x$, $\sin x$, and $\cos x$. Thus the set

$$S = \{x^2 \sin x,\ x^2 \cos x,\ x \sin x,\ x \cos x,\ \sin x,\ \cos x\}$$

is the *UC set* of $x^2 \sin x$.

We now outline the method of undetermined coefficients for finding a particular integral y_p of

$$a_0 \frac{d^n y}{dx^n} + a_1 \frac{d^{n-1} y}{dx^{n-1}} + \cdots + a_{n-1} \frac{dy}{dx} + a_n y = F(x),$$

where F is a finite linear combination

$$F = A_1 u_1 + A_2 u_2 + \cdots + A_m u_m$$

of UC functions $u_1, u_2, \ldots, u_m$, the A_i being known constants. Assuming the complementary function y_c has already been obtained, we proceed as follows:

1. For *each* of the UC functions

$$u_1, \ldots, u_m$$

of which F is a linear combination, form the corresponding UC set, thus obtaining the respective sets

$$S_1, S_2, \ldots, S_m.$$

2. Suppose that one of the UC sets so formed, say S_j, is identical with or completely included in another, say S_k. In this case, we omit the (identical or smaller) set S_j from further consideration (retaining the set S_k).

3. We now consider in turn each of the UC sets which still remain after Step 2. Suppose now that one of these UC sets, say S_l, includes one or more members which are solutions of the corresponding homogeneous differential equation. If this is the case, we multiply *each* member of S_l by the lowest positive integral power of x so that the resulting revised set will contain no members which are solutions of the corresponding homogeneous differential equation. We now replace S_l by this revised set, so obtained. Note that here we consider one UC set at a time and perform the indicated multiplication, if needed, only upon the members of the one UC set under consideration at the moment.

4. In general there now remains:

   (i) certain of the original UC sets, which were neither omitted in Step 2 nor needed revision in Step 3, and
   (ii) certain revised sets resulting from the needed revision in Step 3.

Now form a linear combination of *all* of the elements of *all* of the sets of these two categories, with unknown constant coefficients (*undetermined coefficients*).

5. Determine these unknown coefficients by substituting the linear combination formed in Step 4 into the differential equation and demanding that it identically satisfy the differential equation (that is, that it be a particular solution).

We frankly admit that this outline of procedure may seem unnecessarily complicated. Once it is understood, however, it frees one from the need of considering separately all of the special cases which it covers.

B. Examples

A few illustrative examples, with reference to the above outline, should make the procedure clear. Our first example will be a simple one in which the situations of Steps 2 and 3 do not occur.

▶ Example 4.31

$$\frac{d^2y}{dx^2} - 2\frac{dy}{dx} - 3y = 2e^x - 10 \sin x.$$

The corresponding homogeneous equation is

$$\frac{d^2y}{dx^2} - 2\frac{dy}{dx} - 3y = 0$$

and the complementary function is

$$y_c = c_1e^{3x} + c_2e^{-x}.$$

The nonhomogeneous term is the linear combination $2e^x - 10 \sin x$ of the two UC functions given by e^x and $\sin x$.

1. Form the UC set for each of these two functions. We find

$$S_1 = \{e^x\},$$
$$S_2 = \{\sin x, \cos x\}.$$

2. Note that neither of these sets is identical with nor included in the other; hence both are retained.

3. Furthermore, by examining the complementary function, we see that none of the functions e^x, $\sin x$, $\cos x$ in either of these sets is a solution of the corresponding homogeneous equation. Hence neither set needs to be revised.

4. Thus the original sets S_1 and S_2 remain intact in this problem, and we form the linear combination

$$Ae^x + B \sin x + C \cos x$$

of the three elements e^x, $\sin x$, $\cos x$ of S_1 and S_2, with the undetermined coefficients A, B, C.

5. We determine these unknown coefficients by substituting the linear combination formed in Step 4 into the differential equation and demanding that it satisfy the differential equation identically. That is, we take

$$y_p = Ae^x + B \sin x + C \cos x$$

as a particular solution. Then

$$y_p' = Ae^x + B \cos x - C \sin x,$$
$$y_p'' = Ae^x - B \sin x - C \cos x.$$

Actually substituting, we find

$$(Ae^x - B\sin x - C\cos x) - 2(Ae^x + B\cos x - C\sin x) - 3(Ae^x + B\sin x + C\cos x) = 2e^x - 10\sin x$$

or

$$-4Ae^x + (-4B + 2C)\sin x + (-4C - 2B)\cos x = 2e^x - 10\sin x.$$

Since the solution is to satisfy the differential equation identically for *all* x on some real interval, this relation must be an identity for all such x and hence the coefficients of like terms on both sides must be respectively equal. Equating coefficients of these like terms, we obtain the equations

$$-4A = 2, \quad -4B + 2C = -10, \quad -4C - 2B = 0.$$

From these equations, we find that

$$A = -\tfrac{1}{2}, \quad B = 2, \quad C = -1,$$

and hence we obtain the particular integral

$$y_p = -\tfrac{1}{2}e^x + 2\sin x - \cos x.$$

Thus the general solution of the differential equation under consideration is

$$y = y_c + y_p = c_1e^{3x} + c_2e^{-x} - \tfrac{1}{2}e^x + 2\sin x - \cos x.$$

▶ Example 4.32

$$\frac{d^2y}{dx^2} - 3\frac{dy}{dx} + 2y = 2x^2 + e^x + 2xe^x + 4e^{3x}.$$

The corresponding homogeneous equation is

$$\frac{d^2y}{dx^2} - 3\frac{dy}{dx} + 2y = 0$$

and the complementary function is

$$y_c = c_1e^x + c_2e^{2x}.$$

The nonhomogeneous term is the linear combination

$$2x^2 + e^x + 2xe^x + 4e^{3x}$$

of the four UC functions given by x^2, e^x, xe^x, and e^{3x}.

1. Form the UC set for each of these functions. We have

$$S_1 = \{x^2, x, 1\},$$
$$S_2 = \{e^x\},$$
$$S_3 = \{xe^x, e^x\},$$
$$S_4 = \{e^{3x}\}.$$

2. We note that S_2 is completely included in S_3, so S_2 is omitted from further consideration, leaving the 3 sets

$$S_1 = \{x^2, x, 1\}, \quad S_3 = \{xe^x, e^x\}, \quad S_4 = \{e^{3x}\}.$$

3. We now observe that $S_3 = \{xe^x, e^x\}$ includes e^x, which is included in the complementary function and so is a solution of the corresponding homogeneous differential equation. Thus we multiply *each* member of S_3 by x to obtain the revised family

$$S_3' = \{x^2e^x, xe^x\},$$

which contains no members which are solutions of the corresponding homogeneous equation.

4. Thus there remain the original UC sets

$$S_1 = \{x^2, x, 1\}$$

and

$$S_4 = \{e^{3x}\}$$

and the revised set

$$S_3' = \{x^2e^x, xe^x\}.$$

These contain the six elements

$$x^2, \quad x, \quad 1, \quad e^{3x}, \quad x^2e^x, \quad xe^x.$$

We form the linear combination

$$Ax^2 + Bx + C + De^{3x} + Ex^2e^x + Fxe^x$$

of these six elements.

5. Thus we take as our particular solution,

$$y_p = Ax^2 + Bx + C + De^{3x} + Ex^2e^x + Fxe^x.$$

From this, we have

$$y_p' = 2Ax + B + 3De^{3x} + Ex^2e^x + 2Exe^x + Fxe^x + Fe^x,$$

$$y_p'' = 2A + 9De^{3x} + Ex^2e^x + 4Exe^x + 2Ee^x + Fxe^x + 2Fe^x.$$

We substitute y_p, y_p', y_p'' into the differential equation for y, dy/dx, d^2y/dx^2, respectively, to obtain:

$$\begin{aligned} &2A + 9De^{3x} + Ex^2e^x + (4E + F)xe^x + (2E + 2F)e^x \\ &- 3[2Ax + B + 3De^{3x} + Ex^2e^x + (2E + F)xe^x + Fe^x] \\ &+ 2(Ax^2 + Bx + C + De^{3x} + Ex^2e^x + Fxe^x) \\ &\qquad = 2x^2 + e^x + 2xe^x + 4e^{3x}, \end{aligned}$$

or

$$(2A - 3B + 2C) + (2B - 6A)x + 2Ax^2 + 2De^{3x} + (-2E)xe^x + (2E - F)e^x = 2x^2 + e^x + 2xe^x + 4e^{3x}.$$

Equating coefficients of like terms, we have:

$$\begin{aligned} 2A - 3B + 2C &= 0, \\ 2B - 6A &= 0, \\ 2A &= 2, \\ 2D &= 4, \\ -2E &= 2, \\ 2E - F &= 1. \end{aligned}$$

From this $A = 1$, $B = 3$, $C = \frac{7}{2}$, $D = 2$, $E = -1$, $F = -3$, and so the particular integral is

$$y_p = x^2 + 3x + \tfrac{7}{2} + 2e^{3x} - x^2e^x - 3xe^x.$$

The general solution is therefore

$$y = y_c + y_p = c_1e^x + c_2e^{2x} + x^2 + 3x + \tfrac{7}{2} + 2e^{3x} - x^2e^x - 3xe^x.$$

▶ Example 4.33

$$\frac{d^4y}{dx^4} + \frac{d^2y}{dx^2} = 3x^2 + 4\sin x - 2\cos x.$$

The corresponding homogeneous equation is

$$\frac{d^4y}{dx^4} + \frac{d^2y}{dx^2} = 0,$$

and the complementary function is

$$y_c = c_1 + c_2x + c_3\sin x + c_4\cos x.$$

The nonhomogeneous term is the linear combination

$$3x^2 + 4\sin x - 2\cos x$$

of the three UC functions given by

$$x^2, \quad \sin x, \quad \text{and} \quad \cos x.$$

1. Form the UC set for each of these three functions. These sets are, respectively,

$$\begin{aligned} S_1 &= \{x^2, x, 1\}, \\ S_2 &= \{\sin x, \cos x\}, \\ S_3 &= \{\cos x, \sin x\}. \end{aligned}$$

2. Observe that S_2 and S_3 are identical and so we retain only one of them, leaving the two sets

$$S_1 = \{x^2, x, 1\}, \qquad S_2 = \{\sin x, \cos x\}.$$

3. Now observe that $S_1 = \{x^2, x, 1\}$ includes 1 and x, which, as the complementary function shows, are both solutions of the corresponding homogeneous

differential equation. Thus we multiply each member of the set S_1 by x^2 to obtain the revised set

$$S_1' = \{x^4, x^3, x^2\},$$

none of whose members are solutions of the homogeneous differential equation. We observe that multiplication by x instead of x^2 would not be sufficient, since the resulting set would be $\{x^3, x^2, x\}$, which still includes the homogeneous solution x. Turning to the set S_2, observe that both of its members, $\sin x$ and $\cos x$, are also solutions of the homogeneous differential equation. Hence we replace S_2 by the revised set

$$S_2' = \{x \sin x, x \cos x\}.$$

4. None of the original UC sets remain here. They have been replaced by the revised sets S_1' and S_2' containing the five elements

$$x^4, \quad x^3, \quad x^2, \quad x \sin x, \quad x \cos x.$$

We form a linear combination of these,

$$Ax^4 + Bx^3 + Cx^2 + Dx \sin x + Ex \cos x,$$

with undetermined coefficients A, B, C, D, E.

5. We now take this as our particular solution

$$y_p = Ax^4 + Bx^3 + Cx^2 + Dx \sin x + Ex \cos x.$$

Then

$$\begin{aligned}
y_p' &= 4Ax^3 + 3Bx^2 + 2Cx + Dx \cos x + D \sin x - Ex \sin x + E \cos x,\\
y_p'' &= 12Ax^2 + 6Bx + 2C - Dx \sin x + 2D \cos x - Ex \cos x - 2E \sin x,\\
y_p''' &= 24Ax + 6B - Dx \cos x - 3D \sin x + Ex \sin x - 3E \cos x,\\
y_p^{(iv)} &= 24A + Dx \sin x - 4D \cos x + Ex \cos x + 4E \sin x.
\end{aligned}$$

Substituting into the differential equation, we obtain

$$\begin{aligned}
24A + Dx \sin x - 4D \cos x + Ex \cos x + 4E \sin x + 12Ax^2 + 6Bx + 2C - Dx \sin x\\
+ 2D \cos x - Ex \cos x - 2E \sin x\\
= 3x^2 + 4 \sin x - 2 \cos x.
\end{aligned}$$

Equating coefficients, we find

$$\begin{aligned}
24A + 2C &= 0\\
6B &= 0\\
12A &= 3\\
-2D &= -2\\
2E &= 4.
\end{aligned}$$

Hence $A = \frac{1}{4}$, $B = 0$, $C = -3$, $D = 1$, $E = 2$, and the particular integral is

$$y_p = \tfrac{1}{4}x^4 - 3x^2 + x \sin x + 2x \cos x.$$

The general solution is

$$y = y_c + y_p = c_1 + c_2x + c_3 \sin x + c_4 \cos x + \tfrac{1}{4}x^4 - 3x^2 + x \sin x + 2x \cos x.$$

▶ Example 4.34. An Initial-Value Problem. We close this section by applying our results to the solution of the initial-value problem

$$\frac{d^2y}{dx^2} - 2\frac{dy}{dx} - 3y = 2e^x - 10 \sin x, \tag{4.30}$$

$$y(0) = 2, \tag{4.31}$$

$$y'(0) = 4. \tag{4.32}$$

By Theorem 4.1, this problem has a unique solution, defined for all x, $-\infty < x < \infty$; let us proceed to find it. In Example 4.31 we found that the general solution of the differential equation (4.30) is

$$y = c_1e^{3x} + c_2e^{-x} - \tfrac{1}{2}e^x + 2 \sin x - \cos x. \tag{4.33}$$

From this, we have

$$\frac{dy}{dx} = 3c_1e^{3x} - c_2e^{-x} - \tfrac{1}{2}e^x + 2 \cos x + \sin x. \tag{4.34}$$

Applying the initial conditions (4.31) and (4.32) to Equations (4.33) and (4.34), respectively, we have

$$2 = c_1e^0 + c_2e^0 - \tfrac{1}{2}e^0 + 2 \sin 0 - \cos 0,$$

$$4 = 3c_1e^0 - c_2e^0 - \tfrac{1}{2}e^0 + 2 \cos 0 + \sin 0.$$

These equations simplify at once to the following:

$$c_1 + c_2 = \tfrac{7}{2}, \qquad 3c_1 - c_2 = \tfrac{5}{2}.$$

From these two equations we obtain

$$c_1 = \tfrac{3}{2}, \qquad c_2 = 2.$$

Substituting these values for c_1 and c_2 into Equation (4.33) we obtain the unique solution of the given initial-value problem in the form

$$y = \tfrac{3}{2}e^{3x} + 2e^{-x} - \tfrac{1}{2}e^x + 2 \sin x - \cos x.$$

Exercises

Find the general solution of each of the differential equations in Exercises 1 through 20.

1. $\dfrac{d^2y}{dx^2} - 3\dfrac{dy}{dx} + 2y = 4x^2.$

2. $\dfrac{d^2y}{dx^2} - 2\dfrac{dy}{dx} - 8y = 4e^{2x} - 21e^{-3x}.$

3. $\dfrac{d^2y}{dx^2} + 2\dfrac{dy}{dx} + 5y = 6 \sin 2x + 7 \cos 2x.$

4. $\frac{d^2y}{dx^2} + 2\frac{dy}{dx} + 2y = 10 \sin 4x.$

5. $\frac{d^2y}{dx^2} + 2\frac{dy}{dx} + 4y = \cos 4x.$

6. $\frac{d^3y}{dx^3} + 2\frac{d^2y}{dx^2} - 3\frac{dy}{dx} - 10y = 8xe^{-2x}.$

7. $\frac{d^3y}{dx^3} + \frac{d^2y}{dx^2} + 3\frac{dy}{dx} - 5y = 5 \sin 2x + 10x^2 - 3x + 7.$

8. $4\frac{d^3y}{dx^3} - 4\frac{d^2y}{dx^2} - 5\frac{dy}{dx} + 3y = 3x^3 - 8x.$

9. $\frac{d^2y}{dx^2} + \frac{dy}{dx} - 6y = 10e^{2x} - 18e^{3x} - 6x - 11.$

10. $\frac{d^2y}{dx^2} + \frac{dy}{dx} - 2y = 6e^{-2x} + 3e^{x} - 4x^2.$

11. $\frac{d^3y}{dx^3} - 3\frac{d^2y}{dx^2} + 4y = 4e^{x} - 18e^{-x}.$

12. $\frac{d^3y}{dx^3} - 2\frac{d^2y}{dx^2} - \frac{dy}{dx} + 2y = 9e^{2x} - 8e^{3x}.$

13. $\frac{d^3y}{dx^3} + \frac{dy}{dx} = 2x^2 + 4 \sin x.$

14. $\frac{d^4y}{dx^4} - 3\frac{d^3y}{dx^3} + 2\frac{d^2y}{dx^2} = 3e^{-x} + 6e^{2x} - 6x.$

15. $\frac{d^3y}{dx^3} - 6\frac{d^2y}{dx^2} + 11\frac{dy}{dx} - 6y = xe^{x} - 4e^{2x} + 6e^{4x}.$

16. $\frac{d^3y}{dx^3} - 4\frac{d^2y}{dx^2} + 5\frac{dy}{dx} - 2y = 3x^2e^{x} - 7e^{x}.$

17. $\frac{d^2y}{dx^2} + y = x \sin x.$

18. $\frac{d^2y}{dx^2} + 4y = 12x^2 - 16x \cos 2x.$

19. $\frac{d^4y}{dx^4} + 2\frac{d^3y}{dx^3} - 3\frac{d^2y}{dx^2} = 18x^2 + 16xe^{x} + 4e^{3x} - 9.$

20. $\frac{d^4y}{dx^4} - 5\frac{d^3y}{dx^3} + 7\frac{d^2y}{dx^2} - 5\frac{dy}{dx} + 6y = 5 \sin x - 12 \sin 2x.$

D. HERR

Solve the initial-value problems in Exercises 21 through 28.

21. $\dfrac{d^2y}{dx^2} - 4\dfrac{dy}{dx} + 3y = 9x^2 + 4, \quad y(0) = 6, \quad y'(0) = 8.$

22. $\dfrac{d^2y}{dx^2} + 4\dfrac{dy}{dx} + 13y = 5 \sin 2x, \quad y(0) = 1, \quad y'(0) = -2.$

23. $\dfrac{d^2y}{dx^2} - \dfrac{dy}{dx} - 6y = 8e^{2x} - 5e^{3x}, \quad y(0) = 3, \quad y'(0) = 5.$

24. $\dfrac{d^2y}{dx^2} + y = 3x^2 - 4 \sin x, \quad y(0) = 0, \quad y'(0) = 1.$

25. $\dfrac{d^2y}{dx^2} - 4\dfrac{dy}{dx} + 13y = 8 \sin 3x, \quad y(0) = 1, \quad y'(0) = 2.$

26. $\dfrac{d^2y}{dx^2} - y = 3x^2e^x, \quad y(0) = 1, \quad y'(0) = 2.$

27. $\dfrac{d^3y}{dx^3} - 4\dfrac{d^2y}{dx^2} + \dfrac{dy}{dx} + 6y = 3xe^x + 2e^x - \sin x,$

$$y(0) = \frac{33}{40}, \quad y'(0) = 0, \quad y''(0) = 0.$$

28. $\dfrac{d^3y}{dx^3} - 6\dfrac{d^2y}{dx^2} + 9\dfrac{dy}{dx} - 4y = 8x^2 + 3 - 6e^{2x},$

$$y(0) = 1, \quad y'(0) = 7, \quad y''(0) = 10.$$

For each of the differential equations in Exercises 29 through 42 *set up* the correct linear combination of functions with undetermined literal coefficients to use in finding a particular integral by the method of undetermined coefficients. (Do not actually find the particular integrals.)

29. $\dfrac{d^2y}{dx^2} - 6\dfrac{dy}{dx} + 8y = x^3 + x + e^{-2x}.$

30. $\dfrac{d^2y}{dx^2} + 9y = e^{3x} + e^{-3x} + e^{3x} \sin 3x.$

31. $\dfrac{d^2y}{dx^2} + 4\dfrac{dy}{dx} + 5y = e^{-2x}(1 + \cos x).$

32. $\dfrac{d^2y}{dx^2} - 6\dfrac{dy}{dx} + 9y = x^4e^x + x^3e^{2x} + x^2e^{3x}.$

33. $\dfrac{d^2y}{dx^2} + 6\dfrac{dy}{dx} + 13y = xe^{-3x} \sin 2x + x^2e^{-2x} \sin 3x.$

34. $\dfrac{d^3y}{dx^3} - 3\dfrac{d^2y}{dx^2} + 2\dfrac{dy}{dx} = x^2e^x + 3xe^{2x} + 5x^2.$

35. $\frac{d^3y}{dx^3} - 6\frac{d^2y}{dx^2} + 12\frac{dy}{dx} - 8y = xe^{2x} + x^2e^{3x}$.

36. $\frac{d^4y}{dx^4} + 3\frac{d^3y}{dx^3} + 4\frac{d^2y}{dx^2} + 3\frac{dy}{dx} + y = x^2e^{-x} + 3e^{-x/2}\cos\frac{\sqrt{3}}{2}x$.

37. $\frac{d^4y}{dx^4} - 16y = x^2 \sin 2x + x^4e^{2x}$.

38. $\frac{d^6y}{dx^6} + 2\frac{d^5y}{dx^5} + 5\frac{d^4y}{dx^4} = x^3 + x^2e^{-x} + e^{-x}\sin 2x$.

39. $\frac{d^4y}{dx^4} + 2\frac{d^2y}{dx^2} + y = x^2 \cos x$.

40. $\frac{d^4y}{dx^4} + 16y = xe^{\sqrt{2}x}\sin\sqrt{2}\,x + e^{-\sqrt{2}x}\cos\sqrt{2}\,x$.

41. $\frac{d^4y}{dx^4} + 3\frac{d^2y}{dx^2} - 4y = \cos^2 x - \cosh x$.

42. $\frac{d^4y}{dx^4} + 10\frac{d^2y}{dx^2} + 9y = \sin x \sin 2x$.

4.4 Variation of Parameters

A. *The Method*

While the process of carrying out the method of undetermined coefficients is actually quite straightforward (involving only techniques of college algebra and differentiation), the method applies in general to a rather small class of problems. For example, it would not apply to the apparently simple equation

$$\frac{d^2y}{dx^2} + y = \tan x.$$

We thus seek a method of finding a particular integral which applies in all cases (including variable coefficients) in which the complementary function is known. Such a method is the method of *variation of parameters*, which we now consider.

We shall develop this method in connection with the general second-order linear differential equation with variable coefficients

$$a_0(x)\frac{d^2y}{dx^2} + a_1(x)\frac{dy}{dx} + a_2(x)y = F(x). \tag{4.35}$$

Suppose that y_1 and y_2 are linearly independent solutions of the corresponding homogeneous equation

$$a_0(x)\frac{d^2y}{dx^2} + a_1(x)\frac{dy}{dx} + a_2(x)y = 0. \tag{4.36}$$

Then the complementary function of Equation (4.35) is

$$c_1y_1(x) + c_2y_2(x),$$

where y_1 and y_2 are linearly independent solutions of (4.36) and c_1 and c_2 are arbitrary constants. The procedure in the method of variation of parameters is to replace the arbitrary constants c_1 and c_2 in the complementary function by respective *functions* v_1 and v_2 which will be determined so that the resulting function which is defined by

$$v_1(x)y_1(x) + v_2(x)y_2(x) \tag{4.37}$$

will be a particular integral of Equation (4.35) (hence the name, *variation* of parameters).

We have at our disposal the *two functions* v_1 and v_2 with which to satisfy the *one condition* that (4.37) be a solution of (4.35). Since we have *two* functions but only *one* condition on them, we are thus free to impose a second condition, provided this second condition does not violate the first one. We shall see when and how to impose this additional condition as we proceed.

We thus assume a solution of the form (4.37) and write

$$y_p(x) = v_1(x)y_1(x) + v_2(x)y_2(x). \tag{4.38}$$

Differentiating (4.38), we have

$$y_p'(x) = v_1(x)y_1'(x) + v_2(x)y_2'(x) + v_1'(x)y_1(x) + v_2'(x)y_2(x), \tag{4.39}$$

where we use primes to denote differentiations. At this point we impose the aforementioned second condition; we simplify y_p' by demanding that

$$v_1'(x)y_1(x) + v_2'(x)y_2(x) = 0. \tag{4.40}$$

With this condition imposed, (4.39) reduces to

$$y_p'(x) = v_1(x)y_1'(x) + v_2(x)y_2'(x). \tag{4.41}$$

Now differentiating (4.41), we obtain

$$y_p''(x) = v_1(x)y_1''(x) + v_2(x)y_2''(x) + v_1'(x)y_1'(x) + v_2'(x)y_2'(x). \tag{4.42}$$

We now impose the basic condition that (4.38) be a solution of Equation (4.35). Thus we substitute (4.38), (4.41), and (4.42) for y, $\frac{dy}{dx}$, and $\frac{d^2y}{dx^2}$, respectively, in Equation (4.35) and obtain the identity

$$a_0(x)[v_1(x)y_1''(x) + v_2(x)y_2''(x) + v_1'(x)y_1'(x) + v_2'(x)y_2'(x)]$$
$$+ a_1(x)[v_1(x)y_1'(x) + v_2(x)y_2'(x)] + a_2(x)[v_1(x)y_1(x) + v_2(x)y_2(x)] = F(x).$$

This can be written as

$$v_1(x)[a_0(x)y_1''(x) + a_1(x)y_1'(x) + a_2(x)y_1(x)]$$
$$+ v_2(x)[a_0(x)y_2''(x) + a_1(x)y_2'(x) + a_2(x)y_2(x)]$$
$$+ a_0(x)[v_1'(x)y_1'(x) + v_2'(x)y_2'(x)] = F(x). \tag{4.43}$$

Since y_1 and y_2 are solutions of the corresponding homogeneous differential equation (4.36), the expressions in the first two brackets in (4.43) are identically zero. This leaves merely

$$v_1'(x)y_1'(x) + v_2'(x)y_2'(x) = \frac{F(x)}{a_0(x)}. \tag{4.44}$$

This is actually what the basic condition demands. Thus the two imposed conditions require that the functions v_1 and v_2 be chosen such that the system of equations

$$\begin{aligned} y_1(x)v_1'(x) + y_2(x)v_2'(x) &= 0, \\ y_1'(x)v_1'(x) + y_2'(x)v_2'(x) &= \frac{F(x)}{a_0(x)}, \end{aligned} \tag{4.45}$$

is satisfied. The determinant of coefficients of this system is precisely

$$W[y_1(x), y_2(x)] = \begin{vmatrix} y_1(x) & y_2(x) \\ y_1'(x) & y_2'(x) \end{vmatrix}.$$

Since y_1 and y_2 are linearly independent solutions of the corresponding homogeneous differential equation (4.36), we know that $W[y_1(x), y_2(x)] \neq 0$. Hence the system (4.45) has a unique solution. Actually solving this system, we obtain

$$v_1'(x) = \frac{\begin{vmatrix} 0 & y_2(x) \\ \dfrac{F(x)}{a_0(x)} & y_2'(x) \end{vmatrix}}{\begin{vmatrix} y_1(x) & y_2(x) \\ y_1'(x) & y_2'(x) \end{vmatrix}} = -\frac{F(x)y_2(x)}{a_0(x)W[y_1(x), y_2(x)]},$$

$$v_2'(x) = \frac{\begin{vmatrix} y_1(x) & 0 \\ y_1'(x) & \dfrac{F(x)}{a_0(x)} \end{vmatrix}}{\begin{vmatrix} y_1(x) & y_2(x) \\ y_1'(x) & y_2'(x) \end{vmatrix}} = \frac{F(x)y_1(x)}{a_0(x)W[y_1(x), y_2(x)]}.$$

Thus we obtain the functions v_1 and v_2 defined by

$$\begin{aligned} v_1(x) &= -\int^x \frac{F(t)y_2(t)\,dt}{a_0(t)W[y_1(t), y_2(t)]}, \\ v_2(x) &= \int^x \frac{F(t)y_1(t)\,dt}{a_0(t)W[y_1(t), y_2(t)]}. \end{aligned} \tag{4.46}$$

Therefore a particular integral y_p of Equation (4.35) is defined by

$$y_p(x) = v_1(x)y_1(x) + v_2(x)y_2(x),$$

where v_1 and v_2 are defined by (4.46).

B. Examples

▶ Example 4.35. Consider the differential equation

$$\frac{d^2y}{dx^2} + y = \tan x. \tag{4.47}$$

The complementary function is defined by

$$y_c(x) = c_1 \sin x + c_2 \cos x.$$

We assume
$$y_p(x) = v_1(x) \sin x + v_2(x) \cos x, \tag{4.48}$$

where the functions v_1 and v_2 will be determined such that this is a particular integral of the differential equation (4.47). Then

$$y_p'(x) = v_1(x) \cos x - v_2(x) \sin x + v_1'(x) \sin x + v_2'(x) \cos x.$$

We impose the condition

$$v_1'(x) \sin x + v_2'(x) \cos x = 0, \tag{4.49}$$

leaving
$$y_p'(x) = v_1(x) \cos x - v_2(x) \sin x.$$

From this

$$y_p''(x) = -v_1(x) \sin x - v_2(x) \cos x + v_1'(x) \cos x - v_2'(x) \sin x. \tag{4.50}$$

Substituting (4.48) and (4.50) into (4.47) we obtain

$$v_1'(x) \cos x - v_2'(x) \sin x = \tan x. \tag{4.51}$$

Thus we have the two equations (4.49) and (4.51) from which to determine $v_1'(x)$, $v_2'(x)$:

$$v_1'(x) \sin x + v_2'(x) \cos x = 0,$$
$$v_1'(x) \cos x - v_2'(x) \sin x = \tan x.$$

Solving we find:

$$v_1'(x) = \frac{\begin{vmatrix} 0 & \cos x \\ \tan x & -\sin x \end{vmatrix}}{\begin{vmatrix} \sin x & \cos x \\ \cos x & -\sin x \end{vmatrix}} = \frac{-\cos x \tan x}{-1} = \sin x,$$

$$v_2'(x) = \frac{\begin{vmatrix} \sin x & 0 \\ \cos x & \tan x \end{vmatrix}}{\begin{vmatrix} \sin x & \cos x \\ \cos x & -\sin x \end{vmatrix}} = \frac{\sin x \tan x}{-1} = \frac{-\sin^2 x}{\cos x}$$

$$= \frac{\cos^2 x - 1}{\cos x} = \cos x - \sec x.$$

Integrating we find:

$$v_1(x) = -\cos x + c_3, \qquad v_2(x) = \sin x - \ln |\sec x + \tan x| + c_4. \tag{4.52}$$

Substituting (4.52) into (4.48) we have

$$\begin{aligned} y_p(x) &= (-\cos x + c_3)\sin x + (\sin x - \ln|\sec x + \tan x| + c_4)\cos x \\ &= -\sin x \cos x + c_3 \sin x + \sin x \cos x \\ &\quad - \ln|\sec x + \tan x|(\cos x) + c_4 \cos x \\ &= c_3 \sin x + c_4 \cos x - (\cos x)(\ln|\sec x + \tan x|). \end{aligned}$$

Since a particular integral is a solution free of arbitrary constants, we may assign any particular values A and B to c_3 and c_4, respectively, and the result will be the particular integral

$$A \sin x + B \cos x - (\cos x)(\ln|\sec x + \tan x|).$$

Thus $y = y_c + y_p$ becomes

$$y = c_1 \sin x + c_2 \cos x + A \sin x + B \cos x - (\cos x)(\ln|\sec x + \tan x|),$$

which we may write as

$$y = C_1 \sin x + C_2 \cos x - (\cos x)(\ln|\sec x + \tan x|),$$

where $C_1 = c_1 + A$, $C_2 = c_2 + B$.

Thus we see that we might as well have chosen the constants c_3 and c_4 both equal to 0 in (4.52), for essentially the same result,

$$y = c_1 \sin x + c_2 \cos x - (\cos x)(\ln|\sec x + \tan x|),$$

would have been obtained. This is the general solution of the differential equation (4.47).

The method of variation of parameters extends to higher-order linear equations. We now illustrate the extension to a third-order equation in Example 4.36, although we hasten to point out that the equation of this example can be solved more readily by the method of undetermined coefficients.

▶ **Example 4.36.** Consider the differential equation

$$\frac{d^3y}{dx^3} - 6\frac{d^2y}{dx^2} + 11\frac{dy}{dx} - 6y = e^x. \tag{4.53}$$

The complementary function is

$$y_c(x) = c_1e^x + c_2e^{2x} + c_3e^{3x}.$$

We assume as a particular integral

$$y_p(x) = v_1(x)e^x + v_2(x)e^{2x} + v_3(x)e^{3x}. \tag{4.54}$$

Since we have *three* functions v_1, v_2, v_3 at our disposal in this case, we can apply three conditions. We have:

$$y_p'(x) = v_1(x)e^x + 2v_2(x)e^{2x} + 3v_3(x)e^{3x} + v_1'(x)e^x + v_2'(x)e^{2x} + v_3'(x)e^{3x}.$$

Proceeding in a manner analogous to that of the second-order case, we impose the condition

$$v_1'(x)e^x + v_2'(x)e^{2x} + v_3'(x)e^{3x} = 0, \tag{4.55}$$

leaving

$$y_p'(x) = v_1(x)e^x + 2v_2(x)e^{2x} + 3v_3(x)e^{3x}. \tag{4.56}$$

Then

$$y_p''(x) = v_1(x)e^x + 4v_2(x)e^{2x} + 9v_3(x)e^{3x} + v_1'(x)e^x + 2v_2'(x)e^{2x} + 3v_3'(x)e^{3x}.$$

We now impose the condition

$$v_1'(x)e^x + 2v_2'(x)e^{2x} + 3v_3'(x)e^{3x} = 0, \tag{4.57}$$

leaving

$$y_p''(x) = v_1(x)e^x + 4v_2(x)e^{2x} + 9v_3(x)e^{3x}. \tag{4.58}$$

From this,

$$y_p'''(x) = v_1(x)e^x + 8v_2(x)e^{2x} + 27v_3(x)e^{3x} + v_1'(x)e^x + 4v_2'(x)e^{2x} + 9v_3'(x)e^{3x}. \tag{4.59}$$

We substitute (4.54), (4.56), (4.58), and (4.59) into the differential equation (4.53), obtaining:

$$\begin{aligned} &v_1(x)e^x + 8v_2(x)e^{2x} + 27v_3(x)e^{3x} + v_1'(x)e^x + 4v_2'(x)e^{2x} + 9v_3'(x)e^{3x} \\ &\quad - 6v_1(x)e^x - 24v_2(x)e^{2x} - 54v_3(x)e^{3x} + 11v_1(x)e^x + 22v_2(x)e^{2x} + 33v_3(x)e^{3x} \\ &\quad\quad - 6v_1(x)e^x - 6v_2(x)e^{2x} - 6v_3(x)e^{3x} = e^x \end{aligned}$$

or

$$v_1'(x)e^x + 4v_2'(x)e^{2x} + 9v_3'(x)e^{3x} = e^x. \tag{4.60}$$

Thus we have the three equations (4.55), (4.57), (4.60) from which to determine $v_1'(x)$, $v_2'(x)$, $v_3'(x)$:

$$\begin{aligned} v_1'(x)e^x + v_2'(x)e^{2x} + v_3'(x)e^{3x} &= 0, \\ v_1'(x)e^x + 2v_2'(x)e^{2x} + 3v_3'(x)e^{3x} &= 0, \\ v_1'(x)e^x + 4v_2'(x)e^{2x} + 9v_3'(x)e^{3x} &= e^x. \end{aligned}$$

Solving, we find

$$v_1'(x) = \frac{\begin{vmatrix} 0 & e^{2x} & e^{3x} \\ 0 & 2e^{2x} & 3e^{3x} \\ e^x & 4e^{2x} & 9e^{3x} \end{vmatrix}}{\begin{vmatrix} e^x & e^{2x} & e^{3x} \\ e^x & 2e^{2x} & 3e^{3x} \\ e^x & 4e^{2x} & 9e^{3x} \end{vmatrix}} = \frac{e^{6x}\begin{vmatrix} 1 & 1 \\ 2 & 3 \end{vmatrix}}{e^{6x}\begin{vmatrix} 1 & 1 & 1 \\ 1 & 2 & 3 \\ 1 & 4 & 9 \end{vmatrix}} = \frac{1}{2},$$

$$v_2'(x) = \frac{\begin{vmatrix} e^x & 0 & e^{3x} \\ e^x & 0 & 3e^{3x} \\ e^x & e^x & 9e^{3x} \end{vmatrix}}{\begin{vmatrix} e^x & e^{2x} & e^{3x} \\ e^x & 2e^{2x} & 3e^{3x} \\ e^x & 4e^{2x} & 9e^{3x} \end{vmatrix}} = \frac{-e^{5x}\begin{vmatrix} 1 & 1 \\ 1 & 3 \end{vmatrix}}{2e^{6x}} = -e^{-x},$$

$$v_3'(x) = \frac{\begin{vmatrix} e^x & e^{2x} & 0 \\ e^x & 2e^{2x} & 0 \\ e^x & 4e^{2x} & e^x \end{vmatrix}}{\begin{vmatrix} e^x & e^{2x} & e^{3x} \\ e^x & 2e^{2x} & 3e^{3x} \\ e^x & 4e^{2x} & 9e^{3x} \end{vmatrix}} = \frac{e^{4x}\begin{vmatrix} 1 & 1 \\ 1 & 2 \end{vmatrix}}{2e^{6x}} = \frac{1}{2}e^{-2x}.$$

We now integrate, choosing all the constants of integration to be zero (as the previous example showed was possible). We find:

$$v_1(x) = \tfrac{1}{2}x, \qquad v_2(x) = e^{-x}, \qquad v_3(x) = -\tfrac{1}{4}e^{-2x}.$$

Thus

$$y_p(x) = \tfrac{1}{2}xe^x + e^{-x}e^{2x} - \tfrac{1}{4}e^{-2x}e^{3x} = \tfrac{1}{2}xe^x + \tfrac{3}{4}e^x.$$

Thus the general solution of Equation (4.53) is

$$y = y_c + y_p = c_1e^x + c_2e^{2x} + c_3e^{3x} + \tfrac{1}{2}xe^x + \tfrac{3}{4}e^x$$

or

$$y = c_1'e^x + c_2e^{2x} + c_3e^{3x} + \tfrac{1}{2}xe^x,$$

where $c_1' = c_1 + \tfrac{3}{4}$.

In Examples 4.35 and 4.36 the coefficients in the differential equation were constants. The general discussion at the beginning of this section shows that the method applies equally well to linear differential equations with variable coefficients, once the complementary function y_c is known. We now illustrate its application to such an equation in Example 4.37.

▶ Example 4.37. Consider the differential equation

$$(x^2 + 1)\frac{d^2y}{dx^2} - 2x\frac{dy}{dx} + 2y = 6(x^2 + 1)^2. \tag{4.61}$$

In Example 4.15 we solved the corresponding homogeneous equation

$$(x^2 + 1)\frac{d^2y}{dx^2} - 2x\frac{dy}{dx} + 2y = 0.$$

From the results of that example, we see that the complementary function of equation (4.61) is

$$y_c(x) = c_1x + c_2(x^2 - 1).$$

To find a particular integral of Equation (4.61), we therefore let

$$y_p(x) = v_1(x)x + v_2(x)(x^2 - 1). \tag{4.62}$$

Then

$$y_p'(x) = v_1(x)\cdot 1 + v_2(x)\cdot 2x + v_1'(x)x + v_2'(x)(x^2 - 1).$$

We impose the condition

$$v_1'(x)x + v_2'(x)(x^2 - 1) = 0, \tag{4.63}$$

leaving

$$y_p'(x) = v_1(x)\cdot 1 + v_2(x)\cdot 2x. \tag{4.64}$$

From this, we find

$$y_p''(x) = v_1'(x) + 2v_2(x) + v_2'(x)\cdot 2x. \tag{4.65}$$

Substituting (4.62), (4.64), and (4.65) into (4.61) we obtain

$$(x^2 + 1)[v_1'(x) + 2v_2(x) + 2xv_2'(x)] - 2x[v_1(x) + 2xv_2(x)] + 2[v_1(x)x + v_2(x)(x^2 - 1)] = 6(x^2 + 1)^2$$

or

$$(x^2 + 1)[v_1'(x) + 2xv_2'(x)] = 6(x^2 + 1)^2. \tag{4.66}$$

Thus we have the two equations (4.63) and (4.66) from which to determine $v_1'(x)$ and $v_2'(x)$; that is, $v_1'(x)$ and $v_2'(x)$ satisfy the system

$$v_1'(x)x + v_2'(x)[x^2 - 1] = 0,$$
$$v_1'(x) + v_2'(x)[2x] = 6(x^2 + 1).$$

Solving this system, we find

$$v_1'(x) = \frac{\begin{vmatrix} 0 & x^2 - 1 \\ 6(x^2+1) & 2x \end{vmatrix}}{\begin{vmatrix} x & x^2 - 1 \\ 1 & 2x \end{vmatrix}} = \frac{-6(x^2+1)(x^2-1)}{x^2+1} = -6(x^2 - 1),$$

$$v_2'(x) = \frac{\begin{vmatrix} x & 0 \\ 1 & 6(x^2+1) \end{vmatrix}}{\begin{vmatrix} x & x^2 - 1 \\ 1 & 2x \end{vmatrix}} = \frac{6x(x^2+1)}{x^2+1} = 6x.$$

Integrating, we obtain

$$v_1(x) = -2x^3 + 6x, \qquad v_2(x) = 3x^2, \tag{4.67}$$

where we have chosen both constants of integration to be zero. Substituting (4.67) into (4.62), we have

$$y_p(x) = (-2x^3 + 6x)x + 3x^2(x^2 - 1)$$
$$= x^4 + 3x^2.$$

Therefore the general solution of Equation (4.61) may be expressed in the form

$$y = y_c + y_p$$
$$= c_1x + c_2(x^2 - 1) + x^4 + 3x^2.$$

Exercises

Find the general solution of each of the differential equations in Exercises 1 through 15.

1. $\dfrac{d^2y}{dx^2} + y = \cot x.$

2. $\dfrac{d^2y}{dx^2} + y = \tan^2 x.$

3. $\dfrac{d^2y}{dx^2} + y = \sec x.$

4. $\dfrac{d^2y}{dx^2} + y = \sec^3 x.$

5. $\dfrac{d^2y}{dx^2} + 4y = \sec^2 2x.$

6. $\dfrac{d^2y}{dx^2} + 6\dfrac{dy}{dx} + 9y = \dfrac{e^{-3x}}{x^3}.$

7. $\dfrac{d^2y}{dx^2} + 4\dfrac{dy}{dx} + 5y = e^{-2x}\sec x.$

8. $\dfrac{d^2y}{dx^2} - 2\dfrac{dy}{dx} + y = x\ln x \quad (x > 0).$

9. $\frac{d^2y}{dx^2} - 2\frac{dy}{dx} + y = xe^x \ln x \quad (x > 0).$

10. $\frac{d^2y}{dx^2} - 2\frac{dy}{dx} + y = (\ln x)^2 \quad (x > 0).$

11. $\frac{d^2y}{dx^2} - 2\frac{dy}{dx} + y = e^x \sin^{-1} x.$

12. $\frac{d^2y}{dx^2} + 3\frac{dy}{dx} + 2y = \frac{e^{-x}}{x}.$

13. $\frac{d^2y}{dx^2} + 3\frac{dy}{dx} + 2y = \frac{1}{1 + e^x}.$

14. $\frac{d^2y}{dx^2} + 3\frac{dy}{dx} + 2y = \frac{1}{1 + e^{2x}}.$

15. $\frac{d^2y}{dx^2} + y = \frac{1}{1 + \sin x}.$

16. Find the general solution by two methods:

$$\frac{d^3y}{dx^3} - 3\frac{d^2y}{dx^2} - \frac{dy}{dx} + 3y = x^2e^x.$$

17. Find the general solution of

$$x^2\frac{d^2y}{dx^2} - x(x + 2)\frac{dy}{dx} + (x + 2)y = x^3,$$

given that $y = x$ and $y = xe^x$ are linearly independent solutions of the corresponding homogeneous equation.

18. Find the general solution of

$$x(x - 2)\frac{d^2y}{dx^2} - (x^2 - 2)\frac{dy}{dx} + 2(x - 1)y = 3x^2(x - 2)^2e^x,$$

given that $y = e^x$ and $y = x^2$ are linearly independent solutions of the corresponding homogeneous equation.

19. Find the general solution of

$$(2x + 1)(x + 1)\frac{d^2y}{dx^2} + 2x\frac{dy}{dx} - 2y = (2x + 1)^2,$$

given that $y = x$ and $y = (x + 1)^{-1}$ are linearly independent solutions of the corresponding homogeneous equation.

20. Find the general solution of

$$(\sin^2 x)\frac{d^2y}{dx^2} - 2\sin x \cos x\frac{dy}{dx} + (\cos^2 x + 1)y = \sin^3 x,$$

given that $y = \sin x$ and $y = x \sin x$ are linearly independent solutions of the corresponding homogeneous equation.

4.5 The Cauchy-Euler Equation

A. *The Equation and the Method of Solution*

In the preceding sections we have seen how to obtain the general solution of the nth-order linear differential equation with *constant* coefficients. We have seen that in such cases the form of the complementary function may be readily determined. The general nth-order linear equation with *variable* coefficients is quite a different matter, however, and only in certain special cases can the complementary function be obtained explicitly in closed form. One special case of considerable practical importance for which it is fortunate that this can be done is the so-called *Cauchy-Euler equation* (or *equidimensional* equation). This is an equation of the form

$$a_0x^n \frac{d^ny}{dx^n} + a_1x^{n-1}\frac{d^{n-1}y}{dx^{n-1}} + \cdots + a_{n-1}x\frac{dy}{dx} + a_ny = F(x), \tag{4.68}$$

where $a_0, a_1, \ldots, a_{n-1}, a_n$ are constants. Note the characteristic feature of this equation: each term in the left member is a constant multiple of an expression of the form

$$x^k \frac{d^ky}{dx^k}.$$

How should one proceed to solve such an equation? About the only hopeful thought that comes to mind at this stage of our study is to attempt a transformation. But what transformation should we attempt and where will it lead us? While it is certainly worthwhile to stop for a moment and consider what sort of transformation we might use in solving a "new" type of equation when we first encounter it, it is certainly not worthwhile to spend a great deal of time looking for clever devices which mathematicians have known about for many years. The facts are stated in the following theorem.

THEOREM 4.13
The transformation $x = e^t$ reduces the equation

$$a_0x^n \frac{d^ny}{dx^n} + a_1x^{n-1}\frac{d^{n-1}y}{dx^{n-1}} + \cdots + a_{n-1}x\frac{dy}{dx} + a_ny = F(x) \tag{4.68}$$

to a linear differential equation with constant coefficients.

This is what we need! We shall prove this theorem for the case of the *second*-order Cauchy-Euler differential equation

$$a_0x^2\frac{d^2y}{dx^2} + a_1x\frac{dy}{dx} + a_2y = F(x). \tag{4.69}$$

The proof in the general nth-order case proceeds in a similar fashion. Letting $x = e^t$, assuming $x > 0$, we have $t = \ln x$. Then

$$\frac{dy}{dx} = \frac{dy}{dt}\frac{dt}{dx} = \frac{1}{x}\frac{dy}{dt}$$

and

$$\frac{d^2y}{dx^2} = \frac{1}{x}\frac{d}{dx}\left(\frac{dy}{dt}\right) + \frac{dy}{dt}\frac{d}{dx}\left(\frac{1}{x}\right) = \frac{1}{x}\left(\frac{d^2y}{dt^2}\frac{dt}{dx}\right) - \frac{1}{x^2}\frac{dy}{dt}$$

$$= \frac{1}{x}\left(\frac{d^2y}{dt^2}\frac{1}{x}\right) - \frac{1}{x^2}\frac{dy}{dt} = \frac{1}{x^2}\left(\frac{d^2y}{dt^2} - \frac{dy}{dt}\right).$$

Thus

$$x\frac{dy}{dx} = \frac{dy}{dt} \quad \text{and} \quad x^2\frac{d^2y}{dx^2} = \frac{d^2y}{dt^2} - \frac{dy}{dt}.$$

Substituting into Equation (4.69) we obtain

$$a_0\left(\frac{d^2y}{dt^2} - \frac{dy}{dt}\right) + a_1\frac{dy}{dt} + a_2 y = F(e^t)$$

or

$$A_0\frac{d^2y}{dt^2} + A_1\frac{dy}{dt} + A_2 y = G(t), \tag{4.70}$$

where

$$A_0 = a_0, \quad A_1 = a_1 - a_0, \quad A_2 = a_2, \quad G(t) = F(e^t).$$

This is a second-order linear differential equation with *constant* coefficients, which was what we wished to show.

Remarks. 1. Note that the leading coefficient a_0x^n in Equation (4.68) is zero for $x = 0$. Thus the basic interval $a \leq x \leq b$, referred to in the general theorems of Section 4.1, does *not* include $x = 0$.

2. Observe that in the above proof we assumed that $x > 0$. If $x < 0$, the substitution $x = -e^t$ is actually the correct one. Unless the contrary is explicitly stated, we shall assume $x > 0$ when finding the general solution of a Cauchy-Euler differential equation.

B. Examples

▶ Example 4.38

$$x^2\frac{d^2y}{dx^2} - 2x\frac{dy}{dx} + 2y = x^3. \tag{4.71}$$

Let $x = e^t$. Then, assuming $x > 0$, we have $t = \ln x$, and

$$\frac{dy}{dx} = \frac{dy}{dt}\frac{dt}{dx} = \frac{1}{x}\frac{dy}{dt},$$

$$\frac{d^2y}{dx^2} = \frac{1}{x}\left(\frac{d^2y}{dt^2}\frac{dt}{dx}\right) - \frac{1}{x^2}\frac{dy}{dt} = \frac{1}{x^2}\left(\frac{d^2y}{dt^2} - \frac{dy}{dt}\right).$$

Thus Equation (4.71) becomes

$$\frac{d^2y}{dt^2} - \frac{dy}{dt} - 2\frac{dy}{dt} + 2y = e^{3t}$$

or

$$\frac{d^2y}{dt^2} - 3\frac{dy}{dt} + 2y = e^{3t}. \tag{4.72}$$

The complementary function of this equation is $y_c = c_1e^t + c_2e^{2t}$. We find a particular integral by the method of undetermined coefficients. We assume $y_p = Ae^{3t}$. Then $y_p' = 3Ae^{3t}$, $y_p'' = 9Ae^{3t}$, and substituting into Equation (4.72) we obtain

$$2Ae^{3t} = e^{3t}.$$

Thus $A = \frac{1}{2}$ and we have $y_p = \frac{1}{2}e^{3t}$. The general solution of Equation (4.72) is then

$$y = c_1e^t + c_2e^{2t} + \tfrac{1}{2}e^{3t}.$$

But we are not yet finished! We must return to the original independent variable x. Since $e^t = x$, we find

$$y = c_1x + c_2x^2 + \tfrac{1}{2}x^3.$$

This is the general solution of Equation (4.71).

Remarks. 1. Note carefully that under the transformation $x = e^t$ the right member of (4.71), x^3, transforms into e^{3t}. The student should be careful to transform *both* sides of the equation if he intends to obtain a particular integral of the given equation by finding a particular integral of the transformed equation, as we have done here.

2. We hasten to point out that the following alternative procedure may be used. After finding the complementary function of the transformed equation one can immediately write the complementary function of the original given equation and then proceed to obtain a particular integral of the original equation by variation of parameters. In Example 4.38, upon finding the complementary function $c_1e^t + c_2e^{2t}$ of Equation (4.72), one can immediately write the complementary function $c_1x + c_2x^2$ of Equation (4.71), then assume the particular integral $y_p(x) = v_1(x)x + v_2(x)x^2$, and from here proceed by the method of variation of parameters. However, when the nonhomogeneous function F transforms into a linear combination of UC functions, as it does in this example, the procedure illustrated is generally simpler.

▶ **Example 4.39**

$$x^3\frac{d^3y}{dx^3} - 4x^2\frac{d^2y}{dx^2} + 8x\frac{dy}{dx} - 8y = 4\ln x. \tag{4.73}$$

Assuming $x > 0$, we let $x = e^t$. Then $t = \ln x$, and

$$\frac{dy}{dx} = \frac{1}{x}\frac{dy}{dt},$$

$$\frac{d^2y}{dx^2} = \frac{1}{x^2}\left(\frac{d^2y}{dt^2} - \frac{dy}{dt}\right)$$

Now we must consider $\dfrac{d^3y}{dx^3}$.

$$\begin{aligned}\frac{d^3y}{dx^3} &= \frac{1}{x^2}\frac{d}{dx}\left(\frac{d^2y}{dt^2} - \frac{dy}{dt}\right) - \frac{2}{x^3}\left(\frac{d^2y}{dt^2} - \frac{dy}{dt}\right)\\ &= \frac{1}{x^2}\left(\frac{d^3y}{dt^3}\frac{dt}{dx} - \frac{d^2y}{dt^2}\frac{dt}{dx}\right) - \frac{2}{x^3}\left(\frac{d^2y}{dt^2} - \frac{dy}{dt}\right)\\ &= \frac{1}{x^3}\left(\frac{d^3y}{dt^3} - \frac{d^2y}{dt^2}\right) - \frac{2}{x^3}\left(\frac{d^2y}{dt^2} - \frac{dy}{dt}\right)\\ &= \frac{1}{x^3}\left(\frac{d^3y}{dt^3} - 3\frac{d^2y}{dt^2} + 2\frac{dy}{dt}\right).\end{aligned}$$

Thus, substituting into Equation (4.73), we obtain

$$\left(\frac{d^3y}{dt^3} - 3\frac{d^2y}{dt^2} + 2\frac{dy}{dt}\right) - 4\left(\frac{d^2y}{dt^2} - \frac{dy}{dt}\right) + 8\left(\frac{dy}{dt}\right) - 8y = 4t$$

or

$$\frac{d^3y}{dt^3} - 7\frac{d^2y}{dt^2} + 14\frac{dy}{dt} - 8y = 4t. \tag{4.74}$$

The complementary function of the transformed equation (4.74) is

$$y_c = c_1e^t + c_2e^{2t} + c_3e^{4t}.$$

We proceed to obtain a particular integral of Equation (4.74) by the method of undetermined coefficients. We assume $y_p = At + B$. Then $y_p' = A$, $y_p'' = y_p''' = 0$. Substituting into Equation (4.74), we find

$$14A - 8At - 8B = 4t.$$

Thus

$$-8A = 4, \qquad 14A - 8B = 0,$$

and so $A = -\frac{1}{2}$, $B = -\frac{7}{8}$. Thus the general solution of Equation (4.74) is

$$y = c_1e^t + c_2e^{2t} + c_3e^{4t} - \tfrac{1}{2}t - \tfrac{7}{8},$$

and so the general solution of Equation (4.73) is

$$y = c_1x + c_2x^2 + c_3x^4 - \tfrac{1}{2}\ln x - \tfrac{7}{8}.$$

Remark. In solving the Cauchy-Euler equations of the preceding examples, we observe that the transformation $x = e^t$ reduces

$$x\frac{dy}{dx} \text{ to } \frac{dy}{dt}, \qquad x^2\frac{d^2y}{dx^2} \text{ to } \frac{d^2y}{dt^2} - \frac{dy}{dt},$$

and

$$x^3\frac{d^3y}{dx^3} \text{ to } \frac{d^3y}{dt^3} - 3\frac{d^2y}{dt^2} + 2\frac{dy}{dt}.$$

We now show (without proof) how to find the expression into which the general term

$$x^n\frac{d^ny}{dx^n},$$

where n is an *arbitrary* positive integer, reduces under the transformation $x = e^t$. We present this as the following formal four-step procedure.

1. For the given positive integer n, determine

$$r(r-1)(r-2)\cdots[r-(n-1)].$$

2. Expand the preceding as a polynomial of degree n in r.
3. Replace r^k by $\dfrac{d^k y}{dt^k}$, for each $k = 1, 2, 3, \ldots, n$.
4. Equate $x^n \dfrac{d^n y}{dx^n}$ to the result in Step 3.

For example, when $n = 3$, we have the following illustration.

1. Since $n = 3$, $n - 1 = 2$ and we determine $r(r-1)(r-2)$.
2. Expanding the preceding, we obtain $r^3 - 3r^2 + 2r$.
3. Replacing r^3 by $\dfrac{d^3y}{dt^3}$, r^2 by $\dfrac{d^2y}{dt^2}$, and r by $\dfrac{dy}{dt}$, we have

$$\frac{d^3y}{dt^3} - 3\frac{d^2y}{dt^2} + 2\frac{dy}{dt}.$$

4. Equating $x^3 \dfrac{d^3y}{dx^3}$ to this, we have the relation

$$x^3\frac{d^3y}{dx^3} = \frac{d^3y}{dt^3} - 3\frac{d^2y}{dt^2} + 2\frac{dy}{dt}.$$

Note that this is precisely the relation we found in Example 4.39 and stated above.

Exercises

Find the general solution of each of the differential equations in Exercises 1 through 15.

1. $x^2\dfrac{d^2y}{dx^2} - 3x\dfrac{dy}{dx} + 3y = 0.$
2. $x^2\dfrac{d^2y}{dx^2} + x\dfrac{dy}{dx} - 4y = 0.$
3. $4x^2\dfrac{d^2y}{dx^2} - 4x\dfrac{dy}{dx} + 3y = 0.$
4. $x^2\dfrac{d^2y}{dx^2} - 3x\dfrac{dy}{dx} + 4y = 0.$
5. $x^2\dfrac{d^2y}{dx^2} + x\dfrac{dy}{dx} + 4y = 0.$
6. $x^2\dfrac{d^2y}{dx^2} - 3x\dfrac{dy}{dx} + 13y = 0.$
7. $x^3\dfrac{d^3y}{dx^3} - 3x^2\dfrac{d^2y}{dx^2} + 6x\dfrac{dy}{dx} - 6y = 0.$
8. $x^3\dfrac{d^3y}{dx^3} + 2x^2\dfrac{d^2y}{dx^2} - 10x\dfrac{dy}{dx} - 8y = 0.$
9. $x^3\dfrac{d^3y}{dx^3} - x^2\dfrac{d^2y}{dx^2} - 6x\dfrac{dy}{dx} + 18y = 0.$
10. $x^2\dfrac{d^2y}{dx^2} - 4x\dfrac{dy}{dx} + 6y = 4x - 6.$

11. $x^2 \dfrac{d^2y}{dx^2} - 5x \dfrac{dy}{dx} + 8y = 2x^3.$

12. $x^2 \dfrac{d^2y}{dx^2} + 4x \dfrac{dy}{dx} + 2y = 4 \ln x \quad (x > 0).$

13. $x^2 \dfrac{d^2y}{dx^2} + x \dfrac{dy}{dx} + 4y = 2x \ln x \quad (x > 0).$

14. $x^2 \dfrac{d^2y}{dx^2} + x \dfrac{dy}{dx} + y = \dfrac{1}{1 + x}.$

15. $x^3 \dfrac{d^3y}{dx^3} - x^2 \dfrac{d^2y}{dx^2} + 2x \dfrac{dy}{dx} - 2y = x^3.$

Solve the initial-value problems in Exercises 16 through 20.

16. $x^2 \dfrac{d^2y}{dx^2} - 2x \dfrac{dy}{dx} - 10y = 0, \quad y(1) = 5, \quad y'(1) = 4.$

17. $x^2 \dfrac{d^2y}{dx^2} - 4x \dfrac{dy}{dx} + 6y = 0, \quad y(2) = 0, \quad y'(2) = 4.$

18. $x^2 \dfrac{d^2y}{dx^2} - 5x \dfrac{dy}{dx} + 8y = 2x^3, \quad y(-2) = 1, \quad y'(-2) = 7.$

19. $x^2 \dfrac{d^2y}{dx^2} - 6y = \ln x \quad (x > 0), \quad y(1) = \frac{1}{6}, \quad y'(1) = -\frac{1}{6}.$

20. Solve:

$$(x + 2)^2 \frac{d^2y}{dx^2} - (x + 2) \frac{dy}{dx} - 3y = 0.$$

21. Solve:

$$(2x - 3)^2 \frac{d^2y}{dx^2} - 6(2x - 3) \frac{dy}{dx} + 12y = 0.$$

4.6 Statements and Proofs of Theorems on the Second-Order Homogeneous Linear Equation

Having considered the most fundamental methods of solving higher-order linear differential equations, we now return briefly to the theoretical side of the subject and present detailed statements and proofs of the basic theorems concerning the second-order homogeneous equation. The corresponding results for the general nth-order equation were introduced in Section 4.1B and employed frequently thereafter. By restricting attention here to the second-order case we shall be able to present proofs, which are completely explicit in every detail. However, we point out that each of these proofs may be extended in a straightforward manner to provide a proof of the corresponding theorem for the general nth-order case. For these general proofs, we again refer to Chapter 11 of the author's *Differential Equations*.

We thus consider the second-order homogeneous linear differential equation

$$a_0(x)\frac{d^2y}{dx^2} + a_1(x)\frac{dy}{dx} + a_2(x)y = 0. \tag{4.75}$$

where a_0, a_1, and a_2 are continuous real functions on a real interval $a \le x \le b$ and $a_0(x) \ne 0$ for any x on $a \le x \le b$.

In order to obtain the basic results concerning this equation, we shall need to make use of the following special case of Theorem 4.1 and its corollary.

THOEREM A

Hypothesis. *Consider the second-order homogeneous linear equation (4.75), where a_0, a_1, and a_2 are continuous real functions on a real interval $a \le x \le b$ and $a_0(x) \ne 0$ for any x on $a \le x \le b$. Let x_0 be any point of $a \le x \le b$; and let c_0 and c_1 be any two real constants.*

Conclusion 1. *Then there exists a unique solution f of Equation (4.75) such that $f(x_0) = c_0$ and $f'(x_0) = c_1$, and this solution f is defined over the entire interval $a \le x \le b$.*

Conclusion 2. *In particular, the unique solution f of Equation (4.75), which is such that $f(x_0) = 0$ and $f'(x_0) = 0$, is the function f such that $f(x) = 0$ for all x on $a \le x \le b$.*

Besides this result, we shall also need the following two theorems from algebra.

THEOREM B

Two homogeneous linear algebraic equations in two unknowns have a nontrivial solution if and only if the determinant of coefficients of the system is equal to zero.

THEOREM C

Two linear algebraic equations in two unknowns have a unique solution if and only if the determinant of coefficients of the system is unequal to zero.

We shall now proceed to obtain the basic results concerning Equation (4.75). Since each of the concepts involved has already been introduced and illustrated in Section 4.1, we shall state and prove the various theorems without further comments or examples.

THEOREM 4.14

Hypothesis. *Let the functions f_1 and f_2 be any two solutions of the homogeneous linear differential equation (4.75) on $a \le x \le b$, and let c_1 and c_2 be any two arbitrary constants.*

Conclusion. *Then the linear combination $c_1f_1 + c_2f_2$ of f_1 and f_2 is also a solution of Equation (4.75) on $a \le x \le b$.*

Proof. We must show that the function f defined by

$$f(x) = c_1 f_1(x) + c_2 f_2(x), \qquad a \le x \le b, \tag{4.76}$$

satisfies the differential equation (4.75) on $a \le x \le b$. From (4.76), we see that

$$f'(x) = c_1 f_1'(x) + c_2 f_2'(x), \qquad a \le x \le b, \tag{4.77}$$

and

$$f''(x) = c_1 f_1''(x) + c_2 f_2''(x), \qquad a \le x \le b. \tag{4.78}$$

Substituting $f(x)$ given by (4.76), $f'(x)$ given by (4.77), and $f''(x)$ given by (4.78) for y, $\dfrac{dy}{dx}$, and $\dfrac{d^2y}{dx^2}$, respectively, in the left member of differential equation (4.75), we obtain

$$a_0(x)[c_1 f_1''(x) + c_2 f_2''(x)] + a_1(x)[c_1 f_1'(x) + c_2 f_2'(x)] + a_2(x)[c_1 f_1(x) + c_2 f_2(x)]. \tag{4.79}$$

By rearranging terms, we express this as

$$c_1[a_0(x) f_1''(x) + a_1(x) f_1'(x) + a_2(x) f_1(x)] + c_2[a_0(x) f_2''(x) + a_1(x) f_2'(x) + a_2(x) f_2(x)]. \tag{4.80}$$

Since by hypothesis, f_1 and f_2 are solutions of differential equation (4.75) on $a \le x \le b$, we have, respectively,

$$a_0(x) f_1''(x) + a_1(x) f_1'(x) + a_2(x) f_1(x) = 0$$

and

$$a_0(x) f_2''(x) + a_1(x) f_2'(x) + a_2(x) f_2(x) = 0$$

for all x on $a \le x \le b$.

Thus the expression (4.80) is equal to zero for all x on $a \le x \le b$, and therefore so is the expression (4.79). That is, we have

$$a_0(x)[c_1 f_1''(x) + c_2 f_2''(x)] + a_1(x)[c_1 f_1'(x) + c_2 f_2'(x)] + a_2(x)[c_1 f_1(x) + c_2 f_2(x)] = 0$$

for all x on $a \le x \le b$, and so the function $c_1 f_1 + c_2 f_2$ is also a solution of differential equation (4.75) on this interval. *Q.E.D.*

THEOREM 4.15

Hypothesis. *Consider the second-order homogeneous linear differential equation* (4.75), *where* a_0, a_1, *and* a_2 *are continuous on* $a \le x \le b$ *and* $a_0(x) \ne 0$ *on* $a \le x \le b$.

Conclusion. *There exists a set of two solutions of Equation* (4.75) *which are linearly independent on* $a \le x \le b$.

Proof. We prove this theorem by actually exhibiting such a set of solutions. Let x_0 be a point of the interval $a \le x \le b$. Then by Theorem A, Conclusion 1, there exists a unique solution f_1 of Equation (4.75) such that

$$f_1(x_0) = 1 \quad \text{and} \quad f_1'(x_0) = 0 \tag{4.81}$$

and a unique solution f_2 of Equation (4.75) such that

$$f_2(x_0) = 0 \quad \text{and} \quad f_2'(x_0) = 1. \tag{4.82}$$

We now show that these two solutions f_1 and f_2 are indeed linearly independent. Suppose they were not. Then they would be linear *dependent;* and so by the definition of linear dependence, there would exist constants c_1 and c_2, not both zero, such that

$$c_1 f_1(x) + c_2 f_2(x) = 0 \tag{4.83}$$

for all x such that $a \leq x \leq b$. Then also

$$c_1 f_1'(x) + c_2 f_2'(x) = 0 \tag{4.84}$$

for all x such that $a \leq x \leq b$. The identities (4.83) and (4.84) hold at $x = x_0$, giving

$$c_1 f_1(x_0) + c_2 f_2(x_0) = 0, \qquad c_1 f_1'(x_0) + c_2 f_2'(x_0) = 0.$$

Now apply conditions (4.81) and (4.82) to this set of equations. They reduce to

$$c_1(1) + c_2(0) = 0, \qquad c_1(0) + c_2(1) = 0.$$

or simply $c_1 = c_2 = 0$, which is a contradiction (since c_1 and c_2 are not both zero). Thus the solutions f_1 and f_2 defined respectively by (4.81) and (4.82) are linearly independent on $a \leq x \leq b$. *Q.E.D*

THEOREM 4.16

Two solutions f_1 and f_2 of the second-order homogeneous linear differential equation (4.75) are linear independent on $a \leq x \leq b$ if and only if the value of the Wronskian of f_1 and f_2 is different from zero for some x on the interval $a \leq x \leq b$.

Method of Proof. We prove this theorem by proving the following equivalent theorem.

THEOREM 4.17

Two solutions f_1 and f_2 of the second-order homogeneous linear differential equation (4.75) are linearly dependent on $a \leq x \leq b$ if and only if the value of the Wronskian of f_1 and f_2 is zero for all x on $a \leq x \leq b$:

$$\begin{vmatrix} f_1(x) & f_2(x) \\ f_1'(x) & f_2'(x) \end{vmatrix} = 0 \quad \textit{for all } x \textit{ on } a \leq x \leq b.$$

Proof. Part 1. We must show that if the value of the Wronskian of f_1 and f_2 is zero for all x on $a \leq x \leq b$, then f_1 and f_2 are linearly dependent on $a \leq x \leq b$. We thus assume that

$$\begin{vmatrix} f_1(x) & f_2(x) \\ f_1'(x) & f_2'(x) \end{vmatrix} = 0$$

for all x such that $a \leq x \leq b$. Then at any particular x_0 such that $a \leq x_0 \leq b$, we have

$$\begin{vmatrix} f_1(x_0) & f_2(x_0) \\ f_1'(x_0) & f_2'(x_0) \end{vmatrix} = 0.$$

Thus, by Theorem B, there exist constants c_1 and c_2, not both zero, such that

$$\begin{aligned} c_1 f_1(x_0) + c_2 f_2(x_0) &= 0, \\ c_1 f_1'(x_0) + c_2 f_2'(x_0) &= 0. \end{aligned} \tag{4.85}$$

Now consider the function f defined by

$$f(x) = c_1 f_1(x) + c_2 f_2(x), \qquad a \leq x \leq b.$$

By Theorem 4.14, since f_1 and f_2 are solutions of differential equation (4.75), this function f is also a solution of Equation (4.75). From (4.85), we have

$$f(x_0) = 0 \quad \text{and} \quad f'(x_0) = 0.$$

Thus by Theorem A, Conclusion 2, we know that

$$f(x) = 0 \quad \text{for all } x \text{ on } a \leq x \leq b.$$

That is,

$$c_1 f_1(x) + c_2 f_2(x) = 0$$

for all x on $a \leq x \leq b$, where c_1 and c_2 are not both zero. Therefore the solutions f_1 and f_2 are linearly dependent on $a \leq x \leq b$.

Part 2. We must now show that if f_1 and f_2 are linearly dependent on $a \leq x \leq b$, then their Wronskian has the value zero for all x on this interval. We thus assume that f_1 and f_2 are linearly dependent on $a \leq x \leq b$. Then there exist constants c_1 and c_2, not both zero, such that

$$c_1 f_1(x) + c_2 f_2(x) = 0 \tag{4.86}$$

for all x on $a \leq x \leq b$. From (4.86), we also have

$$c_1 f_1'(x) + c_2 f_2'(x) = 0 \tag{4.87}$$

for all x on $a \leq x \leq b$. Now let $x = x_0$ be an arbitrary point of the interval $a \leq x \leq b$. Then (4.86) and (4.87) hold at $x = x_0$. That is,

$$\begin{aligned} c_1 f_1(x_0) + c_2 f_2(x_0) &= 0, \\ c_1 f_1'(x_0) + c_2 f_2'(x_0) &= 0, \end{aligned}$$

where c_1 and c_2 are not both zero. Thus, by Theorem B, we have

$$\begin{vmatrix} f_1(x_0) & f_2(x_0) \\ f_1'(x_0) & f_2'(x_0) \end{vmatrix} = 0.$$

But this determinant is the value of the Wronskian of f_1 and f_2 at $x = x_0$, and x_0 is an *arbitrary* point of $a \leq x \leq b$. Thus we have

$$\begin{vmatrix} f_1(x) & f_2(x) \\ f_1'(x) & f_2'(x) \end{vmatrix} = 0$$

for *all* x on $a \leq x \leq b$. *Q.E.D.*

THEOREM 4.18

The value of the Wronskian of two solutions f_1 and f_2 of differential equation (4.75) *either is zero for all x on $a \leq x \leq b$ or is zero for no x on $a \leq x \leq b$.*

Proof. If f_1 and f_2 are linearly dependent on $a \le x \le b$, then by Theorem 4.17, the value of the Wronskian of f_1 and f_2 is zero for all x on $a \le x \le b$.

Now let f_1 and f_2 be linearly independent on $a \le x \le b$; and let W denote the Wronskian of f_1 and f_2, so that

$$W(x) = \begin{vmatrix} f_1(x) & f_2(x) \\ f_1'(x) & f_2'(x) \end{vmatrix}.$$

Differentiating this, we obtain

$$W'(x) = \begin{vmatrix} f_1'(x) & f_2'(x) \\ f_1'(x) & f_2'(x) \end{vmatrix} + \begin{vmatrix} f_1(x) & f_2(x) \\ f_1''(x) & f_2''(x) \end{vmatrix},$$

and this reduces at once to

$$W'(x) = \begin{vmatrix} f_1(x) & f_2(x) \\ f_1''(x) & f_2''(x) \end{vmatrix}. \tag{4.88}$$

Since f_1 and f_2 are solutions of differential equation (4.75), we have, respectively,

$$a_0(x)f_1''(x) + a_1(x)f_1'(x) + a_2(x)f_1(x) = 0,$$

$$a_0(x)f_2''(x) + a_1(x)f_2'(x) + a_2(x)f_2(x) = 0,$$

and hence

$$f_1''(x) = -\frac{a_1(x)}{a_0(x)} f_1'(x) - \frac{a_2(x)}{a_0(x)} f_1(x),$$

$$f_2''(x) = -\frac{a_1(x)}{a_0(x)} f_2'(x) - \frac{a_2(x)}{a_0(x)} f_2(x)$$

on $a \le x \le b$. Substituting these expressions into (4.88), we obtain

$$W'(x) = \begin{vmatrix} f_1(x) & f_2(x) \\ -\dfrac{a_1(x)}{a_0(x)} f_1'(x) - \dfrac{a_2(x)}{a_0(x)} f_1(x) & -\dfrac{a_1(x)}{a_0(x)} f_2'(x) - \dfrac{a_2(x)}{a_0(x)} f_2(x) \end{vmatrix}.$$

This reduces at once to

$$W'(x) = \begin{vmatrix} f_1(x) & f_2(x) \\ -\dfrac{a_1(x)}{a_0(x)} f_1'(x) & -\dfrac{a_1(x)}{a_0(x)} f_2'(x) \end{vmatrix} + \begin{vmatrix} f_1(x) & f_2(x) \\ -\dfrac{a_2(x)}{a_0(x)} f_1(x) & -\dfrac{a_2(x)}{a_0(x)} f_2(x) \end{vmatrix},$$

and since the last determinant has two proportional rows, this in turn reduces to

$$W'(x) = -\frac{a_1(x)}{a_0(x)} \begin{vmatrix} f_1(x) & f_2(x) \\ f_1'(x) & f_2'(x) \end{vmatrix},$$

which is simply

$$W'(x) = -\frac{a_1(x)}{a_0(x)} W(x).$$

Thus the Wronskian W satisfies the first-order homogeneous linear differential equation

$$\frac{dW}{dx} + \frac{a_1(x)}{a_0(x)} W = 0.$$

Integrating this from x_0 to x, where x_0 is an arbitrary point of $a \le x \le b$, we obtain

$$W(x) = c \exp\left[-\int_{x_0}^{x} \frac{a_1(t)}{a_0(t)}\, dt\right].$$

Letting $x = x_0$, we find that $c = W(x_0)$. Hence we obtain the identity

$$W(x) = W(x_0) \exp\left[-\int_{x_0}^{x} \frac{a_1(t)}{a_0(t)}\, dt\right], \tag{4.89}$$

valid for all x on $a \le x \le b$, where x_0 is an arbitrary point of this interval.

Now assume that $W(x_0) = 0$. Then by identity (4.89), we have $W(x) = 0$ for all x on $a \le x \le b$. Thus by Theorem 4.17, the solutions f_1 and f_2 must be linearly dependent on $a \le x \le b$. This is a contradiction, since f_1 and f_2 are linearly independent. Therefore the assumption that $W(x_0) = 0$ is false, and so $W(x_0) \ne 0$. But x_0 is an *arbitrary* point of $a \le x \le b$. Thus $W(x)$ is zero for no x on $a \le x \le b$.

Q.E.D

THEOREM 4.19

Hypothesis. *Let f_1 and f_2 be any two linearly independent solutions of differential equation (4.75) on $a \le x \le b$.*

Conclusion. *Then every solution f of differential equation (4.75) can be expressed as a suitable linear combination*

$$c_1 f_1 + c_2 f_2$$

of these two linear independent solutions.

Proof. Let x_0 be an arbitrary point of the interval $a \le x \le b$, and consider the following system of two linear algebraic equations in the two unknowns k_1 and k_2:

$$\begin{aligned} k_1 f_1(x_0) + k_2 f_2(x_0) &= f(x_0), \\ k_1 f_1'(x_0) + k_2 f_2'(x_0) &= f'(x_0). \end{aligned} \tag{4.90}$$

Since f_1 and f_2 are linearly independent on $a \le x \le b$, we know by Theorem 4.16 that the value of the Wronskian of f_1 and f_2 is different from zero at some point of this interval. Then by Theorem 4.18 the value of the Wronskian is zero for no x on $a \le x \le b$ and hence its value at x_0 is not zero. That is,

$$\begin{vmatrix} f_1(x_0) & f_2(x_0) \\ f_1'(x_0) & f_2'(x_0) \end{vmatrix} \ne 0.$$

Thus by Theorem C, the algebraic system (4.90) has a unique solution $k_1 = c_1$ and $k_2 = c_2$. Thus for $k_1 = c_1$ and $k_2 = c_2$, each left member of system (4.90) is the same number as the corresponding right member of (4.90). That is, the number $c_1 f_1(x_0) + c_2 f_2(x_0)$ is equal to the number $f(x_0)$, and the number $c_1 f_1'(x_0) + c_2 f_2'(x_0)$ is equal to the number $f'(x_0)$. But the numbers $c_1 f_1(x_0) + c_2 f_2(x_0)$ and $c_1 f_1'(x_0) + c_2 f_2'(x_0)$ are the values of the solution $c_1 f_1 + c_2 f_2$ and its first derivative, respectively, at x_0; and the numbers $f(x_0)$ and $f'(x_0)$ are the values of the solution f and its first derivative, respectively, at x_0. Thus the two solutions $c_1 f_1 + c_2 f_2$ and

f have equal values and their first derivatives also have equal values at x_0. Hence by Theorem A, Conclusion 1, we know that these two solutions are identical throughout the interval $a \leq x \leq b$. That is,

$$f(x) = c_1 f_1(x) + c_2 f_2(x)$$

for all x on $a \leq x \leq b$, and so f is expressed as a linear combination of f_1 and f_2.

Q.E.D.

Exercises

1. Consider the second-order homogeneous linear differential equation

$$\frac{d^2y}{dx^2} - 3\frac{dy}{dx} + 2y = 0.$$

(a) Find the two linearly independent solutions f_1 and f_2 of this equation which are such that

$$f_1(0) = 1 \quad \text{and} \quad f_1'(0) = 0$$

and

$$f_2(0) = 0 \quad \text{and} \quad f_2'(0) = 1.$$

(b) Express the solution

$$3e^x + 2e^{2x}$$

as a linear combination of the two linearly independent solutions f_1 and f_2 defined in part (a).

2. Consider the second-order homogeneous linear differential equation

$$a_0(x)\frac{d^2y}{dx^2} + a_1(x)\frac{dy}{dx} + a_2(x)y = 0, \tag{A}$$

where a_0, a_1, and a_2 are continuous on a real interval $a \leq x \leq b$, and $a_0(x) \neq 0$ for all x on this interval. Let f_1 and f_2 be two distinct solutions of differential equation (A) on $a \leq x \leq b$, and suppose $f_2(x) \neq 0$ for all x on this interval. Let $W[f_1(x), f_2(x)]$ be the value of the Wronskian of f_1 and f_2 at x.

(a) Show that

$$\frac{d}{dx}\left[\frac{f_1(x)}{f_2(x)}\right] = -\frac{W[f_1(x), f_2(x)]}{[f_2(x)]^2}$$

for all x on $a \leq x \leq b$.

(b) Use the result of part (a) to show that if $W[f_1(x), f_2(x)] = 0$ for all x such that $a \leq x \leq b$, then the solutions f_1 and f_2 are linearly dependent on this interval.

(c) Suppose the solutions f_1 and f_2 are linearly independent on $a \leq x \leq b$, and let f be the function defined by $f(x) = f_1(x)/f_2(x)$, $a \leq x \leq b$. Show that f is a monotonic function on $a \leq x \leq b$.

3. Let f_1 and f_2 be two solutions of the second-order homogeneous linear differential equation (A) of Exercise 2.

(a) Show that if f_1 and f_2 have a common zero at a point x_0 of the interval $a \leq x \leq b$, then f_1 and f_2 are linearly dependent on $a \leq x \leq b$.

(b) Show that if f_1 and f_2 have relative maxima at a common point x_0 of the interval $a \leq x \leq b$, then f_1 and f_2 are linearly dependent on $a \leq x \leq b$.

4. Consider the second-order homogeneous linear differential equation (A) of Exercise 2.

(a) Let f_1 and f_2 be two solutions of this equation. Show that if f_1 and f_2 are linearly independent on $a \leq x \leq b$ and A_1, A_2, B_1, and B_2 are constants such that $A_1B_2 - A_2B_1 \neq 0$, then the solutions $A_1f_1 + A_2f_2$ and $B_1f_1 + B_2f_2$ of Equation (A) are also linearly independent on $a \leq x \leq b$.

(b) Let $\{f_1, f_2\}$ be one set of two linearly independent solutions of Equation (A) on $a \leq x \leq b$, and let $\{g_1, g_2\}$ be another set of two linearly independent solutions of Equation (A) on this interval. Let $W[f_1(x), f_2(x)]$ denote the value of the Wronskian of f_1 and f_2 at x, and let $W[g_1(x), g_2(x)]$ denote the value of the Wronskian of g_1 and g_2 at x. Show that there exists a constant $c \neq 0$ such that

$$W[f_1(x), f_2(x)] = cW[g_1(x), g_2(x)]$$

for all x on $a \leq x \leq b$.

5. Let f_1 and f_2 be two solutions of the second-order homogeneous linear differential equation (A) of Exercise 2. Show that if f_1 and f_2 are linearly independent on $a \leq x \leq b$ and are such that $f_1''(x_0) = f_2''(x_0) = 0$ at some point x_0 of this interval, then $a_1(x_0) = a_2(x_0) = 0$.

Suggested Reading

AGNEW, R., *Differential Equations*, 2nd ed. (McGraw-Hill, New York, 1960).
BOYCE, W., and R. DIPRIMA, *Elementary Differential Equations*, 2nd ed. (Wiley, New York, 1969).
BRAUER, F., and J. NOHEL, *Ordinary Differential Equations: A First Course* (Benjamin, New York, 1967).
CODDINGTON, E., *An Introduction to Ordinary Differential Equations* (Prentice-Hall, Englewood Cliffs, N.J., 1961).
FORD, L., *Differential Equations*, 2nd ed. (McGraw-Hill, New York, 1955).
KAPLAN, W., *Ordinary Differential Equations* (Addison-Wesley, Reading, Mass., 1958).
KREIDER, D., R. KULLER, and D. OSTBERG, *Elementary Differential Equations* (Addison-Wesley, Reading, Mass., 1968).
LEIGHTON, W., *Ordinary Differential Equations*, 3rd ed. (Wadsworth, Belmont, Cal., 1970).
RAINVILLE, E., and P. BEDIENT, *Elementary Differential Equations*, 4th ed. (Macmillan, New York, 1969).
RITGER, P., and N. ROSE, *Differential Equations with Applications* (McGraw-Hill, New York, 1968).

Applications of Second-Order Linear Differential Equations with Constant Coefficients

Higher-order linear differential equations, which were introduced in the previous chapter, are equations having a great variety of important applications. In particular, second-order linear differential equations with constant coefficients have numerous applications in physics and in electrical and mechanical engineering. Two of these applications will be considered in the present chapter. In Sections 5.1 through 5.5 we shall discuss the motion of a mass vibrating up and down at the end of a spring, while in Section 5.6 we shall consider problems in electric circuit theory.

5.1 The Differential Equation of the Vibrations of a Mass on a Spring

The Basic Problem

A coil spring is suspended vertically from a fixed point on a ceiling, beam, or other similar object. A mass is attached to its lower end and allowed to come to rest in an equilibrium position. The system is then set in motion either (1) by pulling the mass down a distance below its equilibrium position (or pushing it up a distance above it) and subsequently releasing it with an initial velocity (zero or nonzero, downward or upward) at $t = 0$; or (2) by forcing the mass out of its equilibrium position by giving it a nonzero initial velocity (downward or upward) at $t = 0$. Our problem is to determine the resulting motion of the mass on the spring. In order to do so we must also consider certain other phenomena which may be present. For one thing, assuming the system is located in some sort of medium (say "ordinary" air or perhaps water), this medium produces a resistance force which tends to retard the motion. Also, certain external forces may be present. For example, a magnetic force from outside the system may be acting upon the mass. Let us then attempt to determine the motion of the mass on the spring, taking into account both the resistance of the medium and possible external forces. We shall do this by first obtaining and then solving the differential equation for the motion.

In order to set up the differential equation for this problem we shall need two laws of physics: Newton's second law and Hooke's law. Newton's second law was encountered in Chapter 3, and we shall not go into a further discussion of it here. Let us then recall the other law which we shall need.

Hooke's Law

The magnitude of the force needed to produce a certain elongation of a spring is directly proportional to the amount of this elongation, provided this elongation is not too great. In mathematical form,

$$|F| = ks,$$

where F is the magnitude of the force, s is the amount of elongation, and k is a constant of proportionality which we shall call the *spring constant.*

The spring constant k depends upon the spring under consideration and is a measure of its stiffness. For example, if a 30-lb weight stretches a spring 2 ft, then Hooke's law gives $30 = (k)(2)$; thus for this spring $k = 15$ lb/ft.

When a mass is hung upon a spring of spring constant k and thus produces an elongation of amount s, the force F of the mass upon the spring therefore has magnitude ks. The spring at the same time exerts a force upon the mass called the *restoring force* of the spring. This force is equal in magnitude but opposite in sign to F and hence has magnitude $-ks$.

Let us formulate the problem systematically. Let the coil spring have natural (unstretched) length L. The mass m is attached to its lower end and comes to rest in its equilibrium position, thereby stretching the spring an amount l so that its stretched length is $L + l$. We choose the axis along the line of the spring, with the origin O at the equilibrium position and the positive direction downward. Thus, letting x denote the displacement of the mass from O along this line, we see that x is positive, zero, or negative according to whether the mass is below, at, or above its equilibrium position. (See Figure 5.1.)

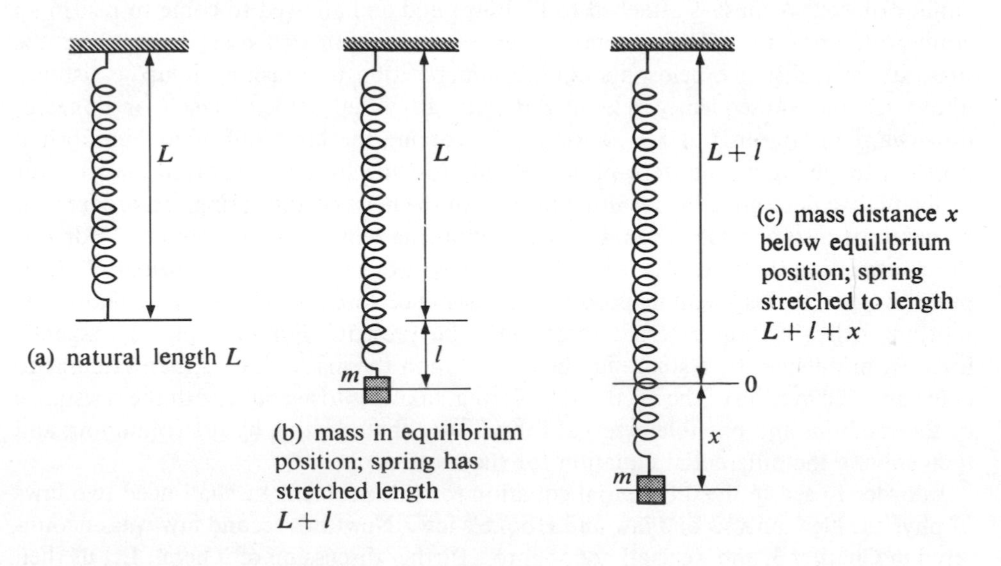

FIGURE 5.1

Forces Acting Upon the Mass

We now enumerate the various forces which act upon the mass. Forces tending to pull the mass downward are positive, while those tending to pull it upward are negative. The forces are:

1. F_1, the *force of gravity*, of magnitude mg, where g is the acceleration due to gravity. Since this acts in the downward direction, it is positive, and so

$$F_1 = mg. \tag{5.1}$$

2. F_2, the *restoring force* of the spring. Since $x + l$ is the total amount of elongation, by Hooke's law the magnitude of this force is $k(x + l)$. When the mass is *below* the end of the unstretched spring, this force acts in the *upward* direction and so is *negative*. Also, for the mass in such a position, $x + l$ is *positive*. Thus, when the mass is *below* the end of the unstretched spring, the restoring force is given by

$$F_2 = -k(x + l). \tag{5.2}$$

This also gives the restoring force when the mass is *above* the end of the unstretched spring, as one can see by replacing each italicized word in the three preceding sentences by its opposite. When the mass is at rest in its equilibrium position the restoring force F_2 is equal in magnitude but opposite in direction to the force of gravity and so is given by $-mg$. Since in this position $x = 0$, Equation (5.2) gives

$$-mg = -k(0 + l)$$

or

$$mg = kl.$$

Replacing kl by mg in Equation (5.2) we see that the restoring force can thus be written as

$$F_2 = -kx - mg. \tag{5.3}$$

3. F_3, the *resisting force* of the medium, called the *damping force*. Although the magnitude of this force is not known *exactly*, it is known that for small velocities it is *approximately* proportional to the magnitude of the velocity:

$$|F_3| = a\left|\frac{dx}{dt}\right|, \tag{5.4}$$

where $a > 0$ is called the *damping constant*. When the mass is moving *downward*, F_3 acts in the *upward* direction (opposite to that of the motion) and so $F_3 < 0$. Also, since m is moving *downward*, x is *increasing* and dx/dt is *positive*. Thus, assuming Equation (5.4) to hold, when the mass is moving *downward*, the damping force is given by

$$F_3 = -a\frac{dx}{dt} \quad (a > 0). \tag{5.5}$$

This also gives the damping force when the mass is moving *upward*, as one may see by replacing each italicized word in the three preceding sentences by its opposite.

4. F_4, any *external impressed forces* which act upon the mass. Let us denote the resultant of all such external forces at time t simply by $F(t)$ and write

$$F_4 = F(t). \tag{5.6}$$

We now apply Newton's second law, $F = ma$, where $F = F_1 + F_2 + F_3 + F_4$. Using (5.1), (5.3), (5.5), and (5.6), we find

$$m \frac{d^2x}{dt^2} = mg - kx - mg - a\frac{dx}{dt} + F(t)$$

or

$$m \frac{d^2x}{dt^2} + a\frac{dx}{dt} + kx = F(t). \tag{5.7}$$

This we take as the differential equation for the motion of the mass on the spring. Observe that it is a nonhomogeneous second-order linear differential equation with constant coefficients. If $a = 0$ the motion is called *undamped;* otherwise it is called *damped.* If there are no external impressed forces, $F(t) = 0$ for all t and the motion is called *free;* otherwise it is called *forced.* In the following sections we consider the solution of (5.7) in each of these cases.

5.2 Free, Undamped Motion

We now consider the special case of *free, undamped motion*, that is, the case in which both $a = 0$ and $F(t) = 0$ for all t. The differential equation (5.7) then reduces to

$$m \frac{d^2x}{dt^2} + kx = 0, \tag{5.8}$$

where m (> 0) is the mass and k (> 0) is the spring constant. Dividing through by m and letting $k/m = \lambda^2$, we write (5.8) in the form

$$\frac{d^2x}{dt^2} + \lambda^2 x = 0. \tag{5.9}$$

The auxiliary equation

$$r^2 + \lambda^2 = 0$$

has roots $r = \pm\lambda i$ and hence the general solution of (5.8) can be written

$$x = c_1 \sin \lambda t + c_2 \cos \lambda t, \tag{5.10}$$

where c_1 and c_2 are arbitrary constants.

Let us now assume that the mass was initially displaced a distance x_0 from its equilibrium position and released from that point with initial velocity v_0. Then, in addition to the differential equation (5.8) [or (5.9)], we have the initial conditions

$$x(0) = x_0, \tag{5.11}$$

$$x'(0) = v_0. \tag{5.12}$$

Differentiating (5.10) with respect to t, we have

$$\frac{dx}{dt} = c_1\lambda \cos \lambda t - c_2\lambda \sin \lambda t. \tag{5.13}$$

Applying conditions (5.11) and (5.12) to Equations (5.10) and (5.13), respectively, we see at once that

$$c_2 = x_0,$$

$$c_1\lambda = v_0.$$

Substituting the values of c_1 and c_2 so determined into Equation (5.10) gives the particular solution of the differential equation (5.8) satisfying the conditions (5.11) and (5.12) in the form

$$x = \frac{v_0}{\lambda} \sin \lambda t + x_0 \cos \lambda t.$$

We put this in an alternative form by first writing it as

$$x = c\left[\frac{(v_0/\lambda)}{c} \sin \lambda t + \frac{x_0}{c} \cos \lambda t\right], \tag{5.14}$$

where

$$c = \sqrt{\left(\frac{v_0}{\lambda}\right)^2 + x_0^2} > 0. \tag{5.15}$$

Then, letting

$$\begin{aligned} \frac{(v_0/\lambda)}{c} &= -\sin \phi, \\ \frac{x_0}{c} &= \cos \phi, \end{aligned} \tag{5.16}$$

Equation (5.14) reduces at once to

$$x = c \cos (\lambda t + \phi), \tag{5.17}$$

where c is given by Equation (5.15) and ϕ is determined by Equations (5.16). Since $\lambda = \sqrt{k/m}$, we now write the solution (5.17) in the form

$$x = c \cos \left(\sqrt{\frac{k}{m}}\, t + \phi\right). \tag{5.18}$$

This, then, gives the displacement x of the mass from the equilibrium position O as a function of the time t ($t > 0$). We see at once that the free, undamped motion of the mass is a *simple harmonic motion*. The constant c is called the *amplitude* of the motion and gives the maximum (positive) displacement of the mass from its equilibrium position. The motion is a *periodic* motion, and the mass oscillates back and forth between $x = c$ and $x = -c$. We have $x = c$ if and only if

$$\sqrt{\frac{k}{m}}\, t + \phi = \pm 2n\pi,$$

$n = 0, 1, 2, 3, \ldots; t > 0$. Thus the maximum (positive) displacement occurs if and only if

$$t = \sqrt{\frac{m}{k}}(\pm 2n\pi - \phi) > 0, \tag{5.19}$$

where $n = 0, 1, 2, 3, \ldots$.

The time interval between two successive maxima is called the *period* of the motion. Using (5.19), we see that it is given by

$$\frac{2\pi}{\sqrt{k/m}} = \frac{2\pi}{\lambda}. \tag{5.20}$$

The reciprocal of the period, which gives the number of oscillations per second, is called the *natural frequency* (or simply *frequency*) of the motion. The number ϕ is called the *phase constant* (or *phase angle*). The graph of this motion is shown in Figure 5.2.

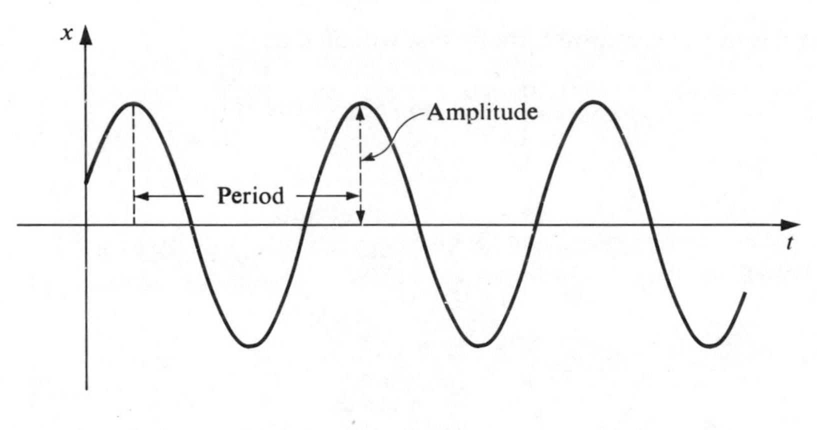

FIGURE 5.2

▶ Example 5.1. An 8-lb weight is placed upon the lower end of a coil spring suspended from the ceiling. The weight comes to rest in its equilibrium position, thereby stretching the spring 6 in. The weight is then pulled down 3 in. below its equilibrium position and released at $t = 0$ with an initial velocity of 1 ft/sec, directed downward. Neglecting the resistance of the medium and assuming that no external forces are present, determine the amplitude, period, and frequency of the resulting motion.

Formulation. This is clearly an example of free, undamped motion and hence Equation (5.8) applies. Since the 8-lb weight stretches the spring 6 in. $= \frac{1}{2}$ ft, Hooke's law $F = ks$ gives $8 = k(\frac{1}{2})$ and so $k = 16$ lb/ft. Also, $m = w/g = \frac{8}{32}$ (slugs), and so Equation (5.8) gives

$$\frac{8}{32}\frac{d^2x}{dt^2} + 16x = 0$$

or

$$\frac{d^2x}{dt^2} + 64x = 0. \tag{5.21}$$

Since the weight was released with a downward initial velocity of 1 ft/sec from a point 3 in. (= $\frac{1}{4}$ ft) below its equilibrium position, we also have the initial conditions

$$x(0) = \tfrac{1}{4}, \qquad x'(0) = 1. \tag{5.22}$$

Solution. The auxiliary equation corresponding to Equation (5.21) is $r^2 + 64 = 0$, and hence $r = \pm 8i$. Thus the general solution of the differential equation (5.21) may be written

$$x = c_1 \sin 8t + c_2 \cos 8t, \tag{5.23}$$

where c_1 and c_2 are arbitrary constants. Applying the first of conditions (5.22) to this, we find $c_2 = \frac{1}{4}$. Differentiating (5.23), we have

$$\frac{dx}{dt} = 8c_1 \cos 8t - 8c_2 \sin 8t.$$

Applying the second of conditions (5.22) to this, we have $8c_1 = 1$ and hence $c_1 = \frac{1}{8}$. Thus the solution of the differential equation (5.21) satisfying the conditions (5.22) is

$$x = \tfrac{1}{8} \sin 8t + \tfrac{1}{4} \cos 8t. \tag{5.24}$$

Let us put this in the form (5.18). We find

$$\sqrt{\left(\frac{1}{8}\right)^2 + \left(\frac{1}{4}\right)^2} = \frac{\sqrt{5}}{8}$$

and thus write

$$x = \frac{\sqrt{5}}{8}\left(\frac{\sqrt{5}}{5} \sin 8t + \frac{2\sqrt{5}}{5} \cos 8t\right).$$

Thus, letting

$$\begin{aligned} \cos \phi &= \frac{2\sqrt{5}}{5}, \\ \sin \phi &= -\frac{\sqrt{5}}{5}, \end{aligned} \tag{5.25}$$

we write the solution (5.24) in the form

$$x = \frac{\sqrt{5}}{8} \cos (8t + \phi), \tag{5.26}$$

where ϕ is determined by Equations (5.25). From these equations we find that $\phi \approx -0.46$ radians. Taking $\sqrt{5} \approx 2.236$, the solution (5.26) is thus given approximately by

$$x = 0.280 \cos (8t - 0.46).$$

The amplitude of the motion is $\sqrt{5}/8 \approx 0.280$ (ft). By formula (5.20), the period is $2\pi/8 = \pi/4$ (sec), and the frequency is $4/\pi$ oscillations/sec. The graph is shown in Figure 5.3.

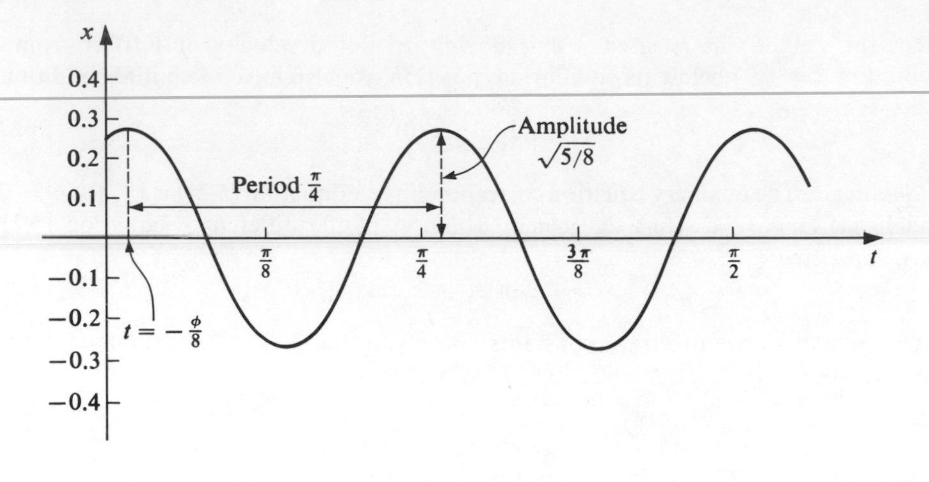

FIGURE 5.3

Before leaving this problem, let us be certain that we can set up initial conditions correctly. Let us replace the third sentence in the statement of the problem by the following: "The weight is then *pushed up* 4 in. *above* its equilibrium position and released at $t = 0$, with an initial velocity of 2 ft/sec, directed *upward*." The initial conditions (5.22) would then have been replaced by

$$x(0) = -\tfrac{1}{3},$$

$$x'(0) = -2.$$

The minus sign appears before the $\frac{1}{3}$ because the initial position is 4 in. $= \frac{1}{3}$ foot *above* the equilibrium position and hence is *negative*. The minus sign before the 2 is due to the fact that the initial velocity is directed *upward*, that is, in the *negative* direction.

Exercises

Note: In Exercises 1–7 neglect the resistance of the medium and assume that no external forces are present.

1. A 12-lb weight is placed upon the lower end of a coil spring suspended from the ceiling. The weight comes to rest in its equilibrium position, thereby stretching the spring 1.5 in. The weight is then pulled down 2 in. below its equilibrium position and released from rest at $t = 0$. Find the displacement of the weight as a function of the time; determine the amplitude, period, and frequency of the resulting motion; and graph the displacement as a function of the time.

2. A 16-lb weight is placed upon the lower end of a coil spring suspended vertically from a fixed support. The weight comes to rest in its equilibrium position, thereby stretching the spring 6 in. Determine the resulting displacement as a function of time in each of the following cases:
   (a) If the weight is then pulled down 4 in. below its equilibrium position and released at $t = 0$ with an initial velocity of 2 ft/sec, directed downward.

(b) If the weight is then pulled down 4 in. below its equilibrium position and released at $t = 0$ with an initial velocity of 2 ft/sec, directed upward.

(c) If the weight is then pushed up 4 in. above its equilibrium position and released at $t = 0$ with an initial velocity of 2 ft/sec, directed downward.

3. A 4-lb. weight is attached to the lower end of a coil spring suspended from the ceiling. The weight comes to rest in its equilibrium position, thereby stretching the spring 6 in. At time $t = 0$ the weight is then struck so as to set it into motion with an initial velocity of 2 ft/sec, directed downward.
   (a) Determine the resulting displacement and velocity of the weight as functions of the time.
   (b) Find the amplitude, period, and frequency of the motion.
   (c) Determine the times at which the weight is 1.5 in. below its equilibrium position and moving downward.
   (d) Determine the times at which it is 1.5 in. below its equilibrium position and moving upward.

4. A 64-lb weight is placed upon the lower end of a coil spring suspended from a rigid beam. The weight comes to rest in its equilibrium position, thereby stretching the spring 2 ft. The weight is then pulled down 1 ft below its equilbrium position and released from rest at $t = 0$.
   (a) What is the position of the weight at $t = 5\pi/12$? How fast and which way is it moving at the time?
   (b) At what time is the weight 6 in. above its equilibrium position and moving downward? What is its velocity at such time?

5. A coil spring is such that a 25-lb weight would stretch it 6 in. The spring is suspended from the ceiling, a 16-lb weight is attached to the end of it, and the weight then comes to rest in its equilibrium position. It is then pulled down 4 in. below its equilibrium position and released at $t = 0$ with an initial velocity of 2 ft/sec, directed upward.
   (a) Determine the resulting displacement of the weight as a function of the time.
   (b) Find the amplitude, period, and frequency of the resulting motion.
   (c) At what time does the weight first pass through its equilibrium position and what is its velocity at this instant?

6. An 8-lb weight is attached to the end of a coil spring suspended from a beam and comes to rest in its equilibrium position. The weight is then pulled down A feet below its equilibrium position and released at $t = 0$ with an initial velocity of 3 ft/sec, directed downward. Determine the spring constant k and the constant A if the amplitude of the resulting motion is $\sqrt{\frac{10}{2}}$ and the period is $\pi/2$.

7. An 8-lb weight is placed at the end of a coil spring suspended from the ceiling. After coming to rest in its equilibrium position, the weight is set into vertical motion and the period of the resulting motion is 4 sec. After a time this motion is stopped, and the 8-lb weight is replaced by another weight. After this other weight has come to

rest in its equilibrium position, it is set into vertical motion. If the period of this new motion is 6 sec, how heavy is the second weight?

8. A simple pendulum is composed of a mass m (the bob) at the end of a straight wire of negligible mass and length l. It is suspended from a fixed point S (its point of support) and is free to vibrate in a vertical plane (see Figure 5.4). Let SP denote the straight wire; and let θ denote the angle which SP makes with the vertical SP_0 at time t, positive when measured counterclockwise. We neglect air resistance and assume that only two forces act on the mass m: F_1, the tension in the wire; and F_2, the force due to gravity, which acts vertically downward and is of magnitude mg. We write $F_2 = F_T + F_N$, where F_T is the component of F_2 along the tangent to the path of m and F_N is the component of F_2 normal to F_T. Then $F_N = -F_1$ and $F_T = -mg \sin \theta$, and so the net force acting on m is $F_1 + F_2 = F_1 + F_T + F_N = -mg \sin \theta$, along the arc P_0P. Letting s denote the length of the arc P_0P, the acceleration along this arc is d^2s/dt^2. Hence applying Newton's second law, we have $m\, d^2s/dt^2 = -mg \sin \theta$. But since $s = l\theta$, this reduces to the differential equation

$$ml \frac{d^2\theta}{dt^2} = -mg \sin \theta \quad \text{or} \quad \frac{d^2\theta}{dt^2} + \frac{g}{l} \sin \theta = 0.$$

(a) The equation

$$\frac{d^2\theta}{dt^2} + \frac{g}{l} \sin \theta = 0$$

is a *nonlinear* second-order differential equation. Now recall that

$$\sin \theta = \theta - \frac{\theta^3}{3!} + \frac{\theta^5}{5!} - \cdots.$$

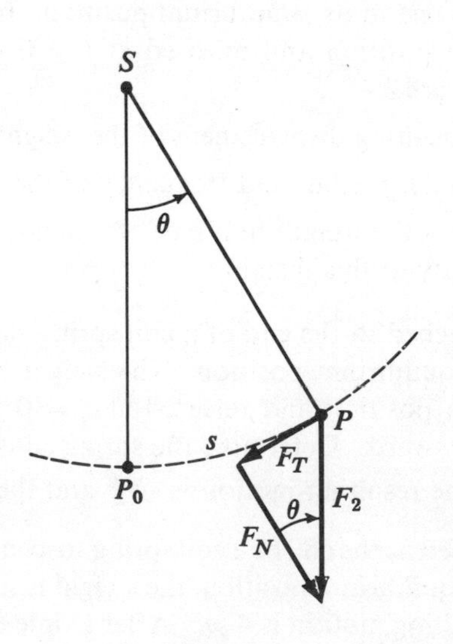

FIGURE 5.4

Hence if θ is sufficiently small, we may replace $\sin\theta$ by θ and consider the *approximate linear* equation

$$\frac{d^2\theta}{dt^2} + \frac{g}{l}\,\theta = 0.$$

Assume that $\theta = \theta_0$ and $d\theta/dt = 0$ when $t = 0$. Obtain the solution of this approximate equation which satisfies these initial conditions and find the amplitude and period of the resulting solution. Observe that this period is independent of the initial displacement.

(b) Now return to the nonlinear equation

$$\frac{d^2\theta}{dt^2} + \frac{g}{l}\sin\theta = 0.$$

Multiply through by $2\,d\theta/dt$, integrate, and apply the initial condition $\theta = \theta_0$, $d\theta/dt = 0$. Then separate variables in the resulting equation to obtain

$$\frac{d\theta}{\sqrt{\cos\theta - \cos\theta_0}} = \pm\sqrt{\frac{2g}{l}}\,dt.$$

From this equation determine the angular velocity $d\theta/dt$ as a function of θ. Note that the left member cannot be integrated in terms of elementary functions to obtain the exact solution $\theta(t)$ of the nonlinear differential equation.

5.3 Free, Damped Motion

We now consider the effect of the resistance of the medium upon the mass on the spring. Still assuming that no external forces are present, this is then the case of *free, damped motion.* Hence with the damping coefficient $a > 0$ and $F(t) = 0$ for all t, the basic differential equation (5.7) reduces to

$$m\frac{d^2x}{dt^2} + a\frac{dx}{dt} + kx = 0. \tag{5.27}$$

Dividing through by m and putting $k/m = \lambda^2$ and $a/m = 2b$ (for convenience) we have the differential equation (5.27) in the form

$$\frac{d^2x}{dt^2} + 2b\frac{dx}{dt} + \lambda^2 x = 0. \tag{5.28}$$

Observe that since a is positive, b is also positive. The auxiliary equation is

$$r^2 + 2br + \lambda^2 = 0. \tag{5.29}$$

Using the quadratic formula we find that the roots of (5.29) are

$$\frac{-2b \pm \sqrt{4b^2 - 4\lambda^2}}{2} = -b \pm \sqrt{b^2 - \lambda^2}. \tag{5.30}$$

Three distinct cases occur, depending upon the nature of these roots, which in turn depends upon the sign of $b^2 - \lambda^2$.

Case 1. Damped, Oscillatory Motion. Here we consider the case in which $b < \lambda$, which implies that $b^2 - \lambda^2 < 0$. Then the roots (5.30) are the conjugate complex numbers $-b \pm \sqrt{\lambda^2 - b^2}\, i$ and the general solution of Equation (5.28) in thus

$$x = e^{-bt}(c_1 \sin \sqrt{\lambda^2 - b^2}\, t + c_2 \cos \sqrt{\lambda^2 - b^2}\, t), \tag{5.31}$$

where c_1 and c_2 are arbitrary constants. We may write this in the alternative form

$$x = ce^{-bt} \cos (\sqrt{\lambda^2 - b^2}\, t + \phi), \tag{5.32}$$

where $c = \sqrt{c_1^2 + c_2^2} > 0$ and ϕ is determined by the equations

$$\frac{c_1}{\sqrt{c_1^2 + c_2^2}} = -\sin \phi,$$

$$\frac{c_2}{\sqrt{c_1^2 + c_2^2}} = \cos \phi.$$

The right member of Equation (5.32) consists of two factors,

$$ce^{-bt} \quad \text{and} \quad \cos (\sqrt{\lambda^2 - b^2}\, t + \phi).$$

The factor ce^{-bt} is called the *damping factor*, or *time-varying amplitude*. Since $c > 0$, it is positive; and since $b > 0$, it tends to zero monotonically as $t \to \infty$. In other words, as time goes on this positive factor becomes smaller and smaller and eventually becomes negligible. The remaining factor, $\cos (\sqrt{\lambda^2 - b^2}\, t + \phi)$, is, of course, of a periodic, oscillatory character; indeed it represents a simple harmonic motion. The product of these two factors, which is precisely the right member of Equation (5.32), therefore represents an oscillatory motion in which the oscillations become successively smaller and smaller. The oscillations are said to be "damped out," and the motion is described as *damped, oscillatory motion.* Of course, the motion is no longer periodic, but the time interval between two successive (positive) maximum displacements is still referred to as the *period.* This is given by

$$\frac{2\pi}{\sqrt{\lambda^2 - b^2}}.$$

The graph of such a motion is shown in Figure 5.5, in which the damping factor ce^{-bt} and its negative are indicated by dashed curves.

The ratio of the amplitude at any time T to that at time $T - (2\pi/\sqrt{\lambda^2 - b^2})$ one period before T is the constant

$$\exp\left(-\frac{2\pi b}{\sqrt{\lambda^2 - b^2}}\right).$$

Thus the quantity $2\pi b/\sqrt{\lambda^2 - b^2}$ is the decrease in the logarithm of the amplitude ce^{-bt} over a time interval of one period. It is called the *logarithmic decrement.*

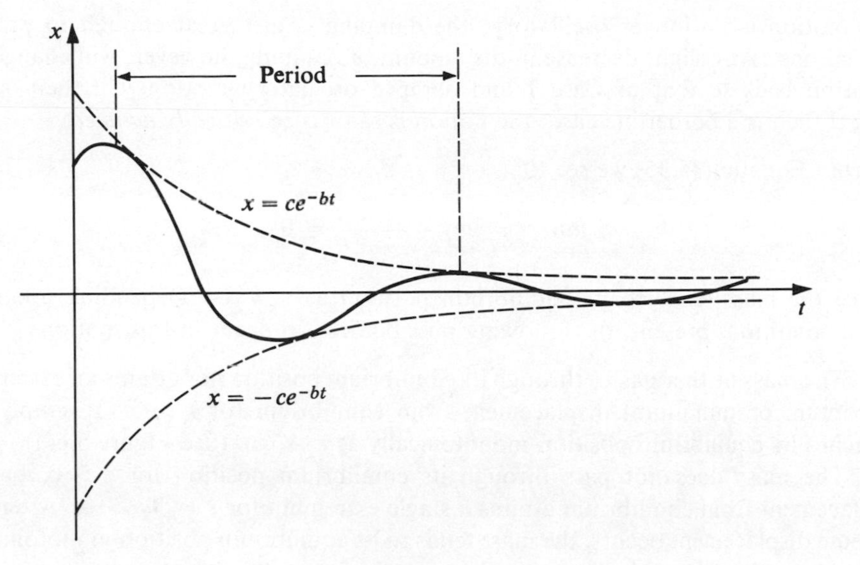

FIGURE 5.5

If we now return to the original notation of the differential equation (5.27), we see from Equation (5.32) that in terms of the original constants m, a, and k, the general solution of (5.27) is

$$x = ce^{-(a/2m)t} \cos\left(\sqrt{\frac{k}{m} - \frac{a^2}{4m^2}}\, t + \phi\right). \tag{5.33}$$

Since $b < \lambda$ is equivalent to $a/2m < \sqrt{k/m}$, we can say that the general solution of (5.27) is given by (5.33) and that damped, oscillatory motion occurs when $a < 2\sqrt{km}$. The frequency of the oscillations

$$\cos\left(\sqrt{\frac{k}{m} - \frac{a^2}{4m^2}}\, t + \phi\right) \tag{5.34}$$

is

$$\frac{1}{2\pi}\sqrt{\frac{k}{m} - \frac{a^2}{4m^2}}.$$

If damping were not present, a would equal zero and the natural frequency of an undamped system would be $(1/2\pi)\sqrt{k/m}$. Thus the frequency of the oscillations (5.34) in the damped oscillatory motion (5.33) is less than the natural frequency of the corresponding undamped system.

Case 2. Critical Damping. This is the case in which $b = \lambda$, which implies that $b^2 - \lambda^2 = 0$. The roots (5.30) are thus both equal to the real negative number $-b$, and the general solution of Equation (5.28) is thus

$$x = (c_1 + c_2 t)e^{-bt}. \tag{5.35}$$

The motion is no longer oscillatory; the damping is just great enough to prevent oscillations. Any slight decrease in the amount of damping, however, will change the situation back to that of Case 1 and damped oscillatory motion will then occur. Case 2 then is a borderline case; the notion is said to be *critically damped.*

From Equation (5.35) we see that

$$\lim_{t\to\infty} x = \lim_{t\to\infty} \frac{c_1 + c_2 t}{e^{bt}} = 0.$$

Hence the mass tends to its equilibrium position as $t \to \infty$. Depending upon the initial conditions present, the following possibilities can occur in this motion:

1. The mass neither passes through its equilibrium position nor attains an extremum (maximum or minimum) displacement from equilibrium for $t > 0$. It simply approaches its equilibrium position monotonically as $t \to \infty$. (See Figure 5.6(a).)
2. The mass does not pass through its equilibrium position for $t > 0$, but its displacement from equilibrium attains a single extremum for $t = T_1 > 0$. After this extreme displacement occurs, the mass tends to its equilibrium position monotonically as $t \to \infty$. (See Figure 5.6(b).)
3. The mass passes through its equilibrium position once at $t = T_2 > 0$ and then attains an extreme displacement at $t = T_3 > T_2$, following which it tends to its equilibrium position monotonically as $t \to \infty$. (See Figure 5.6(c).)

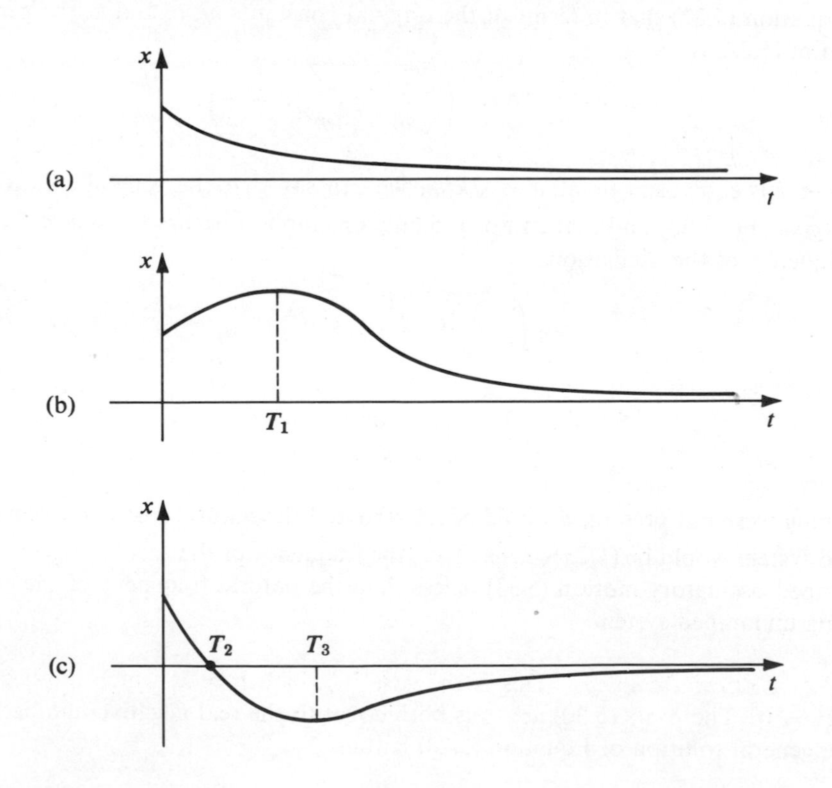

FIGURE 5.6

Case 3. Overcritical Damping. Finally, we consider here the case in which $b > \lambda$, which implies that $b^2 - \lambda^2 > 0$. Here the roots (5.30) are the distinct, real negative numbers

$$r_1 = -b + \sqrt{b^2 - \lambda^2}$$

and

$$r_2 = -b - \sqrt{b^2 - \lambda^2}.$$

The general solution of (5.28) in this case is

$$x = c_1 e^{r_1 t} + c_2 e^{r_2 t}. \tag{5.36}$$

The damping is now so great that no oscillations can occur. Further, we can no longer say that *every* decrease in the amount of damping will result in oscillations, as we could in Case 2. The motion here is said to be *overcritically damped* (or simply *overdamped*).

Equation (5.36) shows us at once that the displacement x approaches zero as $t \to \infty$. As in Case 2 this approach to zero is monotonic for t sufficiently large. Indeed, the three possible motions in Cases 2 and 3 are qualitatively the same. Thus the three motions illustrated in Figure 5.6 can also serve to illustrate the three types of motion possible in Case 3.

▶ Example 5.2. A 32-lb weight is attached to the lower end of a coil spring suspended from the ceiling. The weight comes to rest in its equilibrium position, thereby stretching the spring 2 ft. The weight is then pulled down 6 in. below its equilibrium position and released at $t = 0$. No external forces are present; but the resistance of the medium in pounds is numerically equal to $4(dx/dt)$, where dx/dt is the instantaneous velocity in feet per second. Determine the resulting motion of the weight on the spring.

Formulation. This is a free, damped motion and Equation (5.27) applies. Since the 32-lb weight stretches the spring 2 ft, Hooke's law, $F = ks$, gives $32 = k(2)$ and so $k = 16$ lb/ft. Thus, since $m = w/g = \frac{32}{32} = 1$ (slug), and the damping constant $a = 4$, Equation (5.27) becomes

$$\frac{d^2x}{dt^2} + 4\frac{dx}{dt} + 16x = 0. \tag{5.37}$$

The initial conditions are

$$\begin{aligned} x(0) &= \tfrac{1}{2}, \\ x'(0) &= 0. \end{aligned} \tag{5.38}$$

Solution. The auxiliary equation of Equation (5.37) is

$$r^2 + 4r + 16 = 0.$$

Its roots are the conjugate complex numbers $-2 \pm 2\sqrt{3}\,i$. Thus the general solution of (5.37) may be written

$$x = e^{-2t}(c_1 \sin 2\sqrt{3}\,t + c_2 \cos 2\sqrt{3}\,t), \tag{5.39}$$

where c_1 and c_2 are arbitrary constants. Differentiating (5.39) with respect to t we obtain

$$\frac{dx}{dt} = e^{-2t}[(-2c_1 - 2\sqrt{3}\,c_2)\sin 2\sqrt{3}\,t + (2\sqrt{3}\,c_1 - 2c_2)\cos 2\sqrt{3}\,t]. \tag{5.40}$$

Applying the initial conditions (5.38) to Equations (5.39) and (5.40), we obtain

$$c_2 = \tfrac{1}{2},$$

$$2\sqrt{3}\,c_1 - 2c_2 = 0.$$

Thus $c_1 = \sqrt{3}/6$, $c_2 = \frac{1}{2}$ and the solution is

$$x = e^{-2t}\left(\frac{\sqrt{3}}{6}\sin 2\sqrt{3}\,t + \frac{1}{2}\cos 2\sqrt{3}\,t\right). \tag{5.41}$$

Let us put this in the alternative form (5.32). We have

$$\frac{\sqrt{3}}{6}\sin 2\sqrt{3}\,t + \frac{1}{2}\cos 2\sqrt{3}\,t = \frac{\sqrt{3}}{3}\left[\frac{1}{2}\sin 2\sqrt{3}\,t + \frac{\sqrt{3}}{2}\cos 2\sqrt{3}\,t\right]$$

$$= \frac{\sqrt{3}}{3}\cos\left(2\sqrt{3}\,t - \frac{\pi}{6}\right).$$

Thus the solution (5.41) may be written

$$x = \frac{\sqrt{3}}{3}e^{-2t}\cos\left(2\sqrt{3}\,t - \frac{\pi}{6}\right). \tag{5.42}$$

Interpretation. This a *damped oscillatory motion* (Case 1). The damping factor is $(\sqrt{3}/3)e^{-2t}$, the "period" is $2\pi/2\sqrt{3} = \sqrt{3}\pi/3$, and the logarithmic decrement is $2\sqrt{3}\pi/3$. The graph of the solution (5.42) is shown in Figure 5.7, where the curves $x = \pm(\sqrt{3}/3)e^{-2t}$ are drawn dashed.

▶ Example 5.3. Determine the motion of the weight on the spring described in Example 5.2 if the resistance of the medium in pounds is numerically equal to $8(dx/dt)$ instead of $4(dx/dt)$ (as stated there), all other circumstances being the same as stated in Example 5.2.

Formulation. Once again Equation 5.27 applies, and exactly as in Example 5.2 we find that $m = 1$ (slug) and $k = 16$ lb/ft. But now the damping has increased, and we have $a = 8$. Thus Equation (5.27) now becomes

$$\frac{d^2x}{dt^2} + 8\frac{dx}{dt} + 16x = 0. \tag{5.43}$$

The initial conditions

$$x(0) = \tfrac{1}{2},$$

$$x'(0) = 0, \tag{5.44}$$

are, of course, unchanged from Example 5.2.

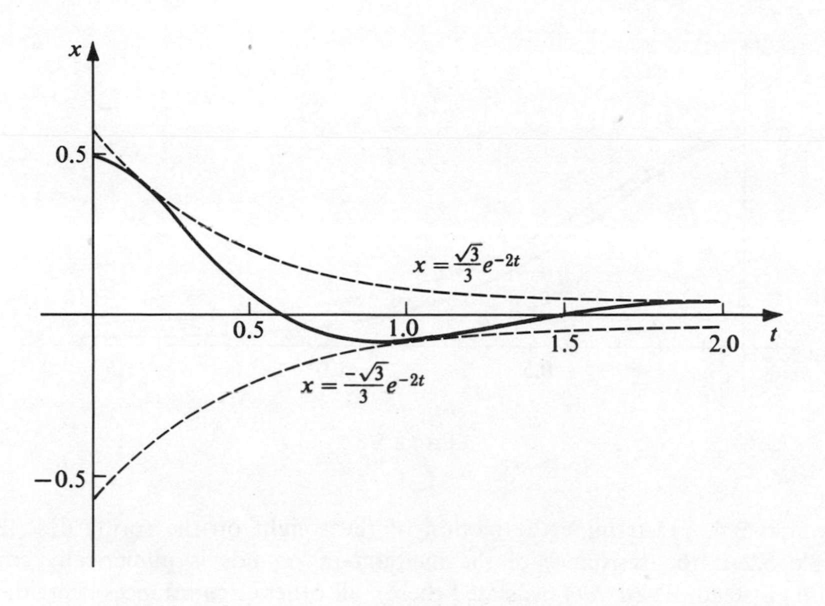

FIGURE 5.7

Solution. The auxiliary equation is now

$$r^2 + 8r + 16 = 0$$

and has the equal roots $r = -4, -4$. The general solution of Equation (5.43) is thus

$$x = (c_1 + c_2 t)e^{-4t}, \tag{5.45}$$

where c_1 and c_2 are arbitrary constants. Differentiating (5.45) with respect to t, we have

$$\frac{dx}{dt} = (c_2 - 4c_1 - 4c_2 t)e^{-4t}. \tag{5.46}$$

Applying the initial conditions (5.44) to Equations (5.45) and (5.46) we obtain

$$c_1 = \tfrac{1}{2},$$
$$c_2 - 4c_1 = 0.$$

Thus $c_1 = \frac{1}{2}$, $c_2 = 2$ and the solution is

$$x = (\tfrac{1}{2} + 2t)e^{-4t}. \tag{5.47}$$

Interpretation. The motion is critically damped. Using (5.47), we see that $x = 0$ if and only if $t = -\frac{1}{4}$. Thus $x \neq 0$ for $t > 0$ and the weight does not pass through its equilibrium position. Differentiating (5.47) one finds at once that $dx/dt < 0$ for all $t > 0$. Thus the displacement of the weight from its equilibrium position is a decreasing function of t for all $t > 0$. In other words, the weight starts to move back toward its equilibrium position at once and $x \to 0$ monotonically as $t \to \infty$. The motion is therefore an example of possibility 1 described in the general discussion of Case 2 above. The graph of the solution (5.47) is shown as the solid curve in Figure 5.8.

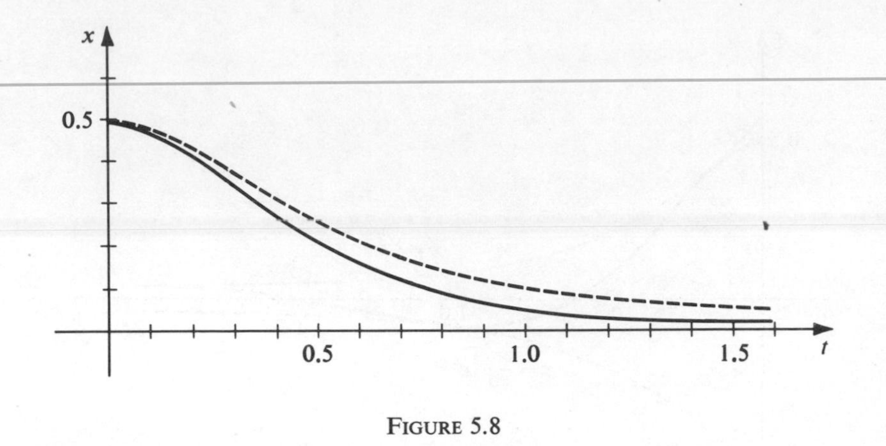

FIGURE 5.8

▶ Example 5.4. Determine the motion of the weight on the spring described in Example 5.2 if the resistance of the medium in pounds is numerically equal to $10(dx/dt)$ instead of $4(dx/dt)$ (as stated there), all other circumstances being the same as stated in Example 5.2.

Formulation. The only difference between this and the two previous examples is in the damping constant. In Example 5.2, $a = 4$; in Example 5.3, $a = 8$; and now we have even greater damping, for here $a = 10$. As before $m = 1$ (slug) and $k = 16$ lb/ft. The differential equation (5.27) thus becomes

$$\frac{d^2x}{dt^2} + 10\frac{dx}{dt} + 16x = 0. \tag{5.48}$$

The initial conditions (5.44) or (5.38) still hold.

Solution. The auxiliary equation is now

$$r^2 + 10r + 16 = 0$$

and its roots are $r = -2, -8$. Thus the general solution of Equation (5.48) is

$$x = c_1e^{-2t} + c_2e^{-8t},$$

where c_1 and c_2 are arbitrary constants. Differentiating this with respect to t to obtain

$$\frac{dx}{dt} = -2c_1e^{-2t} - 8c_2e^{-8t}$$

and applying the initial conditions (5.44), we find the following equations for the determination of c_1 and c_2:

$$c_1 + c_2 = \tfrac{1}{2},$$
$$-2c_1 - 8c_2 = 0.$$

The solution of this system is $c_1 = \frac{2}{3}$, $c_2 = -\frac{1}{6}$; thus the solution of the problem is

$$x = \tfrac{2}{3}e^{-2t} - \tfrac{1}{6}e^{-8t}. \tag{5.49}$$

Interpretation. Clearly the motion described by Equation (5.49) is an example of the overdamped case (Case 3). Qualitatively the motion is the same as that of the solution (5.47) of Example 5.3. Here, however, due to the increased damping, the weight returns to its equilibrium position at a slower rate. The graph of (5.49) is shown as the dashed curve in Figure 5.8. Note that in each of Examples 5.2, 5.3, and 5.4, all circumstances (the weight, the spring, and the initial conditions) were exactly the same, *except* for the damping. In Example 5.2, the damping constant $a = 4$, and the resulting motion was the damped oscillatory motion shown in Figure 5.7. In Example 5.3 the damping was increased to such an extent ($a = 8$) that oscillations no longer occurred, the motion being shown by the solid curve of Figure 5.8. Finally in Example 5.4 the damping was further increased ($a = 10$) and the resulting motion, indicated by the dashed curve of Figure 5.8, was similar to but slower than that of Example 5.3.

Exercises

1. An 8-lb weight is attached to the lower end of a coil spring suspended from the ceiling and comes to rest in its equilibrium position, thereby stretching the spring 0.4 ft. The weight is then pulled down 6 in. below its equilibrium position and released at $t = 0$. The resistance of the medium in pounds is numerically equal to $2\, dx/dt$, where dx/dt is the instantaneous velocity in feet per second.
   (a) Set up the differential equation for the motion and list the initial conditions.
   (b) Solve the initial-value problem set up in part (a) to determine the displacement of the weight as a function of the time.
   (c) Express the solution found in step (b) in the alternative form (5.32) of the text.
   (d) What is the so-called "period" of the motion?
   (e) Graph the displacement as a function of the time.

2. A 16-lb weight is placed upon the lower end of a coil spring suspended from the ceiling and comes to rest in its equilibrium position, thereby stretching the spring 8 in. At time $t = 0$ the weight is then struck so as to set it into motion with an initial velocity of 2 ft/sec, directed downward. The medium offers a resistance in pounds numerically equal to $6\, dx/dt$, where dx/dt is the instantaneous velocity in feet per second. Determine the resulting displacement of the weight as a function of time and graph this displacement.

3. An 8-lb weight is attached to the lower end of a coil spring suspended from a fixed support. The weight comes to rest in its equilibrium position, thereby stretching the spring 6 in. The weight is then pulled down 9 in. below its equilibrium position and released at $t = 0$. The medium offers a resistance in pounds numerically equal to $4\, dx/dt$, where dx/dt is the instantaneous velocity in feet per second. Determine the displacement of the weight as a function of the time and graph this displacement.

4. A 16-lb weight is attached to the lower end of a coil spring suspended from a fixed support. The weight comes to rest in its equilibrium position, thereby

stretching the spring 6 in. The weight is then pulled down 3 in. below its equilibrium position and released at $t = 0$. The medium offers a resistance in pounds numerically equal to $10\, dx/dt$, where dx/dt is the instantaneous velocity in feet per second. Determine the displacement of the weight as a function of the time and graph this displacement.

5. A spring is such that a force of 20 lb would stretch it 6 in. The spring hangs vertically and a 4-lb weight is attached to the end of it. After this 4-lb weight comes to rest in its equilibrium position it is pulled down 8 in. below this position and then released at $t = 0$. The medium offers a resistance in pounds numerically equal to $2\, dx/dt$, where dx/dt is the instantaneous velocity in feet per second.
   (a) Determine the displacement of the weight as a function of the time and express this displacement in the alternative form (5.32) of the text.
   (b) Find the so-called "period" and determine the logarithmic decrement.
   (c) At what time does the weight first pass through its equilibrium position?

6. A 4-lb weight is hung upon the lower end of a coil spring hanging vertically from a fixed support. The weight comes to rest in its equilibrium position, thereby stretching the spring 8 in. The weight is then pulled down a certain distance below this equilibrium position and released at $t = 0$. The medium offers a resistance in pounds numerically equal to $a\, dx/dt$, where $a > 0$ and dx/dt is the instantaneous velocity in feet per second. Show that the motion is oscillatory if $a < \sqrt{3}$, critically damped if $a = \sqrt{3}$, and overdamped if $a > \sqrt{3}$.

7. A 4-lb weight is attached to the lower end of a coil spring which hangs vertically from a fixed support. The weight comes to rest in its equilibrium position, thereby stretching the spring 6 in. The weight is then pulled down 3 in. below this equilibrium position and released at $t = 0$. The medium offers a resistance in pounds numerically equal to $a\, dx/dt$, where $a > 0$ and dx/dt is the instantaneous velocity in feet per second.
   (a) Determine the value of a such that the resulting motion would be critically damped and determine the displacement for this critical value of a.
   (b) Determine the displacement if a is equal to one-half the critical value found in step (a).
   (c) Determine the displacement if a is equal to twice the critical value found in step (a).

8. A 10-lb weight is attached to the lower end of a coil spring suspended from the ceiling, the spring constant being 20 lb/ft. The weight comes to rest in its equilibrium position. It is then pulled down 6 in. below this position and released at $t = 0$ with an initial velocity of 1 ft/sec, directed downward. The resistance of the medium in pounds is numerically equal to $a\, dx/dt$, where $a > 0$ and dx/dt is the instantaneous velocity in feet per second.
   (a) Determine the smallest value of the damping coefficient a for which the motion is not oscillatory.
   (b) Using the value of a found in part (a) find the displacement of the weight as a function of the time.

(c) Show that the weight attains a single extreme displacement from its equilibrium position at time $t = \frac{1}{40}$, determine this extreme displacement, and show that the weight then tends monotonically to its equilibrium position as $t \to \infty$.

(d) Graph the displacement found in step (b).

9. A 32-lb weight is attached to the lower end of a coil spring suspended from the ceiling. After the weight comes to rest in its equilibrium position, it is then pulled down a certain distance below this position and released at $t = 0$. If the medium offered no resistance, the natural frequency of the resulting undamped motion would be $4/\pi$ cycles per second. However, the medium does offer a resistance in pounds numerically equal to $a\, dx/dt$, where $a > 0$ and dx/dt is the instantaneous velocity in feet per second; and the frequency of the resulting damped oscillatory motion is only half as great as the natural frequency.

 (a) Determine the spring constant k.

 (b) Find the value of the damping coefficient a.

10. The differential equation for the vertical motion of a mass m on a coil spring of spring constant k in a medium in which the damping is proportional to the instantaneous velocity is given by Equation (5.27). In the case of damped oscillatory motion the solution of this equation is given by (5.33). Show that the displacement x so defined attains an extremum (maximum or minimum) at the times t_n ($n = 0, 1, 2, \ldots$) given by

$$t_n = \frac{1}{\omega_1}\left[\arc\tan\left(-\frac{a}{2m\omega_1}\right) + n\pi - \phi\right],$$

where

$$\omega_1 = \sqrt{\frac{k}{m} - \frac{a^2}{4m^2}}.$$

11. The differential equation for the vertical motion of a unit mass on a certain coil spring in a certain medium is

$$\frac{d^2x}{dt^2} + 2b\frac{dx}{dt} + b^2 x = 0,$$

where $b > 0$. The initial displacement of the mass is A feet and its initial velocity is B feet per second.

 (a) Show that the motion is critically damped and that the displacement is given by

$$x = (A + Bt + Abt)e^{-bt}.$$

 (b) If A and B are such that

$$-\frac{A}{B + Ab} \quad \text{and} \quad \frac{B}{b(B + Ab)}$$

 are both negative, show that the mass approaches its equilibrium position monotonically as $t \to \infty$ without either passing through this equilibrium position or attaining an extreme displacement from it for $t > 0$.

(c) If A and B are such that $-\dfrac{A}{B + Ab}$ is negative but $\dfrac{B}{b(B + Ab)}$ is positive, show that the mass does not pass through its equilibrium position for $t > 0$, that its displacement from this position attains a single extremum at $t = \dfrac{B}{b(B + Ab)}$, and that thereafter the mass tends to its equilibrium position monotonically as $t \to \infty$.

(d) If A and B are such that $-\dfrac{A}{B + Ab}$ is positive, show that the mass passes through its equilibrium position at $t = -\dfrac{A}{B + Ab}$, attains an extreme displacement at $t = \dfrac{B}{b(B + Ab)}$, and thereafter tends to its equilibrium position monotonically as $t \to \infty$.

5.4 Forced Motion

We now consider an important special case of *forced motion*. That is, we not only consider the effect of damping upon the mass on the spring but also the effect upon it of a periodic external impressed force F defined by $F(t) = F_1 \cos \omega t$ for all $t \geq 0$, where F_1 and ω are constants. Then the basic differential equation (5.7) becomes

$$m \frac{d^2x}{dt^2} + a \frac{dx}{dt} + kx = F_1 \cos \omega t. \tag{5.50}$$

Dividing through by m and letting

$$\frac{a}{m} = 2b, \quad \frac{k}{m} = \lambda^2, \quad \text{and} \quad \frac{F_1}{m} = E_1,$$

this takes the more convenient form

$$\frac{d^2x}{dt^2} + 2b \frac{dx}{dt} + \lambda^2 x = E_1 \cos \omega t. \tag{5.51}$$

We shall assume that the positive damping constant a is small enough so that the damping is less than critical. In other words we assume that $b < \lambda$. Hence by Equation (5.32) the complementary function of Equation (5.51) can be written

$$x_c = ce^{-bt} \cos (\sqrt{\lambda^2 - b^2}\, t + \phi). \tag{5.52}$$

We shall now find a particular integral of (5.51) by the method of undetermined coefficients. We let

$$x_p = A \cos \omega t + B \sin \omega t. \tag{5.53}$$

Then

$$\frac{dx_p}{dt} = -\omega A \sin \omega t + \omega B \cos \omega t,$$

$$\frac{d^2x_p}{dt^2} = -\omega^2 A \cos \omega t - \omega^2 B \sin \omega t.$$

Substituting into Equation (5.51), we have

$$[-2b\omega A + (\lambda^2 - \omega^2)B] \sin \omega t + [(\lambda^2 - \omega^2)A + 2b\omega B] \cos \omega t = E_1 \cos \omega t.$$

Thus, we have the following two equations from which to determine A and B:

$$-2b\omega A + (\lambda^2 - \omega^2)B = 0,$$
$$(\lambda^2 - \omega^2)A + 2b\omega B = E_1.$$

Solving these, we obtain

$$A = \frac{E_1(\lambda^2 - \omega^2)}{(\lambda^2 - \omega^2)^2 + 4b^2\omega^2}, \quad B = \frac{2b\omega E_1}{(\lambda^2 - \omega^2)^2 + 4b^2\omega^2}. \tag{5.54}$$

Substituting these values into Equation (5.53), we obtain a particular integral in the form

$$x_p = \frac{E_1}{(\lambda^2 - \omega^2)^2 + 4b^2\omega^2}[(\lambda^2 - \omega^2) \cos \omega t + 2b\omega \sin \omega t].$$

We now put this in the alternative "phase angle" form; we write

$$(\lambda^2 - \omega^2) \cos \omega t + 2b\omega \sin \omega t$$
$$= \sqrt{(\lambda^2 - \omega^2)^2 + 4b^2\omega^2}\left[\frac{\lambda^2 - \omega^2}{\sqrt{(\lambda^2 - \omega^2)^2 + 4b^2\omega^2}} \cos \omega t + \frac{2b\omega}{\sqrt{(\lambda^2 - \omega^2)^2 + 4b^2\omega^2}} \sin \omega t\right]$$
$$= \sqrt{(\lambda^2 - \omega^2)^2 + 4b^2\omega^2}\,[\cos \omega t \cos \theta + \sin \omega t \sin \theta]$$
$$= \sqrt{(\lambda^2 - \omega^2)^2 + 4b^2\omega^2}\,\cos (\omega t - \theta),$$

where

$$\cos \theta = \frac{\lambda^2 - \omega^2}{\sqrt{(\lambda^2 - \omega^2)^2 + 4b^2\omega^2}}, \quad \sin \theta = \frac{2b\omega}{\sqrt{(\lambda^2 - \omega^2)^2 + 4b^2\omega^2}}. \tag{5.55}$$

Thus the particular integral appears in the form

$$x_p = \frac{E_1}{\sqrt{(\lambda^2 - \omega^2)^2 + 4b^2\omega^2}} \cos (\omega t - \theta), \tag{5.56}$$

where θ is determined from Equations (5.55). Thus, using (5.52) and (5.56) the general solution of Equation (5.51) is

$$x = x_c + x_p = ce^{-bt} \cos (\sqrt{\lambda^2 - b^2}\, t + \phi) + \frac{E_1}{\sqrt{(\lambda^2 - \omega^2)^2 + 4b^2\omega^2}} \cos (\omega t - \theta). \tag{5.57}$$

Observe that this solution is the sum of two terms. The first term, $ce^{-bt}\cos(\sqrt{\lambda^2 - b^2}\,t + \phi)$, represents the damped oscillation which would be the entire motion of the system if the external force $F_1 \cos \omega t$ were not present. The second term,

$$\frac{E_1}{\sqrt{(\lambda^2 - \omega^2)^2 + 4b^2\omega^2}} \cos(\omega t - \theta),$$

which results from the presence of the external force, represents a simple harmonic motion of period $2\pi/\omega$. Because of the damping factor ce^{-bt} the contribution of the first term will become smaller and smaller as time goes on and will eventually become negligible. The first term is thus called the *transient* term. The second term, however, being a cosine term of constant amplitude, continues to contribute to the motion in a periodic, oscillatory manner. Eventually, the transient term having become relatively small, the entire motion will consist essentially of that given by this second term. This second term is thus called the *steady-state* term.

▶ Example 5.5. A 16-lb weight is attached to the lower end of a coil spring suspended from the ceiling, the spring constant of the spring being 10 lb/ft. The weight comes to rest in its equilibrium position. Beginning at $t = 0$ an external force given by $F(t) = 5\cos 2t$ is applied to the system. Determine the resulting motion if the damping force in pounds is numerically equal to $2(dx/dt)$, where dx/dt is the instantaneous velocity in feet per second.

Formulation. The basic differential equation for the motion is

$$m\frac{d^2x}{dt^2} + a\frac{dx}{dt} + kx = F(t). \tag{5.58}$$

Here $m = w/g = \frac{16}{32} = \frac{1}{2}$ (slug), $a = 2$, $k = 10$, and $F(t) = 5\cos 2t$. Thus Equation (5.58) becomes

$$\frac{1}{2}\frac{d^2x}{dt^2} + 2\frac{dx}{dt} + 10x = 5\cos 2t$$

or

$$\frac{d^2x}{dt^2} + 4\frac{dx}{dt} + 20x = 10\cos 2t. \tag{5.59}$$

The initial conditions are

$$x(0) = 0, \quad x'(0) = 0. \tag{5.60}$$

Solution. The auxiliary equation of the homogeneous equation corresponding to (5.59) is $r^2 + 4r + 20 = 0$; its roots are $-2 \pm 4i$. Thus the complementary function of Equation (5.59) is

$$x_c = e^{-2t}(c_1 \sin 4t + c_2 \cos 4t),$$

where c_1 and c_2 are arbitrary constants. Using the method of undetermined coefficients to obtain a particular integral, we let

$$x_p = A \cos 2t + B \sin 2t.$$

Upon differentiating and substituting into (5.59), we find the following equations for the determination of A and B.

$$-8A + 16B = 0,$$

$$16A + 8B = 10.$$

Solving these, we find

$$A = \tfrac{1}{2}, \qquad B = \tfrac{1}{4}.$$

Thus a particular integral is

$$x_p = \tfrac{1}{2} \cos 2t + \tfrac{1}{4} \sin 2t$$

and the general solution of (5.59) is

$$x = x_c + x_p = e^{-2t}(c_1 \sin 4t + c_2 \cos 4t) + \tfrac{1}{2} \cos 2t + \tfrac{1}{4} \sin 2t. \tag{5.61}$$

Differentiating (5.61) with respect to t, we obtain

$$\frac{dx}{dt} = e^{-2t}[(-2c_1 - 4c_2) \sin 4t + (-2c_2 + 4c_1) \cos 4t] - \sin 2t + \tfrac{1}{2} \cos 2t. \tag{5.62}$$

Applying the initial conditions (5.60) to Equations (5.61) and (5.62), we see that

$$c_2 + \tfrac{1}{2} = 0,$$

$$4c_1 - 2c_2 + \tfrac{1}{2} = 0.$$

From these equations we find that

$$c_1 = -\tfrac{3}{8}, \qquad c_2 = -\tfrac{1}{2}.$$

Hence the solution is

$$x = e^{-2t}(-\tfrac{3}{8} \sin 4t - \tfrac{1}{2} \cos 4t) + \tfrac{1}{2} \cos 2t + \tfrac{1}{4} \sin 2t. \tag{5.63}$$

Let us write this in the "phase angle" form. We have first

$$3 \sin 4t + 4 \cos 4t = 5(\tfrac{3}{5} \sin 4t + \tfrac{4}{5} \cos 4t) = 5 \cos (4t - \phi),$$

where

$$\cos \phi = \tfrac{4}{5}, \qquad \sin \phi = \tfrac{3}{5}. \tag{5.64}$$

Also, we have

$$2 \cos 2t + \sin 2t = \sqrt{5}\left(\frac{2}{\sqrt{5}} \cos 2t + \frac{1}{\sqrt{5}} \sin 2t\right) = \sqrt{5} \cos (2t - \theta),$$

where

$$\cos \theta = \frac{2}{\sqrt{5}}, \qquad \sin \theta = \frac{1}{\sqrt{5}}. \tag{5.65}$$

Thus we may write the solution (5.63) as

$$x = -\frac{5e^{-2t}}{8} \cos (4t - \phi) + \frac{\sqrt{5}}{4} \cos (2t - \theta), \tag{5.66}$$

where ϕ and θ are determined by Equations (5.64) and (5.65), respectively. We find that $\phi \approx 0.64$ (rad) and $\theta \approx 0.46$ (rad). Thus the solution (5.66) is given approximately by

$$x = -0.63e^{-2t} \cos (4t - 0.64) + 0.56 \cos (2t - 0.46).$$

Interpretation. The term

$$-\frac{5e^{-2t}}{8} \cos (4t - \phi) \approx -0.63e^{-2t} \cos (4t - 0.64)$$

is the *transient* term, representing a damped oscillatory motion. It becomes negligible in a short time; for example, for $t > 3$, its numerical value is less than 0.002. Its graph is shown in Figure 5.9(a). The term

$$\frac{\sqrt{5}}{4} \cos (2t - \theta) \approx 0.56 \cos (2t - 0.46)$$

is the *steady-state* term, representing a simple harmonic motion of amplitude

$$\frac{\sqrt{5}}{4} \approx 0.56$$

and period π. Its graph appears in Figure 5.9(b). The graph in Figure 5.9(c) is that of the complete solution (5.66). It is clear from this that the effect of the transient term soon becomes negligible, and that after a short time the contribution of the steady-state term is essentially all that remains.

Exercises

1. A 6-lb weight is attached to the lower end of a coil spring suspended from the ceiling, the spring constant of the spring being 27 lb/ft. The weight comes to rest in its equilibrium position, and beginning at $t = 0$ an external force given by $F(t) = 12 \cos 20t$ is applied to the system. Determine the resulting displacement as a function of the time, assuming damping is negligible.

2. A 16-lb weight is attached to the lower end of a coil spring suspended from the ceiling. The weight comes to rest in its equilibrium position, thereby stretching the spring 0.4 ft. Then, beginning at $t = 0$, an external force given by $F(t) = 40 \cos 16t$ is applied to the system. The medium offers a resistance in pounds numerically equal to $4\, dx/dt$, where dx/dt is the instantaneous velocity in feet per second.
   (a) Find the displacement of the weight as a function of the time.
   (b) Graph separately the transient and steady-state terms of the motion found in step (a) and then use the curves so obtained to graph the entire displacement itself.

3. A 10-lb weight is hung on the lower end of a coil spring suspended from the ceiling, the spring constant of the spring being 20 lb/ft. The weight comes to rest in its equilibrium position, and beginning at $t = 0$ an external force given by

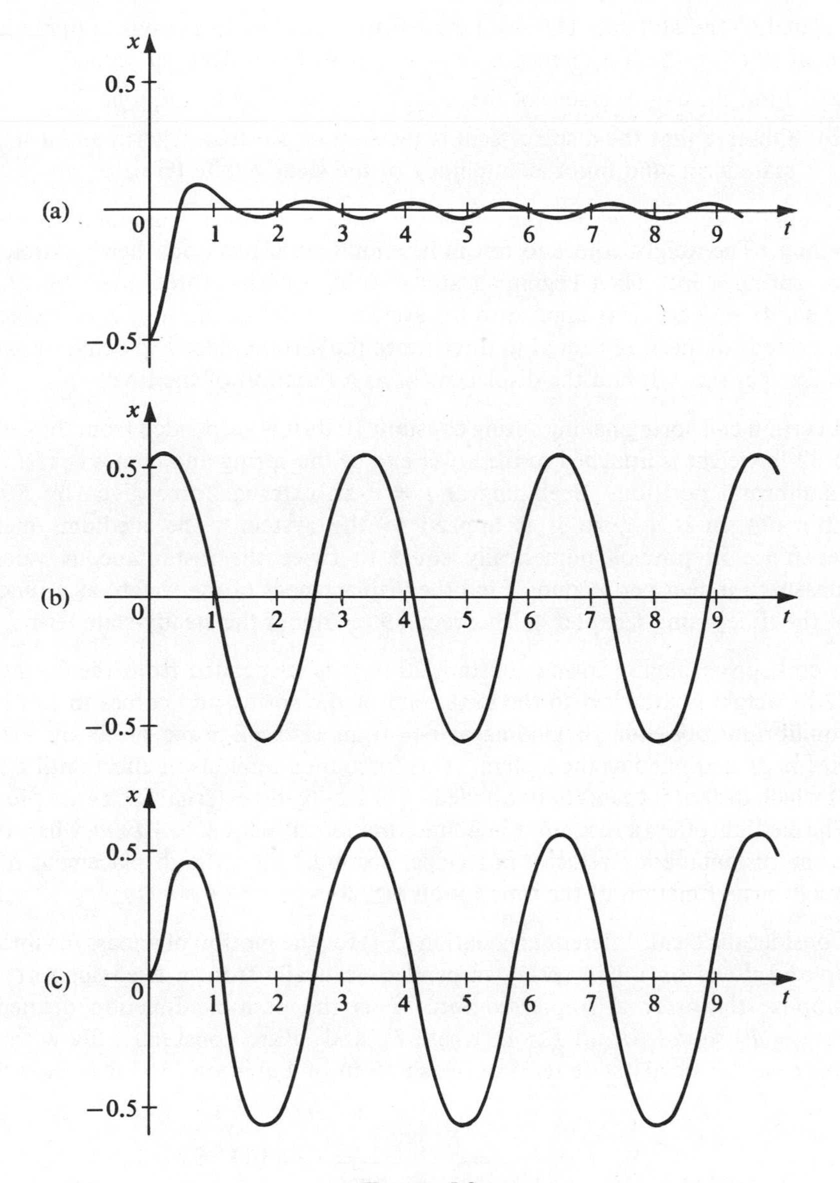

FIGURE 5.9

$F(t) = 10 \cos 8t$ is applied to the system. The medium offers a resistance in pounds numerically equal to $5\, dx/dt$, where dx/dt is the instantaneous velocity in feet per second. Find the displacement of the weight as a function of the time.

4. A 4-lb weight is hung on the lower end of a coil spring suspended from a beam. The weight comes to rest in its equilibrium position, thereby stretching the spring 3 in. The weight is then pulled down 6 in. below this position and released at $t = 0$. At this instant an external force given by $F(t) = 13 \sin 4t$ is

applied to the system. The resistance of the medium in pounds is numerically equal to twice the instantaneous velocity, measured in feet per second.

(a) Find the displacement of the weight as a function of the time.

(b) Observe that the displacement is the sum of a transient term and a steady-state term, and find the amplitude of the steady-state term.

5. A 6-lb weight is hung on the lower end of a coil spring suspended from the ceiling. The weight comes to rest in its equilibrium position, thereby stretching the spring 4 in. Then beginning at $t = 0$ an external force given by $F(t) = 27 \sin 4t - 3 \cos 4t$ is applied to the system. If the medium offers a resistance in pounds numerically equal to three times the instantaneous velocity, measured in feet per second, find the displacement as a function of the time.

6. A certain coil spring having spring constant 10 lb/ft is suspended from the ceiling. A 32-lb weight is attached to the lower end of the spring and comes to rest in its equilibrium position. Beginning at $t = 0$ an external force given by $F(t) = \sin t + \frac{1}{4} \sin 2t + \frac{1}{9} \sin 3t$ is applied to the system. The medium offers a resistance in pounds numerically equal to twice the instantaneous velocity, measured in feet per second. Find the displacement of the weight as a function of the time, using Chapter 4, Theorem 4.9 to obtain the steady-state term.

7. A coil spring having spring constant 20 lb/ft is suspended from the ceiling. A 32-lb weight is attached to the lower end of the spring and comes to rest in its equilibrium position. Beginning at $t = 0$ an external force given by $F(t) = 40 \cos 2t$ is applied to the system. This force then remains in effect until $t = \pi$, at which instant it ceases to be applied. For $t > \pi$, no external forces are present. The medium offers a resistance in pounds numerically equal to $4\,dx/dt$, where dx/dt is the instantaneous velocity in feet per second. Find the displacement of the weight as a function of the time for all $t \geq 0$.

8. Consider the basic differential equation (5.7) for the motion of a mass m vibrating up and down on a coil spring suspended vertically from a fixed support; and suppose the external impressed force F is the periodic function defined by $F(t) = F_2 \sin \omega t$ for all $t \geq 0$, where F_2 and ω are constants. Show that in this case the steady-state term in the solution of Equation (5.7) may be written

$$x_p = \frac{E_2}{\sqrt{(\lambda^2 - \omega^2)^2 + 4b^2\omega^2}} \sin (\omega t - \theta),$$

where $b = a/2m$, $\lambda^2 = k/m$, $E_2 = F_2/m$, and θ is determined from Equations (5.55).

9. A 32-lb weight is attached to the lower end of a coil spring suspended vertically from a fixed support and comes to rest in its equilibrium position, thereby stretching the spring 6 in. Beginning at $t = 0$ an external force given by $F(t) = 15 \cos 7t$ is applied to the system. Assume that the damping is negligible.

(a) Find the displacement of the weight as a function of the time.

(b) Show that this displacement may be expressed as $x = A(t) \sin (15t/2)$, where $A(t) = 2 \sin \frac{1}{2}t$. The function $A(t)$ may be regarded as the "slowly varying" amplitude of the more rapid oscillation $\sin (15t/2)$. When a phenomenon involving such fluctuations in maximum amplitude takes place in acoustical applications, *beats* are said to occur.

(c) Carefully graph the slowly varying amplitude $A(t) = 2 \sin (t/2)$ and its negative $-A(t)$ and then use these "bounding curves" to graph the displacement $x = A(t) \sin (15t/2)$.

10. A 16-lb weight is attached to the lower end of a coil spring which is suspended vertically from a support and for which the spring constant k is 10 lb/ft. The weight comes to rest in its equilibrium position and is then pulled down 6 in. below this position and released at $t = 0$. At this instant the support of the spring begins a vertical oscillation such that its distance from its initial position is given by $\frac{1}{2} \sin 2t$ for $t \geq 0$. The resistance of the medium in pounds is numerically equal to $2\, dx/dt$, where dx/dt is the instantaneous velocity of the moving weight in feet per second.

(a) Show that the differential equation for the displacement of the weight from its equilibrium position is

$$\frac{1}{2}\frac{d^2x}{dt^2} = -10(x - y) - 2\frac{dx}{dt}, \qquad \text{where } y = \frac{1}{2}\sin 2t,$$

and hence that this differential equation may be written

$$\frac{d^2x}{dt^2} + 4\frac{dx}{dt} + 20x = 10 \sin 2t.$$

(b) Solve the differential equation obtained in step (a), apply the relevant initial conditions, and thus obtain the displacement x as a function of time.

5.5 Resonance Phenomena

We now consider the amplitude of the steady-state vibration which results from the periodic external force defined for all t by $F(t) = F_1 \cos \omega t$, where we assume that $F_1 > 0$. For fixed b, λ, and E_1 we see from Equation (5.56) that this is the function f of ω defined by

$$f(\omega) = \frac{E_1}{\sqrt{(\lambda^2 - \omega^2)^2 + 4b^2\omega^2}}. \tag{5.67}$$

If $\omega = 0$, the force $F(t)$ is the constant F_1 and the amplitude $f(\omega)$ has the value $E_1/\lambda^2 > 0$. Also, as $\omega \to \infty$, we see from (5.67) that $f(\omega) \to 0$. Let us consider the function f for $0 < \omega < \infty$. Calculating the derivative $f'(\omega)$ we find that this derivative equals zero only if

$$4\omega[2b^2 - (\lambda^2 - \omega^2)] = 0$$

and hence only if $\omega = 0$ or $\omega = \sqrt{\lambda^2 - 2b^2}$. If $\lambda^2 < 2b^2$, $\sqrt{\lambda^2 - 2b^2}$ is a complex number. Hence in this case f has no extremum for $0 < \omega < \infty$, but rather f decreases monotonically for $0 < \omega < \infty$ from the value E_1/λ^2 at $\omega = 0$ and approaches zero as $\omega \to \infty$. Let us assume that $\lambda^2 > 2b^2$. Then the function f has a relative maximum at $\omega_1 = \sqrt{\lambda^2 - 2b^2}$, and this maximum value is given by

$$f(\omega_1) = \frac{E_1}{\sqrt{(2b^2)^2 + 4b^2(\lambda^2 - 2b^2)}} = \frac{E_1}{2b\sqrt{\lambda^2 - b^2}}.$$

When the frequency of the forcing function $F_1 \cos \omega t$ is such that $\omega = \omega_1$, then the forcing function is said to be in *resonance* with the system. In other words, the forcing function defined by $F_1 \cos \omega t$ is in resonance with the system when ω assumes the value ω_1 at which $f(\omega)$ is a maximum. The value $\omega_1/2\pi$ is called the *resonance frequency* of the system. Note carefully that resonance can occur only if $\lambda^2 > 2b^2$. Since then $\lambda^2 > b^2$, the damping must be less than critical in such a case.

We now return to the original notation of Equation (5.50). In terms of the quantities m, a, k, and F_1 of that equation, the function f is given by

$$f(\omega) = \frac{\dfrac{F_1}{m}}{\sqrt{\left(\dfrac{k}{m} - \omega^2\right)^2 + \left(\dfrac{a}{m}\right)^2 \omega^2}}. \tag{5.68}$$

In this original notation the resonance frequency is

$$\frac{1}{2\pi}\sqrt{\frac{k}{m} - \frac{a^2}{2m^2}}. \tag{5.69}$$

Since the frequency of the corresponding free, damped oscillation is

$$\frac{1}{2\pi}\sqrt{\frac{k}{m} - \frac{a^2}{4m^2}},$$

we see that the resonance frequency is less than that of the corresponding free, damped oscillation.

The graph of $f(\omega)$ is called the *resonance curve* of the system. For a given system with fixed m, k, and F_1, there is a resonance curve corresponding to each value of the damping coefficient $a \geq 0$. Let us choose $m = k = F_1 = 1$, for example, and graph the resonance curves corresponding to certain selected values of a. In this case we have

$$f(\omega) = \frac{1}{\sqrt{(1 - \omega^2)^2 + a^2\omega^2}}$$

and the resonance frequency is given by $(1/2\pi)\sqrt{1 - a^2/2}$. The graphs appear in Figure 5.10.

Observe that resonance occurs in this case only if $a < \sqrt{2}$. As a decreases from $\sqrt{2}$ to 0, the value ω_1 at which resonance occurs increases from 0 to 1 and the corresponding maximum value of $f(\omega)$ becomes larger and larger. In the limiting case $a = 0$, the

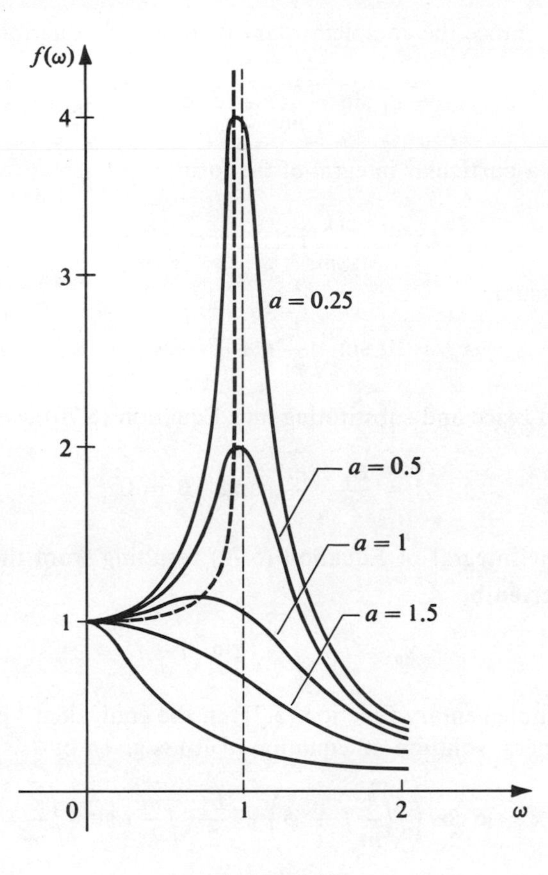

FIGURE 5.10

maximum has disappeared and an infinite discontinuity occurs at $\omega = 1$. In this case our solution actually breaks down, for then

$$f(\omega) = \frac{1}{\sqrt{(1 - \omega^2)^2}} = \frac{1}{1 - \omega^2}$$

and $f(1)$ is undefined. This limiting case is an example of *undamped resonance*, a phenomenon which we shall now investigate.

Undamped resonance occurs when there is no damping and the frequency of the impressed force is equal to the natural frequency of the system. Since in this case $a = 0$ and the frequency $\omega/2\pi$ equals the natural frequency $(1/2\pi)\sqrt{k/m}$, the differential equation (5.50) reduces to

$$m \frac{d^2x}{dt^2} + kx = F_1 \cos \sqrt{\frac{k}{m}}\, t$$

or

$$\frac{d^2x}{dt^2} + \frac{k}{m} x = E_1 \cos \sqrt{\frac{k}{m}}\, t, \tag{5.70}$$

where $E_1 = F_1/m$. Since the complementary function of Equation (5.70) is

$$x_c = c_1 \sin \sqrt{\frac{k}{m}}\, t + c_2 \cos \sqrt{\frac{k}{m}}\, t, \tag{5.71}$$

we cannot assume a particular integral of the form

$$A \sin \sqrt{\frac{k}{m}}\, t + B \cos \sqrt{\frac{k}{m}}\, t.$$

Rather we must assume

$$x_p = At \sin \sqrt{\frac{k}{m}}\, t + Bt \cos \sqrt{\frac{k}{m}}\, t.$$

Differentiating this twice and substituting into Equation (5.70), we find that

$$A = \frac{E_1}{2}\sqrt{\frac{m}{k}} \quad \text{and} \quad B = 0.$$

Thus the particular integral of Equation (5.70) resulting from the forcing function $E_1 \cos \sqrt{k/m}\, t$ is given by

$$x_p = \frac{E_1}{2}\sqrt{\frac{m}{k}}\, t \sin \sqrt{\frac{k}{m}}\, t.$$

Expressing the complementary function (5.71) in the equivalent "phase-angle" form, we see that the general solution of Equation (5.70) is given by

$$x = c \cos\left(\sqrt{\frac{k}{m}}\, t + \phi\right) + \frac{E_1}{2}\sqrt{\frac{m}{k}}\, t \sin \sqrt{\frac{k}{m}}\, t. \tag{5.72}$$

The motion defined by (5.72) is thus the sum of a periodic term and an oscillatory term whose amplitude $(E_1/2)\sqrt{m/k}\, t$ increases with t. The graph of the function defined by this latter term,

$$\frac{E_1}{2}\sqrt{\frac{m}{k}}\, t \sin \sqrt{\frac{k}{m}}\, t,$$

appears in Figure 5.11. As t increases, this term clearly dominates the entire motion. One might argue that Equation (5.72) informs us that as $t \to \infty$ the oscillations will become infinite. However, common sense intervenes and convinces us that before this exciting phenomenon can occur the system will break down and then Equation (5.72) will no longer apply!

▶ **Example 5.6.** A 64-lb weight is attached to the lower end of a coil spring suspended from the ceiling, the spring constant being 18 lb/ft. The weight comes to rest in its equilibrium position. It is then pulled down 6 in. below its equilibrium position and released at $t = 0$. At this instant an external force given by $F(t) = 3 \cos \omega t$ is applied to the system.

1. Assuming the damping force in pounds is numerically equal to $4(dx/dt)$, where dx/dt is the instantaneous velocity in feet per second, determine the resonance frequency of the resulting motion.

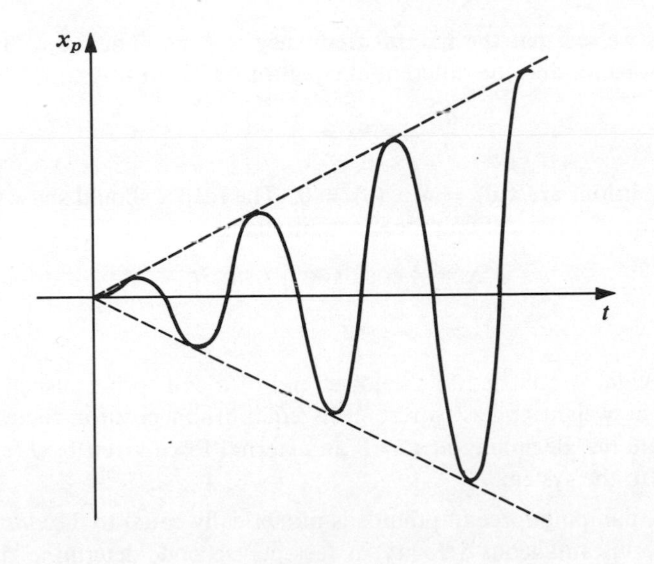

FIGURE 5.11

2. Assuming there is no damping, determine the value of ω which gives rise to undamped resonance.

Solution. Since $m = w/g = \frac{64}{32} = 2$ (slugs), $k = 18$, and $F(t) = 3 \cos \omega t$, the differential equation is

$$2 \frac{d^2x}{dt^2} + a \frac{dx}{dt} + 18x = 3 \cos \omega t,$$

where a is the damping coefficient. In Part 1, $a = 4$ and so in this case the differential equation reduces to

$$2 \frac{d^2x}{dt^2} + 4 \frac{dx}{dt} + 18x = 3 \cos \omega t.$$

Here we are not asked to solve the differential equation but merely to determine the resonance frequency. Using formula (5.69) we find that this is

$$\frac{1}{2\pi} \sqrt{\frac{18}{2} - \frac{1}{2}\left(\frac{16}{4}\right)} = \frac{1}{2\pi} \sqrt{7} \approx 0.42 \text{ (cycles/sec).}$$

Thus resonance occurs when $\omega = \sqrt{7} \approx 2.65$.

In Part 2, $a = 0$ and so the differential equation reduces to

$$\frac{d^2x}{dt^2} + 9x = \frac{3}{2} \cos \omega t. \tag{5.73}$$

Undamped resonance occurs when the frequency $\omega/2\pi$ of the impressed force is equal to the natural frequency. The complementary function of Equation (5.73) is

$$x_c = c_1 \sin 3t + c_2 \cos 3t,$$

and from this we see that the natural frequency is $3/2\pi$. Thus $\omega = 3$ gives rise to undamped resonance and the differential equation (5.73) in this case becomes

$$\frac{d^2x}{dt^2} + 9x = \tfrac{3}{2}\cos 3t. \tag{5.74}$$

The initial conditions are $x(0) = \frac{1}{2}$, $x'(0) = 0$. The reader should show that the solution of Equation (5.74) satisfying these conditions is

$$x = \tfrac{1}{2}\cos 3t + \tfrac{1}{4}t\sin 3t.$$

Exercises

1. A 12-lb weight is attached to the lower end of a coil spring suspended from the ceiling. The weight comes to rest in its equilibrium position thereby stretching the spring 6 in. Beginning at $t = 0$ an external force given by $F(t) = 2\cos\omega t$ is applied to the system.

   (a) If the damping force in pounds is numerically equal to $3\,dx/dt$, where dx/dt is the instantaneous velocity in feet per second, determine the resonance frequency of the resulting motion and find the displacement as a function of the time when the forcing function is in resonance with the system.

   (b) Assuming there is no damping, determine the value of ω which gives rise to undamped resonance and find the displacement as a function of the time in this case.

2. A 20-lb weight is attached to the lower end of a coil spring suspended from the ceiling. The weight comes to rest in its equilibrium position, thereby stretching the spring 6 in. Various external forces of the form $F(t) = \cos\omega t$ are applied to the system and it is found that the resonance frequency is 0.5 cycles/sec. Assuming that the resistance of the medium in pounds is numerically equal to $a\,dx/dt$, where dx/dt is the instantaneous velocity in feet per second, determine the damping coefficient a.

3. The differential equation for the motion of a unit mass on a certain coil spring under the action of an external force of the form $F(t) = 30\cos\omega t$ is

   $$\frac{d^2x}{dt^2} + a\frac{dx}{dt} + 24x = 30\cos\omega t,$$

   where $a \geq 0$ is the damping coefficient.

   (a) Graph the resonance curves of the system for $a = 0, 2, 4, 6$, and $4\sqrt{3}$.

   (b) If $a = 4$, find the resonance frequency and determine the amplitude of the steady-state vibration when the forcing function is in resonance with the system.

   (c) Proceed as in part (b) if $a = 2$.

5.6 Electric Circuit Problems

In this section we consider the application of differential equations to series circuits containing (1) an electromotive force, and (2) resistors, inductors, and capacitors. We assume that the reader is somewhat familiar with these items and so we shall avoid an extensive discussion. Let us simply recall that the electromotive force (for example, a battery or generator) produces a flow of current in a closed circuit and that this current produces a so-called *voltage drop* across each resistor, inductor, and capacitor. Further, the following three laws concerning the voltage drops across these various elements are known to hold:

1. The voltage drop across a resistor is given by

$$E_R = Ri, \tag{5.75}$$

where R is a constant of proportionality called the *resistance*, and i is the current.

2. The voltage drop across an inductor is given by

$$E_L = L\frac{di}{dt}, \tag{5.76}$$

where L is a constant of proportionality called the *inductance*, and i again denotes the current.

3. The voltage drop across a capacitor is given by

$$E_C = \frac{1}{C}q, \tag{5.77}$$

where C is a constant of proportionality called the *capacitance* and q is the instantaneous charge on the capacitor. Since $i = dq/dt$, this is often written as

$$E_C = \frac{1}{C}\int i\,dt.$$

The units in common use are listed in Table 5.1.

TABLE 5.1

Quantity and symbol	*Unit*
emf or voltage E	volt (V)
current i	ampere
charge q	coulomb
resistance R	ohm (Ω)
inductance L	henry (H)
capacitance C	farad

The fundamental law in the study of electric circuits is the following:

Kirchhoff's Voltage Law (*Form 1*). The algebraic sum of the instantaneous voltage drops around a closed circuit in a specific direction is zero.

Since voltage drops across resistors, inductors, and capacitors have the opposite sign from voltages arising from electromotive forces, we may state this law in the following alternative form:

Kirchhoff's Voltage Law (Form 2). The sum of the voltage drops across resistors, inductors, and capacitors is equal to the total electromotive force in a closed circuit.

We now consider the circuit shown in Figure 5.12.

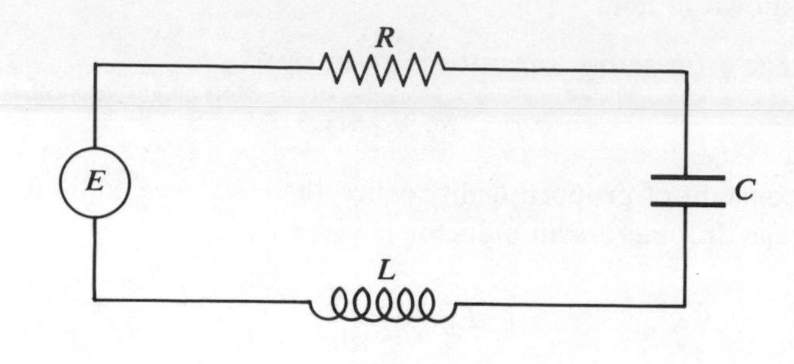

FIGURE 5.12

Here and in later diagrams the following conventional symbols are employed:

E Electromotive force (battery or generator)

R Resistor

L Inductor

C Capacitor

Let us apply Kirchhoff's law to the circuit of Figure 5.12. Letting E denote the electromotive force, and using the laws 1, 2, and 3 for voltage drops which were given above, we are led at once to the equation

$$L\frac{di}{dt} + Ri + \frac{1}{C}q = E. \tag{5.78}$$

This equation contains *two* dependent variables i and q. However, we recall that these two variables are related to each other by the equation

$$i = \frac{dq}{dt}. \tag{5.79}$$

Using this we may eliminate i from Equation (5.78) and write it in the form

$$L\frac{d^2q}{dt^2} + R\frac{dq}{dt} + \frac{1}{C}q = E. \tag{5.80}$$

Equation (5.80) is a second-order linear differential equation in the single dependent variable q. On the other hand, if we differentiate Equation (5.78) with respect to t and make use of (5.79), we may eliminate q from Equation (5.78) and write

$$L\frac{d^2i}{dt^2} + R\frac{di}{dt} + \frac{1}{C}i = \frac{dE}{dt}. \tag{5.81}$$

This is a second-order linear differential equation in the single dependent variable i.

Thus we have the two second-order linear differential equations (5.80) and (5.81) for the charge q and current i, respectively. Further observe that in two very simple cases the problem reduces to a *first*-order linear differential equation. If the circuit contains no capacitor, Equation (5.78) itself reduces directly to

$$L\frac{di}{dt} + Ri = E;$$

while if no inductor is present, Equation (5.80) reduces to

$$R\frac{dq}{dt} + \frac{1}{C}q = E.$$

Before considering examples, we observe an interesting and useful analogy. The differential equation (5.80) for the charge is exactly the same as the differential equation (5.7) of Section 5.1 for the vibrations of a mass on a coil spring, except for the notations used. That is, the electrical system described by Equation (5.80) is analogous to the mechanical system described by Equation (5.7) of Section 5.1. This analogy is brought out by Table 5.2.

TABLE 5.2

Mechanical system	*Electrical system*
mass m	inductance L
damping constant a	resistance R
spring constant k	reciprocal of capacitance $\frac{1}{C}$
impressed force $F(t)$	impressed voltage or emf E
displacement x	charge q
velocity $v = \frac{dx}{dt}$	current $i = \frac{dq}{dt}$

▶ Example 5.7. A circuit has in series an electromotive force given by $E = 100 \sin 40t$ V, a resistor of 10 Ω and an inductor of 0.5 H. If the initial current is 0, find the current at time $t > 0$.

Formulation. The circuit diagram is shown in Figure 5.13. Let i denote the current in amperes at time t. The total electromotive force is $100 \sin 40t$ V. Using the laws 1 and 2, we find that the voltage drops are as follows:

1. across the resistor: $E_R = Ri = 10i$.
2. across the inductor: $E_L = L\frac{di}{dt} = \frac{1}{2}\frac{di}{dt}$.

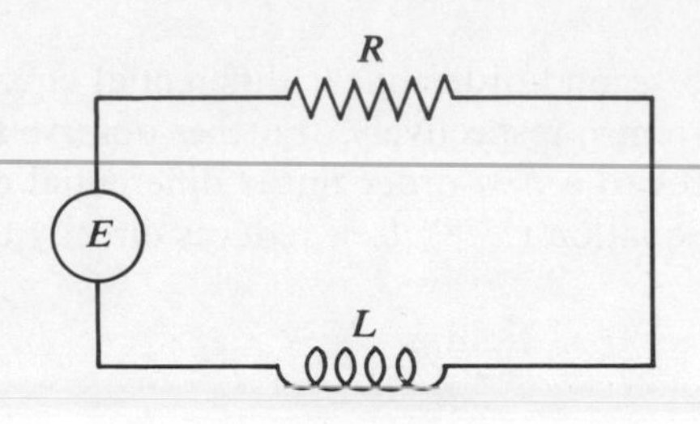

FIGURE 5.13

Applying Kirchhoff's law, we have the differential equation

$$\frac{1}{2}\frac{di}{dt} + 10i = 100 \sin 40t,$$

or

$$\frac{di}{dt} + 20i = 200 \sin 40t. \tag{5.82}$$

Since the initial current is 0, the initial condition is

$$i(0) = 0. \tag{5.83}$$

Solution. Equation (5.82) is a first-order linear equation. An integrating factor is

$$e^{\int 20\, dt} = e^{20t}.$$

Multiplying (5.82) by this, we obtain

$$e^{20t}\frac{di}{dt} + 20e^{20t}i = 200e^{20t} \sin 40t$$

or

$$\frac{d}{dt}(e^{20t}i) = 200e^{20t} \sin 40t.$$

Integrating and simplifying, we find

$$i = 2(\sin 40t - 2 \cos 40t) + Ce^{-20t}.$$

Applying the condition (5.83), $i = 0$ when $t = 0$, we find $C = 4$. Thus the solution is

$$i = 2(\sin 40t - 2 \cos 40t) + 4e^{-20t}.$$

Expressing the trigonometric terms in a "phase-angle" form, we have

$$i = 2\sqrt{5}\left(\frac{1}{\sqrt{5}} \sin 40t - \frac{2}{\sqrt{5}} \cos 40t\right) + 4e^{-20t}$$

or

$$i = 2\sqrt{5} \sin (40t + \phi) + 4e^{-20t}, \tag{5.84}$$

where ϕ is determined by the equation

$$\phi = \arccos \frac{1}{\sqrt{5}} = \arcsin \left(-\frac{2}{\sqrt{5}} \right).$$

We find $\phi \approx -1.11$ rad, and hence the current is given approximately by

$$i = 4.47 \sin (40t - 1.11) + 4e^{-20t}.$$

Interpretation. The current is clearly expressed as the sum of a sinusoidal term and an exponential. The exponential becomes so very small in a short time that its effect is soon practically negligible; it is the *transient* term. Thus, after a short time, essentially all that remains is the sinusoidal term; it is the *steady-state current.* Observe that its *period* $\pi/20$ is the same as that of the electromotive force. However, the *phase angle* $\phi \approx -1.11$ indicates that the electromotive force leads the steady-state current by approximately $\frac{1}{40}$ (1.11). The graph of the current as a function of time appears in Figure 5.14.

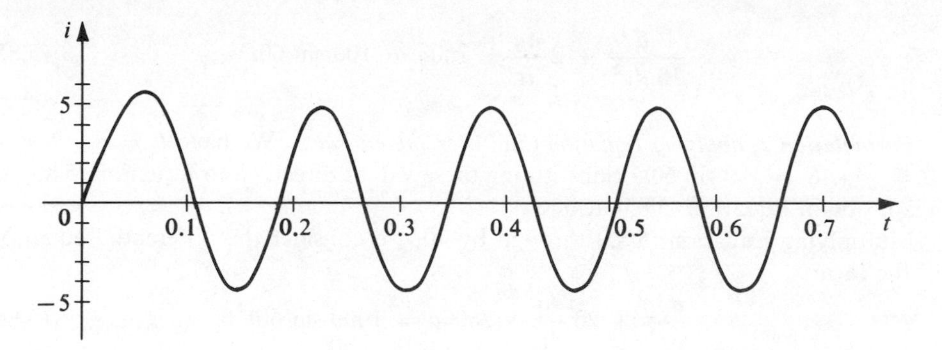

FIGURE 5.14

▶ **Example 5.8.** A circuit has in series an electromotive force given by $E = 100 \sin 60t$ V, a resistor of 2 Ω, an inductor of 0.1 H, and a capacitor of $\frac{1}{260}$ farads. (See Figure 5.15.) If the initial current and the initial charge on the capacitor are both zero, find the charge on the capacitor at any time $t > 0$.

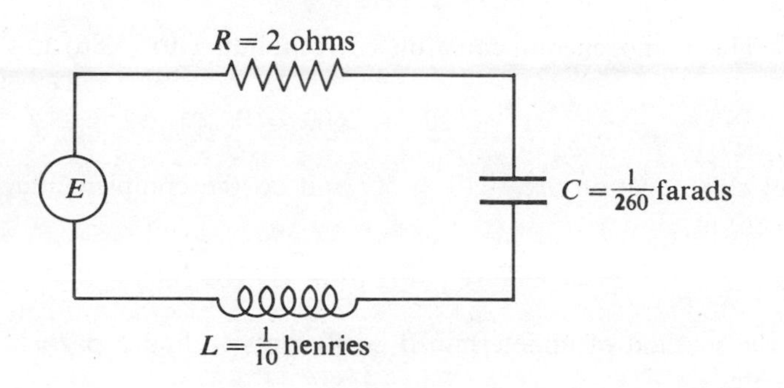

FIGURE 5.15

***Formulation 1**, by directly applying Kirchhoff's law:* Let i denote the current and q the charge on the capacitor at time t. The total electromotive force is 100 sin $60t$ (volts). Using the voltage drop laws 1, 2, and 3 we find that the voltage drops are as follows:

1. Across the resistor: $E_R = Ri = 2i$.
2. Across the inductor: $E_L = L\dfrac{di}{dt} = \dfrac{1}{10}\dfrac{di}{dt}$.
3. Across the capacitor: $E_c = \dfrac{1}{C}q = 260q$.

Now applying Kirchhoff's law we have at once:

$$\frac{1}{10}\frac{di}{dt} + 2i + 260q = 100 \sin 60t.$$

Since $i = dq/dt$, this reduces to

$$\frac{1}{10}\frac{d^2q}{dt^2} + 2\frac{dq}{dt} + 260q = 100 \sin 60t. \tag{5.85}$$

***Formulation 2**, applying Equation (5.80) for the charge:* We have $L = \frac{1}{10}$, $R = 2$, $C = \frac{1}{260}$, $E = 100 \sin 60t$. Substituting these values directly into Equation (5.80) we again obtain Equation (5.85) at once.

Multiplying Equation (5.85) through by 10, we consider the differential equation in the form

$$\frac{d^2q}{dt^2} + 20\frac{dq}{dt} + 2600q = 1000 \sin 60t. \tag{5.86}$$

Since the charge q is initially zero, we have as a first initial condition

$$q(0) = 0. \tag{5.87}$$

Since the current i is also initially zero and $i = dq/dt$, we take the second initial condition in the form

$$q'(0) = 0. \tag{5.88}$$

Solution. The homogeneous equation corresponding to (5.86) has the auxiliary equation

$$r^2 + 20r + 2600 = 0.$$

The roots of this equation are $-10 \pm 50i$ and so the complementary function of Equation (5.86) is

$$q_c = e^{-10t}(C_1 \sin 50t + C_2 \cos 50t).$$

Employing the method of undetermined coefficients to find a particular integral of (5.86), we write

$$q_p = A \sin 60t + B \cos 60t.$$

Differentiating twice and substituting into Equation (5.86) we find that

$$A = -\tfrac{25}{61} \quad \text{and} \quad B = -\tfrac{30}{61},$$

and so the general solution of Equation (5.86) is

$$q = e^{-10t}(C_1 \sin 50t + C_2 \cos 50t) - \tfrac{25}{61} \sin 60t - \tfrac{30}{61} \cos 60t. \tag{5.89}$$

Differentiating (5.89), we obtain

$$\frac{dq}{dt} = e^{-10t}[(-10C_1 - 50C_2) \sin 50t + (50C_1 - 10C_2) \cos 50t] - \tfrac{1500}{61} \cos 60t + \tfrac{1800}{61} \sin 60t. \tag{5.90}$$

Applying condition (5.87) to Equation (5.89) and condition (5.88) to Equation (5.90), we have

$$C_2 - \tfrac{30}{61} = 0 \quad \text{and} \quad 50C_1 - 10C_2 - \tfrac{1500}{61} = 0.$$

From these equations, we find that

$$C_1 = \tfrac{36}{61} \quad \text{and} \quad C_2 = \tfrac{30}{61}.$$

Thus the solution of the problem is

$$q = \frac{6e^{-10t}}{61}(6 \sin 50t + 5 \cos 50t) - \frac{5}{61}(5 \sin 60t + 6 \cos 60t)$$

or

$$q = \frac{6\sqrt{61}}{61} e^{-10t} \cos (50t - \phi) - \frac{5\sqrt{61}}{61} \cos (60t - \theta),$$

where $\cos \phi = 5/\sqrt{61}$, $\sin \phi = 6/\sqrt{61}$ and $\cos \theta = 6/\sqrt{61}$, $\sin \theta = 5/\sqrt{61}$. From these equations we determine $\phi \approx 0.88$ (radians) and $\theta \approx 0.69$ (radians). Thus our solution is given approximately by

$$q = 0.77e^{-10t} \cos (50t - 0.88) - 0.64 \cos (60t - 0.69).$$

Interpretation. The first term in the above solution clearly becomes negligible after a relatively short time; it is the *transient* term. After a sufficient time essentially all that remains is the periodic second term; this is the *steady-state term.* The graphs of these two components and that of their sum (the complete solution) are shown in Figure 5.16.

Exercises

1. A circuit has in series a constant electromotive force of 40 V, a resistor of 10 Ω and an inductor of 0.2 H. If the initial current is 0, find the current at time $t > 0$.

2. Solve Problem 1 if the electromotive force is given by $E(t) = 150 \cos 200t$ V instead of the constant electromotive force given in that problem.

3. A circuit has in series a constant electromotive force of 100 V, a resistor of 10 Ω, and a capacitor of 2×10^{-4} farads. The switch is closed at time $t = 0$, and the

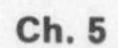

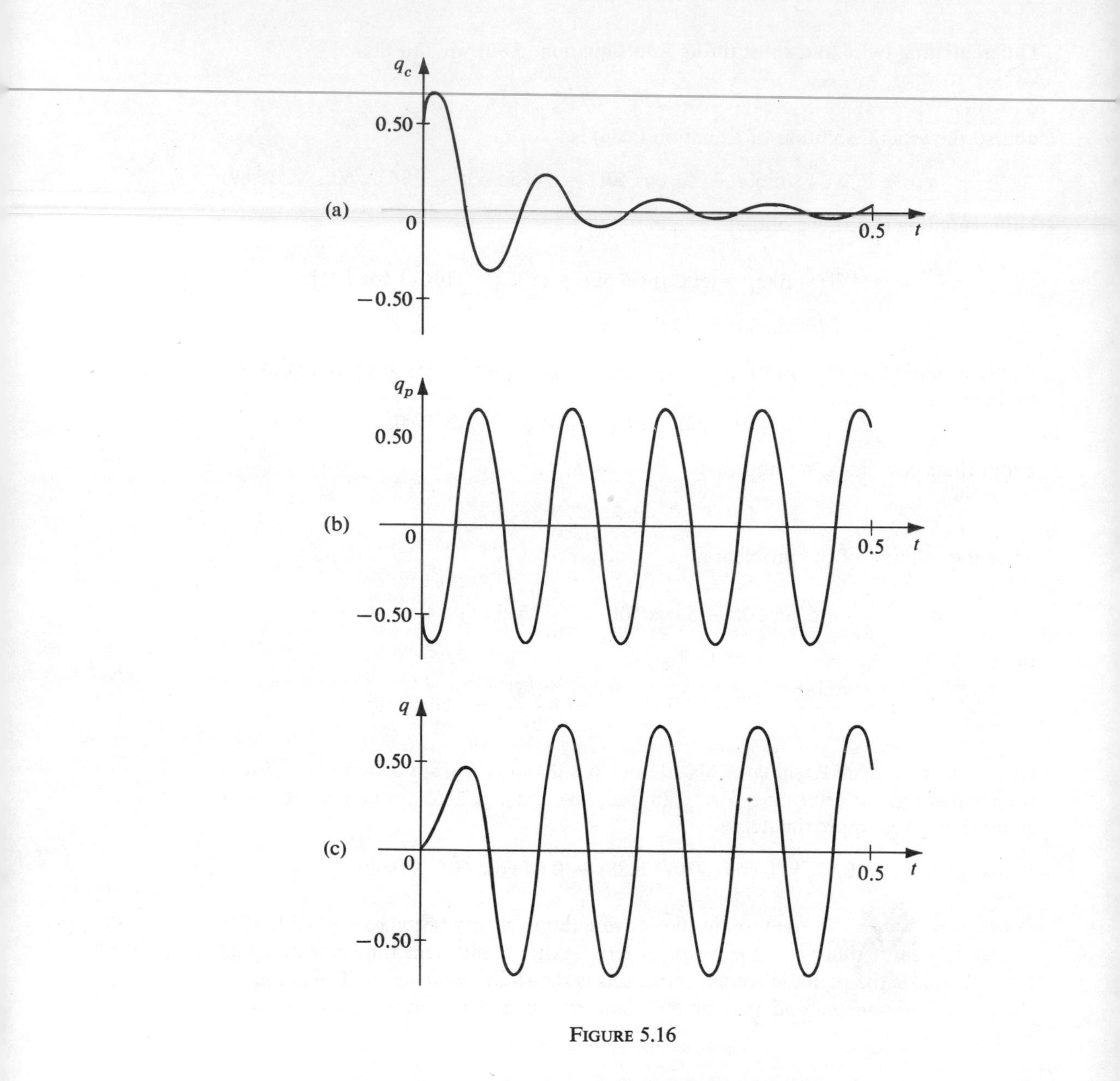

FIGURE 5.16

charge on the capacitor at this instant is zero. Find the charge and the current at time $t > 0$.

4. A circuit has in series an electromotive force given by $E(t) = 5 \sin 100t$ V, a resistor of 10 Ω, an inductor of 0.05 H, and a capacitor of 2×10^{-4} farads. If the initial current and the initial charge on the capacitor are both zero, find the charge on the capacitor at any time $t > 0$.

5. A circuit has in series an electromotive force given by $E(t) = 100 \sin 200t$ V, a resistor of 40 Ω, an inductor of 0.25 H, and a capacitor of 4×10^{-4} farads. If the initial current is zero, and the initial charge on the capacitor is 0.01 coulombs, find the current at any time $t > 0$.

6. A circuit has in series an electromotive force given by $E(t) = 200e^{-100t}$ V, a resistor of 80 Ω, an inductor of 0.2 H, and a capacitor of 5×10^{-6} farads. If the initial current and the initial charge on the capacitor are zero, find the current at any time $t > 0$.

7. A circuit has in series a resistor R Ω, an inductor of L H, and a capacitor of C farads. The initial current is zero and the initial charge on the capacitor is Q_0 coulombs.
   (a) Show that the charge and the current are damped oscillatory functions of time if and only if $R < 2\sqrt{L/C}$, and find the expressions for the charge and the current in this case.
   (b) If $R \geq 2\sqrt{L/C}$, discuss the nature of the charge and the current as functions of time.

8. A circuit has in series an electromotive force given by $E(t) = E_0 \sin \omega t$ V, a resistor of R Ω, and inductor of L H, and a capacitor of C farads.
   (a) Show that the steady-state current is

$$i = \frac{E_0}{Z}\left(\frac{R}{Z} \sin \omega t - \frac{X}{Z} \cos \omega t\right),$$

   where $X = L\omega - 1/C\omega$ and $Z = \sqrt{X^2 + R^2}$. The quantity X is called the *reactance* of the circuit and Z is called the *impedance*.
   (b) Using the result of part (a) show that the steady-state current may be written

$$i = \frac{E_0}{Z} \sin (\omega t - \phi),$$

   where ϕ is determined by the equations

$$\cos \phi = \frac{R}{Z}, \qquad \sin \phi = \frac{X}{Z}$$

   Thus show that the steady-state current attains its maximum absolute value E_0/Z at times $t_n + \phi/\omega$, where

$$t_n = \frac{1}{\omega}\left[\frac{(2n - 1)\pi}{2}\right] \qquad (n = 1, 2, 3, \ldots),$$

   are the times at which the electromotive force attains its maximum absolute value E_0.

(c) Show that the amplitude of the steady-state current is a maximum when

$$\omega = \frac{1}{\sqrt{LC}}.$$

For this value of ω *electrical resonance* is said to occur.

(d) If $R = 20$, $L = \frac{1}{4}$, $C = 10^{-4}$, and $E_0 = 100$, find the value of ω which gives rise to electrical resonance and determine the amplitude of the steady-state current in this case.

Suggested Reading

AGNEW, R., *Differential Equations*, 2nd ed. (McGraw-Hill, New York, 1960).
BOYCE, W., and R. DIPRIMA, *Elementary Differential Equations*, 2nd ed. (Wiley, New York, 1969).
KAPLAN, W., *Ordinary Differential Equations* (Addison-Wesley, Reading, Mass., 1958).
RAINVILLE, E., and P. BEDIENT, *Elementary Differential Equations*, 4th ed. (Macmillan, New York, 1969).
RITGER, P., and N. ROSE, *Differential Equations with Applications* (McGraw-Hill, New York, 1968).
SPIEGEL, M., *Applied Differential Equations*, 2nd ed. (Prentice-Hall, Englewood Cliffs, N.J., 1967).

Series Solutions of Linear Differential Equations

In Chapter 4 we learned that certain types of higher-order linear differential equations (for example, those with constant coefficients) have solutions which can be expressed as finite linear combinations of known elementary functions. In general, however, higher-order linear equations have no solutions which can be expressed in such a simple manner. Thus we must seek other means of expression for the solutions of these equations. One such means of expression is furnished by infinite series representations, and the present chapter is devoted to methods of obtaining solutions in infinite series form.

6.1 Power Series Solutions About an Ordinary Point

A. Basic Concepts and Results

Consider the second-order homogeneous linear differential equation

$$a_0(x)\frac{d^2y}{dx^2} + a_1(x)\frac{dy}{dx} + a_2(x)y = 0, \tag{6.1}$$

and suppose that this equation has no solution which is expressible as a finite linear combination of known elementary functions. Let us assume, however, that it does have a solution which can be expressed in the form of an infinite series. Specifically, we assume that it has a solution expressible in the form

$$c_0 + c_1(x - x_0) + c_2(x - x_0)^2 + \cdots = \sum_{n=0}^{\infty} c_n(x - x_0)^n, \tag{6.2}$$

where $c_0, c_1, c_2, \ldots$ are constants. An expression of the form (6.2) is called a *power series* in $x - x_0$. We have thus assumed that the differential equation (6.1) has a so-called *power series solution* of the form (6.2). Assuming that this assumption is valid, we can proceed to determine the coefficients $c_0, c_1, c_2, \ldots$ in (6.2) in such a manner that the expression (6.2) does indeed satisfy the Equation (6.1).

But under what conditions is this assumption actually valid? That is, under what conditions can we be certain that the differential equation (6.1) actually *does* have a solution of the form (6.2)? This is a question of considerable importance; for it would be quite absurd to actually try to find a "solution" of the form (6.2) if there were really no such solution to be found! In order to answer this important question concerning the existence of a solution of the form (6.2), we shall first introduce certain basic definitions. For this purpose let us write the differential equation (6.1) in the equivalent normalized form

$$\frac{d^2y}{dx^2} + P_1(x)\frac{dy}{dx} + P_2(x)y = 0, \tag{6.3}$$

where

$$P_1(x) = \frac{a_1(x)}{a_0(x)} \quad \text{and} \quad P_2(x) = \frac{a_2(x)}{a_0(x)}.$$

DEFINITION

A function f is said to be analytic *at x_0 if its Taylor series about x_0,*

$$\sum_{n=0}^{\infty} \frac{f^{(n)}(x_0)}{n!}(x - x_0)^n,$$

exists and converges to $f(x)$ for all x in some open interval including x_0.

We note that all polynomial functions are analytic everywhere; so also are the functions with values e^x, $\sin x$, and $\cos x$. A rational function is analytic except at those values of x at which its denominator is zero. For example, the rational function defined by $1/(x^2 - 3x + 2)$ is analytic except at $x = 1$ and $x = 2$.

DEFINITION

The point x_0 is called an ordinary point *of the differential equation (6.1) if* both *of the functions P_1 and P_2 in the equivalent normalized equation (6.3) are analytic at x_0. If either (or both) of these functions is* not *analytic at x_0, then x_0 is called a* singular point *of the differential equation (6.1).*

▶ **Example 6.1.** Consider the differential equation

$$\frac{d^2y}{dx^2} + x\frac{dy}{dx} + (x^2 + 2)y = 0. \tag{6.4}$$

Here $P_1(x) = x$ and $P_2(x) = x^2 + 2$. Both of the functions P_1 and P_2 are polynomial functions and so they are analytic everywhere. Thus all points are ordinary points of this differential equation.

▶ **Example 6.2.** Consider the differential equation

$$(x - 1)\frac{d^2y}{dx^2} + x\frac{dy}{dx} + \frac{1}{x}y = 0. \tag{6.5}$$

Equation (6.5) has not been written in the normalized form (6.3). We must first express (6.5) in the normalized form, thereby obtaining

$$\frac{d^2y}{dx^2} + \frac{x}{x - 1}\frac{dy}{dx} + \frac{1}{x(x - 1)}y = 0.$$

Here

$$P_1(x) = \frac{x}{x - 1} \quad \text{and} \quad P_2(x) = \frac{1}{x(x - 1)}.$$

The function P_1 is analytic, except at $x = 1$, and P_2 is analytic except at $x = 0$ and $x = 1$. Thus $x = 0$ and $x = 1$ are singular points of the differential equation under consideration. All other points are ordinary points. Note clearly that $x = 0$ is a singular point, even though P_1 *is* analytic at $x = 0$. We mention this fact to emphasize that *both* P_1 and P_2 must be analytic at x_0 in order for x_0 to be an ordinary point.

We are now in a position to state a theorem concerning the existence of power series solutions of the form (6.2).

THEOREM 6.1

Hypothesis. *The point x_0 is an ordinary point of the differential equation (6.1).*

Conclusion. *The differential equation (6.1) has two nontrivial linearly independent power series solutions of the form*

$$\sum_{n=0}^{\infty} c_n(x - x_0)^n, \tag{6.2}$$

and these power series converge in some interval $|x - x_0| < R$ (where $R > 0$) about x_0.

This theorem gives us a sufficient condition for the existence of power series solutions of the differential equation (6.1). It states that if x_0 is an ordinary point of equation (6.1), then this equation has *two* power series solutions in powers of $x - x_0$ and that these two power series solutions are *linearly independent*. Thus if x_0 is an ordinary point of (6.1), we may obtain the *general solution* of (6.1) as a linear combination of these two linearly independent power series. We shall omit the proof of this important theorem.

▶ Example 6.3. In Example 6.1 we noted that all points are ordinary points of the differential equation (6.4). Thus this differential equation has two linearly independent solutions of the form (6.2) about *any* point x_0.

▶ Example 6.4. In Example 6.2 we observed that $x = 0$ and $x = 1$ are the only singular points of the differential equation (6.5). Thus this differential equation has two linearly independent solutions of the form (6.2) about any point $x_0 \neq 0$ or 1. For example, the equation has two linearly independent solutions of the form

$$\sum_{n=0}^{\infty} c_n(x - 2)^n$$

about the ordinary point 2. However, we are *not* assured that there exists any solution of the form

$$\sum_{n=0}^{\infty} c_n x^n$$

about the singular point 0 or any solution of the form

$$\sum_{n=0}^{\infty} c_n(x - 1)^n$$

about the singular point 1.

B. The Method of Solution

Now that we are assured that under appropriate hypotheses Equation (6.1) actually does have power series solutions of the form (6.2), how do we proceed to find these solutions? In other words, how do we determine the coefficients $c_0, c_1, c_2, \ldots$ in the expression

$$\sum_{n=0}^{\infty} c_n(x - x_0)^n \tag{6.2}$$

so that this expression actually does satisfy Equation (6.1)? We shall first give a brief outline of the procedure for finding these coefficients and shall then illustrate the procedure in detail by considering specific examples.

Assuming that x_0 is an ordinary point of the differential equation (6.1), so that solutions in powers of $x - x_0$ actually do exist, we denote such a solution by

$$y = c_0 + c_1(x - x_0) + c_2(x - x_0)^2 + \cdots = \sum_{n=0}^{\infty} c_n(x - x_0)^n. \tag{6.6}$$

Since the series in (6.6) converges on an interval $|x - x_0| < R$ about x_0, it may be differentiated term by term on this interval twice in succession to obtain

$$\frac{dy}{dx} = c_1 + 2c_2(x - x_0) + 3c_3(x - x_0)^2 + \cdots = \sum_{n=1}^{\infty} nc_n(x - x_0)^{n-1} \tag{6.7}$$

and

$$\frac{d^2y}{dx^2} = 2c_2 + 6c_3(x - x_0) + 12c_4(x - x_0)^2 + \cdots = \sum_{n=2}^{\infty} n(n - 1)c_n(x - x_0)^{n-2}, \tag{6.8}$$

respectively. We now substitute the series in the right members of (6.6), (6.7), and (6.8) for y and its first two derivatives, respectively, in the differential equation (6.1). We then simplify the resulting expression so that it takes the form

$$K_0 + K_1(x - x_0) + K_2(x - x_0)^2 + \cdots = 0, \tag{6.9}$$

where the coefficients K_i $(i = 0, 1, 2, \ldots)$ are functions of certain coefficients c_n of the solution (6.6). In order that (6.9) be valid for all x in the interval of convergence $|x - x_0| < R$, we must set

$$K_0 = K_1 = K_2 = \cdots = 0.$$

In other words, we must equate to zero the coefficient of each power of $(x - x_0)$ in the left member of (6.9). This leads to a set of conditions which must be satisfied by the various coefficients c_n in the series (6.6) in order that (6.6) be a solution of the differential equation (6.1). If the c_n are chosen to satisfy the set of conditions which thus occurs, then the resulting series (6.6) is the desired solution of the differential equation (6.1). We shall illustrate this procedure in detail in the two examples which follow.

▶ Example 6.5. Find the power series solution of the differential equation

$$\frac{d^2y}{dx^2} + x\frac{dy}{dx} + (x^2 + 2)y = 0 \tag{6.4}$$

in powers of x (that is, about $x_0 = 0$).

Solution. We have already observed that $x_0 = 0$ is an ordinary point of the differential equation (6.4) and that two linearly independent solutions of the desired type actually exist. Our procedure will yield both of these solutions at once.

We thus assume a solution of the form (6.6) with $x_0 = 0$. That is, we assume

$$y = \sum_{n=0}^{\infty} c_n x^n. \tag{6.10}$$

Differentiating term by term we obtain

$$\frac{dy}{dx} = \sum_{n=1}^{\infty} nc_n x^{n-1} \tag{6.11}$$

and

$$\frac{d^2y}{dx^2} = \sum_{n=2}^{\infty} n(n-1)c_n x^{n-2}. \tag{6.12}$$

Substituting the series (6.10), (6.11), and (6.12) into the differential equation (6.4), we obtain

$$\sum_{n=2}^{\infty} n(n-1)c_n x^{n-2} + x\sum_{n=1}^{\infty} nc_n x^{n-1} + x^2\sum_{n=0}^{\infty} c_n x^n + 2\sum_{n=0}^{\infty} c_n x^n = 0.$$

Since x is independent of the index of summation n, we may rewrite this as

$$\sum_{n=2}^{\infty} n(n-1)c_n x^{n-2} + \sum_{n=1}^{\infty} nc_n x^n + \sum_{n=0}^{\infty} c_n x^{n+2} + 2\sum_{n=0}^{\infty} c_n x^n = 0. \tag{6.13}$$

In order to put the left member of Equation (6.13) in the form (6.9), we shall rewrite the first and third summations in (6.13) so that x in each of these summations will have the exponent n. Let us consider the first summation

$$\sum_{n=2}^{\infty} n(n-1)c_n x^{n-2} \tag{6.14}$$

in (6.13). To rewrite the summation (6.14) so that x will have the desired exponent n, we first replace the present exponent $n - 2$ in (6.14) by a new variable m. That is, we let $m = n - 2$ in (6.14). Then $n = m + 2$, and since $m = 0$ for $n = 2$, the summation (6.14) takes the form

$$\sum_{m=0}^{\infty} (m+2)(m+1)c_{m+2}x^m. \tag{6.15}$$

Now since the variable of summation is merely a "dummy" variable, we may replace m by n in (6.15) to write the first summation in (6.13) as

$$\sum_{n=0}^{\infty} (n + 2)(n + 1)c_{n+2}x^n. \tag{6.16}$$

In like manner, letting $m = n + 2$, the third summation

$$\sum_{n=0}^{\infty} c_n x^{n+2} \tag{6.17}$$

in (6.13) first takes the form

$$\sum_{m=2}^{\infty} c_{m-2}x^m. \tag{6.18}$$

Then replacing m by n in (6.18), the third summation in (6.13) may be written as

$$\sum_{n=2}^{\infty} c_{n-2}x^n. \tag{6.19}$$

Thus replacing (6.14) by its equivalent (6.16) and (6.17) by its equivalent (6.19), Equation (6.13) may be written

$$\sum_{n=0}^{\infty} (n + 2)(n + 1)c_{n+2}x^n + \sum_{n=1}^{\infty} nc_n x^n + \sum_{n=2}^{\infty} c_{n-2}x^n + 2\sum_{n=0}^{\infty} c_n x^n = 0. \tag{6.20}$$

Although x has the same exponent n in each summation in (6.20), the ranges of the various summations are not all the same. In the first and fourth summations n ranges from 0 to ∞, in the second n ranges from 1 to ∞, and in the third the range is from 2 to ∞. The common range is from 2 to ∞. We now write out individually the terms in each summation which do *not* belong to this common range, and we continue to employ the "sigma" notation to denote the remainder of each such summation. For example, in the first summation

$$\sum_{n=0}^{\infty} (n + 2)(n + 1)c_{n+2}x^n$$

of (6.20) we write out individually the terms corresponding to $n = 0$ and $n = 1$ and denote the remainder of this summation by

$$\sum_{n=2}^{\infty} (n + 2)(n + 1)c_{n+2}x^n.$$

We thus rewrite

$$\sum_{n=0}^{\infty} (n + 2)(n + 1)c_{n+2}x^n$$

in (6.20) as

$$2c_2 + 6c_3 x + \sum_{n=2}^{\infty} (n + 2)(n + 1)c_{n+2}x^n.$$

In like manner, we write

$$\sum_{n=1}^{\infty} nc_n x^n$$

in (6.20) as

$$c_1x + \sum_{n=2}^{\infty} nc_nx^n$$

and

$$2\sum_{n=0}^{\infty} c_nx^n$$

in (6.20) as

$$2c_0 + 2c_1x + 2\sum_{n=2}^{\infty} c_nx^n.$$

Thus Equation (6.20) is now written as

$$2c_2 + 6c_3x + \sum_{n=2}^{\infty}(n+2)(n+1)c_{n+2}x^n + c_1x + \sum_{n=2}^{\infty} nc_nx^n$$
$$+ \sum_{n=2}^{\infty} c_{n-2}x^n + 2c_0 + 2c_1x + 2\sum_{n=2}^{\infty} c_nx^n = 0.$$

We can now combine like powers of x and write this equation as

$$(2c_0 + 2c_2) + (3c_1 + 6c_3)x + \sum_{n=2}^{\infty}[(n+2)(n+1)c_{n+2} + (n+2)c_n + c_{n-2}]x^n = 0. \tag{6.21}$$

Equation (6.21) is in the desired form (6.9). For (6.21) to be valid for all x in the interval of convergence $|x - x_0| < R$, the coefficient of each power of x in the left member of (6.21) must be equated to zero. This leads immediately to the conditions

$$2c_0 + 2c_2 = 0, \tag{6.22}$$

$$3c_1 + 6c_3 = 0, \tag{6.23}$$

and

$$(n+2)(n+1)c_{n+2} + (n+2)c_n + c_{n-2} = 0, \quad n \geq 2. \tag{6.24}$$

The condition (6.22) enables us to express c_2 in terms of c_0. Doing so, we find that

$$c_2 = -c_0. \tag{6.25}$$

The condition (6.23) enables us to express c_3 in terms of c_1. This leads to

$$c_3 = -\tfrac{1}{2}c_1. \tag{6.26}$$

The condition (6.24) is called a *recurrence formula.* It enables us to express each coefficient c_{n+2} for $n \geq 2$ in terms of the previous coefficients c_n and c_{n-2}, thus giving

$$c_{n+2} = -\frac{(n+2)c_n + c_{n-2}}{(n+1)(n+2)}, \quad n \geq 2. \tag{6.27}$$

For $n = 2$, formula (6.27) is

$$c_4 = -\frac{4c_2 + c_0}{12}.$$

Now using (6.25), this reduces to

$$c_4 = \tfrac{1}{4}c_0, \tag{6.28}$$

which expresses c_4 in terms of c_0. For $n = 3$, formula (6.27) is

$$c_5 = -\frac{5c_3 + c_1}{20}.$$

Now using (6.26), this reduces to

$$c_5 = \tfrac{3}{40}c_1, \tag{6.29}$$

which expresses c_5 in terms of c_1. In the same way we may express each even coefficient in terms of c_0 and each odd coefficient in terms of c_1.

Substituting the values of c_2, c_3, c_4, and c_5, given by (6.25), (6.26), (6.28), and (6.29), respectively, into the assumed solution (6.10), we have

$$y = c_0 + c_1x - c_0x^2 - \tfrac{1}{2}c_1x^3 + \tfrac{1}{4}c_0x^4 + \tfrac{3}{40}c_1x^5 + \cdots.$$

Collecting terms in c_0 and c_1, we have finally

$$y = c_0(1 - x^2 + \tfrac{1}{4}x^4 + \cdots) + c_1(x - \tfrac{1}{2}x^3 + \tfrac{3}{40}x^5 + \cdots), \tag{6.30}$$

which gives the solution of the differential equation (6.4) in powers of x through terms in x^5. The two series in parentheses in (6.30) are the power series expansions of two linearly independent solutions of (6.4), and the constants c_0 and c_1 are arbitrary constants. Thus (6.30) represents the general solution of (6.4) in powers of x (through terms in x^5).

▶ Example 6.6. Find a power series solution of the initial-value problem

$$(x^2 - 1)\frac{d^2y}{dx^2} + 3x\frac{dy}{dx} + xy = 0, \tag{6.31}$$

$$y(0) = 4, \tag{6.32}$$

$$y'(0) = 6. \tag{6.33}$$

Solution. We first observe that all points except $x = \pm 1$ are ordinary points for the differential equation (6.31). Thus we could assume solutions of the form (6.6) for any $x_0 \neq \pm 1$. However, since the initial values of y and its first derivative are prescribed at $x = 0$, we shall choose $x_0 = 0$ and seek solutions in powers of x. Thus we assume

$$y = \sum_{n=0}^{\infty} c_nx^n. \tag{6.34}$$

Differentiating term by term we obtain

$$\frac{dy}{dx} = \sum_{n=1}^{\infty} nc_nx^{n-1} \tag{6.35}$$

and

$$\frac{d^2y}{dx^2} = \sum_{n=2}^{\infty} n(n-1)c_nx^{n-2}. \tag{6.36}$$

Substituting the series (6.34), (6.35), and (6.36) into the differential equation (6.31), we obtain

$$\sum_{n=2}^{\infty} n(n-1)c_n x^n - \sum_{n=2}^{\infty} n(n-1)c_n x^{n-2} + 3\sum_{n=1}^{\infty} nc_n x^n + \sum_{n=0}^{\infty} c_n x^{n+1} = 0. \quad (6.37)$$

We now rewrite the second and fourth summations in (6.37) so that x in each of these summations has the exponent n. Doing this, Equation (6.37) takes the form

$$\sum_{n=2}^{\infty} n(n-1)c_n x^n - \sum_{n=0}^{\infty} (n+2)(n+1)c_{n+2} x^n + 3\sum_{n=1}^{\infty} nc_n x^n + \sum_{n=1}^{\infty} c_{n-1} x^n = 0. \quad (6.38)$$

The common range of the four summations in (6.38) is from 2 to ∞. We can write out the individual terms in each summation which do *not* belong to this common range and thus express (6.38) in the form

$$\sum_{n=2}^{\infty} n(n-1)c_n x^n - 2c_2 - 6c_3 x - \sum_{n=2}^{\infty} (n+2)(n+1)c_{n+2} x^n$$
$$+ 3c_1 x + 3\sum_{n=2}^{\infty} nc_n x^n + c_0 x + \sum_{n=2}^{\infty} c_{n-1} x^n = 0.$$

Combining like powers of x, this takes the form

$$-2c_2 + (c_0 + 3c_1 - 6c_3)x$$
$$+ \sum_{n=2}^{\infty} [-(n+2)(n+1)c_{n+2} + n(n+2)c_n + c_{n-1}]x^n = 0. \quad (6.39)$$

For (6.39) to be valid for all x in the interval of convergence $|x - x_0| < R$, the coefficient of each power of x in the left member of (6.39) must be equated to zero. In doing this, we are led to the relations

$$-2c_2 = 0, \quad (6.40)$$

$$c_0 + 3c_1 - 6c_3 = 0, \quad (6.41)$$

and

$$-(n+2)(n+1)c_{n+2} + n(n+2)c_n + c_{n-1} = 0, \quad n \geq 2. \quad (6.42)$$

From (6.40), we find that $c_2 = 0$; and from (6.41), $c_3 = \frac{1}{6}c_0 + \frac{1}{2}c_1$. The recurrence formula (6.42) gives

$$c_{n+2} = \frac{n(n+2)c_n + c_{n-1}}{(n+1)(n+2)}, \quad n \geq 2.$$

Using this, we find successively

$$c_4 = \frac{8c_2 + c_1}{12} = \frac{1}{12}c_1,$$

$$c_5 = \frac{15c_3 + c_2}{20} = \frac{1}{8}c_0 + \frac{3}{8}c_1,$$

Substituting these values of $c_2, c_3, c_4, c_5, \ldots$ into the assumed solution (6.34), we have

$$y = c_0 + c_1 x + \left(\frac{c_0}{6} + \frac{c_1}{2}\right)x^3 + \frac{c_1}{12}x^4 + \left(\frac{c_0}{8} + \frac{3c_1}{8}\right)x^5 + \cdots$$

or

$$y = c_0(1 + \tfrac{1}{6}x^3 + \tfrac{1}{8}x^5 + \cdots) + c_1(x + \tfrac{1}{2}x^3 + \tfrac{1}{12}x^4 + \tfrac{3}{8}x^5 + \cdots). \quad (6.43)$$

The solution (6.43) is the general solution of the differential equation (6.31) in powers of x (through terms in x^5).

We must now apply the initial conditions (6.32) and (6.33). Applying (6.32) to (6.43), we immediately find that

$$c_0 = 4.$$

Differentiating (6.43), we have

$$\frac{dy}{dx} = c_0(\tfrac{1}{2}x^2 + \tfrac{5}{8}x^4 + \cdots) + c_1(1 + \tfrac{3}{2}x^2 + \tfrac{1}{3}x^3 + \tfrac{15}{8}x^4 + \cdots). \quad (6.44)$$

Applying (6.33) to (6.44) we find that

$$c_1 = 6.$$

Thus the solution of the given initial-value problem in powers of x (through terms in x^5) is

$$y = 4(1 + \tfrac{1}{6}x^3 + \tfrac{1}{8}x^5 + \cdots) + 6(x + \tfrac{1}{2}x^3 + \tfrac{1}{12}x^4 + \tfrac{3}{8}x^5 + \cdots)$$

or

$$y = 4 + 6x + \tfrac{11}{3}x^3 + \tfrac{1}{2}x^4 + \tfrac{11}{4}x^5 + \cdots.$$

Remark 1. Suppose the initial values of y and its first derivative in conditions (6.32) and (6.33) of Example 6.6 are prescribed at $x = 2$, instead of $x = 0$. Then we have the initial-value problem

$$(x^2 - 1)\frac{d^2y}{dx^2} + 3x\frac{dy}{dx} + xy = 0,$$
$$y(2) = 4, \qquad y'(2) = 6. \quad (6.45)$$

Since the initial values in this problem are prescribed at $x = 2$, we would seek solutions in powers of $x - 2$. That is, in this case we would seek solutions of the form

$$y = \sum_{n=0}^{\infty} c_n(x - 2)^n. \quad (6.46)$$

The simplest procedure for obtaining a solution of the form (6.46) is first to make the substitution $t = x - 2$. This replaces the initial-value problem (6.45) by the equivalent problem

$$(t^2 + 4t + 3)\frac{d^2y}{dt^2} + (3t + 6)\frac{dy}{dt} + (t + 2)y = 0,$$
$$y(0) = 4, \qquad y'(0) = 6, \quad (6.47)$$

in which t is the independent variable and the initial values are prescribed at $t = 0$. One then seeks a solution of the problem (6.47) in powers of t,

$$y = \sum_{n=0}^{\infty} c_n t^n. \tag{6.48}$$

Differentiating (6.48) and substituting into the differential equation in (6.47), one determines the c_n as in Examples 6.5 and 6.6. The initial conditions in (6.47) are then applied. Replacing t by $x - 2$ in the resulting solution (6.48), one obtains the desired solution (6.46) of the original problem (6.45).

Remark 2. In Examples 6.5 and 6.6 we obtained power series solutions of the differential equations under consideration but made no attempt to discuss the convergence of these solutions. According to Theorem 6.1, if x_0 is an ordinary point of the differential equation

$$a_0(x)\frac{d^2y}{dx^2} + a_1(x)\frac{dy}{dx} + a_2(x)y = 0, \tag{6.1}$$

then the power series solutions of the form (6.2) converge in some interval $|x - x_0| < R$ (where $R > 0$) about x_0. Let us again write (6.1) in the normalized from

$$\frac{d^2y}{dx^2} + P_1(x)\frac{dy}{dx} + P_2(x)y = 0, \tag{6.3}$$

where

$$P_1(x) = \frac{a_1(x)}{a_0(x)} \quad \text{and} \quad P_2(x) = \frac{a_2(x)}{a_0(x)}.$$

If x_0 is an ordinary point of (6.1), the functions P_1 and P_2 have Taylor series expansions about x_0 which converge in intervals $|x - x_0| < R_1$ and $|x - x_0| < R_2$, respectively, about x_0. It can be proved that the interval of convergence $|x - x_0| < R$ of a series solution (6.2) of (6.1) is at least as great as the smaller of the intervals $|x - x_0| < R_1$ and $|x - x_0| < R_2$.

In the differential equation (6.4) of Example 6.5, $P_1(x) = x$ and $P_2(x) = x^2 + 2$. Thus in this example the Taylor series expansions for P_1 and P_2 about the ordinary point $x_0 = 0$ converge for all x. Hence the series solutions (6.30) of (6.4) also converge for all x.

In the differential equation (6.31) of Example 6.6,

$$P_1(x) = \frac{3x}{x^2 - 1} \quad \text{and} \quad P_2(x) = \frac{x}{x^2 - 1}.$$

In this example the Taylor series for P_1 and P_2 about $x_0 = 0$ both converge for $|x| < 1$. Thus the solutions (6.43) of (6.31) converge at least for $|x| < 1$.

Exercises

Find power series solutions in powers of x of each of the differential equations in Exercises 1 through 10.

1. $\dfrac{d^2y}{dx^2} + x\dfrac{dy}{dx} + y = 0.$

2. $\dfrac{d^2y}{dx^2} + 8x\dfrac{dy}{dx} - 4y = 0.$

3. $\dfrac{d^2y}{dx^2} + x\dfrac{dy}{dx} + (2x^2 + 1)y = 0.$

4. $\dfrac{d^2y}{dx^2} + x\dfrac{dy}{dx} + (x^2 - 4)y = 0.$

5. $\dfrac{d^2y}{dx^2} + x\dfrac{dy}{dx} + (3x + 2)y = 0.$

6. $\dfrac{d^2y}{dx^2} - x\dfrac{dy}{dx} + (3x - 2)y = 0.$

7. $(x^2 + 1)\dfrac{d^2y}{dx^2} + x\dfrac{dy}{dx} + xy = 0.$

8. $(x - 1)\dfrac{d^2y}{dx^2} - (3x - 2)\dfrac{dy}{dx} + 2xy = 0.$

9. $(x^3 - 1)\dfrac{d^2y}{dx^2} + x^2\dfrac{dy}{dx} + xy = 0.$

10. $(x + 3)\dfrac{d^2y}{dx^2} + (x + 2)\dfrac{dy}{dx} + y = 0.$

Find the power series solution of each of the initial-value problems in Exercises 11 through 14.

11. $\dfrac{d^2y}{dx^2} - x\dfrac{dy}{dx} - y = 0, \quad y(0) = 1, \quad y'(0) = 0.$

12. $\dfrac{d^2y}{dx^2} + x\dfrac{dy}{dx} - 2y = 0, \quad y(0) = 0, \quad y'(0) = 1.$

13. $(x^2 + 1)\dfrac{d^2y}{dx^2} + x\dfrac{dy}{dx} + 2xy = 0, \quad y(0) = 2, \quad y'(0) = 3.$

14. $(2x^2 - 3)\dfrac{d^2y}{dx^2} - 2x\dfrac{dy}{dx} + y = 0, \quad y(0) = -1, \quad y'(0) = 5.$

Find power series solutions in powers of $(x - 1)$ of each of the differential equations in Exercises 15 and 16.

15. $x^2\dfrac{d^2y}{dx^2} + x\dfrac{dy}{dx} + y = 0.$

16. $x^2\dfrac{d^2y}{dx^2} + 3x\dfrac{dy}{dx} - y = 0.$

17. Find the power series solution in powers of $(x - 1)$ of the initial-value problem

$$x\frac{d^2y}{dx^2} + \frac{dy}{dx} + 2y = 0, \quad y(1) = 2, \quad y'(1) = 4.$$

18. The differential equation

$$(1 - x^2)\frac{d^2y}{dx^2} - 2x\frac{dy}{dx} + n(n + 1)y = 0,$$

where n is a constant, is called *Legendre's differential equation.*

(a) Show that $x = 0$ is an ordinary point of this differential equation, and find two linearly independent power series solutions in powers of x.

(b) Show that if n is a nonnegative integer, then one of the two solutions found in part (a) is a polynomial of degree n.

6.2 Solutions About Singular Points; the Method of Frobenius

A. Regular Singular Points

We again consider the homogeneous linear differential equation

$$a_0(x)\frac{d^2y}{dx^2} + a_1(x)\frac{dy}{dx} + a_2(x)y = 0, \tag{6.1}$$

and we assume that x_0 is a *singular* point of (6.1). Then Theorem 6.1 does *not* apply at the point x_0, and we are *not* assured of a power series solution

$$y = \sum_{n=0}^{\infty} c_n(x - x_0)^n \tag{6.2}$$

of (6.1) in powers of $x - x_0$. Indeed an equation of the form (6.1) with a singular point at x_0 does *not*, in general, have a solution of the form (6.2). Clearly we must seek a different type of solution in such a case, but what type of solution can we expect? It happens that under certain conditions we are justified in assuming a solution of the form

$$y = |x - x_0|^r \sum_{n=0}^{\infty} c_n(x - x_0)^n, \tag{6.49}$$

where r is a certain (real or complex) constant. Such a solution is clearly a power series in $x - x_0$ multiplied by a certain *power* of $|x - x_0|$. In order to state conditions under which a solution of this form is assured, we proceed to classify singular points.

We again write the differential equation (6.1) in the equivalent normalized form

$$\frac{d^2y}{dx^2} + P_1(x)\frac{dy}{dx} + P_2(x)y = 0, \tag{6.3}$$

where

$$P_1(x) = \frac{a_1(x)}{a_0(x)} \quad \text{and} \quad P_2(x) = \frac{a_2(x)}{a_0(x)}.$$

DEFINITION

Consider the differential equation (6.1), and assume that at least one of the functions P_1 and P_2 in the equivalent normalized equation (6.3) is not *analytic at x_0, so that x_0 is a singular point of (6.1). If the functions defined by the products*

$$(x - x_0)P_1(x) \quad \textit{and} \quad (x - x_0)^2P_2(x) \tag{6.50}$$

are both analytic at x_0, *then* x_0 *is called a* regular singular point *of the differential equation (6.1). If either (or both) of the functions defined by the products (6.50) is not analytic at* x_0, *then* x_0 *is called an* irregular singular point *of (6.1).*

▶ **Example 6.7.** Consider the differential equation

$$2x^2 \frac{d^2y}{dx^2} - x\frac{dy}{dx} + (x - 5)y = 0. \tag{6.51}$$

Writing this in the normalized form (6.3), we have

$$\frac{d^2y}{dx^2} - \frac{1}{2x}\frac{dy}{dx} + \frac{x-5}{2x^2}y = 0.$$

Here $P_1(x) = -1/2x$ and $P_2(x) = (x - 5)/2x^2$. Since both P_1 and P_2 fail to be analytic at $x = 0$, we conclude that $x = 0$ is a singular point of (6.51). We now consider the functions defined by the products

$$xP_1(x) = -\frac{1}{2} \quad \text{and} \quad x^2P_2(x) = \frac{x-5}{2}$$

of the form (6.50). Both of these product functions are analytic at $x = 0$, and so $x = 0$ is a *regular* singular point of the differential equation (6.51).

▶ **Example 6.8.** Consider the differential equation

$$x^2(x-2)^2\frac{d^2y}{dx^2} + 2(x-2)\frac{dy}{dx} + (x+1)y = 0. \tag{6.52}$$

In the normalized form (6.3), this is

$$\frac{d^2y}{dx^2} + \frac{2}{x^2(x-2)}\frac{dy}{dx} + \frac{x+1}{x^2(x-2)^2}y = 0.$$

Here

$$P_1(x) = \frac{2}{x^2(x-2)} \quad \text{and} \quad P_2(x) = \frac{x+1}{x^2(x-2)^2}.$$

Clearly the singular points of the differential equation (6.52) are $x = 0$ and $x = 2$. We investigate them one at a time.

Consider $x = 0$ first, and form the functions defined by the products

$$xP_1(x) = \frac{2}{x(x-2)} \quad \text{and} \quad x^2P_2(x) = \frac{x+1}{(x-2)^2}$$

of the form (6.50). The product function defined by $x^2P_2(x)$ is analytic at $x = 0$, but that defined by $xP_1(x)$ is *not*. Thus $x = 0$ is an *irregular* singular point of (6.52).

Now consider $x = 2$. Forming the products (6.50) for this point, we have

$$(x-2)P_1(x) = \frac{2}{x^2} \quad \text{and} \quad (x-2)^2P_2(x) = \frac{x+1}{x^2}.$$

Both of the product functions thus defined are analytic at $x = 2$, and hence $x = 2$ is a *regular* singular point of (6.52).

Now that we can distinguish between regular and irregular singular points, we shall state a basic theorem concerning solutions of the form (6.49) about regular singular points. We shall later give a more complete theorem on this topic.

THEOREM 6.2

Hypothesis. *The point x_0 is a regular singular point of the differential equation (6.1).*

Conclusion. *The differential equation (6.1) has* at least one *nontrivial solution of the form*

$$|x - x_0|^r \sum_{n=0}^{\infty} c_n(x - x_0)^n, \tag{6.49}$$

where r is a definite (real or complex) constant which may be determined, and this solution is valid in some deleted interval $0 < |x - x_0| < R$ (where $R > 0$) about x_0.

▶ **Example 6.9.** In Example 6.7 we saw that $x = 0$ is a regular singular point of the differential equation

$$2x^2 \frac{d^2y}{dx^2} - x \frac{dy}{dx} + (x - 5)y = 0. \tag{6.51}$$

By Theorem 6.2 we conclude that this equation has at least one nontrivial solution of the form

$$|x|^r \sum_{n=0}^{\infty} c_n x^n,$$

valid in some deleted interval $0 < |x| < R$ about $x = 0$.

▶ **Example 6.10.** In Example 6.8 we saw that $x = 2$ is a regular singular point of the differential equation

$$x^2(x - 2)^2 \frac{d^2y}{dx^2} + 2(x - 2) \frac{dy}{dx} + (x + 1)y = 0. \tag{6.52}$$

Thus we know that this equation has at least one nontrivial solution of the form

$$|x - 2|^r \sum_{n=0}^{\infty} c_n(x - 2)^n,$$

valid in some deleted interval $0 < |x - 2| < R$ about $x = 2$.

We also observed that $x = 0$ is a singular point of Equation (6.52). However, this singular point is irregular and so Theorem 6.2 does not apply to it. We are *not* assured that the differential equation (6.52) has a solution of the form

$$|x|^r \sum_{n=0}^{\infty} c_n x^n$$

in any deleted interval about $x = 0$.

B. The Method of Frobenius

Now that we are assured of at least one solution of the form (6.49) about a regular singular point x_0 of the differential equation (6.1), how do we proceed to determine the coefficients c_n and the number r in this solution? The procedure is similar to that introduced in Section 6.1 and is commonly called the *method of Frobenius.* We shall briefly outline the method and then illustrate it by applying it to the differential equation (6.51). In this outline and the illustrative example which follows we shall seek solutions valid in some interval $0 < x - x_0 < R$. Note that for all such x, $|x - x_0|$ is simply $x - x_0$. To obtain solutions valid for $-R < x - x_0 < 0$, simply replace $x - x_0$ by $-(x - x_0) > 0$ and proceed as in the outline.

Outline of the Method of Frobenius

1. Let x_0 be a regular singular point of the differential equation (6.1), seek solutions valid in some interval $0 < x - x_0 < R$, and assume a solution

$$y = (x - x_0)^r \sum_{n=0}^{\infty} c_n(x - x_0)^n$$

of the form (6.49), where $c_0 \neq 0$. We write this solution in the form

$$y = \sum_{n=0}^{\infty} c_n(x - x_0)^{n+r}, \tag{6.53}$$

where $c_0 \neq 0$.

2. Assuming term-by-term differentiation of (6.53) is valid, we obtain

$$\frac{dy}{dx} = \sum_{n=0}^{\infty} (n + r)c_n(x - x_0)^{n+r-1} \tag{6.54}$$

and

$$\frac{d^2y}{dx^2} = \sum_{n=0}^{\infty} (n + r)(n + r - 1)c_n(x - x_0)^{n+r-2}. \tag{6.55}$$

We now substitute the series (6.53), (6.54), and (6.55) for y and its first two derivatives, respectively, into the differential equation (6.1).

3. We now proceed (essentially as in Section 6.1) to simplify the resulting expression so that it takes the form

$$K_0(x - x_0)^{r+k} + K_1(x - x_0)^{r+k+1} + K_2(x - x_0)^{r+k+2} + \cdots = 0, \tag{6.56}$$

where k is a certain integer and the coefficients K_i ($i = 0, 1, 2, \ldots$) are functions of r and certain of the coefficients c_n of the solution (6.53).

4. In order that (6.56) be valid for all x in the deleted interval $0 < x - x_0 < R$, we must set

$$K_0 = K_1 = K_2 = \cdots = 0.$$

5. Upon equating to zero the coefficient K_0 of the *lowest* power $r + k$ of $(x - x_0)$, we obtain a quadratic equation in r, called the *indicial equation* of the differential equation (6.1). The two roots of this quadratic equation are often called the *exponents* of the differential equation (6.1) and are the only possible values for the constant r

in the assumed solution (6.53). Thus at this stage the "unknown" constant r is determined. We denote the roots of the indicial equation by r_1 and r_2, where $\text{Re}(r_1) \geq \text{Re}(r_2)$. Here $\text{Re}(r_j)$ denotes the real part of r_j $(j = 1, 2)$; and of course if r_j is real, then $\text{Re}(r_j)$ is simply r_j itself.

6. We now equate to zero the remaining coefficients $K_1, K_2, \ldots$ in (6.56). We are thus led to a set of conditions, involving the constant r, which must be satisfied by the various coefficients c_n in the series (6.53).

7. We now substitute the root r_1 for r into the conditions obtained in Step 6, and then choose the c_n to satisfy these conditions. If the c_n are so chosen, the resulting series (6.53) with $r = r_1$ is a solution of the desired form. Note that if r_1 and r_2 are real and unequal, then r_1 is the *larger* root.

8. If $r_2 \neq r_1$, we may repeat the procedure of Step 7 using the root r_2 instead of r_1. In this way a second solution of the desired form (6.53) may be obtained. Note that if r_1 and r_2 are real and unequal, then r_2 is the *smaller* root. However, in the case in which r_1 and r_2 are real and unequal, the second solution of the desired form (6.53) obtained in this step may not be linearly independent of the solution obtained in Step 7. Also, in the case in which r_1 and r_2 are real and equal, the solution obtained in this step is clearly identical with the one obtained in Step 7. We shall consider these "exceptional" situations after we have considered an example.

▶ **Example 6.11.** Use the method of Frobenius to find solutions of the differential equation

$$2x^2 \frac{d^2y}{dx^2} - x\frac{dy}{dx} + (x - 5)y = 0 \tag{6.51}$$

in some interval $0 < x < R$.

Solution. Since $x = 0$ is a regular singular point of the differential equation (6.51) and we seek solution for $0 < x < R$, we assume

$$y = \sum_{n=0}^{\infty} c_n x^{n+r}, \tag{6.57}$$

where $c_0 \neq 0$. Then

$$\frac{dy}{dx} = \sum_{n=0}^{\infty} (n + r)c_n x^{n+r-1} \tag{6.58}$$

and

$$\frac{d^2y}{dx^2} = \sum_{n=0}^{\infty} (n + r)(n + r - 1)c_n x^{n+r-2}. \tag{6.59}$$

Substituting the series (6.57), (6.58), and (6.59) into (6.51), we have

$$2\sum_{n=0}^{\infty} (n + r)(n + r - 1)c_n x^{n+r} - \sum_{n=0}^{\infty} (n + r)c_n x^{n+r} + \sum_{n=0}^{\infty} c_n x^{n+r+1} - 5\sum_{n=0}^{\infty} c_n x^{n+r} = 0.$$

Simplifying, as in the examples of Section 6.1, we may write this as

$$\sum_{n=0}^{\infty} [2(n + r)(n + r - 1) - (n + r) - 5]c_n x^{n+r} + \sum_{n=1}^{\infty} c_{n-1} x^{n+r} = 0$$

or

$$[2r(r - 1) - r - 5]c_0 x^r$$

$$+ \sum_{n=1}^{\infty} \{[2(n + r)(n + r - 1) - (n + r) - 5]c_n + c_{n-1}\}x^{n+r} = 0. \quad (6.60)$$

This is of the form (6.56), where $k = 0$.

Equating to zero the coefficient of the lowest power of x (that is, the coefficient of x^r) in (6.60), we are led to the quadratic equation

$$2r(r - 1) - r - 5 = 0$$

(since we have assumed that $c_0 \neq 0$). This is the *indicial* equation of the differential equation (6.51). We write it in the form

$$2r^2 - 3r - 5 = 0$$

and observe that its roots are

$$r_1 = \tfrac{5}{2} \quad \text{and} \quad r_2 = -1.$$

These are the so-called *exponents* of the differential equation (6.51) and are the only possible values for the previously unknown constant r in the solution (6.57). Note that they are real and unequal.

Equating to zero the coefficients of the higher powers of x in (6.60), we obtain the *recurrence formula*

$$[2(n + r)(n + r - 1) - (n + r) - 5]c_n + c_{n-1} = 0, \qquad n \geq 1. \quad (6.61)$$

Letting $r = r_1 = \frac{5}{2}$ in (6.61), we obtain the recurrence formula

$$[2(n + \tfrac{5}{2})(n + \tfrac{3}{2}) - (n + \tfrac{5}{2}) - 5]c_n + c_{n-1} = 0, \qquad n \geq 1,$$

corresponding to the larger root $\frac{5}{2}$ of the indicial equation. This simplifies to the form

$$n(2n + 7)c_n + c_{n-1} = 0, \qquad n \geq 1,$$

or, finally,

$$c_n = -\frac{c_{n-1}}{n(2n + 7)}, \qquad n \geq 1. \quad (6.62)$$

Using (6.62) we find that

$$c_1 = -\frac{c_0}{9}, \quad c_2 = -\frac{c_1}{22} = \frac{c_0}{198}, \quad c_3 = -\frac{c_2}{39} = -\frac{c_0}{7722}, \ldots .$$

Letting $r = \frac{5}{2}$ in (6.57), and using these values of $c_1, c_2, c_3, \ldots$, we obtain the solution

$$\begin{aligned} y &= c_0(x^{5/2} - \tfrac{1}{9}x^{7/2} + \tfrac{1}{198}x^{9/2} - \tfrac{1}{7722}x^{11/2} + \cdots) \\ &= c_0 x^{5/2}(1 - \tfrac{1}{9}x + \tfrac{1}{198}x^2 - \tfrac{1}{7722}x^3 + \cdots), \end{aligned} \quad (6.63)$$

corresponding to the larger root $r_1 = \frac{5}{2}$.

We now let $r = r_2 = -1$ in (6.61) to obtain the recurrence formula

$$[2(n-1)(n-2) - (n-1) - 5]c_n + c_{n-1} = 0, \qquad n \geq 1,$$

corresponding to this smaller root of the indicial equation. This simplifies to the form

$$n(2n-7)c_n + c_{n-1} = 0, \qquad n \geq 1,$$

or, finally

$$c_n = -\frac{c_{n-1}}{n(2n-7)}, \qquad n \geq 1.$$

Using this, we find that

$$c_1 = \tfrac{1}{5}c_0, \qquad c_2 = \tfrac{1}{6}c_1 = \tfrac{1}{30}c_0, \qquad c_3 = \tfrac{1}{3}c_2 = \tfrac{1}{90}c_0, \ldots.$$

Letting $r = -1$ in (6.57), and using these values of $c_1, c_2, c_3, \ldots$ we obtain the solution

$$\begin{aligned} y &= c_0(x^{-1} + \tfrac{1}{5} + \tfrac{1}{30}x + \tfrac{1}{90}x^2 - \cdots) \\ &= c_0x^{-1}(1 + \tfrac{1}{5}x + \tfrac{1}{30}x^2 + \tfrac{1}{90}x^3 - \cdots), \end{aligned} \tag{6.64}$$

corresponding to the smaller exponent $r_2 = -1$.

The two solutions (6.63) and (6.64) corresponding to the two roots $\frac{5}{2}$ and -1, respectively, are linearly independent. Thus the general solution of (6.51) may be written

$$\begin{aligned} y = {} & C_1x^{5/2}(1 - \tfrac{1}{9}x + \tfrac{1}{198}x^2 - \tfrac{1}{7722}x^3 + \cdots) \\ & + C_2x^{-1}(1 + \tfrac{1}{5}x + \tfrac{1}{30}x^2 + \tfrac{1}{90}x^3 - \cdots), \end{aligned}$$

where C_1 and C_2 are arbitrary constants.

Observe that in Example 6.11 *two* linearly independent solutions of the form (6.49) were obtained for $x > 0$. However, in Step 8 of the outline preceding Example 6.11, we indicated that this is not always the case. Thus we are led to ask the following questions:

1. Under what conditions are we assured that the differential equation (6.1) has *two* linearly independent solutions

$$|x - x_0|^r \sum_{n=0}^{\infty} c_n(x - x_0)^n$$

of the form (6.49) about a regular singular point x_0?

2. If the differential equation (6.1) does *not* have two linearly independent solutions of the form (6.49) about a regular singular point x_0, then what is the form of a solution which *is* linearly independent of the basic solution of the form (6.49)?

In answer to these questions we state the following theorem.

THEOREM 6.3

Hypothesis. *Let the point x_0 be a regular singular point of the differential equation (6.1). Let r_1 and r_2 [where* Re $(r_1) \geq$ Re (r_2)*] be the roots of the indicial equation associated with x_0.*

Conclusion 1. *Suppose either $r_1 - r_2 \neq 0$ or $r_1 - r_2 \neq N$, where N is a positive integer. Then the differential equation (6.1) has two nontrivial linearly independent solutions y_1 and y_2 of the form (6.49) given respectively by*

$$y_1(x) = |x - x_0|^{r_1} \sum_{n=0}^{\infty} c_n(x - x_0)^n, \tag{6.65}$$

where $c_0 \neq 0$, and

$$y_2(x) = |x - x_0|^{r_2} \sum_{n=0}^{\infty} c_n^*(x - x_0)^n, \tag{6.66}$$

where $c_0^* \neq 0$.

Conclusion 2. *Suppose $r_1 - r_2 = N$, where N is a positive integer. Then the differential equation (6.1) has two nontrivial linearly independent solutions y_1 and y_2 given respectively by*

$$y_1(x) = |x - x_0|^{r_1} \sum_{n=0}^{\infty} c_n(x - x_0)^n, \tag{6.65}$$

where $c_0 \neq 0$, and

$$y_2(x) = |x - x_0|^{r_2} \sum_{n=0}^{\infty} c_n^*(x - x_0)^n + Cy_1(x) \ln |x - x_0|, \tag{6.67}$$

where $c_0^* \neq 0$ and C is a constant which may or may not be zero.

Conclusion 3. *Suppose $r_1 - r_2 = 0$. Then the differential equation (6.1) has two nontrivial linearly independent solutions y_1 and y_2 given respectively by*

$$y_1(x) = |x - x_0|^{r_1} \sum_{n=0}^{\infty} c_n(x - x_0)^n, \tag{6.65}$$

where $c_0 \neq 0$, and

$$y_2(x) = |x - x_0|^{r_1+1} \sum_{n=0}^{\infty} c_n^*(x - x_0)^n + y_1(x) \ln |x - x_0|. \tag{6.68}$$

The solutions in Conclusions 1, 2, and 3 are valid in some deleted interval $0 < |x - x_0| < R$ about x_0.

In the illustrative examples and exercises which follow, we shall again seek solutions valid in some interval $0 < x - x_0 < R$. We shall therefore discuss the conclusions of Theorem 6.3 for such an interval. Before doing so, we again note that if $0 < x - x_0 < R$, then $|x - x_0|$ is simply $x - x_0$.

From the three conclusions of Theorem 6.3 we see that if x_0 is a regular singular point of (6.1), and $0 < x - x_0 < R$, then there is *always* a solution

$$y_1(x) = (x - x_0)^{r_1} \sum_{n=0}^{\infty} c_n(x - x_0)^n$$

of the form (6.49) for $0 < x - x_0 < R$ corresponding to the root r_1 of the indicial equation associated with x_0. Note again that the root r_1 is the *larger* root if r_1 and r_2 are real and unequal. From Conclusion 1 we see that if $0 < x - x_0 < R$ and the difference $r_1 - r_2$ between the roots of the indicial equation is *not* zero or a positive integer, then there is always a linearly independent solution

$$y_2(x) = (x - x_0)^{r_2} \sum_{n=0}^{\infty} c_n^*(x - x_0)^n$$

of the form (6.49) for $0 < x - x_0 < R$ corresponding to the root r_2. Note that the root r_2 is the *smaller* root if r_1 and r_2 are real and unequal. In particular, observe that if r_1 and r_2 are conjugate complex, then $r_1 - r_2$ is pure imaginary, and there will *always* be a linearly independent solution of the form (6.49) corresponding to r_2. However, from Conclusion 2 we see that if $0 < x - x_0 < R$ and the difference $r_1 - r_2$ *is a positive integer*, then a solution which is linearly independent of the "basic" solution of the form (6.49) for $0 < x - x_0 < R$ is of the generally more complicated form

$$y_2(x) = (x - x_0)^{r_2} \sum_{n=0}^{\infty} c_n^*(x - x_0)^n + Cy_1(x) \ln |x - x_0|$$

for $0 < x - x_0 < R$. Of course, if the constant C in this solution is zero, then it reduces to the simpler type of the form (6.49) for $0 < x - x_0 < R$. Finally, from Conclusion 3, we see that if $r_1 - r_2$ *is zero*, then the linearly independent solution $y_2(x)$ *always* involves the logarithmic term $y_1(x) \ln |x - x_0|$ and is *never* of the simple form (6.49) for $0 < x - x_0 < R$.

We shall now consider several examples which will (1) give further practice in the method of Frobenius, (2) illustrate the conclusions of Theorem 6.3, and (3) indicate how a linearly independent solution of the more complicated form involving the logarithmic term may be found in cases in which it exists. In each example we shall take $x_0 = 0$ and seek solutions valid in some interval $0 < x < R$. Thus note that in each example $|x - x_0| = |x| = x$.

▶ Example 6.12. Use the method of Frobenius to find solutions of the differential equation

$$2x^2 \frac{d^2y}{dx^2} + x \frac{dy}{dx} + (x^2 - 3)y = 0 \qquad (6.69)$$

in some interval $0 < x < R$.

Solution. We observe at once that $x = 0$ is a regular singular point of the differential equation (6.69). Hence, since we seek solutions for $0 < x < R$, we assume a solution

$$y = \sum_{n=0}^{\infty} c_n x^{n+r}, \qquad (6.70)$$

where $c_0 \neq 0$. Differentiating (6.70), we obtain

$$\frac{dy}{dx} = \sum_{n=0}^{\infty} (n + r)c_n x^{n+r-1} \quad \text{and} \quad \frac{d^2y}{dx^2} = \sum_{n=0}^{\infty} (n + r)(n + r - 1)c_n x^{n+r-2}.$$

Upon substituting (6.70) and these derivatives into (6.69), we find

$$2\sum_{n=0}^{\infty}(n+r)(n+r-1)c_n x^{n+r}+\sum_{n=0}^{\infty}(n+r)c_n x^{n+r}$$
$$+\sum_{n=0}^{\infty}c_n x^{n+r+2}-3\sum_{n=0}^{\infty}c_n x^{n+r}=0.$$

Simplifying, as in the previous examples, we write this as

$$\sum_{n=0}^{\infty}[2(n+r)(n+r-1)+(n+r)-3]c_n x^{n+r}+\sum_{n=2}^{\infty}c_{n-2}x^{n+r}=0$$

or

$$[2r(r-1)+r-3]c_0 x^r+[2(r+1)r+(r+1)-3]c_1 x^{r+1}$$
$$+\sum_{n=2}^{\infty}\{[2(n+r)(n+r-1)+(n+r)-3]c_n+c_{n-2}\}x^{n+r}=0. \quad (6.71)$$

This is of the form (6.56), where $k = 0$.

Equating to zero the coefficient of the lowest power of x in (6.71), we obtain the indicial equation

$$2r(r-1)+r-3=0 \quad \text{or} \quad 2r^2-r-3=0.$$

The roots of this equation are

$$r_1=\tfrac{3}{2} \quad \text{and} \quad r_2=-1.$$

Since the difference $r_1 - r_2 = \frac{5}{2}$ between these roots is *not* zero or a positive integer, Conclusion 1 of Theorem 6.3 tells us that Equation (6.69) has *two* linearly independent solutions of the form (6.70), one corresponding to each of the roots r_1 and r_2.

Equating to zero the coefficients of the higher powers of x in (6.71), we obtain the condition

$$[2(r+1)r+(r+1)-3]c_1=0 \quad (6.72)$$

and the recurrence formula

$$[2(n+r)(n+r-1)+(n+r)-3]c_n+c_{n-2}=0, \qquad n\geq 2. \quad (6.73)$$

Letting $r = r_1 = \frac{3}{2}$ in (6.72), we obtain $7c_1 = 0$ and hence $c_1 = 0$. Letting $r = r_1 = \frac{3}{2}$ in (6.73), we obtain (after slight simplifications) the recurrence formula

$$n(2n+5)c_n+c_{n-2}=0, \qquad n\geq 2,$$

corresponding to the larger root $\frac{3}{2}$. Writing this in the form

$$c_n=-\frac{c_{n-2}}{n(2n+5)}, \qquad n\geq 2,$$

we obtain

$$c_2=-\frac{c_0}{18}, \qquad c_3=-\frac{c_1}{33}=0 \text{ (since } c_1=0), \qquad c_4=-\frac{c_2}{52}=\frac{c_0}{936}, \ldots .$$

Note that *all* odd coefficients are zero, since $c_1 = 0$. Letting $r = \frac{3}{2}$ in (6.70) and using these values of $c_1, c_2, c_3, \ldots$, we obtain the solution corresponding to the larger root $r_1 = \frac{3}{2}$. This solution is $y = y_1(x)$, where

$$y_1(x) = c_0 x^{3/2}(1 - \tfrac{1}{18}x^2 + \tfrac{1}{936}x^4 - \cdots). \tag{6.74}$$

Now let $r = r_2 = -1$ in (6.72). We obtain $-3c_1 = 0$ and hence $c_1 = 0$. Letting $r = r_2 = -1$ in (6.73), we obtain the recurrence formula

$$n(2n - 5)c_n + c_{n-2} = 0, \qquad n \geq 2,$$

corresponding to the smaller root -1. Writing this in the form

$$c_n = -\frac{c_{n-2}}{n(2n - 5)}, \qquad n \geq 2,$$

we obtain

$$c_2 = \frac{c_0}{2}, \qquad c_3 = -\frac{c_1}{3} = 0 \text{ (since } c_1 = 0), \qquad c_4 = -\frac{c_2}{12} = -\frac{c_0}{24}, \ldots.$$

In this case also all odd coefficients are zero. Letting $r = -1$ in (6.70) and using these values of $c_1, c_2, c_3, \ldots$, we obtain the solution corresponding to the smaller root $r_2 = -1$. This solution is $y = y_2(x)$, where

$$y_2(x) = c_0 x^{-1}(1 + \tfrac{1}{2}x^2 - \tfrac{1}{24}x^4 + \cdots). \tag{6.75}$$

Since the solutions defined by (6.74) and (6.75) are linearly independent, the general solution of (6.69) may be written

$$y = C_1 y_1(x) + C_2 y_2(x),$$

where C_1 and C_2 are arbitrary constants and $y_1(x)$ and $y_2(x)$ are defined by (6.74) and (6.75), respectively.

▶ **Example 6.13.** Use the method of Frobenius to find solutions of the differential equation

$$x^2 \frac{d^2y}{dx^2} - x\frac{dy}{dx} - (x^2 + \tfrac{5}{4})y = 0 \tag{6.76}$$

in some interval $0 < x < R$.

Solution. We observe that $x = 0$ is a regular singular point of this differential equation and we seek solutions for $0 < x < R$. Hence we assume a solution

$$y = \sum_{n=0}^{\infty} c_n x^{n+r}, \tag{6.77}$$

where $c_0 \neq 0$. Upon differentiating (6.77) twice and substituting into (6.76), we obtain

$$\sum_{n=0}^{\infty} (n + r)(n + r - 1)c_n x^{n+r} - \sum_{n=0}^{\infty} (n + r)c_n x^{n+r} - \sum_{n=0}^{\infty} c_n x^{n+r+2} - \frac{5}{4}\sum_{n=0}^{\infty} c_n x^{n+r} = 0.$$

Simplifying, we write this in the form

$$\sum_{n=0}^{\infty} [(n + r)(n + r - 1) - (n + r) - \tfrac{5}{4}]c_n x^{n+r} - \sum_{n=2}^{\infty} c_{n-2} x^{n+r} = 0$$

or

$$[r(r - 1) - r - \tfrac{5}{4}]c_0 x^r + [(r + 1)r - (r + 1) - \tfrac{5}{4}]c_1 x^{r+1} \tag{6.78}$$
$$+ \sum_{n=2}^{\infty} \{[(n + r)(n + r - 1) - (n + r) - \tfrac{5}{4}]c_n - c_{n-2}\}x^{n+r} = 0.$$

Equating to zero the coefficient of the lowest power of x in (6.78), we obtain the indicial equation

$$r^2 - 2r - \tfrac{5}{4} = 0.$$

The roots of this equation are

$$r_1 = \tfrac{5}{2}, \qquad r_2 = -\tfrac{1}{2}.$$

Although these roots themselves are not integers, the difference $r_1 - r_2$ between them is the positive integer 3. By Conclusion 2 of Theorem 6.3 we know that the differential equation (6.76) has a solution of the assumed form (6.77) corresponding to the larger root $r_1 = \frac{5}{2}$. We proceed to obtain this solution.

Equating to zero the coefficients of the higher powers of x in (6.78), we obtain the condition

$$[(r + 1)r - (r + 1) - \tfrac{5}{4}]c_1 = 0 \tag{6.79}$$

and the recurrence formula

$$[(n + r)(n + r - 1) - (n + r) - \tfrac{5}{4}]c_n - c_{n-2} = 0, \qquad n \geq 2. \tag{6.80}$$

Letting $r = r_1 = \frac{5}{2}$ in (6.79), we obtain

$$4c_1 = 0 \text{ and hence } c_1 = 0.$$

Letting $r = r_1 = \frac{5}{2}$ in (6.80), we obtain the recurrence formula

$$n(n + 3)c_n - c_{n-2} = 0, \qquad n \geq 2,$$

corresponding to the larger root $\frac{5}{2}$. Since $n \geq 2$, we may write this in the form

$$c_n = \frac{c_{n-2}}{n(n + 3)}, \qquad n \geq 2.$$

From this we obtain successively

$$c_2 = \frac{c_0}{2 \cdot 5}, \qquad c_3 = \frac{c_1}{3 \cdot 6} = 0 \text{ (since } c_1 = 0), \qquad c_4 = \frac{c_2}{4 \cdot 7} = \frac{c_0}{2 \cdot 4 \cdot 5 \cdot 7}, \ldots.$$

We note that all odd coefficients are zero. The general even coefficient may be written

$$c_{2n} = \frac{c_0}{[2 \cdot 4 \cdot 6 \cdots (2n)][5 \cdot 7 \cdot 9 \cdots (2n + 3)]}, \qquad n \geq 1.$$

Letting $r = \frac{5}{2}$ in (6.77) and using these values of c_{2n}, we obtain the solution corresponding to the larger root $r_1 = \frac{5}{2}$. This solution is $y = y_1(x)$, where

$$\begin{aligned} y_1(x) &= c_0 x^{5/2}\left[1 + \frac{x^2}{2 \cdot 5} + \frac{x^4}{2 \cdot 4 \cdot 5 \cdot 7} + \cdots\right], \\ &= c_0 x^{5/2}\left[1 + \sum_{n=1}^{\infty} \frac{x^{2n}}{[2 \cdot 4 \cdot 6 \cdots (2n)][5 \cdot 7 \cdot 9 \cdots (2n+3)]}\right]. \end{aligned} \tag{6.81}$$

We now consider the smaller root $r_2 = -\frac{1}{2}$. Theorem 6.3 does *not* assure us that the differential equation (6.76) has a linearly independent solution of the assumed form (6.77) corresponding to this smaller root. Conclusion 2 of that theorem merely tells us that there is a linearly independent solution of the form

$$\sum_{n=0}^{\infty} c_n^* x^{n+r_2} + C y_1(x) \ln x, \tag{6.82}$$

where C may or may not be zero. Of course, if $C = 0$, then the linearly independent solution (6.82) *is* of the assumed form (6.77) and we can let $r = r_2 = -\frac{1}{2}$ in the formula (6.79) and the recurrence formula (6.80) and proceed as in the previous examples. Let us assume (hopefully, but without justification!) that this is indeed the case.

Thus we let $r = r_2 = -\frac{1}{2}$ in (6.79) and (6.80). Letting $r = -\frac{1}{2}$ in (6.79) we obtain $-2c_1 = 0$ and hence $c_1 = 0$. Letting $r = -\frac{1}{2}$ in (6.80) we obtain the recurrence formula

$$n(n-3)c_n - c_{n-2} = 0, \qquad n \geq 2, \tag{6.83}$$

corresponding to the smaller root $-\frac{1}{2}$. For $n \neq 3$, this may be written

$$c_n = \frac{c_{n-2}}{n(n-3)}, \qquad n \geq 2, \quad n \neq 3. \tag{6.84}$$

For $n = 2$, formula (6.84) gives $c_2 = -c_0/2$. For $n = 3$, formula (6.84) does not apply and we must use (6.83). For $n = 3$ formula (6.83) is $0 \cdot c_3 - c_1 = 0$ or simply $0 = 0$ (since $c_1 = 0$). Hence, for $n = 3$, the recurrence formula (6.83) is automatically satisfied with *any* choice of c_3. Thus c_3 is independent of the arbitrary constant c_0; it is a second arbitrary constant! For $n > 3$, we can again use (6.84). Proceeding, we have

$$c_4 = \frac{c_2}{4} = -\frac{c_0}{2 \cdot 4}, \qquad c_5 = \frac{c_3}{2 \cdot 5},$$

$$c_6 = \frac{c_4}{6 \cdot 3} = -\frac{c_0}{2 \cdot 4 \cdot 6 \cdot 3}, \qquad c_7 = \frac{c_5}{4 \cdot 7} = \frac{c_3}{2 \cdot 4 \cdot 5 \cdot 7},$$

$$\ldots .$$

We note that all even coefficients may be expressed in terms of c_0 and that all odd coefficients beyond c_3 may be expressed in terms of c_3. In fact, we may write

$$c_{2n} = -\frac{c_0}{[2 \cdot 4 \cdot 6 \cdots (2n)][3 \cdot 5 \cdot 7 \cdots (2n-3)]}, \qquad n \geq 3$$

(even coefficients c_6 and beyond), and

$$c_{2n+1} = \frac{c_3}{[2 \cdot 4 \cdot 6 \cdots (2n-2)][5 \cdot 7 \cdot 9 \cdots (2n+1)]}, \qquad n \geq 2$$

(odd coefficients c_5 and beyond). Letting $r = -\frac{1}{2}$ in (6.77) and using the values of c_n in terms of c_0 (for even n) and c_3 (for odd n beyond c_3), we obtain the solution corresponding to the smaller root $r_2 = -\frac{1}{2}$. This solution is $y = y_2(x)$, where

$$\begin{aligned} y_2(x) &= c_0 x^{-1/2}\left[1 - \frac{x^2}{2} - \frac{x^4}{2 \cdot 4} - \frac{x^6}{2 \cdot 4 \cdot 6 \cdot 3} - \cdots\right] \\ &\quad + c_3 x^{-1/2}\left[x^3 + \frac{x^5}{2 \cdot 5} + \frac{x^7}{2 \cdot 4 \cdot 5 \cdot 7} + \cdots\right] \\ &= c_0 x^{-1/2}\left[1 - \frac{x^2}{2} - \frac{x^4}{2 \cdot 4} - \sum_{n=3}^{\infty} \frac{x^{2n}}{[2 \cdot 4 \cdot 6 \cdots (2n)][3 \cdot 5 \cdot 7 \cdots (2n-3)]}\right] \\ &\quad + c_3 x^{-1/2}\left[x^3 + \sum_{n=2}^{\infty} \frac{x^{2n+1}}{[2 \cdot 4 \cdot 6 \cdots (2n-2)][5 \cdot 7 \cdot 9 \cdots (2n+1)]}\right], \end{aligned} \tag{6.85}$$

and c_0 and c_3 are arbitrary constants.

If we now let $c_0 = 1$ in (6.81) we obtain the particular solution $y = y_{11}(x)$, where

$$y_{11}(x) = x^{5/2}\left[1 + \sum_{n=1}^{\infty} \frac{x^{2n}}{[2 \cdot 4 \cdot 6 \cdots (2n)][5 \cdot 7 \cdot 9 \cdots (2n+3)]}\right]$$

corresponding to the larger root $\frac{5}{2}$; and if we let $c_0 = 1$ and $c_3 = 0$ in (6.85) we obtain the particular solution $y = y_{21}(x)$, where

$$y_{21}(x) = x^{-1/2}\left[1 - \frac{x^2}{2} - \frac{x^4}{2 \cdot 4} - \sum_{n=3}^{\infty} \frac{x^{2n}}{[2 \cdot 4 \cdot 6 \cdots (2n)][3 \cdot 5 \cdot 7 \cdots (2n-3)]}\right]$$

corresponding to the smaller root $-\frac{1}{2}$. These two particular solutions, which are both of the assumed form (6.77), are linearly independent. Thus the general solution of the differential equation (6.76) may be written

$$y = C_1 y_{11}(x) + C_2 y_{21}(x), \tag{6.86}$$

where C_1 and C_2 are arbitrary constants.

Now let us examine more carefully the solution y_2 defined by (6.85). The expression

$$x^{-1/2}\left[x^3 + \sum_{n=2}^{\infty} \frac{x^{2n+1}}{[2 \cdot 4 \cdot 6 \cdots (2n-2)][5 \cdot 7 \cdot 9 \cdots (2n+1)]}\right]$$

of which c_3 is the coefficient in (6.85) may be written

$$\begin{aligned} &x^{5/2}\left[1 + \sum_{n=2}^{\infty} \frac{x^{2n-2}}{[2 \cdot 4 \cdot 6 \cdots (2n-2)][5 \cdot 7 \cdot 9 \cdots (2n+1)]}\right] \\ &\qquad = x^{5/2}\left[1 + \sum_{n=1}^{\infty} \frac{x^{2n}}{[2 \cdot 4 \cdot 6 \cdots (2n)][5 \cdot 7 \cdot 9 \cdots (2n+3)]}\right] \end{aligned}$$

and this is precisely $y_{11}(x)$. Thus we may write

$$y_2(x) = c_0 y_{21}(x) + c_3 y_{11}(x), \tag{6.87}$$

where c_0 and c_3 are arbitrary constants. Now compare (6.86) and (6.87). We see that the solution $y = y_2(x)$ by itself is actually the general solution of the differential equation (6.76), even though $y_2(x)$ was obtained using only the smaller root $-\frac{1}{2}$.

We thus observe that if the difference $r_1 - r_2$ between the roots of the indicial equation is a positive integer, it is sometimes possible to obtain the general solution using the smaller root alone, without bothering to find explicitly the solution corresponding to the larger root. Indeed, if the difference $r_1 - r_2$ is a positive integer, it is a worthwhile practice to work with the smaller root first, in the hope that this smaller root by itself may lead directly to the general solution.

▶ **Example 6.14.** Use the method of Frobenius to find solutions of the differential equation

$$x^2 \frac{d^2y}{dx^2} + (x^2 - 3x)\frac{dy}{dx} + 3y = 0 \tag{6.88}$$

in some interval $0 < x < R$.

Solution. We observe that $x = 0$ is a regular singular point of (6.88) and we seek solutions for $0 < x < R$. Hence we assume a solution

$$y = \sum_{n=0}^{\infty} c_n x^{n+r}, \tag{6.89}$$

where $c_0 \neq 0$. Upon differentiating (6.89) twice and substituting into (6.88), we obtain

$$\sum_{n=0}^{\infty} (n + r)(n + r - 1)c_n x^{n+r} + \sum_{n=0}^{\infty} (n + r)c_n x^{n+r+1}$$
$$- 3\sum_{n=0}^{\infty} (n + r)c_n x^{n+r} + 3\sum_{n=0}^{\infty} c_n x^{n+r} = 0.$$

Simplifying, we write this in the form

$$\sum_{n=0}^{\infty} [(n + r)(n + r - 1) - 3(n + r) + 3]c_n x^{n+r} + \sum_{n=1}^{\infty} (n + r - 1)c_{n-1} x^{n+r} = 0$$

or

$$[r(r - 1) - 3r + 3]c_0 x^r + \sum_{n=1}^{\infty} \{[(n + r)(n + r - 1) - 3(n + r) + 3]c_n + (n + r - 1)c_{n-1}\}x^{n+r} = 0. \tag{6.90}$$

Equating to zero the coefficient of the lowest power of x in (6.90) we obtain the indicial equation

$$r^2 - 4r + 3 = 0.$$

The roots of this equation are

$$r_1 = 3, \qquad r_2 = 1.$$

The difference $r_1 - r_2$ between these roots is the positive integer 2. We know from Theorem 6.3 that the differential equation (6.88) has a solution of the assumed form (6.89) corresponding to the larger root $r_1 = 3$. We shall find this first, even though the results of Example 6.13 suggest that we should work first with the smaller root $r_2 = 1$ in the hopes of finding the general solution directly from this smaller root.

Equating to zero the coefficients of the higher powers of x in (6.90), we obtain the recurrence formula

$$[(n + r)(n + r - 1) - 3(n + r) + 3]c_n + (n + r - 1)c_{n-1} = 0, \qquad n \geq 1. \tag{6.91}$$

Letting $r = r_1 = 3$ in (6.91), we obtain the recurrence formula

$$n(n + 2)c_n + (n + 2)c_{n-1} = 0, \qquad n \geq 1,$$

corresponding to the larger root 3. Since $n \geq 1$, we may write this in the form

$$c_n = -\frac{c_{n-1}}{n}, \qquad n \geq 1.$$

From this we find successively

$$c_1 = -c_0, \qquad c_2 = -\frac{c_1}{2} = \frac{c_0}{2!}, \qquad c_3 = -\frac{c_2}{3} = -\frac{c_0}{3!}, \ldots, c_n = \frac{(-1)^n c_0}{n!}, \ldots.$$

Letting $r = 3$ in (6.89) and using these values of c_n, we obtain the solution corresponding to the larger root $r_1 = 3$. This solution is $y = y_1(x)$, where

$$y_1(x) = c_0 x^3 \left[1 - x + \frac{x^2}{2!} - \frac{x^3}{3!} + \cdots + \frac{(-1)^n x^n}{n!} + \cdots\right].$$

We recognize the series in brackets in this solution as the Maclaurin expansion for e^{-x}. Thus we may write

$$y_1(x) = c_0 x^3 e^{-x} \tag{6.92}$$

and express the solution corresponding to r_1 in the closed form

$$y = c_0 x^3 e^{-x},$$

where c_0 is an arbitrary constant.

We now consider the smaller root $r_2 = 1$. As in Example 6.13, we have no assurance that the differential equation has a linearly independent solution of the assumed form (6.89) corresponding to this smaller root. However, as in that example, we shall tentatively assume that such a solution actually does exist and let $r = r_2 = 1$ in (6.91) in the hopes of finding "it." Further, we are now aware that this step by itself *might* even provide us with the *general* solution.

Thus we let $r = r_2 = 1$ in (6.91) to obtain the recurrence formula

$$n(n - 2)c_n + nc_{n-1} = 0, \qquad n \geq 1, \tag{6.93}$$

corresponding to the smaller root 1. For $n \neq 2$, this may be written

$$c_n = -\frac{c_{n-1}}{n - 2}, \qquad n \geq 1, \quad n \neq 2. \tag{6.94}$$

For $n = 1$, formula (6.94) gives $c_1 = c_0$. For $n = 2$, formula (6.94) does not apply and we must use (6.93). For $n = 2$ formula (6.93) is $0 \cdot c_2 + 2c_1 = 0$, and hence we must have $c_1 = 0$. But then, since $c_1 = c_0$, we must have $c_0 = 0$. However, $c_0 \neq 0$ in the assumed solution (6.89). This contradiction shows that there is no solution of the form (6.89), with $c_0 \neq 0$, corresponding to the smaller root 1.

Further, we observe that the use of (6.94) for $n \geq 3$ will only lead us to the solution y_1 already obtained. For, from the condition $0 \cdot c_2 + 2c_1 = 0$ we see that c_2 is arbitrary; and using (6.94) for $n \geq 3$, we obtain successively

$$c_3 = -c_2, \qquad c_4 = -\frac{c_3}{2} = \frac{c_2}{2!}, \qquad c_5 = -\frac{c_4}{3} = -\frac{c_2}{3!}, \ldots, c_{n+2} = \frac{(-1)^n c_2}{n!}, \quad n \geq 1, \ldots.$$

Thus letting $r = 1$ in (6.89) and using these values of c_n, we obtain formally

$$\begin{aligned} y &= c_2 x \left[x^2 - x^3 + \frac{x^4}{2!} - \frac{x^5}{3!} + \cdots + \frac{(-1)^n x^{n+2}}{n!} + \cdots \right] \\ &= c_2 x^3 \left[1 - x + \frac{x^2}{2!} - \frac{x^3}{3!} + \cdots + \frac{(-1)^n x^n}{n!} + \cdots \right] \\ &= c_2 x^3 e^{-x}. \end{aligned}$$

Comparing this with (6.92), we see that it is essentially the solution $y = y_1(x)$.

We now seek a solution of (6.88) which is linearly independent of the solution y_1. From Theorem 6.3 we now know that this solution is of the form

$$\sum_{n=0}^{\infty} c_n^* x^{n+1} + C y_1(x) \ln x, \tag{6.95}$$

where $c_0^* \neq 0$ and $C \neq 0$. Various methods for obtaining such a solution are available; we shall employ the method of reduction of order (Section 4.1). We let $y = f(x)v$, where $f(x)$ is a known solution of (6.88). Choosing for f the known solution y_1 defined by (6.92), with $c_0 = 1$, we thus let

$$y = x^3 e^{-x} v. \tag{6.96}$$

From this we obtain

$$\frac{dy}{dx} = x^3 e^{-x} \frac{dv}{dx} + (3x^2 e^{-x} - x^3 e^{-x}) v \tag{6.97}$$

and

$$\frac{d^2y}{dx^2} = x^3 e^{-x} \frac{d^2v}{dx^2} + 2(3x^2 e^{-x} - x^3 e^{-x}) \frac{dv}{dx} + (x^3 e^{-x} - 6x^2 e^{-x} + 6x e^{-x}) v. \tag{6.98}$$

Substituting (6.96), (6.97), and (6.98) for y and its first two derivatives, respectively, in the differential equation (6.88), after some simplifications we obtain

$$x \frac{d^2v}{dx^2} + (3 - x) \frac{dv}{dx} = 0. \tag{6.99}$$

Letting $w = dv/dx$, this reduces at once to the first-order differential equation

$$x \frac{dw}{dx} + (3 - x)w = 0,$$

a particular solution of which is $w = x^{-3}e^x$. Thus a particular solution of (6.99) is given by

$$v = \int x^{-3}e^x \, dx,$$

and hence $y = y_2(x)$, where

$$y_2(x) = x^3e^{-x} \int x^{-3}e^x \, dx \tag{6.100}$$

is a particular solution of (6.88) which is linearly independent of the solution y_1 defined by (6.92).

We now show that the solution y_2 defined by (6.100) is of the form (6.95). Introducing the Maclaurin series for e^x in (6.100) we have

$$\begin{aligned} y_2(x) &= x^3e^{-x} \int x^{-3}\left(1 + x + \frac{x^2}{2} + \frac{x^3}{6} + \frac{x^4}{24} + \cdots\right) dx \\ &= x^3e^{-x} \int \left(x^{-3} + x^{-2} + \frac{1}{2}x^{-1} + \frac{1}{6} + \frac{x}{24} + \cdots\right) dx. \end{aligned}$$

Integrating term by term, we obtain

$$y_2(x) = x^3e^{-x}\left[-\frac{1}{2x^2} - \frac{1}{x} + \frac{1}{2}\ln x + \frac{1}{6}x + \frac{1}{48}x^2 + \cdots\right].$$

Now introducing the Maclaurin series for e^{-x}, we may write

$$y_2(x) = \left(x^3 - x^4 + \frac{x^5}{2} - \frac{x^6}{6} + \cdots\right)\left(-\frac{1}{2x^2} - \frac{1}{x} + \frac{1}{6}x + \frac{1}{48}x^2 + \cdots\right) + \tfrac{1}{2}x^3e^{-x}\ln x.$$

Finally, multiplying the two series involved, we have

$$y_2(x) = (-\tfrac{1}{2}x - \tfrac{1}{2}x^2 + \tfrac{3}{4}x^3 - \tfrac{1}{4}x^4 + \cdots) + \tfrac{1}{2}x^3e^{-x}\ln x,$$

which is of the form (6.95), where $y_1(x) = x^3e^{-x}$. The general solution of the differential equation (6.88) may thus be written

$$y = C_1y_1(x) + C_2y_2(x),$$

where C_1 and C_2 are arbitrary constants.

In this example it was fortunate that we were able to express the first solution y_1 in closed form, for this simplified the computations involved in finding the second solution y_2. Of course the method of reduction of order may be applied to find the second solution, even if we cannot express the first solution in closed form. In such cases the various steps of the method must be carried out in terms of the series expression for y_1. The computations which result are generally quite complicated.

Examples 6.12, 6.13, and 6.14 illustrate all of the possibilities listed in the conclusions of Theorem 6.3 except the case in which $r_1 - r_2 = 0$ (that is, the case in which the roots of the indicial equation are equal). In this case it is obvious that for $0 < x - x_0 < R$ both roots lead to the *same* solution

$$y_1 = (x - x_0)^r \sum_{n=0}^{\infty} c_n(x - x_0)^n,$$

where r is the common value of r_1 and r_2. Thus, as Conclusion 3 of Theorem 6.3 states, for $0 < x - x_0 < R$, a linearly independent solution is of the form

$$y_2 = (x - x_0)^{r+1} \sum_{n=0}^{\infty} c_n^*(x - x_0)^n + y_1(x) \ln (x - x_0),$$

Once $y_1(x)$ has been found, we may obtain y_2 by the method of reduction of order. This procedure has already been illustrated in finding the second solution of the equation in Example 6.14. A further illustration is provided in Section 6.3 by the solution of Bessel's equation of order zero.

Exercises

Locate and classify the singular points of each of the differential equations in Exercises 1 through 4.

1. $(x^2 - 3x)\dfrac{d^2y}{dx^2} + (x + 2)\dfrac{dy}{dx} + y = 0.$

2. $(x^3 + x^2)\dfrac{d^2y}{dx^2} + (x^2 - 2x)\dfrac{dy}{dx} + 4y = 0.$

3. $(x^4 - 2x^3 + x^2)\dfrac{d^2y}{dx^2} + 2(x - 1)\dfrac{dy}{dx} + x^2y = 0.$

4. $(x^5 + x^4 - 6x^3)\dfrac{d^2y}{dx^2} + x^2\dfrac{dy}{dx} + (x - 2)y = 0.$

Use the method of Frobenius to find solutions near $x = 0$ of each of the differential equations in Exercises 5 through 26.

5. $2x^2\dfrac{d^2y}{dx^2} + x\dfrac{dy}{dx} + (x^2 - 1)y = 0.$

6. $2x^2\dfrac{d^2y}{dx^2} + x\dfrac{dy}{dx} + (2x^2 - 3)y = 0.$

7. $x^2\dfrac{d^2y}{dx^2} - x\dfrac{dy}{dx} + \left(x^2 + \dfrac{8}{9}\right)y = 0.$

8. $x^2\dfrac{d^2y}{dx^2} - x\dfrac{dy}{dx} + \left(2x^2 + \dfrac{5}{9}\right)y = 0.$

9. $x^2 \dfrac{d^2y}{dx^2} + x\dfrac{dy}{dx} + \left(x^2 - \dfrac{1}{9}\right) y = 0.$

10. $2x \dfrac{d^2y}{dx^2} + \dfrac{dy}{dx} + 2y = 0.$

11. $3x \dfrac{d^2y}{dx^2} - (x - 2)\dfrac{dy}{dx} - 2y = 0.$

12. $x \dfrac{d^2y}{dx^2} + 2\dfrac{dy}{dx} + xy = 0.$

13. $x^2 \dfrac{d^2y}{dx^2} + x\dfrac{dy}{dx} + \left(x^2 - \dfrac{1}{4}\right) y = 0.$

14. $x^2 \dfrac{d^2y}{dx^2} + (x^4 + x)\dfrac{dy}{dx} - y = 0.$

15. $x \dfrac{d^2y}{dx^2} - (x^2 + 2)\dfrac{dy}{dx} + xy = 0.$

16. $x^2 \dfrac{d^2y}{dx^2} + x^2\dfrac{dy}{dx} - 2y = 0.$

17. $(2x^2 - x)\dfrac{d^2y}{dx^2} + (2x - 2)\dfrac{dy}{dx} + (-2x^2 + 3x - 2)y = 0.$

18. $x^2 \dfrac{d^2y}{dx^2} - x\dfrac{dy}{dx} + \dfrac{3}{4} y = 0.$

19. $x^2 \dfrac{d^2y}{dx^2} + x\dfrac{dy}{dx} + (x - 1)y = 0.$

20. $x^2 \dfrac{d^2y}{dx^2} + (x^3 - x)\dfrac{dy}{dx} - 3y = 0.$

21. $x^2 \dfrac{d^2y}{dx^2} - x\dfrac{dy}{dx} + 8(x^2 - 1)y = 0.$

22. $x^2 \dfrac{d^2y}{dx^2} + x^2\dfrac{dy}{dx} - \dfrac{3}{4} y = 0.$

23. $x \dfrac{d^2y}{dx^2} + \dfrac{dy}{dx} + 2y = 0.$

24. $2x \dfrac{d^2y}{dx^2} + 6\dfrac{dy}{dx} + y = 0.$

25. $x^2 \dfrac{d^2y}{dx^2} - x\dfrac{dy}{dx} + (x^2 + 1)y = 0.$

26. $x^2 \dfrac{d^2y}{dx^2} - x\dfrac{dy}{dx} + (x^2 - 3)y = 0.$

6.3 Bessel's Equation and Bessel Functions

A. Bessel's Equation of Order Zero

The differential equation

$$x^2 \frac{d^2y}{dx^2} + x \frac{dy}{dx} + (x^2 - p^2)y - 0, \tag{6.101}$$

where p is a parameter, is called *Bessel's equation of order p*. Any solution of Bessel's equation of order p is called a *Bessel function of order p*. Bessel's equation and Bessel functions occur in connection with many problems of physics and engineering, and there is an extensive literature dealing with the theory and application of this equation and its solutions.

If $p = 0$, Equation (6.101) is equivalent to the equation

$$x \frac{d^2y}{dx^2} + \frac{dy}{dx} + xy = 0, \tag{6.102}$$

which is called *Bessel's equation of order zero*. We shall seek solutions of this equation which are valid in an interval $0 < x < R$. We observe at once that $x = 0$ is a regular singular point of (6.102); and hence, since we seek solutions for $0 < x < R$, we assume a solution

$$y = \sum_{n=0}^{\infty} c_n x^{n+r}, \tag{6.103}$$

where $c_0 \neq 0$. Upon differentiating (6.103) twice and substituting into (6.102), we obtain

$$\sum_{n=0}^{\infty} (n + r)(n + r - 1)c_n x^{n+r-1} + \sum_{n=0}^{\infty} (n + r)c_n x^{n+r-1} + \sum_{n=0}^{\infty} c_n x^{n+r+1} = 0.$$

Simplifying, we write this in the form

$$\sum_{n=0}^{\infty} (n + r)^2 c_n x^{n+r-1} + \sum_{n=2}^{\infty} c_{n-2} x^{n+r-1} = 0$$

or

$$r^2 c_0 x^{r-1} + (1 + r)^2 c_1 x^r + \sum_{n=2}^{\infty} [(n + r)^2 c_n + c_{n-2}]x^{n+r-1} = 0. \tag{6.104}$$

Equating to zero the coefficient of the lowest power of x in (6.104), we obtain the indicial equation $r^2 = 0$, which has equal roots $r_1 = r_2 = 0$. Equating to zero the coefficients of the higher powers of x in (6.104) we obtain

$$(1 + r)^2 c_1 = 0 \tag{6.105}$$

and the recurrence formula

$$(n + r)^2 c_n + c_{n-2} = 0, \qquad n \geq 2. \tag{6.106}$$

Letting $r = 0$ in (6.105), we find at once that $c_1 = 0$. Letting $r = 0$ in (6.106) we obtain the recurrence formula in the form

$$n^2 c_n + c_{n-2} = 0, \qquad n \geq 2,$$

or

$$c_n = -\frac{c_{n-2}}{n^2}, \qquad n \geq 2.$$

From this we obtain successively

$$c_2 = -\frac{c_0}{2^2}, \qquad c_3 = -\frac{c_1}{3^2} = 0 \text{ (since } c_1 = 0), \qquad c_4 = -\frac{c_2}{4^2} = \frac{c_0}{2^2 \cdot 4^2}, \ldots.$$

We note that all odd coefficients are zero and that the general even coefficient may be written

$$c_{2n} = \frac{(-1)^n c_0}{2^2 \cdot 4^2 \cdot 6^2 \cdots (2n)^2} = \frac{(-1)^n c_0}{(n!)^2 2^{2n}}, \qquad n \geq 1.$$

Letting $r = 0$ in (6.103) and using these values of c_{2n}, we obtain the solution $y = y_1(x)$, where

$$y_1(x) = c_0 \sum_{n=0}^{\infty} \frac{(-1)^n}{(n!)^2} \left(\frac{x}{2}\right)^{2n}.$$

If we set the arbitrary constant $c_0 = 1$, we obtain an important particular solution of Equation (6.102). This particular solution defines a function denoted by J_0 and called the *Bessel function of the first kind of order zero*. That is, the function J_0 is the particular solution of Equation (6.102) defined by

$$J_0(x) = \sum_{n=0}^{\infty} \frac{(-1)^n}{(n!)^2} \left(\frac{x}{2}\right)^{2n}. \tag{6.107}$$

Writing out the first few terms of this series solution, we see that

$$\begin{aligned} J_0(x) &= 1 - \frac{1}{(1!)^2}\left(\frac{x}{2}\right)^2 + \frac{1}{(2!)^2}\left(\frac{x}{2}\right)^4 - \frac{1}{(3!)^2}\left(\frac{x}{2}\right)^6 + \cdots \\ &= 1 - \frac{x^2}{4} + \frac{x^4}{64} - \frac{x^6}{2304} + \cdots. \end{aligned} \tag{6.108}$$

Since the roots of the indicial equation are equal, we know from Theorem 6.3 that a solution of Equation (6.102) which is linearly independent of J_0 must be of the form

$$y = x \sum_{n=0}^{\infty} c_n^* x^n + J_0(x) \ln x,$$

for $0 < x < R$. Also, we know that such a linearly independent solution can be found by the method of reduction of order (Section 4.1). Indeed from Theorem 4.7 we know that this linearly independent solution y_2 is given by

$$y_2(x) = J_0(x) \int \frac{e^{-\int dx/x}}{[J_0(x)]^2}\, dx$$

and hence by

$$y_2(x) = J_0(x) \int \frac{dx}{x[J_0(x)]^2}.$$

From (6.108) we find that

$$[J_0(x)]^2 = 1 - \frac{x^2}{2} + \frac{3x^4}{32} - \frac{5x^6}{576} + \cdots$$

and hence

$$\frac{1}{[J_0(x)]^2} = 1 + \frac{x^2}{2} + \frac{5x^4}{32} + \frac{23x^6}{576} + \cdots.$$

Thus

$$\begin{aligned} y_2(x) &= J_0(x) \int \left(\frac{1}{x} + \frac{x}{2} + \frac{5x^3}{32} + \frac{23x^5}{576} + \cdots \right) dx \\ &= J_0(x) \left(\ln x + \frac{x^2}{4} + \frac{5x^4}{128} + \frac{23x^6}{3456} + \cdots \right) \\ &= J_0(x) \ln x + \left(1 - \frac{x^2}{4} + \frac{x^4}{64} - \frac{x^6}{2304} + \cdots \right)\left(\frac{x^2}{4} + \frac{5x^4}{128} + \frac{23x^6}{3456} + \cdots \right) \\ &= J_0(x) \ln x + \frac{x^2}{4} - \frac{3x^4}{128} + \frac{11x^6}{13824} + \cdots. \end{aligned}$$

We thus obtain the first few terms of the "second" solution y_2 by the method of reduction of order. However, our computations give no information concerning the general coefficient c_{2n}^* in the above series. Indeed, it seems unlikely that an expression for the general coefficient can be found. However, let us observe that

$$(-1)^2 \frac{1}{2^2(1!)^2}(1) = \frac{1}{2^2} = \frac{1}{4},$$

$$(-1)^3 \frac{1}{2^4(2!)^2}\left(1 + \frac{1}{2}\right) = -\frac{3}{2^4 \cdot 2^2 \cdot 2} = -\frac{3}{128},$$

$$(-1)^4 \frac{1}{2^6(3!)^2}\left(1 + \frac{1}{2} + \frac{1}{3}\right) = \frac{11}{2^6 \cdot 6^2 \cdot 6} = \frac{11}{13824}.$$

Having observed these relations, we may express the solution y_2 in the following more systematic form:

$$y_2(x) = J_0(x) \ln x + \frac{x^2}{2^2} - \frac{x^4}{2^4(2!)^2}\left(1 + \frac{1}{2}\right) + \frac{x^6}{2^6(3!)^2}\left(1 + \frac{1}{2} + \frac{1}{3}\right) + \cdots.$$

Further, we would certainly suspect that the general coefficient c_{2n}^* is given by

$$c_{2n}^* = \frac{(-1)^{n+1}}{2^{2n}(n!)^2}\left(1 + \frac{1}{2} + \frac{1}{3} + \cdots + \frac{1}{n}\right), \qquad n \geq 1.$$

It may be shown (though not without some difficulty) that this is indeed the case. This being true, we may express y_2 in the form

$$y_2(x) = J_0(x) \ln x + \sum_{n=1}^{\infty} \frac{(-1)^{n+1} x^{2n}}{2^{2n}(n!)^2} \left(1 + \frac{1}{2} + \frac{1}{3} + \cdots + \frac{1}{n}\right). \tag{6.109}$$

Since the solution y_2 defined by (6.109) is linearly independent of J_0 we could write the general solution of the differential equation (6.102) as a general linear combination of J_0 and y_2. However, this is not usually done; instead, it has been customary to choose a certain special linear combination of J_0 and y_2 and take this special combination as the "second" solution of Equation (6.102). This special combination is defined by

$$\frac{2}{\pi}[y_2(x) + (\gamma - \ln 2)J_0(x)],$$

where γ is a number called *Euler's constant* and is defined by

$$\gamma = \lim_{n\to\infty}\left(1 + \frac{1}{2} + \frac{1}{3} + \cdots + \frac{1}{n} - \ln n\right) \approx 0.5772.$$

It is called the *Bessel function of the second kind of order zero* (Weber's form) and is commonly denoted by Y_0. Thus the second solution of (6.102) is commonly taken as the function Y_0, where

$$Y_0(x) = \frac{2}{\pi}\left[J_0(x) \ln x + \sum_{n=1}^{\infty} \frac{(-1)^{n+1} x^{2n}}{2^{2n}(n!)^2}\left(1 + \frac{1}{2} + \frac{1}{3} + \cdots + \frac{1}{n}\right) + (\gamma - \ln 2)J_0(x)\right]$$

or

$$Y_0(x) = \frac{2}{\pi}\left[\left(\ln\frac{x}{2} + \gamma\right)J_0(x) + \sum_{n=1}^{\infty} \frac{(-1)^{n+1} x^{2n}}{2^{2n}(n!)^2}\left(1 + \frac{1}{2} + \frac{1}{3} + \cdots + \frac{1}{n}\right)\right]. \tag{6.110}$$

Therefore if we choose Y_0 as the second solution of the differential equation (6.102), the general solution of (6.102) for $0 < x < R$ is given by

$$y = c_1 J_0(x) + c_2 Y_0(x), \tag{6.111}$$

where c_1 and c_2 are arbitrary constants, and J_0 and Y_0 are defined by (6.107) and (6.110), respectively.

The functions J_0 and Y_0 have been studied extensively and tabulated. Many of the interesting properties of these functions are indicated by their graphs, which are shown in Figure 6.1.

B. Bessel's Equation of Order p

We now consider Bessel's equation of order p for $x > 0$, which we have already introduced at the beginning of Section 6.3A, and seek solutions valid for $0 < x < R$. This is the equation

$$x^2 \frac{d^2y}{dx^2} + x\frac{dy}{dx} + (x^2 - p^2)y = 0, \tag{6.101}$$

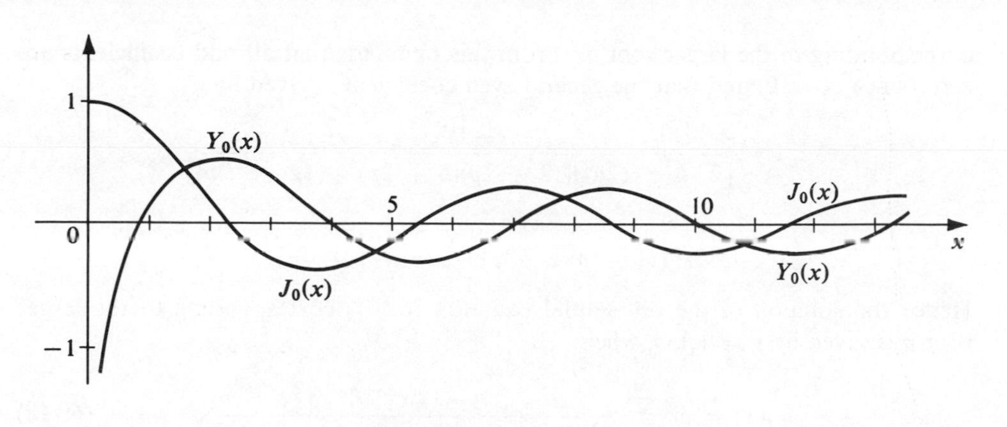

FIGURE 6.1

where we now assume that p is real and positive. We see at once that $x = 0$ is a regular singular point of equation (6.101); and since we seek solutions valid for $0 < x < R$, we may assume a solution

$$y = \sum_{n=0}^{\infty} c_n x^{n+r}, \tag{6.112}$$

where $c_0 \neq 0$. Differentiating (6.112), substituting into (6.101), and simplifying as in our previous examples, we obtain

$$(r^2 - p^2)c_0 x^r + [(r+1)^2 - p^2]c_1 x^{r+1} + \sum_{n=2}^{\infty} \{[(n+r)^2 - p^2]c_n + c_{n-2}\}x^{n+r} = 0. \tag{6.113}$$

Equating to zero the coefficient of each power of x in (6.113), we obtain

$$r^2 - p^2 = 0, \tag{6.114}$$

$$[(r+1)^2 - p^2]c_1 = 0, \tag{6.115}$$

and

$$[(n+r)^2 - p^2]c_n + c_{n-2} = 0, \quad n \geq 2. \tag{6.116}$$

Equation (6.114) is the indicial equation of the differential equation (6.101). Its roots are $r_1 = p > 0$ and $r_2 = -p$. If $r_1 - r_2 = 2p > 0$ is unequal to a positive integer, then from Theorem 6.3 we know that the differential equation (6.101) has *two* linearly independent solutions of the form (6.112). However, if $r_1 - r_2 = 2p$ is equal to a positive integer, we are only certain of a solution of this form corresponding to the *larger* root $r_1 = p$. We shall now proceed to obtain this one solution, the existence of which is always assured.

Letting $r = r_1 = p$ in (6.115), we obtain $(2p + 1)c_1 = 0$. Thus, since $p > 0$, we must have $c_1 = 0$. Letting $r = r_1 = p$ in (6.116), we obtain the recurrence formula

$$n(n + 2p)c_n + c_{n-2} = 0, \quad n \geq 2,$$

or

$$c_n = -\frac{c_{n-2}}{n(n+2p)}, \quad n \geq 2, \tag{6.117}$$

corresponding to the larger root p. From this one finds that all odd coefficients are zero (since $c_1 = 0$) and that the general even coefficient is given by

$$c_{2n} = \frac{(-1)^n c_0}{[2 \cdot 4 \cdots (2n)][(2 + 2p)(4 + 2p) \cdots (2n + 2p)]}$$

$$= \frac{(-1)^n c_0}{2^{2n} n! \, [(1 + p)(2 + p) \cdots (n + p)]}, \qquad n \geq 1.$$

Hence the solution of the differential equation (6.101) corresponding to the larger root p is given by $y = y_1(x)$, where

$$y_1(x) = c_0 \sum_{n=0}^{\infty} \frac{(-1)^n x^{2n+p}}{2^{2n} n! \, [(1 + p)(2 + p) \cdots (n + p)]}. \tag{6.118}$$

If p is a positive integer, we may write this in the form

$$y_1(x) = c_0 2^p p! \sum_{n=0}^{\infty} \frac{(-1)^n}{n! \, (n + p)!} \left(\frac{x}{2}\right)^{2n+p}. \tag{6.119}$$

If p is unequal to a positive integer, we need a generalization of the factorial function in order to express $y_1(x)$ in a form analogous to that given by (6.119). Such a generalization is provided by the so-called *gamma function*, which we now introduce.

For $N > 0$ the gamma function is defined by the convergent improper integral

$$\Gamma(N) = \int_0^{\infty} e^{-x} x^{N-1} \, dx. \tag{6.120}$$

If N is a positive integer, it can be shown that

$$N! = \Gamma(N + 1). \tag{6.121}$$

If N is positive but *not* an integer, we use (6.121) to *define* $N!$ The gamma function has been studied extensively. It can be shown that $\Gamma(N)$ satisfies the recurrence formula

$$\Gamma(N + 1) = N\Gamma(N) \qquad (N > 0). \tag{6.122}$$

Values of $\Gamma(N)$ have been tabulated and are usually given for the range $1 \leq N \leq 2$. Using the tabulated values of $\Gamma(N)$ for $1 \leq N \leq 2$, one can evaluate $\Gamma(N)$ for all $N > 0$ by repeated application of formula (6.122). Suppose, for example, that we wish to evaluate $(\frac{3}{2})!$. From the definition (6.121), we have $(\frac{3}{2})! = \Gamma(\frac{5}{2})$. Then from (6.122), we find that $\Gamma(\frac{5}{2}) = \frac{3}{2}\Gamma(\frac{3}{2})$. From tables one finds that $\Gamma(\frac{3}{2}) \approx 0.8862$, and thus $(\frac{3}{2})! = \Gamma(\frac{5}{2}) \approx 1.3293$.

For $N < 0$ the integral (6.120) diverges, and thus $\Gamma(N)$ is not defined by (6.120) for negative values of N. We extend the definition of $\Gamma(N)$ to values of $N < 0$ by demanding that the recurrence formula (6.122) hold for *negative* (as well as positive) values of N. Repeated use of this formula thus defines $\Gamma(N)$ for every nonintegral negative value of N.

Thus $\Gamma(N)$ is defined for all $N \neq 0, -1, -2, -3, \ldots$. The graph of this function is shown in Figure 6.2. We now define $N!$ for all $N \neq -1, -2, -3, \ldots$ by the formula (6.121).

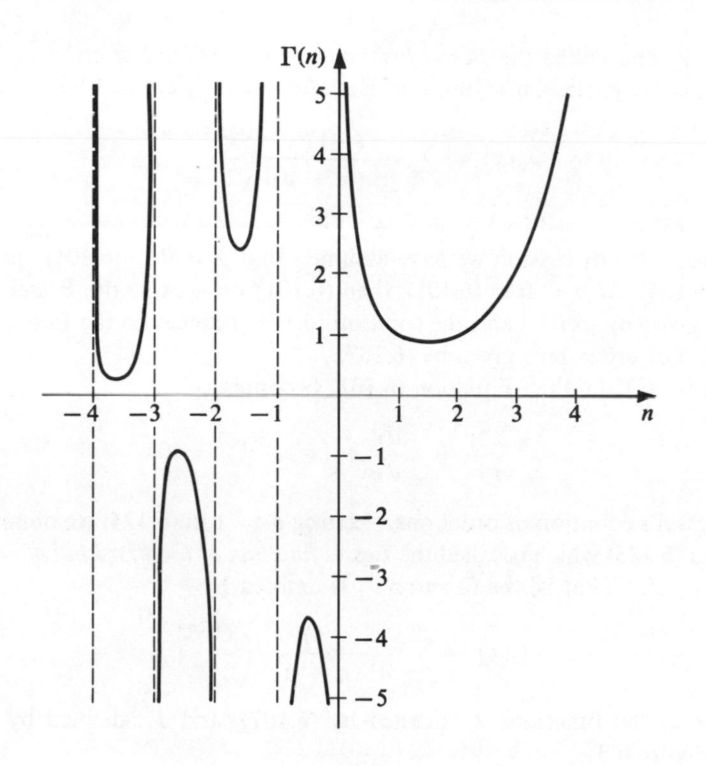

FIGURE 6.2. Graph of $\Gamma(N)$

We now return to the solution y_1 defined by (6.118), for the case in which p is unequal to a positive integer. Applying the recurrence formula (6.122) successively with $N = n + p, n + p - 1, n + p - 2, \ldots, p + 1$, we obtain

$$\Gamma(n + p + 1) = (n + p)(n + p - 1)\cdots(p + 1)\Gamma(p + 1).$$

Thus for p unequal to a positive integer we may write the solution defined by (6.118) in the form

$$\begin{aligned} y_1(x) &= c_0\Gamma(p + 1) \sum_{n=0}^{\infty} \frac{(-1)^n x^{2n+p}}{2^{2n} n!\, \Gamma(n + p + 1)} \\ &= c_0 2^p \Gamma(p + 1) \sum_{n=0}^{\infty} \frac{(-1)^n}{n!\, \Gamma(n + p + 1)} \left(\frac{x}{2}\right)^{2n+p}. \end{aligned} \tag{6.123}$$

Now using (6.121) with $N = p$ and $N = n + p$, we see that (6.123) takes the form (6.119). Thus the solution of the differential equation (6.101) corresponding to the larger root $p > 0$ is given by (6.119), where $p!$ and $(n + p)!$ are defined by $\Gamma(p + 1)$ and $\Gamma(n + p + 1)$, respectively, if p is not a positive integer.

If we set the arbitrary constant c_0 in (6.119) equal to the reciprocal of $2^p p!$, we obtain an important particular solution of (6.101). This particular solution defines a function denoted by J_p and called the *Bessel function of the first kind of order p*. That is, the function J_p is the particular solution of Equation (6.101) defined by

$$J_p(x) = \sum_{n=0}^{\infty} \frac{(-1)^n}{n!\,(n+p)!} \left(\frac{x}{2}\right)^{2n+p}, \tag{6.124}$$

where $(n + p)!$ is defined by $\Gamma(n + p + 1)$ if p is not a positive integer.

Throughout this discussion we have assumed that $p > 0$ in (6.101) and hence that $p > 0$ in (6.124). If $p = 0$ in (6.101), then (6.101) reduces to the Bessel equation of order zero given by (6.102) and the solution (6.124) reduces to the Bessel function of the first kind of order zero given by (6.107).

If $p = 1$ in (6.101), then Equation (6.101) becomes

$$x^2 \frac{d^2y}{dx^2} + x\frac{dy}{dx} + (x^2 - 1)y = 0, \tag{6.125}$$

which is Bessel's equation of order one. Letting $p = 1$ in (6.124) we obtain a solution of Equation (6.125) which is called the *Bessel function of the first kind of order one* and is denoted by J_1. That is, the function J_1 is defined by

$$J_1(x) = \sum_{n=0}^{\infty} \frac{(-1)^n}{n!\,(n+1)!} \left(\frac{x}{2}\right)^{2n+1}. \tag{6.126}$$

The graphs of the functions J_0, defined by (6.107), and J_1, defined by (6.126), are shown in Figure 6.3.

Several interesting properties of the Bessel functions of the first kind are suggested by these graphs. For one thing, they suggest that J_0 and J_1 each have a damped oscillatory behavior and that the positive zeros of J_0 and J_1 separate each other. This is indeed the case. In fact, it may be shown that for every $p \geq 0$ the function J_p has a damped oscillatory behavior as $x \to \infty$ and the positive zeros of J_p and J_{p+1} separate each other.

We now know that for every $p \geq 0$ one solution of Bessel's equation of order p (6.101) is given by (6.124). We now consider briefly the problem of finding a linearly independent solution of (6.101). We have already found such a solution for the case in which $p = 0$; it is given by (6.110). For $p > 0$, we have observed that if $2p$

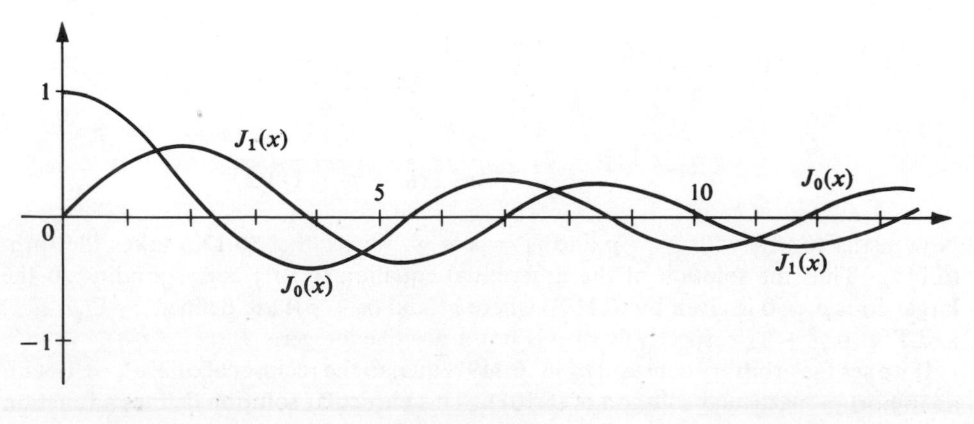

FIGURE 6.3

is unequal to a positive integer, then the differential equation (6.101) has a linearly independent solution of the form (6.112) corresponding to the smaller root $r_2 = -p$. We now proceed to work with this smaller root.

Letting $r = r_2 = -p$ in (6.115), we obtain

$$(-2p + 1)c_1 = 0. \tag{6.127}$$

Letting $r = r_2 = -p$ in (6.116) we obtain the recurrence formula

$$n(n - 2p)c_n + c_{n-2} = 0, \quad n \geq 2, \tag{6.128}$$

or

$$c_n = -\frac{c_{n-2}}{n(n - 2p)}, \quad n \geq 2, \quad n \neq 2p. \tag{6.129}$$

Using (6.127) and (6.128) or (6.129) one finds solutions $y = y_2(x)$, corresponding to the smaller root $-p$. Three distinct cases occur, leading to solutions of the following forms:

1. If $2p \neq$ a positive integer,

$$y_2(x) = c_0 x^{-p}\left(1 + \sum_{n=1}^{\infty} \alpha_{2n} x^{2n}\right), \tag{6.130}$$

where c_0 is an arbitrary constant and the α_{2n} $(n = 1, 2, \ldots)$ are definite constants.

2. If $2p =$ an *odd* positive integer,

$$y_2(x) = c_0 x^{-p}\left(1 + \sum_{n=1}^{\infty} \beta_{2n} x^{2n}\right) + c_{2p} x^{p}\left(1 + \sum_{n=1}^{\infty} \gamma_{2n} x^{2n}\right), \tag{6.131}$$

where c_0 and c_{2p} are arbitrary constants and β_{2n} and γ_{2n} $(n = 1, 2, \ldots)$ are definite constants.

3. If $2p =$ an *even* positive integer,

$$y_2(x) = c_{2p} x^{p}\left(1 + \sum_{n=1}^{\infty} \delta_{2n} x^{2n}\right), \tag{6.132}$$

where c_{2p} is an arbitrary constant and the δ_{2n} $(n = 1, 2, \ldots)$ are definite constants.

In Case 1 the solution defined by (6.130) is linearly independent of J_p. In Case 2 the solution defined by (6.131) with $c_{2p} = 0$ is also linearly independent of J_p. However, in Case 3 the solution defined by (6.132) is merely a constant multiple of $J_p(x)$, and hence this solution is *not* linearly independent of J_p. Thus if $2p$ is unequal to an *even* positive integer, there exists a linearly independent solution of the form (6.112) corresponding to the smaller root $-p$. In other words, if p is unequal to a positive integer, the differential equation (6.101) has a solution of the form $y = y_2(x)$, where

$$y_2(x) = \sum_{n=0}^{\infty} c_{2n} x^{2n-p}, \tag{6.133}$$

and this solution y_2 is linearly independent of J_p.

It is easy to determine the coefficients c_{2n} in (6.133). We observe that the recurrence formula (6.129) corresponding to the smaller root $-p$ is obtained from the recurrence formula (6.117) corresponding to the larger root p simply by replacing p in (6.117) by $-p$. Hence a solution of the form (6.133) may be obtained from (6.124) simply by replacing p in (6.124) by $-p$. This leads at once to the solution denoted by J_{-p} and defined by

$$J_{-p}(x) = \sum_{n=0}^{\infty} \frac{(-1)^n}{n!\,(n-p)!} \left(\frac{x}{2}\right)^{2n-p}, \tag{6.134}$$

where $(n - p)!$ is defined by $\Gamma(n - p + 1)$.

Thus if $p > 0$ is unequal to a positive integer, two linearly independent solutions of the differential equation (6.101) are J_p defined by (6.124), and J_{-p}, defined by (6.134). Hence, if $p > 0$ is unequal to a positive integer, the general solution of Bessel's equation of order p is given by

$$y = C_1 J_p(x) + C_2 J_{-p}(x),$$

where J_p and J_{-p} are defined by (6.124) and (6.134), respectively, and C_1 and C_2 are arbitrary constants.

If p is a positive integer, the corresponding solution defined by (6.132) is *not* linearly independent of J_p, as we have already noted. Hence in this case a solution which is linearly independent of J_p must be given by $y = y_p(x)$, where

$$y_p(x) = x^{-p} \sum_{n=0}^{\infty} c_n^* x^n + CJ_p(x) \ln x,$$

where $C \neq 0$. Such a linearly independent solution y_p may be found by the method of reduction of order. Then the general solution of the differential equation (6.101) may be written as a general linear combination of J_p and y_p. However, as in the case of Bessel's equation of order zero, it is customary to choose a certain special linear combination of J_p and y_p and take this special combination as the "second" solution of equation (6.101). This special combination is denoted by Y_p and defined by

$$Y_p(x) = \frac{2}{\pi}\left\{\left(\ln\frac{x}{2} + \gamma\right) J_p(x) - \frac{1}{2}\sum_{n=0}^{p-1} \frac{(p-n-1)!}{n!}\left(\frac{x}{2}\right)^{2n-p} + \frac{1}{2}\sum_{n=0}^{\infty} (-1)^{n+1}\left(\sum_{k=1}^{n}\frac{1}{k} + \sum_{k=1}^{n+p}\frac{1}{k}\right)\left[\frac{1}{n!\,(n+p)!}\left(\frac{x}{2}\right)^{2n+p}\right]\right\}, \tag{6.135}$$

where γ is Euler's constant. The solution Y_p is called the *Bessel function of the second kind of order p* (Weber's form).

Thus if p is a positive integer, two linearly independent solutions of the differential equation (6.101) are J_p, defined by (6.124), and Y_p, defined by (6.135). Hence if p is a positive integer, the general solution of Bessel's equation of order p is given by

$$y = C_1 J_p(x) + C_2 Y_p(x),$$

where J_p and Y_p are defined by (6.124) and (6.135), respectively, and C_1 and C_2 are arbitrary constants.

Exercises

1. Show that $J_0(kx)$, where k is a constant, satisfies the differential equation

$$x\frac{d^2y}{dx^2} + \frac{dy}{dx} + k^2xy = 0.$$

2. Show that the transformation

$$y = \frac{u(x)}{\sqrt{x}}$$

reduces the Bessel equation of order p, Equation (6.101), to the form

$$\frac{d^2u}{dx^2} + \left[1 + \left(\frac{1}{4} - p^2\right)\frac{1}{x^2}\right]u = 0.$$

3. Use the result of Exercise 2 to obtain a solution of the Bessel equation of order $\frac{1}{2}$.
4. Using the series definition (6.124) for J_p, show that

$$\frac{d}{dx}[x^p J_p(kx)] = kx^p J_{p-1}(kx)$$

and

$$\frac{d}{dx}[x^{-p} J_p(kx)] = -kx^{-p} J_{p+1}(kx),$$

where k is a constant.

5. Use the results of Exercise 4 to show that

$$\frac{d}{dx}[J_p(kx)] = kJ_{p-1}(kx) - \frac{p}{x}J_p(kx),$$

$$\frac{d}{dx}[J_p(kx)] = -kJ_{p+1}(kx) + \frac{p}{x}J_p(kx).$$

Hence show that

$$\frac{d}{dx}[J_p(kx)] = \frac{k}{2}[J_{p-1}(kx) - J_{p+1}(kx)],$$

$$J_p(kx) = \frac{kx}{2p}[J_{p-1}(kx) + J_{p+1}(kx)].$$

6. Using the results of Exercise 5,

(a) express $J_1(x)$ and $\frac{d}{dx}[J_1(x)]$ in terms of $J_0(x)$ and $J_2(x)$;

(b) express $J_{n+1/2}(x)$ in terms of $J_{n-1/2}(x)$ and $J_{n-3/2}(x)$.

Suggested Reading

AGNEW, R., *Differential Equations*, 2nd ed. (McGraw-Hill, New York, 1960).

BOYCE, W., and R. DIPRIMA, *Elementary Differential Equations*, 2nd ed. (Wiley, New York, 1969).

BRAUER, F., and J. NOHEL, *Ordinary Differential Equations: A First Course* (Benjamin, New York, 1967).

CODDINGTON, E., *An Introduction to Ordinary Differential Equations* (Prentice-Hall, Englewood Cliffs, N.J., 1961).

FORD, L., *Differential Equations*, 2nd ed. (McGraw-Hill, New York, 1955).
HILDEBRAND, F., *Advanced Calculus for Applications*, rev. ed. (Prentice-Hall, Englewood Cliffs, N.J., 1962).
KAPLAN, W., *Ordinary Differential Equations* (Addison-Wesley, Reading, Mass., 1958).
KREIDER, D., R. KULLER, and D. OSTBERG, *Elementary Differential Equations* (Addison-Wesley, Reading, Mass., 1968).
RAINVILLE, E., and P. BEDIENT, *Elementary Differential Equations*, 4th ed. (Macmillan, New York, 1969).
RITGER, P., and N. ROSE, *Differential Equations with Applications* (McGraw-Hill, New York, 1968).

Systems of Linear Differential Equations

In the previous chapters we have been concerned with one differential equation in one unknown function. Now we shall consider systems of two differential equations in two unknown functions, and more generally, systems of n differential equations in n unknown functions. We shall restrict our attention to linear systems only, and we shall begin by considering various types of these systems. After this, we shall proceed to introduce differential operators, present an operator method of solving linear systems, and then consider some basic applications of this method. We shall then turn to a study of the fundamental theory and basic method of solution for a standard type of linear system in the special case of two equations in two unknown functions. Finally, we shall outline the most basic material about matrices and vectors and then use this to study the fundamental theory and basic method of solution for the corresponding standard type of linear system in the general case of n equations in n unknown functions.

7.1 Differential Operators and an Operator Method

A. *Types of Linear Systems*

We start by introducing the various types of linear systems which we shall consider. The general linear system of two first-order differential equations in two unknown functions x and y is of the form

$$\begin{aligned} a_1(t)\frac{dx}{dt} + a_2(t)\frac{dy}{dt} + a_3(t)x + a_4(t)y &= F_1(t), \\ b_1(t)\frac{dx}{dt} + b_2(t)\frac{dy}{dt} + b_3(t)x + b_4(t)y &= F_2(t). \end{aligned} \tag{7.1}$$

We shall be concerned with systems of this type which have constant coefficients. An example of such a system is

$$\begin{aligned} 2\frac{dx}{dt} + 3\frac{dy}{dt} - 2x + y &= t^2, \\ \frac{dx}{dt} - 2\frac{dy}{dt} + 3x + 4y &= e^t. \end{aligned}$$

We shall say that a *solution* of system (7.1) is an ordered pair of real functions (f, g) such that $x = f(t)$, $y = g(t)$ simultaneously satisfy both equations of the system (7.1) on some real interval $a \leq t \leq b$.

The general linear system of three first-order differential equations in three unknown functions x, y, and z is of the form

$$\begin{aligned}
a_1(t)\frac{dx}{dt} + a_2(t)\frac{dy}{dt} + a_3(t)\frac{dz}{dt} + a_4(t)x + a_5(t)y + a_6(t)z &= F_1(t),\\
b_1(t)\frac{dx}{dt} + b_2(t)\frac{dy}{dt} + b_3(t)\frac{dz}{dt} + b_4(t)x + b_5(t)y + b_6(t)z &= F_2(t), \qquad (7.2)\\
c_1(t)\frac{dx}{dt} + c_2(t)\frac{dy}{dt} + c_3(t)\frac{dz}{dt} + c_4(t)x + c_5(t)y + c_6(t)z &= F_3(t).
\end{aligned}$$

As in the case of systems of the form (7.1), so also in this case we shall be concerned with systems that have constant coefficients. An example of such a system is

$$\begin{aligned}
\frac{dx}{dt} + \frac{dy}{dt} - 2\frac{dz}{dt} + 2x - 3y + z &= t,\\
2\frac{dx}{dt} - \frac{dy}{dt} + 3\frac{dz}{dt} + x + 4y - 5z &= \sin t,\\
\frac{dx}{dt} + 2\frac{dy}{dt} + \frac{dz}{dt} - 3x + 2y - z &= \cos t.
\end{aligned}$$

We shall say that a solution of system (7.2) is an ordered triple of real functions (f, g, h) such that $x = f(t)$, $y = g(t)$, $z = h(t)$ simultaneously satisfy all three equations of the system (7.2) on some real interval $a \leq t \leq b$.

Systems of the form (7.1) and (7.2) contained only first derivatives, and we now consider the basic linear system involving higher derivatives. This is the general linear system of two second-order differential equations in two unknown functions x and y, and is a system of the form

$$\begin{aligned}
a_1(t)\frac{d^2x}{dt^2} + a_2(t)\frac{d^2y}{dt^2} + a_3(t)\frac{dx}{dt} + a_4(t)\frac{dy}{dt} + a_5(t)x + a_6(t)y &= F_1(t),\\
b_1(t)\frac{d^2x}{dt^2} + b_2(t)\frac{d^2y}{dt^2} + b_3(t)\frac{dx}{dt} + b_4(t)\frac{dy}{dt} + b_5(t)x + b_6(t)y &= F_2(t).
\end{aligned} \qquad (7.3)$$

We shall be concerned with systems having constant coefficients in this case also, and an example is provided by

$$\begin{aligned}
2\frac{d^2x}{dt^2} + 5\frac{d^2y}{dt^2} + 7\frac{dx}{dt} + 3\frac{dy}{dt} + 2y &= 3t + 1,\\
3\frac{d^2x}{dt^2} + 2\frac{d^2y}{dt^2} - 2\frac{dy}{dt} + 4x + y &= 0.
\end{aligned}$$

For given fixed positive integers m and n, we could proceed, in like manner, to exhibit other general linear systems of n mth-order differential equations in n unknown functions and give examples of each such type of system. Instead we proceed to introduce the standard type of linear system referred to in the introductory paragraph at the start of the chapter, and of which we shall make a more systematic study later. We introduce this standard type as a special case of the system (7.1) of two first-order differential equations in two unknowns functions x and y.

We consider the special type of linear system (7.1) which is of the form

$$\begin{aligned}\frac{dx}{dt} &= a_{11}(t)x + a_{12}(t)y + F_1(t),\\ \frac{dy}{dt} &= a_{21}(t)x + a_{22}(t)y + F_2(t).\end{aligned} \tag{7.4}$$

This is the so-called *normal form* in the case of two linear differential equations in two unknown functions. The characteristic feature of such a system is apparent from the manner in which the derivatives appear in it. An example of such a system with variable coefficients is

$$\begin{aligned}\frac{dx}{dt} &= t^2x + (t + 1)y + t^3,\\ \frac{dy}{dt} &= te^tx + t^3y - e^t,\end{aligned}$$

while one with constant coefficients is

$$\begin{aligned}\frac{dx}{dt} &= 5x + 7y + t^2,\\ \frac{dy}{dt} &= 2x - 3y + 2t.\end{aligned}$$

The normal form in the case of a linear system of three differential equations in three unknown functions x, y, and z is

$$\begin{aligned}\frac{dx}{dt} &= a_{11}(t)x + a_{12}(t)y + a_{13}(t)z + F_1(t),\\ \frac{dy}{dt} &= a_{21}(t)x + a_{22}(t)y + a_{23}(t)z + F_2(t),\\ \frac{dz}{dt} &= a_{31}(t)x + a_{32}(t)y + a_{33}(t)z + F_3(t).\end{aligned}$$

An example of such a system is the constant coefficient system

$$\begin{aligned}\frac{dx}{dt} &= 3x + 2y + z + t,\\ \frac{dy}{dt} &= 2x - 4y + 5z - t^2,\\ \frac{dz}{dt} &= 4x + y - 3z + 2t + 1.\end{aligned}$$

The normal form in the general case of a linear system of n differential equations in n unknown functions $x_1, x_2, \ldots, x_n$ is

$$\begin{aligned} \frac{dx_1}{dt} &= a_{11}(t)x_1 + a_{12}(t)x_2 + \cdots + a_{1n}(t)x_n + F_1(t), \\ \frac{dx_2}{dt} &= a_{21}(t)x_1 + a_{22}(t)x_2 + \cdots + a_{2n}(t)x_n + F_2(t), \\ &\vdots \\ \frac{dx_n}{dt} &= a_{n1}(t)x_1 + a_{n2}(t)x_2 + \cdots + a_{nn}(t)x_n + F_n(t). \end{aligned} \tag{7.5}$$

An important fundamental property of a normal linear system (7.5) is its relationship to a single nth-order linear differential equation in one unknown function. Specifically, consider the so-called normalized (meaning, the coefficient of the highest derivative is one) nth-order linear differential equation

$$\frac{d^n x}{dt^n} + a_1(t)\frac{d^{n-1}x}{dt^{n-1}} + \cdots + a_{n-1}(t)\frac{dx}{dt} + a_n(t)x = F(t) \tag{7.6}$$

in the one unknown function x. Let

$$x_1 = x, \quad x_2 = \frac{dx}{dt}, \quad x_3 = \frac{d^2x}{dt^2}, \ldots, x_{n-1} = \frac{d^{n-2}x}{dt^{n-2}}, \quad x_n = \frac{d^{n-1}x}{dt^{n-1}}. \tag{7.7}$$

From (7.7), we have

$$\frac{dx}{dt} = \frac{dx_1}{dt}, \quad \frac{d^2x}{dt^2} = \frac{dx_2}{dt}, \ldots, \frac{d^{n-1}x}{dt^{n-1}} = \frac{dx_{n-1}}{dt}, \quad \frac{d^n x}{dt^n} = \frac{dx_n}{dt}. \tag{7.8}$$

Then using both (7.7) and (7.8), the single nth-order equation (7.6) can be transformed into

$$\begin{aligned} \frac{dx_1}{dt} &= x_2, \\ \frac{dx_2}{dt} &= x_3, \\ &\vdots \\ \frac{dx_{n-1}}{dt} &= x_n, \\ \frac{dx_n}{dt} &= -a_n(t)x_1 - a_{n-1}(t)x_2 - \cdots - a_1(t)x_n + F(t), \end{aligned} \tag{7.9}$$

which is a special case of the normal linear system (7.5) of n equations in n unknown functions. Thus we see that a single nth-order linear differential equation of form (7.6) in one unknown function is indeed intimately related to a normal linear system (7.5) of n first-order differential equations in n unknown functions.

B. *Differential Operators*

In this section we shall present a symbolic operator method for solving linear systems with constant coefficients. This method depends upon the use of so-called *differential operators,* which we now introduce.

Let x be an n-times differentiable function of the independent variable t. We denote the operation of differentiation with respect to t by the symbol D and call D a differential operator. In terms of this differential operator the derivative dx/dt is denoted by Dx. That is,

$$Dx \equiv dx/dt.$$

In like manner, we denote the second derivative of x with respect to t by D^2x. Extending this, we denote the nth derivative of x with respect to t by D^nx. That is,

$$D^n x = \frac{d^n x}{dt^n} \qquad (n = 1, 2, \ldots).$$

Further extending this operator notation, we write

$$(D + c)x \quad \text{to denote} \quad \frac{dx}{dt} + cx$$

and

$$(aD^n + bD^m)x \quad \text{to denote} \quad a\frac{d^n x}{dt^n} + b\frac{d^m x}{dt^m},$$

where a, b, and c are constants.

In this notation the general linear differential expression with constant coefficients $a_0, a_1, \ldots, a_{n-1}, a_n$,

$$a_0\frac{d^n x}{dt^n} + a_1\frac{d^{n-1}x}{dt^{n-1}} + \cdots + a_{n-1}\frac{dx}{dt} + a_n x,$$

is written

$$(a_0 D^n + a_1 D^{n-1} + \cdots + a_{n-1}D + a_n)x.$$

Observe carefully that the operators $D^n, D^{n-1}, \ldots, D$ in this expression do *not* represent quantities which are to be multiplied by the function x, but rather they indicate *operations* (differentiations) which are to be carried out upon this function. The expression

$$a_0 D^n + a_1 D^{n-1} + \cdots + a_{n-1}D + a_n$$

by itself, where $a_0, a_1, \ldots, a_{n-1}, a_n$ are constants, is called a linear differential operator with constant coefficients.

▶ Example 7.1. Consider the linear differential operator

$$3D^2 + 5D - 2.$$

If x is a twice differentiable function of t, then

$$(3D^2 + 5D - 2)x \quad \text{denotes} \quad 3\frac{d^2x}{dt^2} + 5\frac{dx}{dt} - 2x.$$

For example, if $x = t^3$, we have

$$(3D^2 + 5D - 2)t^3 = 3\frac{d^2}{dt^2}(t^3) + 5\frac{d}{dt}(t^3) - 2(t^3) = 18t + 15t^2 - 2t^3.$$

We shall now discuss certain useful properties of the linear differential operator with constant coefficients. In order to facilitate our discussion, we shall let L denote this operator. That is,

$$L \equiv a_0D^n + a_1D^{n-1} + \cdots + a_{n-1}D + a_n,$$

where $a_0, a_1, \ldots, a_{n-1}, a_n$ are constants. Now suppose that f_1 and f_2 are both n-times differentiable functions of t and c_1 and c_2 are constants. Then it can be shown that

$$L[c_1f_1 + c_2f_2] = c_1L[f_1] + c_2L[f_2].$$

For example, if the operator $L \equiv 3D^2 + 5D - 2$ is applied to $3t^2 + 2 \sin t$, then

$$L[3t^2 + 2 \sin t] = 3L[t^2] + 2L[\sin t]$$

or

$$(3D^2 + 5D - 2)(3t^2 + 2 \sin t) = 3(3D^2 + 5D - 2)t^2 + 2(3D^2 + 5D - 2) \sin t.$$

Now let

$$L_1 \equiv a_0D^m + a_1D^{m-1} + \cdots + a_{m-1}D + a_m$$

and

$$L_2 \equiv b_0D^n + b_1D^{n-1} + \cdots + b_{n-1}D + b_n$$

be two linear differential operators with constant coefficients $a_0, a_1, \ldots, a_{m-1}, a_m$, and $b_0, b_1, \ldots, b_{n-1}, b_n$, respectively. Let

$$L_1(r) \equiv a_0r^m + a_1r^{m-1} + \cdots + a_{m-1}r + a_m$$

and

$$L_2(r) \equiv b_0r^n + b_1r^{n-1} + \cdots + b_{n-1}r + b_n$$

be the two polynomials in the quantity r obtained from the operators L_1 and L_2, respectively, by formally replacing D by r, D^2 by $r^2, \ldots, D^k$ by r^k. Let us denote the product of the polynomials $L_1(r)$ and $L_2(r)$ by $L(r)$; that is,

$$L(r) = L_1(r)L_2(r).$$

Then, if f is a function possessing $n + m$ derivatives, it can be shown that

$$L_1L_2f = L_2L_1f = Lf, \tag{7.10}$$

where L is the operator obtained from the "product polynomial" $L(r)$ by formally replacing r by D, r^2 by $D^2, \ldots, r^{m+n}$ by D^{m+n}. Equation (7.10) indicates two important properties of linear differential operators with constant coefficients. First, it states the effect of first operating on f by L_2 and then operating on the resulting function by L_1 is the same as that which results from first operating on f by L_1 and then operating on this resulting function by L_2. Second, Equation (7.10) states that the effect of first operating on f by either L_1 or L_2 and then operating on the resulting function by the other is the same as that which results from operating on f by the "product operator" L. We illustrate these important properties in the following example.

▶ Example 7.2. Let $L_1 \equiv D^2 + 1$, $L_2 \equiv 3D + 2$, $f(t) = t^3$. Then

$$\begin{aligned} L_1L_2 f &= (D^2 + 1)(3D + 2)t^3 = (D^2 + 1)(9t^2 + 2t^3) \\ &= 9(D^2 + 1)t^2 + 2(D^2 + 1)t^3 \\ &= 9(2 + t^2) + 2(6t + t^3) = 2t^3 + 9t^2 + 12t + 18 \end{aligned}$$

and

$$\begin{aligned} L_2L_1 f &= (3D + 2)(D^2 + 1)t^3 = (3D + 2)(6t + t^3) \\ &= 6(3D + 2)t + (3D + 2)t^3 \\ &= 6(3 + 2t) + (9t^2 + 2t^3) = 2t^3 + 9t^2 + 12t + 18. \end{aligned}$$

Finally, $L \equiv 3D^3 + 2D^2 + 3D + 2$ and

$$\begin{aligned} Lf &= (3D^3 + 2D^2 + 3D + 2)t^3 = 3(6) + 2(6t) + 3(3t^2) + 2t^3 \\ &= 2t^3 + 9t^2 + 12t + 18. \end{aligned}$$

Now let $L \equiv a_0D^n + a_1D^{n-1} + \cdots + a_{n-1}D + a_n$, where $a_0, a_1, \ldots, a_{n-1}, a_n$ are constants; and let $L(r) \equiv a_0r^n + a_1r^{n-1} + \cdots + a_{n-1}r + a_n$ be the polynomial in r obtained from L by formally replacing D by r, D^2 by $r^2, \ldots, D^n$ by r^n. Let $r_1, r_2, \ldots, r_n$ be the roots of the polynomial equation $L(r) = 0$. Then $L(r)$ may be written in the factored form

$$L(r) = a_0(r - r_1)(r - r_2)\cdots(r - r_n).$$

Now formally replacing r by D in the right member of this identity, we may express the operator $L \equiv a_0D^n + a_1D^{n-1} + \cdots + a_{n-1}D + a_n$ in the factored form

$$L = a_0(D - r_1)(D - r_2)\cdots(D - r_n).$$

We thus observe that linear differential operators with constant coefficients can be formally multiplied and factored exactly as if they were polynomials in the algebraic quantity D.

C. *An Operator Method for Linear Systems with Constant Coefficients*

We now proceed to explain a symbolic operator method for solving linear systems with constant coefficients. We shall outline the procedure of this method on a strictly formal basis and shall make no attempt to justify it.

We consider a linear system of the form

$$\begin{aligned} L_1x + L_2y &= f_1(t), \\ L_3x + L_4y &= f_2(t), \end{aligned} \tag{7.11}$$

where L_1, L_2, L_3, and L_4 are linear differential operators with constant coefficients. That is, L_1, L_2, L_3, and L_4 are operators of the forms

$$\begin{aligned} L_1 &\equiv a_0D^m + a_1D^{m-1} + \cdots + a_{m-1}D + a_m, \\ L_2 &\equiv b_0D^n + b_1D^{n-1} + \cdots + b_{n-1}D + b_n, \\ L_3 &\equiv \alpha_0D^p + \alpha_1D^{p-1} + \cdots + \alpha_{p-1}D + \alpha_p, \\ L_4 &\equiv \beta_0D^q + \beta_1D^{q-1} + \cdots + \beta_{q-1}D + \beta_q, \end{aligned}$$

where the a's, b's, α's, and β's are constants.

A simple example of a system which may be expressed in the form (7.11) is provided by

$$2\frac{dx}{dt} - 2\frac{dy}{dt} - 3x = t,$$

$$2\frac{dx}{dt} + 2\frac{dy}{dt} + 3x + 8y = 2.$$

Introducing operator notation this system takes the form

$$(2D - 3)x - 2Dy = t,$$

$$(2D + 3)x + (2D + 8)y = 2.$$

This is clearly of the form (7.11), where $L_1 \equiv 2D - 3$, $L_2 \equiv -2D$, $L_3 \equiv 2D + 3$, and $L_4 \equiv 2D + 8$.

Returning now to the general system (7.11), we apply the operator L_4 to the first equation of (7.11) and the operator L_2 to the second equation of (7.11), obtaining

$$L_4L_1x + L_4L_2y = L_4f_1,$$

$$L_2L_3x + L_2L_4y = L_2f_2.$$

We now subtract the second of these equations from the first. Since $L_4L_2y = L_2L_4y$, we obtain

$$L_4L_1x - L_2L_3x = L_4f_1 - L_2f_2,$$

or

$$(L_1L_4 - L_2L_3)x = L_4f_1 - L_2f_2. \tag{7.12}$$

The expression $L_1L_4 - L_2L_3$ in the left member of this equation is itself a linear differential operator with constant coefficients. We assume that it is neither zero nor a nonzero constant and denote it by L_5. If we further assume that the functions f_1 and f_2 are such that the right member $L_4f_1 - L_2f_2$ of (7.12) exists, then this member is some function, say g_1, of t. Then Equation (7.12) may be written

$$L_5x = g_1. \tag{7.13}$$

Equation (7.13) is a linear differential equation with constant coefficients in the single dependent variable x. We thus observe that our procedure has eliminated the other dependent variable y. We now solve the differential equation (7.13) for x using the methods developed in Chapter 4. Suppose Equation (7.13) is of order N. Then the general solution of (7.13) is of the form

$$x = c_1u_1 + c_2u_2 + \cdots + c_Nu_N + U_1, \tag{7.14}$$

where $u_1, u_2, \ldots, u_N$ are N linearly independent solutions of the homogeneous linear equation $L_5x = 0$, $c_1, c_2, \ldots, c_N$ are arbitrary constants, and U_1 is a particular solution of $L_5x = g_1$.

We again return to the system (7.11) and this time apply the operators L_3 and L_1 to the first and second equations, respectively, of the system. We obtain

$$L_3L_1x + L_3L_2y = L_3f_1,$$

$$L_1L_3x + L_1L_4y = L_1f_2.$$

Subtracting the first of these from the second, we obtain

$$(L_1L_4 - L_2L_3)y = L_1f_2 - L_3f_1.$$

Assuming that f_1 and f_2 are such that the right member $L_1f_2 - L_3f_1$ of this equation exists, we may express it as some function, say g_2, of t. Then this equation may be written

$$L_5y = g_2, \tag{7.15}$$

where L_5 denotes the operator $L_1L_4 - L_2L_3$. Equation (7.15) is a linear differential equation with constant coefficients in the single dependent variable y. This time we have eliminated the dependent variable x. Solving the differential equation (7.15) for y, we obtain its general solution in the form

$$y = k_1u_1 + k_2u_2 + \cdots + k_Nu_N + U_2, \tag{7.16}$$

where $u_1, u_2, \ldots, u_N$ are the N linearly independent solutions of $L_5y = 0$ (or $L_5x = 0$) which already appear in (7.14), $k_1, k_2, \ldots, k_N$ are arbitrary constants, and U_2 is a particular solution of $L_5y = g_2$.

We thus see that if x and y satisfy the linear system (7.11), then x satisfies the single linear differential equation (7.13) and y satisfies the single linear differential equation (7.15). Thus if x and y satisfy the system (7.11), then x is of the form (7.14) and y is of the form (7.16). However, the pairs of functions given by (7.14) and (7.16) do *not* satisfy the given system (7.11) for *all* choices of the constants $c_1, c_2, \ldots, c_N, k_1, k_2, \ldots, k_N$. That is, these pairs (7.14) and (7.16) do not simultaneously satisfy both equations of the given system (7.11) for arbitrary choices of the $2N$ constants $c_1, c_2, \ldots, c_N, k_1, k_2, \ldots, k_N$. In other words, in order for x given by (7.14) and y given by (7.16) to satisfy the given system (7.11), the $2N$ constants $c_1, c_2, \ldots, c_N, k_1, k_2, \ldots, k_N$ cannot all be independent but rather certain of them must be dependent on the others. It can be shown that the number of independent constants in the so-called general solution of the linear system (7.11) is equal to the order of the operator $L_1L_4 - L_2L_3$ obtained from the determinant

$$\begin{vmatrix} L_1 & L_2 \\ L_3 & L_4 \end{vmatrix}$$

of the operator "coefficients" of x and y in (7.11), provided that this determinant is not zero. We have assumed that this operator is of order N. Thus in order for the pair (7.14) and (7.16) to satisfy the system (7.11) only N of the $2N$ constants in this pair can be independent. The remaining N constants must depend upon the N which are independent. In order to determine which of these $2N$ constants may be chosen as independent and how the remaining N then relate to the N so chosen, we must substitute x as given by (7.14) and y as given by (7.16) into the system (7.11). This determines the relations which must exist among the constants $c_1, c_2, \ldots, c_N, k_1, k_2, \ldots, k_N$ in order that the pair (7.14) and (7.16) constitute the so-called general solution of (7.11). Once this has been done, appropriate substitutions based on these relations are made in (7.14) and/or (7.16) and then the resulting pair (7.14) and (7.16) contain the required number N of arbitrary constants and so does indeed constitute the so-called general solution of system (7.11).

We now illustrate the above procedure with an example.

▶ Example 7.3. Solve the system

$$2\frac{dx}{dt} - 2\frac{dy}{dt} - 3x = t,$$
$$2\frac{dx}{dt} + 2\frac{dy}{dt} + 3x + 8y = 2. \tag{7.17}$$

We introduce operator notation and write this system in the form

$$(2D - 3)x - 2Dy = t,$$
$$(2D + 3)x + (2D + 8)y = 2. \tag{7.18}$$

We apply the operator $(2D + 8)$ to the first equation of (7.18) and the operator $2D$ to the second equation of (7.18), obtaining

$$(2D + 8)(2D - 3)x - (2D + 8)2Dy = (2D + 8)t,$$
$$2D(2D + 3)x + 2D(2D + 8)y = (2D)2.$$

Adding these two equations, we obtain

$$[(2D + 8)(2D - 3) + 2D(2D + 3)]x = (2D + 8)t + (2D)2$$

or

$$(8D^2 + 16D - 24)x = 2 + 8t + 0$$

or, finally,

$$(D^2 + 2D - 3)x = t + \tfrac{1}{4}. \tag{7.19}$$

The general solution of the differential equation (7.19) is

$$x = c_1e^t + c_2e^{-3t} - \tfrac{1}{3}t - \tfrac{11}{36}. \tag{7.20}$$

We now return to the system (7.18) and apply the operator $(2D + 3)$ to the first equation of (7.18) and the operator $(2D - 3)$ to the second equation of (7.18). We obtain

$$(2D + 3)(2D - 3)x - (2D + 3)2Dy = (2D + 3)t,$$
$$(2D - 3)(2D + 3)x + (2D - 3)(2D + 8)y = (2D - 3)2.$$

Subtracting the first of these equations from the second, we have

$$[(2D - 3)(2D + 8) + (2D + 3)2D]y = (2D - 3)2 - (2D + 3)t$$

or

$$(8D^2 + 16D - 24)y = 0 - 6 - 2 - 3t$$

or, finally,

$$(D^2 + 2D - 3)y = -\tfrac{3}{8}t - 1. \tag{7.21}$$

The general solution of the differential equation (7.21) is

$$y = k_1e^t + k_2e^{-3t} + \tfrac{1}{8}t + \tfrac{5}{12}. \tag{7.22}$$

Thus if x and y satisfy the system (7.17), then x must be of the form (7.20) and y must be of the form (7.22) for some choice of the constants c_1, c_2, k_1, k_2. The determinant of the operator "coefficients" of x and y in (7.18) is

$$\begin{vmatrix} 2D - 3 & -2D \\ 2D + 3 & 2D + 8 \end{vmatrix} = 8D^2 + 16D - 24.$$

Since this is of order two, the number of independent constants in the general solution of the system (7.17) must also be two. Thus in order for the pair (7.20) and (7.22) to satisfy the system (7.17) only two of the four constants c_1, c_2, k_1, and k_2 can be independent. In order to determine the necessary relations which must exist among these constants, we substitute x as given by (7.20) and y as given by (7.22) into the system (7.17). Substituting into the first equation of (7.17), we have

$$[2c_1e^t - 6c_2e^{-3t} - \tfrac{2}{3}] - [2k_1e^t - 6k_2e^{-3t} + \tfrac{1}{4}] - [3c_1e^t + 3c_2e^{-3t} - t - \tfrac{11}{12}] = t$$

or

$$(-c_1 - 2k_1)e^t + (-9c_2 + 6k_2)e^{-3t} = 0.$$

Thus in order that the pair (7.20) and (7.22) satisfy the first equation of the system (7.17) we must have

$$\begin{aligned} -c_1 - 2k_1 &= 0, \\ -9c_2 + 6k_2 &= 0. \end{aligned} \tag{7.23}$$

Substitution of x and y into the second equation of the system (7.17) will lead to relations equivalent to (7.23). Hence in order for the pair (7.20) and (7.22) to satisfy the system (7.17), the relations (7.23) must be satisfied. Two of the four constants in (7.23) must be chosen as independent. If we choose c_1 and c_2 as independent, then we have

$$k_1 = -\tfrac{1}{2}c_1 \quad \text{and} \quad k_2 = \tfrac{3}{2}c_2.$$

Using these values for k_1 and k_2 in (7.22), the resulting pair (7.20) and (7.22) constitute the general solution of the system (7.17). That is, the general solution of (7.17) is given by

$$\begin{aligned} x &= c_1e^t + c_2e^{-3t} - \tfrac{1}{3}t - \tfrac{11}{36}, \\ y &= -\tfrac{1}{2}c_1e^t + \tfrac{3}{2}c_2e^{-3t} + \tfrac{1}{8}t + \tfrac{5}{12}, \end{aligned}$$

where c_1 and c_2 are arbitrary constants. If we had chosen k_1 and k_2 as the independent constants in (7.23), then the general solution of the system (7.17) would have been written

$$\begin{aligned} x &= -2k_1e^t + \tfrac{2}{3}k_2e^{-3t} - \tfrac{1}{3}t - \tfrac{11}{36}, \\ y &= k_1e^t + k_2e^{-3t} + \tfrac{1}{8}t + \tfrac{5}{12}. \end{aligned}$$

An Alternative Procedure. Here we present an alternative procedure for solving a linear system of the form

$$\begin{aligned} L_1x + L_2y &= f_1(t), \\ L_3x + L_4y &= f_2(t), \end{aligned} \tag{7.11}$$

where L_1, L_2, L_3, and L_4 are linear differential operators with constant coefficients. This alternative procedure begins in exactly the same way as the procedure already described. That is, we first apply the operator L_4 to the first equation of (7.11) and the operator L_2 to the second equation of (7.11), obtaining

$$L_4L_1x + L_4L_2y = L_4f_1,$$
$$L_2L_3x + L_2L_4y = L_2f_2.$$

We next subtract the second from the first, obtaining

$$(L_1L_4 - L_2L_3)x = L_4f_1 - L_2f_2, \tag{7.12}$$

which, under the same assumptions as we previously made at this point, may be written

$$L_5x = g_1. \tag{7.13}$$

Then we solve this single linear differential equation with constant coefficients in the single dependent variable x. Assuming its order is N, we obtain its general solution in the form

$$x = c_1u_1 + c_2u_2 + \cdots + c_Nu_N + U_1, \tag{7.14}$$

where $u_1, u_2, \ldots, u_N$ are N linearly independent solutions of the homogeneous linear equation $L_5x = 0$, $c_1, c_2, \ldots, c_N$ are N arbitrary constants, and U_1 is a particular solution of $L_5x = g_1$.

Up to this point, we have indeed proceeded just exactly as before. However, we now return to system (7.11) and attempt to eliminate from it *all* terms which involve the derivatives of the *other* dependent variable y. In other words, we attempt to obtain from system (7.11) a relation R which involves the still unknown y but *none of the derivatives of* y. This relation R will involve x and/or certain of the derivatives of x; but x is given by (7.14) and its derivatives can readily be found from (7.14). Finding these derivatives of x and substituting them and the known x itself into the relation R, we see that the result is merely a single linear *algebraic* equation in the one unknown y. Solving it, we thus determine y without the need to find (7.15) and (7.16) or to relate the arbitrary constants.

As we shall see, this alternative procedure always applies in an easy straightforward manner if the operators L_1, L_2, L_3, and L_4 are all of the first order. However, for systems involving one or more higher-order operators, it is generally difficult to eliminate *all* the derivatives of y.

We now give an explicit presentation of the procedure for finding y when L_1, L_2, L_3, and L_4 are all first-order operators.

Specifically, suppose

$$L_1 \equiv a_0D + a_1,$$
$$L_2 \equiv b_0D + b_1,$$
$$L_3 \equiv \alpha_0D + \alpha_1,$$
$$L_4 \equiv \beta_0D + \beta_1.$$

Then (7.11) is

$$\begin{aligned}(a_0D + a_1)x + (b_0D + b_1)y &= f_1(t),\\ (\alpha_0D + \alpha_1)x + (\beta_0D + \beta_1)y &= f_2(t).\end{aligned} \tag{7.24}$$

Multiplying the first equation of (7.24) by β_0 and the second by $-b_0$ and adding, we obtain

$$[(a_0\beta_0 - b_0\alpha_0)D + (a_1\beta_0 - b_0\alpha_1)]x + (b_1\beta_0 - b_0\beta_1)y = \beta_0 f_1(t) - b_0 f_2(t).$$

Note that this involves y but *none of the derivatives of* y. From this, we at once obtain

$$y = \frac{(b_0\alpha_0 - a_0\beta_0)Dx + (b_0\alpha_1 - a_1\beta_0)x + \beta_0 f_1(t) - b_0 f_2(t)}{b_1\beta_0 - b_0\beta_1}, \tag{7.25}$$

assuming $b_1\beta_0 - b_0\beta_1 \neq 0$. Now x is given by (7.14) and Dx may be found from (7.14) by straightforward differentiation. Then substituting these known expressions for x and Dx into (7.25), we at once obtain y without the need of obtaining (7.15) and (7.16) and hence without having to determine any relations between constants c_i and k_i ($i = 1, 2, \ldots, N$), as in the original procedure.

We illustrate the alternative procedure by applying it to the system of Example 7.3.

▶ **Example 7.4.** Solve the system

$$\begin{aligned} 2\frac{dx}{dt} - 2\frac{dy}{dt} - 3x &= t, \\ 2\frac{dx}{dt} + 2\frac{dy}{dt} + 3x + 8y &= 2. \end{aligned} \tag{7.17}$$

of Example 7.3 by the alternative procedure which we have just described.

Following this alternative procedure, we introduce operator notation and write the system (7.17) in the form

$$\begin{aligned} (2D - 3)x - 2Dy &= t, \\ (2D + 3)x + (2D + 8)y &= 2. \end{aligned} \tag{7.18}$$

Now we eliminate y, obtain the differential equation

$$(D^2 + 2D - 3)x = t + \tfrac{1}{4} \tag{7.19}$$

for x, and find its general solution

$$x = c_1 e^t + c_2 e^{-3t} - \tfrac{1}{3}t - \tfrac{11}{36}, \tag{7.20}$$

exactly as in Example 7.3.

We now proceed using the alternative method. We first obtain from (7.18) a relation which involves the unknown y but *not* the derivative Dy. The system (7.18) of this example is so very simple that we do so by merely adding the equations (7.18). Doing so, we at once obtain

$$4Dx + 8y = t + 2,$$

which does indeed involve y but *not* the derivative Dy, as desired. From this, we at once find

$$y = \tfrac{1}{8}(t + 2 - 4Dx). \tag{7.26}$$

From (7.20), we find

$$Dx = c_1 e^t - 3c_2 e^{-3t} - \tfrac{1}{3}.$$

Substituting into (7.26), we get

$$\begin{aligned} y &= \tfrac{1}{8}(t + 2 - 4c_1e^t + 12c_2e^{-3t} + \tfrac{4}{3}) \\ &= -\tfrac{1}{2}c_1e^t + \tfrac{3}{2}c_2e^{-3t} + \tfrac{1}{8}t + \tfrac{5}{12}. \end{aligned}$$

Thus the general solution of the system may be written

$$\begin{aligned} x &= c_1e^t + c_2e^{-3t} - \tfrac{1}{3}t - \tfrac{11}{36}, \\ y &= -\tfrac{1}{2}c_1e^t + \tfrac{3}{2}c_2e^{-3t} + \tfrac{1}{8}t + \tfrac{5}{12}, \end{aligned}$$

where c_1 and c_2 are arbitrary constants.

Exercises

Use the operator method described in this section to find the general solution of each of the following linear systems.

1. $\dfrac{dx}{dt} + \dfrac{dy}{dt} - 2x - 4y = e^t,$
   $\dfrac{dx}{dt} + \dfrac{dy}{dt} - y = e^{4t}.$

2. $\dfrac{dx}{dt} + \dfrac{dy}{dt} - x = -2t,$
   $\dfrac{dx}{dt} + \dfrac{dy}{dt} - 3x - y = t^2.$

3. $\dfrac{dx}{dt} + \dfrac{dy}{dt} - x - 3y = e^t,$
   $\dfrac{dx}{dt} + \dfrac{dy}{dt} + x = e^{3t}.$

4. $\dfrac{dx}{dt} + \dfrac{dy}{dt} - x - 2y = 2e^t,$
   $\dfrac{dx}{dt} + \dfrac{dy}{dt} - 3x - 4y = e^{2t}.$

5. $2\dfrac{dx}{dt} + \dfrac{dy}{dt} - x - y = e^{-t},$
   $\dfrac{dx}{dt} + \dfrac{dy}{dt} + 2x + y = e^t.$

6. $2\dfrac{dx}{dt} + \dfrac{dy}{dt} - 3x - y = t,$
   $\dfrac{dx}{dt} + \dfrac{dy}{dt} - 4x - y = e^t.$

7. $\dfrac{dx}{dt} + \dfrac{dy}{dt} - x - 6y = e^{3t},$
   $\dfrac{dx}{dt} + 2\dfrac{dy}{dt} - 2x - 6y = t.$

8. $\dfrac{dx}{dt} + \dfrac{dy}{dt} - x - 3y = 3t,$
   $\dfrac{dx}{dt} + 2\dfrac{dy}{dt} - 2x - 3y = 1.$

9. $\dfrac{dx}{dt} + \dfrac{dy}{dt} + 2y = \sin t,$
   $\dfrac{dx}{dt} + \dfrac{dy}{dt} - x - y = 0.$

10. $\dfrac{dx}{dt} - \dfrac{dy}{dt} - 2x + 4y = t,$
    $\dfrac{dx}{dt} + \dfrac{dy}{dt} - x - y = 1.$

11. $2\dfrac{dx}{dt} + \dfrac{dy}{dt} + x + 5y = 4t,$
    $\dfrac{dx}{dt} + \dfrac{dy}{dt} + 2x + 2y = 2.$

12. $\dfrac{dx}{dt} + \dfrac{dy}{dt} - x + 5y = t^2,$
    $\dfrac{dx}{dt} + 2\dfrac{dy}{dt} - 2x + 4y = 2t + 1.$

13. $2\frac{dx}{dt} + \frac{dy}{dt} + x + y = t^2 + 4t,$
$\frac{dx}{dt} + \frac{dy}{dt} + 2x + 2y = 2t^2 - 2t.$

14. $3\frac{dx}{dt} + 2\frac{dy}{dt} - x + y = t - 1,$
$\frac{dx}{dt} + \frac{dy}{dt} - x = t + 2.$

15. $2\frac{dx}{dt} + 4\frac{dy}{dt} + x - y = 3e^t,$
$\frac{dx}{dt} + \frac{dy}{dt} + 2x + 2y = e^t.$

16. $2\frac{dx}{dt} + \frac{dy}{dt} - x - y = -2t,$
$\frac{dx}{dt} + \frac{dy}{dt} + x - y = t^2.$

17. $2\frac{dx}{dt} + \frac{dy}{dt} - x - y = 1,$
$\frac{dx}{dt} + \frac{dy}{dt} + 2x - y = t.$

18. $\frac{d^2x}{dt^2} + \frac{dy}{dt} = e^{2t},$
$\frac{dx}{dt} + \frac{dy}{dt} - x - y = 0.$

19. $\frac{d^2x}{dt^2} + \frac{dy}{dt} - x + y = 1,$
$\frac{d^2y}{dt^2} + \frac{dx}{dt} - x + y = 0.$

20. $\frac{d^2x}{dt^2} - \frac{dy}{dt} = e^t,$
$\frac{dx}{dt} + \frac{dy}{dt} - 4x - y = 2e^t.$

21. $\frac{d^2x}{dt^2} - \frac{dy}{dt} = t + 1,$
$\frac{dx}{dt} + \frac{dy}{dt} - 3x + y = 2t - 1.$

22. $\frac{d^2x}{dt^2} + 4\frac{dy}{dt} + x - 4y = 0,$
$\frac{dx}{dt} + \frac{dy}{dt} - x + 9y = e^{2t}.$

In each of Exercises 23 through 26, transform the single linear differential equation of the form (7.6) into a system of first-order differential equations of the form (7.9).

23. $\frac{d^2x}{dt^2} - 3\frac{dx}{dt} + 2x = t^2.$

24. $\frac{d^3x}{dt^3} + 2\frac{d^2x}{dt^2} - \frac{dx}{dt} - 2x = e^{3t}.$

25. $\frac{d^3x}{dt^3} + t\frac{d^2x}{dt^2} + 2t^3\frac{dx}{dt} - 5t^4 = 0.$

26. $\frac{d^4x}{dt^4} - t^2\frac{d^2x}{dt^2} + 2tx = \cos t.$

7.2 Applications

A. *Applications to Mechanics*

Systems of linear differential equations originate in the mathematical formulation of numerous problems in mechanics. We consider one such problem in the following example. Another mechanics problem leading to a linear system is given in the exercises at the end of this section.

▶ Example 7.5. On a smooth horizontal plane BC (for example, a smooth table top) an object A_1 is connected to a fixed point P by a massless spring S_1 of natural length L_1. An object A_2 is then connected to A_1 by a massless spring S_2 of natural length L_2 in such a way that the fixed point P and the centers of gravity A_1 and A_2 all lie in a straight line (see Figure 7.1).

The object A_1 is then displaced a distance a_1 to the right or left of its equilibrium position O_1, the object A_2 is displaced a distance a_2 to the right or left of its equilibrium position O_2, and at time $t = 0$ the two objects are released (see Figure 7.2). What are the positions of the two objects at any time $t > 0$?

Formulation. We assume first that the plane BC is so smooth that frictional forces may be neglected. We also assume that no external forces act upon the system. Suppose object A_1 has mass m_1 and object A_2 has mass m_2. Further suppose spring S_1 has spring constant k_1 and spring S_2 has spring constant k_2.

Let x_1 denote the displacement of A_1 from its equilibrium position O_1 at time $t \geq 0$ and assume that x_1 is positive when A_1 is to the right of O_1. In like manner, let x_2 denote the displacement of A_2 from its equilibrium position O_2 at time $t \geq 0$ and assume that x_2 is positive when A_2 is to the right of O_2 (see Figure 7.3).

Consider the forces acting on A_1 at time $t > 0$. There are two such forces, F_1 and F_2, where F_1 is exerted by spring S_1 and F_2 is exerted by spring S_2. By Hooke's law

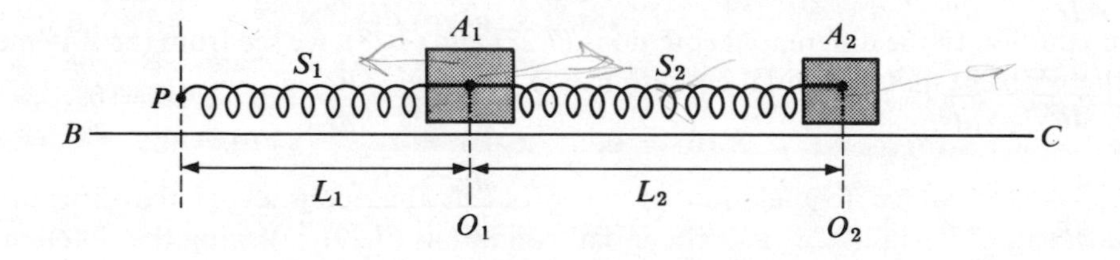

FIGURE 7.1

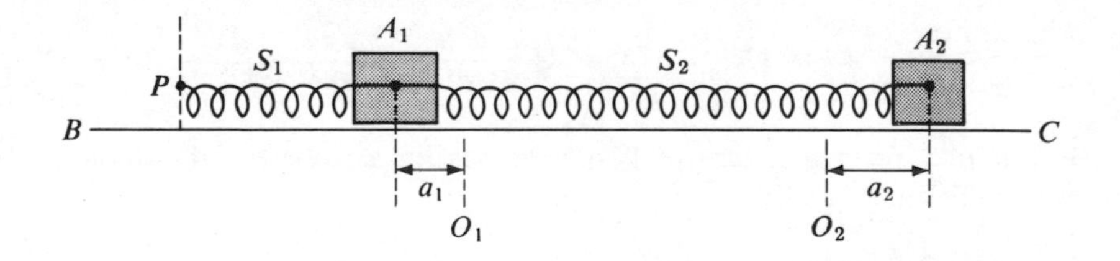

FIGURE 7.2

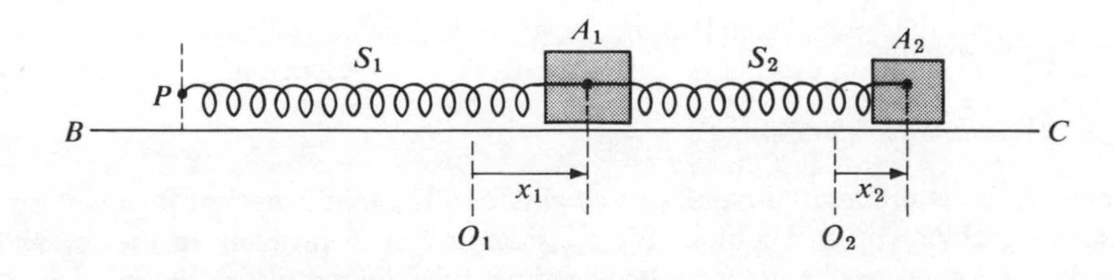

FIGURE 7.3

(Section 5.1) the force F_1 is of magnitude $k_1|x_1|$. Since this force is exerted toward the left when A_1 is to the right of O_1 and toward the right when A_1 is to the left of O_1, we have $F_1 = -k_1x_1$. Again using Hooke's law, the force F_2 is of magnitude k_2s, where s is the elongation of S_2 at time t. Since $s = |x_2 - x_1|$, we see that the magnitude of F_2 is $k_2|x_2 - x_1|$. Further, since this force is exerted toward the left when $x_2 - x_1 < 0$ and toward the right when $x_2 - x_1 > 0$, we see that $F_2 = k_2(x_2 - x_1)$.

Now applying Newton's second law (Section 3.2) to the object A_1, we obtain the differential equation

$$m_1 \frac{d^2x_1}{dt^2} = -k_1x_1 + k_2(x_2 - x_1). \tag{7.27}$$

We now turn to the object A_2 and consider the forces which act upon it at time $t > 0$. There is one such force, F_3, and this is exerted by spring S_2. Applying Hooke's law once again, we observe that this force is also of magnitude $k_2s = k_2|x_2 - x_1|$. Since F_3 is exerted toward the left when $x_2 - x_1 > 0$ and toward the right when $x_2 - x_1 < 0$, we see that $F_3 = -k_2(x_2 - x_1)$. Applying Newton's second law to the object A_2, we obtain the differential equation

$$m_2 \frac{d^2x_2}{dt^2} = -k_2(x_2 - x_1). \tag{7.28}$$

In addition to the differential equations (7.27) and (7.28), we see from the statement of the problem that the initial conditions are given by

$$x_1(0) = a_1, \quad x_1'(0) = 0, \quad x_2(0) = a_2, \quad x_2'(0) = 0. \tag{7.29}$$

The mathematical formulation of the problem thus consists of the differential equations (7.27) and (7.28) and the initial conditions (7.29). Writing the differential equations in the form

$$\begin{aligned} m_1 \frac{d^2x_1}{dt^2} + (k_1 + k_2)x_1 - k_2x_2 &= 0, \\ m_2 \frac{d^2x_2}{dt^2} - k_2x_1 + k_2x_2 &= 0, \end{aligned} \tag{7.30}$$

we see that they form a system of homogeneous linear differential equations with constant coefficients.

Solution of a Specific Case. Rather than solve the general problem consisting of the system (7.30) and conditions (7.29), we shall carry through the solution in a particular case which was chosen to facilitate the work. Suppose the two objects A_1 and A_2 are each of unit mass, so that $m_1 = m_2 = 1$. Further, suppose that the springs S_1 and S_2 have spring constants $k_1 = 3$ and $k_2 = 2$, respectively. Also, we shall take $a_1 = -1$ and $a_2 = 2$. Then the system (7.30) reduces to

$$\begin{aligned} \frac{d^2x_1}{dt^2} + 5x_1 - 2x_2 &= 0, \\ \frac{d^2x_2}{dt^2} - 2x_1 + 2x_2 &= 0, \end{aligned} \tag{7.31}$$

and the initial conditions (7.29) become

$$x_1(0) = -1, \quad x_1'(0) = 0, \quad x_2(0) = 2, \quad x_2'(0) = 0. \tag{7.32}$$

Writing the system (7.31) in operator notation, we have

$$\begin{aligned}(D^2 + 5)x_1 - 2x_2 &= 0,\\ -2x_1 + (D^2 + 2)x_2 &= 0.\end{aligned} \tag{7.33}$$

We apply the operator $(D^2 + 2)$ to the first equation of (7.33), multiply the second equation of (7.33) by 2, and add the two equations to obtain

$$[(D^2 + 2)(D^2 + 5) - 4]x_1 = 0$$

or

$$(D^4 + 7D^2 + 6)x_1 = 0. \tag{7.34}$$

The auxiliary equation corresponding to the fourth-order differential equation (7.34) is

$$m^4 + 7m^2 + 6 = 0 \quad \text{or} \quad (m^2 + 6)(m^2 + 1) = 0.$$

Thus the general solution of the differential equation (7.34) is

$$x_1 = c_1 \sin t + c_2 \cos t + c_3 \sin \sqrt{6}\, t + c_4 \cos \sqrt{6}\, t. \tag{7.35}$$

We now multiply the first equation of (7.33) by 2, apply the operator $(D^2 + 5)$ to the second equation of (7.33), and add to obtain the differential equation

$$(D^4 + 7D^2 + 6)x_2 = 0 \tag{7.36}$$

for x_2. The general solution of (7.36) is clearly

$$x_2 = k_1 \sin t + k_2 \cos t + k_3 \sin \sqrt{6}\, t + k_4 \cos \sqrt{6}\, t. \tag{7.37}$$

The determinant of the operator "coefficients" in the system (7.33) is

$$\begin{vmatrix} D^2 + 5 & -2 \\ -2 & D^2 + 2 \end{vmatrix} = D^4 + 7D^2 + 6.$$

Since this is a fourth-order operator, the general solution of (7.31) must contain four independent constants. We must substitute x_1 given by (7.35) and x_2 given by (7.37) into the equations of the system (7.31) to determine the relations which must exist among the constants c_1, c_2, c_3, c_4, k_1, k_2, k_3, and k_4 in order that the pair (7.35) and (7.37) represent the general solution of (7.31). Substituting, we find that

$$k_1 = 2c_1, \quad k_2 = 2c_2, \quad k_3 = -\tfrac{1}{2}c_3, \quad k_4 = -\tfrac{1}{2}c_4.$$

Thus the general solution of the system (7.31) is given by

$$\begin{aligned}x_1 &= c_1 \sin t + c_2 \cos t + c_3 \sin \sqrt{6}\, t + c_4 \cos \sqrt{6}\, t,\\ x_2 &= 2c_1 \sin t + 2c_2 \cos t - \tfrac{1}{2}c_3 \sin \sqrt{6}\, t - \tfrac{1}{2}c_4 \cos \sqrt{6}\, t.\end{aligned} \tag{7.38}$$

We now apply the initial conditions (7.32). Applying the conditions $x_1 = -1$, $dx_1/dt = 0$ at $t = 0$ to the first of the pair (7.38), we find

$$\begin{aligned} -1 &= c_2 + c_4, \\ 0 &= c_1 + \sqrt{6}\, c_3. \end{aligned} \tag{7.39}$$

Applying the conditions $x_2 = 2$, $dx_2/dt = 0$ at $t = 0$ to the second of the pair (7.38), we obtain

$$\begin{aligned} 2 &= 2c_2 - \tfrac{1}{2}c_4, \\ 0 &= 2c_1 - \frac{\sqrt{6}}{2} c_3. \end{aligned} \tag{7.40}$$

From Equations (7.39) and (7.40), we find that

$$c_1 = 0, \quad c_2 = \tfrac{3}{5}, \quad c_3 = 0, \quad c_4 = -\tfrac{8}{5}.$$

Thus the particular solution of the specific problem consisting of the system (7.31) and the conditions (7.32) is

$$\begin{aligned} x_1 &= \tfrac{3}{5}\cos t - \tfrac{8}{5}\cos\sqrt{6}\, t, \\ x_2 &= \tfrac{6}{5}\cos t + \tfrac{4}{5}\cos\sqrt{6}\, t. \end{aligned}$$

B. *Applications to Electric Circuits*

In Section 5.6 we considered the application of differential equations to electric circuits consisting of a single closed path. A closed path in an electrical network is called a *loop*. We shall now consider electrical networks which consist of several loops. For example, consider the network shown in Figure 7.4.

This network consists of the three loops *ABMNA*, *BJKMB*, and *ABJKMNA*. Points such as *B* and *M* at which two or more circuits join are called *junction points* or *branch points*. The direction of current flow has been arbitrarily assigned and indicated by arrows.

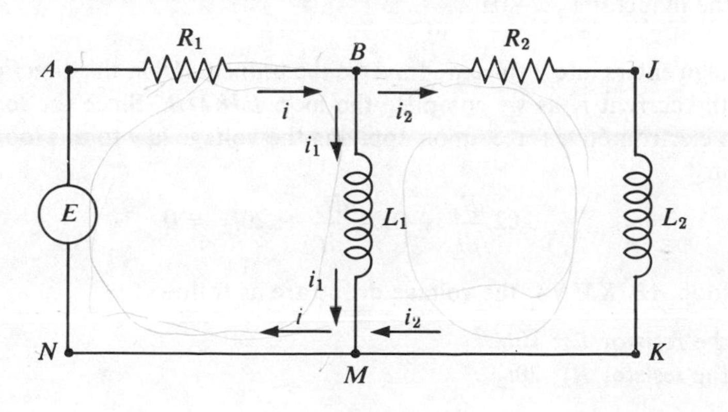

FIGURE 7.4

In order to solve problems involving multiple loop networks we shall need two fundamental laws of circuit theory. One of these is Kirchhoff's voltage law, which we have already stated and applied in Section 5.6. The other basic law which we shall employ is the following:

Kirchhoff's Current Law. In an electrical network the total current flowing into a junction point is equal to the total current flowing away from the junction point.

As an application of these laws we consider the following problem dealing with the circuit of Figure 7.4.

▶ Example 7.6. Determine the currents in the electrical network of Figure 7.4, if E is an electromotive force of 30 V, R_1 is a resistor of 10 Ω, R_2 is a resistor of 20 Ω, L_1 is an inductor of 0.02 H, L_2 is an inductor of 0.04 H, and the currents are initially zero.

Formulation. The current flowing in the branch $MNAB$ is denoted by i, that flowing on the branch BM by i_1, and that flowing on the branch $BJKM$ by i_2.

We now apply Kirchhoff's voltage law (Section 5.6) to each of the three loops $ABMNA$, $BJKMB$, and $ABJKMNA$.

For the loop $ABMNA$ the voltage drops are as follows:

1. Across the resistor R_1: $10i$.
2. Across the inductor L_1: $0.02\dfrac{di_1}{dt}$.

Thus applying the voltage law to the loop $ABMNA$, we have the equation

$$0.02\frac{di_1}{dt} + 10i = 30. \tag{7.41}$$

For the loop $BJKMB$, the voltage drops are as follows:

1. Across the resistor R_2: $20i_2$.
2. Across the inductor L_2: $0.04\dfrac{di_2}{dt}$.
3. Across the inductor L_1: $-0.02\dfrac{di_1}{dt}$.

The minus sign enters into 3 since we traverse the branch MB in the direction opposite to that of the current i_1 as we complete the loop $BJKMB$. Since the loop $BJKMB$ contains no electromotive force, upon applying the voltage law to this loop we obtain the equation

$$-0.02\frac{di_1}{dt} + 0.04\frac{di_2}{dt} + 20i_2 = 0. \tag{7.42}$$

For the loop $ABJKMNA$, the voltage drops are as follows:

1. Across the resistor R_1: $10i$.
2. Across the resistor R_2: $20i_2$.
3. Across the inductor L_2: $0.04\dfrac{di_2}{dt}$.

Applying the voltage law to this loop, we obtain the equation

$$10i + 0.04\frac{di_2}{dt} + 20i_2 = 30. \tag{7.43}$$

We observe that the three equations (7.41), (7.42), and (7.43) are not all independent. For example, we note that (7.42) may be obtained by subtracting (7.41) from (7.43). Thus we need to retain only the two equations (7.41) and (7.43).

We now apply Kirchhoff's current law to the junction point B. From this we see at once that

$$i = i_1 + i_2. \tag{7.44}$$

In accordance with this we replace i by $i_1 + i_2$ in (7.41) and (7.43) and thus obtain the linear system

$$\begin{aligned} 0.02\frac{di_1}{dt} + 10i_1 + 10i_2 &= 30, \\ 10i_1 + 0.04\frac{di_2}{dt} + 30i_2 &= 30. \end{aligned} \tag{7.45}$$

Since the currents are initially zero, we have the initial conditions

$$i_1(0) = 0 \quad \text{and} \quad i_2(0) = 0. \tag{7.46}$$

Solution. We introduce operator notation and write the system (7.45) in the form

$$\begin{aligned} (0.02D + 10)i_1 + 10i_2 &= 30, \\ 10i_1 + (0.04D + 30)i_2 &= 30. \end{aligned} \tag{7.47}$$

We apply the operator $(0.04D + 30)$ to the first equation of (7.47), multiply the second by 10, and subtract to obtain

$$[(0.04D + 30)(0.02D + 10) - 100]i_1 = (0.04D + 30)30 - 300$$

or

$$(0.0008D^2 + D + 200)i_1 = 600$$

or, finally,

$$(D^2 + 1250D + 250{,}000)i_1 = 750{,}000. \tag{7.48}$$

We now solve the differential equation (7.48) for i_1. The auxiliary equation is

$$m^2 + 1250m + 250{,}000 = 0$$

or

$$(m + 250)(m + 1000) = 0.$$

Thus the complementary function of Equation (7.48) is

$$i_{1,c} = c_1e^{-250t} + c_2e^{-1000t},$$

and a particular integral is obviously $i_{1,p} = 3$. Hence the general solution of the differential equation (7.48) is

$$i_1 = c_1e^{-250t} + c_2e^{-1000t} + 3. \tag{7.49}$$

Now returning to the system (7.47), we multiply the first equation of the system by 10, apply the operator $(0.02 + 10)$ to the second equation, and subtract the first from the second. After simplifications we obtain the differential equation

$$(D^2 + 1250D + 250{,}000)i_2 = 0$$

for i_2. The general solution of this differential equation is clearly

$$i_2 = k_1 e^{-250t} + k_2 e^{-1000t}. \tag{7.50}$$

Since the determinant of the operator "coefficients" in the system (7.47) is a second-order operator, the general solution of the system (7.45) must contain two independent constants. We must substitute i_1 given by (7.49) and i_2 given by (7.50) into the equations of the system (7.45) to determine the relations which must exist among the constants c_1, c_2, k_1, k_2 in order that the pair (7.49) and (7.50) represent the general solution of (7.45). Substituting, we find that

$$k_1 = -\tfrac{1}{2}c_1, \qquad k_2 = c_2. \tag{7.51}$$

Thus the general solution of the system (7.45) is given by

$$\begin{aligned} i_1 &= c_1 e^{-250t} + c_2 e^{-1000t} + 3, \\ i_2 &= -\tfrac{1}{2}c_1 e^{-250t} + c_2 e^{-1000t}. \end{aligned} \tag{7.52}$$

Now applying the initial conditions (7.46), we find that $c_1 + c_2 + 3 = 0$ and $-\frac{1}{2}c_1 + c_2 = 0$ and hence $c_1 = -2$ and $c_2 = -1$. Thus the solution of the linear system (7.45) which satisfies the conditions (7.46) is

$$i_1 = -2e^{-250t} - e^{-1000t} + 3,$$

$$i_2 = e^{-250t} - e^{-1000t}.$$

Finally, using (7.44) we find that

$$i = -e^{-250t} - 2e^{-1000t} + 3.$$

We observe that the current i_2 rapidly approaches zero. On the other hand, the currents, i_1 and $i = i_1 + i_2$ rapidly approach the value 3.

Exercises

1. Solve the problem of Example 7.5 for the case in which the object A_1 has mass $m_1 = 2$, the object A_2 has mass $m_2 = 1$, the spring S_1 has spring constant $k_1 = 4$, the spring S_2 has spring constant $k_2 = 2$, and the initial conditions are $x_1(0) = 1$, $x_1'(0) = 0$, $x_2(0) = 5$, and $x_2'(0) = 0$.

2. A projectile of mass m is fired into the air from a gun which is inclined at an angle θ with the horizontal, and suppose the initial velocity of the projectile is v_0 feet

per second. Neglect all forces except that of gravity and the air resistance, and assume that this latter force (in pounds) is numerically equal to k times the velocity (in feet/second).

(a) Taking the origin at the position of the gun, with the x-axis horizontal and the y-axis vertical, show that the differential equations of the resulting motion are

$$m\frac{d^2x}{dt^2} + k\frac{dx}{dt} = 0,$$

$$m\frac{d^2y}{dt^2} + k\frac{dy}{dt} + mg = 0.$$

(b) Find the solution of the system of differential equations of part (a).

3. Determine the currents in the electrical network of Figure 7.5 if E is an electromotive force of 100 V, R_1 is a resistor of 20 Ω, R_2 is a resistor of 40 Ω, L_1 is an inductor of 0.01 H, L_2 is an inductor of 0.02 H, and the currents are initially zero.

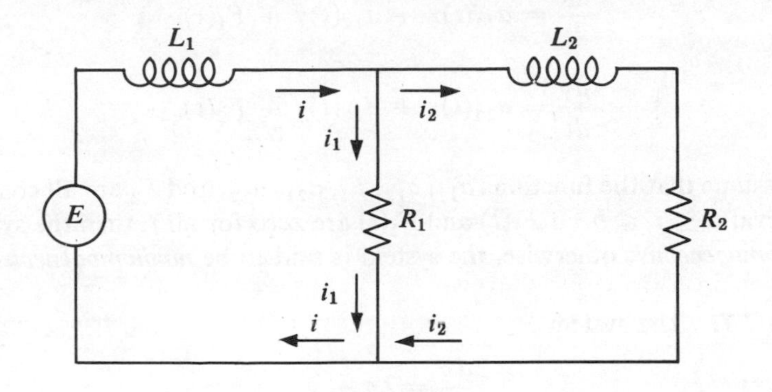

FIGURE 7.5

4. Set up differential equations for the currents in each of the electrical networks shown in Figure 7.6. Do not solve the equations.

(a) For the network in Figure 7.6(a) assume that E is an electromotive force of 15 V, R is a resistor of 20 Ω, L is an inductor of 0.02 H, and C is a capacitor of 10^{-4} farads.

(b) For the network in Figure 7.6(b) assume that E is an electromotive force of $100 \sin 130t$ V, R_1 is a resistor of 20 Ω, R_2 is a resistor of 30 Ω, and L is an inductor of 0.05 H.

(c) For the network of Figure 7.6(c), assume that E is an electromotive force of 100 V, R_1 is a resistor of 20 Ω, R_2 is a resistor of 10 Ω, C_1 is a capacitor of 10^{-4} farads, and C_2 is a capacitor of 2×10^{-4} farads.

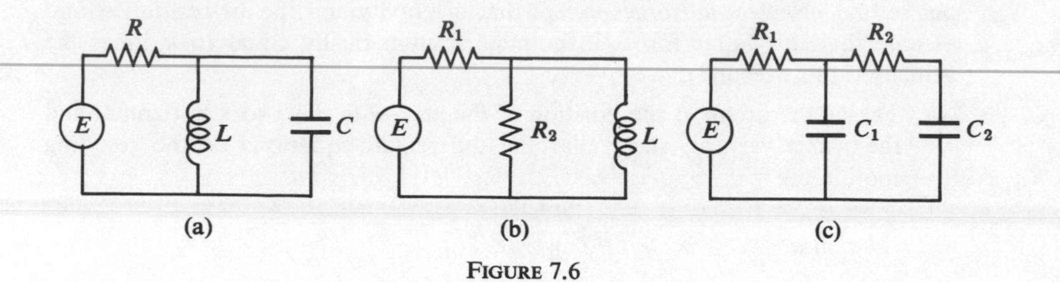

FIGURE 7.6

7.3 Basic Theory of Linear Systems in Normal Form: Two Equations in Two Unknown Functions

A. Introduction

We shall begin by considering a basic type of system of two linear differential equations in two unknown functions. This system is of the form

$$\begin{aligned} \frac{dx}{dt} &= a_{11}(t)x + a_{12}(t)y + F_1(t), \\ \frac{dy}{dt} &= a_{21}(t)x + a_{22}(t)y + F_2(t). \end{aligned} \tag{7.53}$$

We shall assume that the functions a_{11}, a_{12}, F_1, a_{21}, a_{22}, and F_2 are all continuous on a real interval $a \leq t \leq b$. If $F_1(t)$ and $F_2(t)$ are zero for all t, then the system (7.53) is called *homogeneous*; otherwise, the system is said to be *nonhomogeneous*.

▶ Example 7.7. The system

$$\begin{aligned} \frac{dx}{dt} &= 2x - y, \\ \frac{dy}{dt} &= 3x + 6y, \end{aligned} \tag{7.54}$$

is homogeneous; the system

$$\begin{aligned} \frac{dx}{dt} &= 2x - y - 5t, \\ \frac{dy}{dt} &= 3x + 6y - 4, \end{aligned} \tag{7.55}$$

is nonhomogeneous.

DEFINITION

By a solution *of the system (7.53) we shall mean an ordered pair of real functions*

$$(f, g), \tag{7.56}$$

each having a continuous derivative on the real interval $a \leq t \leq b$, such that

$$\frac{df(t)}{dt} = a_{11}(t)f(t) + a_{12}(t)g(t) + F_1(t),$$

$$\frac{dg(t)}{dt} = a_{21}(t)f(t) + a_{22}(t)g(t) + F_2(t),$$

for all t such that $a \leq t \leq b$. In other words,

$$\begin{aligned} x &= f(t), \\ y &= g(t), \end{aligned} \qquad (7.57)$$

simultaneously satisfy both equations of the system (7.53) identically for $a \leq t \leq b$.

Notation. We shall use the notation

$$\begin{aligned} x &= f(t), \\ y &= g(t), \end{aligned} \qquad (7.57)$$

to denote a solution of the system (7.53) and shall speak of "the solution

$$\begin{aligned} x &= f(t), \\ y &= g(t)." \end{aligned}$$

Whenever we do this, we must remember that the solution thus referred to is really the ordered pair of functions (f, g) such that (7.57) simultaneously satisfy both equations of the system (7.53) identically on $a \leq t \leq b$.

▶ Example 7.8. The ordered pair of functions defined for all t by $(e^{5t}, -3e^{5t})$, which we denote by

$$\begin{aligned} x &= e^{5t}, \\ y &= -3e^{5t}, \end{aligned} \qquad (7.58)$$

is a solution of the system (7.54). That is,

$$\begin{aligned} x &= e^{5t}, \\ y &= -3e^{5t}, \end{aligned} \qquad (7.58)$$

simultaneously satisfy both equations of the system (7.54). Let us verify this by directly substituting (7.58) into (7.54). We have

$$\frac{d}{dt}(e^{5t}) = 2(e^{5t}) - (-3e^{5t}),$$

$$\frac{d}{dt}(-3e^{5t}) = 3(e^{5t}) + 6(-3e^{5t}),$$

or

$$5e^{5t} = 2e^{5t} + 3e^{5t},$$

$$-15e^{5t} = 3e^{5t} - 18e^{5t}.$$

Hence (7.58) is indeed a solution of the system (7.54). The reader should verify that the ordered pair of functions defined for all t by $(e^{3t}, -e^{3t})$, which we denote by

$$x = e^{3t},$$

$$y = -e^{3t},$$

is also a solution of the system (7.54)

We shall now proceed to survey the basic theory of linear systems. We shall observe a close analogy between this theory and that introduced in Section 4.1 for the single linear equation of higher order. Theorem 7.1 is the basic existence theorem dealing with the system (7.53).

THEOREM 7.1

Hypothesis. *Let the functions a_{11}, a_{12}, F_1, a_{21}, a_{22}, and F_2 in the system (7.53) all be continuous on the interval $a \leq t \leq b$. Let t_0 be any point of the interval $a \leq t \leq b$; and let c_1 and c_2 be two arbitrary constants.*

Conclusion. *There exists a unique solution*

$$x = f(t),$$

$$y = g(t),$$

of the system (7.53) such that

$$f(t_0) = c_1 \quad \text{and} \quad g(t_0) = c_2,$$

and this solution is defined on the entire interval $a \leq t \leq b$.

▶ Example 7.9. Let us consider the system (7.55). The continuity requirements of the hypothesis of Theorem 7.1 are satisfied on every closed interval $a \leq t \leq b$. Hence, given any point t_0 and any two constants c_1 and c_2, there exists a unique solution $x = f(t)$, $y = g(t)$ of the system (7.55) which satisfies the conditions $f(t_0) = c_1$, $g(t_0) = c_2$. For example, there exists one and only one solution $x = f(t)$, $y = g(t)$ such that $f(2) = 5$, $g(2) = -7$.

B. Homogeneous Linear Systems

We shall now assume that $F_1(t)$ and $F_2(t)$ in the system (7.53) are both zero for all t and consider the basic theory of the resulting *homogeneous* linear system

$$\frac{dx}{dt} = a_{11}(t)x + a_{12}(t)y,$$
$$\frac{dy}{dt} = a_{21}(t)x + a_{22}(t)y. \tag{7.59}$$

We shall see that this theory is analogous to that of the single nth-order homogeneous linear differential equation presented in Section 4.1B. Our first result concerning the system (7.59) is the following.

THEOREM 7.2

Hypothesis. *Let*

$$\begin{aligned} x &= f_1(t), \\ y &= g_1(t), \end{aligned} \quad \text{and} \quad \begin{aligned} x &= f_2(t), \\ y &= g_2(t), \end{aligned} \tag{7.60}$$

be two solutions of the homogeneous linear system (7.59). Let c_1 and c_2 be two arbitrary constants.

Conclusion. *Then*

$$\begin{aligned} x &= c_1 f_1(t) + c_2 f_2(t), \\ y &= c_1 g_1(t) + c_2 g_2(t), \end{aligned} \tag{7.61}$$

is also a solution of the system (7.59).

DEFINITION

The solution (7.61) is called a linear combination *of the solutions (7.60). This definition enables us to express Theorem 7.2 in the following alternative form.*

THEOREM 7.2 restated

Any linear combination of two solutions of the homogeneous linear system (7.59) is itself a solution of the system (7.59).

▶ Example 7.10. We have already observed that

$$\begin{aligned} x &= e^{5t}, \\ y &= -3e^{5t}, \end{aligned} \quad \text{and} \quad \begin{aligned} x &= e^{3t}, \\ y &= -e^{3t}, \end{aligned}$$

are solutions of the homogeneous linear system (7.54). Theorem 7.2 tells us that

$$\begin{aligned} x &= c_1 e^{5t} + c_2 e^{3t}, \\ y &= -3c_1 e^{5t} - c_2 e^{3t}, \end{aligned}$$

where c_1 and c_2 are arbitrary constants, is also a solution of the system (7.54). For example, if $c_1 = 4$ and $c_2 = -2$, we have the solution

$$\begin{aligned} x &= 4e^{5t} - 2e^{3t}, \\ y &= -12e^{5t} + 2e^{3t}. \end{aligned}$$

DEFINITION

Let

$$\begin{aligned} x &= f_1(t), \\ y &= g_1(t), \end{aligned} \quad \text{and} \quad \begin{aligned} x &= f_2(t), \\ y &= g_2(t), \end{aligned}$$

be two solutions of the homogeneous linear system (7.59). These two solutions are linearly dependent *on the interval* $a \leq t \leq b$ *if there exist constants* c_1 *and* c_2, not both zero, *such that*

$$c_1 f_1(t) + c_2 f_2(t) = 0,$$
$$c_1 g_1(t) + c_2 g_2(t) = 0, \tag{7.62}$$

for all t *such that* $a \leq t \leq b$.

DEFINITION

Let

$$x = f_1(t), \quad y = g_1(t), \qquad \textit{and} \qquad x = f_2(t), \quad y = g_2(t),$$

be two solutions of the homogeneous linear system (7.59). These two solutions are linearly independent *on* $a \leq t \leq b$ *if they are not linearly dependent on* $a \leq t \leq b$. *That is, the solutions* $x = f_1(t)$, $y = g_1(t)$ *and* $x = f_2(t)$, $y = g_2(t)$ *are linearly independent on* $a \leq t \leq b$ *if*

$$c_1 f_1(t) + c_2 f_2(t) = 0,$$
$$c_1 g_1(t) + c_2 g_2(t) = 0, \tag{7.63}$$

for all t *such that* $a \leq t \leq b$ *implies that*

$$c_1 = c_2 = 0.$$

▶ Example 7.11. The solutions

$$x = e^{5t}, \quad y = -3e^{5t}, \qquad \text{and} \qquad x = 2e^{5t}, \quad y = -6e^{5t},$$

of the system (7.54) are linearly dependent on every interval $a \leq t \leq b$. For in this case the conditions (7.62) become

$$c_1 e^{5t} + 2c_2 e^{5t} = 0,$$
$$-3c_1 e^{5t} - 6c_2 e^{5t} = 0, \tag{7.64}$$

and clearly there exist constants c_1 and c_2, not both zero, such that the conditions (7.64) hold on $a \leq t \leq b$. For example, let $c_1 = 2$ and $c_2 = -1$.

On the other hand, the solutions

$$x = e^{5t}, \quad y = -3e^{5t}, \qquad \text{and} \qquad x = e^{3t}, \quad y = -e^{3t},$$

of system (7.54) are linearly independent on $a \leq t \leq b$. For in this case the conditions (7.63) are

$$c_1 e^{5t} + c_2 e^{3t} = 0,$$
$$-3c_1 e^{5t} - c_2 e^{3t} = 0.$$

If these conditions hold for all t such that $a \leq t \leq b$, then we must have $c_1 = c_2 = 0$.

We now state the following basic theorem concerning sets of linearly independent solutions of the homogeneous linear system (7.59).

THEOREM 7.3

There exist sets of two linearly independent solutions of the homogeneous linear system (7.59). Every solution of the system (7.59) can be written as a linear combination of any two linearly independent solutions of (7.59).

▶ Example 7.12. We have seen that

$$x = e^{5t}, \quad y = -3e^{5t}, \qquad \text{and} \qquad x = e^{3t}, \quad y = -e^{3t},$$

constitute a pair of linearly independent solutions of the system (7.54). This illustrates the first part of Theorem 7.3. The second part of the theorem tells us that every solution of the system (7.54) can be written in the form

$$x = c_1e^{5t} + c_2e^{3t},$$

$$y = -3c_1e^{5t} - c_2e^{3t},$$

where c_1 and c_2 are suitably chosen constants.

Recall that in Section 4.1 in connection with the single nth-order homogeneous linear differential equation, we defined the general solution of such an equation to be a linear combination of n linearly independent solutions. As a result of Theorems 7.2 and 7.3 we now give an analogous definition of general solution for the homogeneous linear system (7.59).

DEFINITION

Let

$$x = f_1(t), \quad y = g_1(t), \qquad \textit{and} \qquad x = f_2(t), \quad y = g_2(t),$$

be two linearly independent solutions of the homogeneous linear system (7.59). Let c_1 and c_2 be two arbitrary constants. Then the solution

$$x = c_1f_1(t) + c_2f_2(t),$$

$$y = c_1g_1(t) + c_2g_2(t),$$

is called a general solution *of the system (7.59).*

▶ Example 7.13. Since

$$x = e^{5t}, \quad y = -3e^{5t}, \qquad \text{and} \qquad x = e^{3t}, \quad y = -e^{3t},$$

are linearly independent solutions of the system (7.54), we may write the general solution of (7.54) in the form

$$x = c_1e^{5t} + c_2e^{3t},$$

$$y = -3c_1e^{5t} - c_2e^{3t},$$

where c_1 and c_2 are arbitrary constants.

As a criterion for the linear independence of two solutions of the system (7.59), we state the following theorem, which is analogous to the now familiar "Wronskian theorem" of Section 4.1.

THEOREM 7.4

Let

$$x = f_1(t), \qquad x = f_2(t),$$
$$\textit{and}$$
$$y = g_1(t), \qquad y = g_2(t),$$

be two solutions of the homogeneous linear system (7.59). A necessary and sufficient condition that these two solutions be linearly independent on $a \le t \le b$ is that the determinant

$$\Delta(t) = \begin{vmatrix} f_1(t) & f_2(t) \\ g_1(t) & g_2(t) \end{vmatrix} \tag{7.65}$$

be different from zero for all t such that $a \le t \le b$.

Concerning this determinant, we also state the following result.

THEOREM 7.5

The determinant $\Delta(t)$ of Theorem 7.4 either is identically zero or vanishes for no value of t on the interval $a \le t \le b$.

▶ Example 7.14. Let us employ Theorem 7.4 to verify the linear independence of the solutions

$$x = e^{5t}, \qquad x = e^{3t},$$
$$\text{and}$$
$$y = -3e^{5t}, \qquad y = -e^{3t},$$

of the system (7.54). We have

$$\Delta(t) = \begin{vmatrix} e^{5t} & e^{3t} \\ -3e^{5t} & -e^{3t} \end{vmatrix} = 2e^{8t} \neq 0$$

on every closed interval $a \le t \le b$. Thus by Theorem 7.4 the two solutions are indeed linearly independent on $a \le t \le b$.

C. *Nonhomogeneous Linear Systems*

Let us now return briefly to the nonhomogeneous system (7.53). A theorem and a definition, illustrated by a simple example, will suffice for our purposes here.

THEOREM 7.6

Hypothesis. *Let*

$$x = f_0(t),$$
$$y = g_0(t),$$

be any solution of the nonhomogeneous system (7.53), and let

$$x = f(t),$$
$$y = g(t),$$

be any solution of the corresponding homogeneous system (7.59).

Conclusion. *Then*

$$x = f(t) + f_0(t),$$
$$y = g(t) + g_0(t),$$

is also a solution of the nonhomogeneous system (7.53).

DEFINITION

Let

$$x = f_0(t),$$
$$y = g_0(t),$$

be any solution of the nonhomogeneous system (7.53), and let

$$x = f_1(t), \quad x = f_2(t),$$
$$\text{and}$$
$$y = g_1(t), \quad y = g_2(t),$$

be two linearly independent solutions of the corresponding homogeneous system (7.59). Let c_1 and c_2 be two arbitrary constants. Then the solution

$$x = c_1 f_1(t) + c_2 f_2(t) + f_0(t),$$
$$y = c_1 g_1(t) + c_2 g_2(t) + g_0(t),$$

will be called a general solution *of the nonhomogeneous system (7.53).*

▶ Example 7.15. The student may verify that

$$x = 2t + 1,$$
$$y = -t,$$

is a solution of the nonhomogeneous system (7.55). The corresponding homogeneous system is the system (7.54), and we have already seen that

$$x = e^{5t}, \quad x = e^{3t},$$
$$\text{and}$$
$$y = -3e^{5t}, \quad y = -e^{3t},$$

are linearly independent solutions of this homogeneous system. Theorem 7.6 tells us, for example, that

$$x = e^{5t} + 2t + 1,$$
$$y = -3e^{5t} - t,$$

is a solution of the nonhomogeneous system (7.55). From the preceding definition we see that the general solution of (7.55) may be written in the form

$$x = c_1 e^{5t} + c_2 e^{3t} + 2t + 1,$$
$$y = -3c_1 e^{5t} - c_2 e^{3t} - t,$$

where c_1 and c_2 are arbitrary constants.

Exercises

1. Consider the linear system

$$\frac{dx}{dt} = 3x + 4y,$$
$$\frac{dy}{dt} = 2x + y.$$

   (a) Show that

$$x = 2e^{5t}, \quad y = e^{5t}, \qquad \text{and} \qquad x = e^{-t}, \quad y = -e^{-t},$$

   are solutions of this system.

   (b) Show that the two solutions of part (a) are linearly independent on every interval $a \le t \le b$, and write the general solution of the system.

   (c) Find the solution

$$x = f(t),$$
$$y = g(t),$$

   of the system which is such that $f(0) = 1$ and $g(0) = 2$. Why is this solution unique? Over what interval is it defined?

2. Consider the linear system

$$\frac{dx}{dt} = 5x + 3y,$$
$$\frac{dy}{dt} = 4x + y.$$

   (a) Show that

$$x = 3e^{7t}, \quad y = 2e^{7t}, \qquad \text{and} \qquad x = e^{-t}, \quad y = -2e^{-t},$$

   are solutions of this system.

   (b) Show that the two solutions of part (a) are linearly independent on every interval $a \le t \le b$, and write the general solution of the system.

   (c) Find the solution

$$x = f(t),$$
$$y = g(t),$$

   of the system which is such that $f(0) = 0$ and $g(0) = 8$.

3. (a) Show that

$$x = 2e^{2t}, \quad\quad x = e^{7t},$$
$$\text{and}$$
$$y = -3e^{2t}, \quad\quad y = e^{7t},$$

are solutions of the homogeneous linear system

$$\frac{dx}{dt} = 5x + 2y,$$

$$\frac{dy}{dt} = 3x + 4y.$$

(b) Show that the two solutions defined in part (a) are linearly independent on every interval $a \leq t \leq b$, and write the general solution of the homogeneous system of part (a).

(c) Show that

$$x = t + 1,$$
$$y = -5t - 2,$$

is a particular solution of the nonhomogeneous linear system

$$\frac{dx}{dt} = 5x + 2y + 5t,$$

$$\frac{dy}{dt} = 3x + 4y + 17t,$$

and write the general solution of this system.

4. Let

$$x = f_1(t), \quad\quad x = f_2(t),$$
$$\text{and}$$
$$y = g_1(t), \quad\quad y = g_2(t),$$

be two linearly independent solutions of the homogeneous linear system (7.59), and let $\Delta(t)$ be the determinant defined by (7.65). Show that Δ satisfies the first-order differential equation

$$\frac{d\Delta}{dt} = [a_{11}(t) + a_{22}(t)]\,\Delta.$$

5. Prove Theorem 7.2.

6. Prove Theorem 7.6.

7.4 Homogeneous Linear Systems with Constant Coefficients: Two Equations in Two Unknown Functions

A. *Introduction*

In this section we shall be concerned with the homogeneous linear system

$$\frac{dx}{dt} = a_1 x + b_1 y,$$

$$\frac{dy}{dt} = a_2 x + b_2 y, \tag{7.66}$$

where the coefficients a_1, b_1, a_2, and b_2 are real constants. We seek solutions of this system; but how shall we proceed? Recall that in Section 4.2 we sought and found exponential solutions of the single nth-order linear equation with constant coefficients. Remembering the analogy which exists between linear systems and single higher-order linear equations, we might now attempt to find exponential solutions of the system (7.66). Let us therefore attempt to determine a solution of the form

$$\begin{aligned} x &= Ae^{\lambda t}, \\ y &= Be^{\lambda t}, \end{aligned} \tag{7.67}$$

where A, B, and λ are constants. If we substitute (7.67) into (7.66) we obtain

$$\begin{aligned} A\lambda e^{\lambda t} &= a_1 Ae^{\lambda t} + b_1 Be^{\lambda t}, \\ B\lambda e^{\lambda t} &= a_2 Ae^{\lambda t} + b_2 Be^{\lambda t}. \end{aligned}$$

These equations lead at once to the system

$$\begin{aligned} (a_1 - \lambda)A + b_1 B &= 0, \\ a_2 A + (b_2 - \lambda)B &= 0, \end{aligned} \tag{7.68}$$

in the unknowns A and B. This system obviously has the trivial solution $A = B = 0$. But this would only lead to the trivial solution $x = 0$, $y = 0$ of the system (7.66). Thus we seek nontrivial solutions of the system (7.68). A necessary and sufficient condition (see Section 4.6, Theorem B) that this system have a nontrivial solution is that the determinant

$$\begin{vmatrix} a_1 - \lambda & b_1 \\ a_2 & b_2 - \lambda \end{vmatrix} = 0. \tag{7.69}$$

Expanding this determinant we are led at once to the quadratic equation

$$\lambda^2 - (a_1 + b_2)\lambda + (a_1 b_2 - a_2 b_1) = 0 \tag{7.70}$$

in the unknown λ. This equation is called the *characteristic equation* associated with the system (7.66). Its roots λ_1 and λ_2 are called the *characteristic roots.* If the pair (7.67) is to be a solution of the system (7.66), then λ in (7.67) must be one of these roots. Suppose $\lambda = \lambda_1$. Then substituting $\lambda = \lambda_1$ into the algebraic system (7.68), we may obtain a nontrivial solution A_1, B_1, of this algebraic system. With these values A_1, B_1 we obtain the nontrivial solution

$$\begin{aligned} x &= A_1 e^{\lambda_1 t}, \\ y &= B_1 e^{\lambda_1 t}, \end{aligned}$$

of the given system (7.66).

Three cases must now be considered:

1. The roots λ_1 and λ_2 are real and distinct.
2. The roots λ_1 and λ_2 are real and equal.
3. The roots λ_1 and λ_2 are conjugate complex.

B. *Case 1. The Roots of the Characteristic Equation (7.70) are Real and Distinct*

If the roots λ_1 and λ_2 of the characteristic equation (7.70) are real and distinct, it appears that we should expect two distinct solutions of the form (7.67), one corresponding to each of the two distinct roots. This is indeed the case. Furthermore, these two distinct solutions are linearly independent. We summarize this case in the following theorem.

THEOREM 7.7

Hypothesis. *The roots λ_1 and λ_2, of the characteristic equation (7.70) associated with the system (7.66) are real and distinct.*

Conclusion. *The system (7.66) has two nontrivial linearly independent solutions of the form*

$$x = A_1 e^{\lambda_1 t}, \quad y = B_1 e^{\lambda_1 t}, \quad \textit{and} \quad x = A_2 e^{\lambda_2 t}, \quad y = B_2 e^{\lambda_2 t},$$

where A_1, B_1, A_2, and B_2 are definite constants. The general solution of the system (7.66) may thus be written

$$x = c_1 A_1 e^{\lambda_1 t} + c_2 A_2 e^{\lambda_2 t},$$

$$y = c_1 B_1 e^{\lambda_1 t} + c_2 B_2 e^{\lambda_2 t},$$

where c_1 and c_2 are arbitrary constants.

▶ Example 7.16

$$\frac{dx}{dt} = 6x - 3y,$$

$$\frac{dy}{dt} = 2x + y. \tag{7.71}$$

We assume a solution of the form (7.67):

$$x = Ae^{\lambda t},$$

$$y = Be^{\lambda t}. \tag{7.72}$$

Substituting (7.72) into (7.71) we obtain

$$A\lambda e^{\lambda t} = 6Ae^{\lambda t} - 3Be^{\lambda t},$$

$$B\lambda e^{\lambda t} = 2Ae^{\lambda t} + Be^{\lambda t},$$

and this leads at once to the algebraic system

$$(6 - \lambda)A - 3B = 0,$$

$$2A + (1 - \lambda)B = 0, \tag{7.73}$$

in the unknown λ. For nontrivial solutions of this system we must have

$$\begin{vmatrix} 6-\lambda & -3 \\ 2 & 1-\lambda \end{vmatrix} = 0.$$

Expanding this we obtain the characteristic equation

$$\lambda^2 - 7\lambda + 12 = 0.$$

Solving this, we find the roots $\lambda_1 = 3$, $\lambda_2 = 4$.

Setting $\lambda = \lambda_1 = 3$ in (7.73), we obtain

$$3A - 3B = 0,$$
$$2A - 2B = 0.$$

A simple nontrivial solution of this system is obviously $A = B = 1$. With these values of A, B, and λ we find the nontrivial solution

$$x = e^{3t}, \quad y = e^{3t}. \tag{7.74}$$

Now setting $\lambda = \lambda_2 = 4$ in (7.73), we find

$$2A - 3B = 0,$$
$$2A - 3B = 0.$$

A simple nontrivial solution of this system is $A = 3$, $B = 2$. Using these values of A, B, and λ we find the nontrivial solution

$$x = 3e^{4t}, \quad y = 2e^{4t}. \tag{7.75}$$

By Theorem 7.7 the solutions (7.74) and (7.75) are linearly independent (one may check this using Theorem 7.4) and the general solution of the system (7.71) may be written

$$x = c_1e^{3t} + 3c_2e^{4t},$$
$$y = c_1e^{3t} + 2c_2e^{4t},$$

where c_1 and c_2 are arbitrary constants.

C. *Case 2. The Roots of the Characteristic Equation (7.70) are Real and Equal*

If the two roots of the characteristic equation (7.70) are real and equal, it would appear that we could find only one solution of the form (7.67). Except in the special subcase in which $a_1 = b_2 \neq 0$, $a_2 = b_1 = 0$ (see Exercise 27 at the end of this section) this is indeed true. In general, how shall we then proceed to find a second, linearly independent solution? Recall the analogous situation in which the auxiliary equation corresponding to a single nth-order linear equation has a double root. This would lead us to expect a second solution of the form

$$x = Ate^{\lambda t},$$
$$y = Bte^{\lambda t}.$$

However, the situation here is not quite so simple (see Exercise 28 at the end of this section). We must actually seek a second solution of the form

$$\begin{aligned} x &= (A_1 t + A_2)e^{\lambda t}, \\ y &= (B_1 t + B_2)e^{\lambda t}. \end{aligned} \tag{7.76}$$

We shall illustrate this in Example 7.17. We first summarize Case 2 in the following theorem.

THEOREM 7.8

Hypothesis. *The roots λ_1 and λ_2 of the characteristic equation (7.70) associated with the system (7.66) are real and equal. Let λ denote their common value. Further assume that system (7.66) is* not *such that $a_1 = b_2 \neq 0$, $a_2 = b_1 = 0$.*

Conclusion. *The system (7.66) has two linearly independent solutions of the form*

$$x = Ae^{\lambda t}, \quad y = Be^{\lambda t}, \qquad \text{and} \qquad x = (A_1 t + A_2)e^{\lambda t}, \quad y = (B_1 t + B_2)e^{\lambda t},$$

where A, B, A_1, A_2, B_1, and B_2 are definite constants, A_1 and B_1 are not both zero, and $B_1/A_1 = B/A$. The general solution of the system (7.66) may thus be written

$$x = c_1 Ae^{\lambda t} + c_2(A_1 t + A_2)e^{\lambda t},$$
$$y = c_1 Be^{\lambda t} + c_2(B_1 t + B_2)e^{\lambda t},$$

where c_1 and c_2 are arbitrary constants.

▶ Example 7.17.

$$\begin{aligned} \frac{dx}{dt} &= 4x - y, \\ \frac{dy}{dt} &= x + 2y. \end{aligned} \tag{7.77}$$

We assume a solution of the form (7.67):

$$\begin{aligned} x &= Ae^{\lambda t}, \\ y &= Be^{\lambda t}. \end{aligned} \tag{7.78}$$

Substituting (7.78) into (7.77) we obtain

$$A\lambda e^{\lambda t} = 4Ae^{\lambda t} - Be^{\lambda t},$$
$$B\lambda e^{\lambda t} = Ae^{\lambda t} + 2Be^{\lambda t},$$

and this leads at once to the algebraic system

$$\begin{aligned} (4-\lambda)A - B &= 0, \\ A + (2-\lambda)B &= 0, \end{aligned} \tag{7.79}$$

in the unknown λ. For nontrivial solutions of this system we must have

$$\begin{vmatrix} 4-\lambda & -1 \\ 1 & 2-\lambda \end{vmatrix} = 0.$$

Expanding this we obtain the characteristic equation

$$\lambda^2 - 6\lambda + 9 = 0 \tag{7.80}$$

or

$$(\lambda - 3)^2 = 0.$$

Thus the characteristic equation (7.80) has the real and equal roots 3, 3.

Setting $\lambda = 3$ in (7.79), we obtain

$$\begin{aligned} A - B &= 0, \\ A - B &= 0. \end{aligned}$$

A simple nontrivial solution of this system being $A = B = 1$, we obtain the nontrivial solution

$$\begin{aligned} x &= e^{3t}, \\ y &= e^{3t}, \end{aligned} \tag{7.81}$$

of the given system (7.77).

Since the roots of the characteristic equation are both equal to 3, we must seek a second solution of the form (7.76), with $\lambda = 3$. That is, we must determine A_1, A_2, B_1, and B_2 (with A_1 and B_1 not both zero) such that

$$\begin{aligned} x &= (A_1 t + A_2)e^{3t}, \\ y &= (B_1 t + B_2)e^{3t}, \end{aligned} \tag{7.82}$$

is a solution of the system (7.77). Substituting (7.82) into (7.77), we obtain

$$\begin{aligned} (3A_1 t + 3A_2 + A_1)e^{3t} &= 4(A_1 t + A_2)e^{3t} - (B_1 t + B_2)e^{3t}, \\ (3B_1 t + 3B_2 + B_1)e^{3t} &= (A_1 t + A_2)e^{3t} + 2(B_1 t + B_2)e^{3t}. \end{aligned}$$

These equations reduce at once to

$$\begin{aligned} (A_1 - B_1)t + (A_2 - A_1 - B_2) &= 0, \\ (A_1 - B_1)t + (A_2 - B_1 - B_2) &= 0. \end{aligned}$$

In order for these equations to be identities we must have

$$\begin{aligned} A_1 - B_1 &= 0, \quad A_2 - A_1 - B_2 = 0, \\ A_1 - B_1 &= 0, \quad A_2 - B_1 - B_2 = 0. \end{aligned} \tag{7.83}$$

Thus in order for (7.82) to be a solution of the system (7.77), the constants A_1, A_2, B_1, and B_2 must be chosen to satisfy the equations (7.83). From the equations $A_1 - B_1 = 0$, we see that $A_1 = B_1$. The other two equations of (7.83) show that A_2 and B_2 must satisfy

$$A_2 - B_2 = A_1 = B_1. \tag{7.84}$$

We may choose any convenient nonzero value for A_1 and B_1. We choose $A_1 = B_1 = 1$. Then (7.84) reduces to $A_2 - B_2 = 1$, and we can choose any convenient values for A_2 and B_2 which will satisfy this equation. We choose $A_2 = 1$, $B_2 = 0$. We are thus led to the solution

$$\begin{aligned} x &= (t + 1)e^{3t}, \\ y &= te^{3t}. \end{aligned} \tag{7.85}$$

By Theorem 7.8 the solutions (7.81) and (7.85) are linearly independent. We may thus write the general solution of the system (7.77) in the form

$$\begin{aligned} x &= c_1e^{3t} + c_2(t + 1)e^{3t}, \\ y &= c_1e^{3t} + c_2te^{3t}, \end{aligned}$$

where c_1 and c_2 are arbitrary constants.

D. Case 3. The Roots of the Characteristic Equation (7.70) are Conjugate Complex

If the roots λ_1 and λ_2 of the characteristic equation (7.70) are the conjugate complex numbers $a + bi$ and $a - bi$, then we obtain two distinct solutions

$$\begin{aligned} x &= A_1^*e^{(a+bi)t}, \\ y &= B_1^*e^{(a+bi)t}, \end{aligned} \quad \text{and} \quad \begin{aligned} x &= A_2^*e^{(a-bi)t}, \\ y &= B_2^*e^{(a-bi)t}, \end{aligned} \tag{7.86}$$

of the form (7.67), one corresponding to each of the complex roots. However, the solutions (7.86) are *complex* solutions. In order to obtain *real* solutions in this case we consider the first of the two solutions (7.86) and proceed as follows: We first express the complex constants A_1^* and B_1^* in this solution in the forms $A_1^* = A_1 + iA_2$ and $B_1^* = B_1 + iB_2$, where A_1, A_2, B_1, and B_2, are real. We then apply Euler's formula $e^{i\theta} = \cos\theta + i\sin\theta$ and express the first solution (7.86) in the form

$$\begin{aligned} x &= (A_1 + iA_2)e^{at}(\cos bt + i\sin bt), \\ y &= (B_1 + iB_2)e^{at}(\cos bt + i\sin bt). \end{aligned}$$

Rewriting this, we have

$$\begin{aligned} x &= e^{at}[(A_1\cos bt - A_2\sin bt) + i(A_2\cos bt + A_1\sin bt)], \\ y &= e^{at}[(B_1\cos bt - B_2\sin bt) + i(B_2\cos bt + B_1\sin bt)]. \end{aligned} \tag{7.87}$$

It can be shown that a pair $[f_1(t) + if_2(t), g_1(t) + ig_2(t)]$ of complex functions is a solution of the system (7.66) if and only if both the pair $[f_1(t), g_1(t)]$ consisting of their real parts and the pair $[f_2(t), g_2(t)]$ consisting of their imaginary parts are solutions of (7.66). Thus both the real part

$$\begin{aligned} x &= e^{at}(A_1 \cos bt - A_2 \sin bt), \\ y &= e^{at}(B_1 \cos bt - B_2 \sin bt), \end{aligned} \tag{7.88}$$

and the imaginary part

$$\begin{aligned} x &= e^{at}(A_2 \cos bt + A_1 \sin bt), \\ y &= e^{at}(B_2 \cos bt + B_1 \sin bt), \end{aligned} \tag{7.89}$$

of the solution (7.87) of the system (7.66) are also solutions of (7.66). Furthermore, the solutions (7.88) and (7.89) are linearly independent. We verify this by evaluating the determinant (7.65) for these solutions. We find

$$\begin{aligned} \Delta(t) &= \begin{vmatrix} e^{at}(A_1 \cos bt - A_2 \sin bt) & e^{at}(A_2 \cos bt + A_1 \sin bt) \\ e^{at}(B_1 \cos bt - B_2 \sin bt) & e^{at}(B_2 \cos bt + B_1 \sin bt) \end{vmatrix} \\ &= e^{2at}(A_1B_2 - A_2B_1). \end{aligned} \tag{7.90}$$

Now, the constant B_1^* is a *nonreal* multiple of the constant A_1^*. If we assume that $A_1B_2 - A_2B_1 = 0$, then it follows that B_1^* is a *real* multiple of A_1^*, which contradicts the result stated in the previous sentence. Thus $A_1B_2 - A_2B_1 \neq 0$ and so the determinant $\Delta(t)$ in (7.90) is unequal to zero. Thus by Theorem 7.4 the solutions (7.88) and (7.89) are indeed linearly independent. Hence a linear combination of these two real solutions provides the general solution of the system (7.66) in this case. There is no need to consider the second of the two solutions (7.86). We summarize the above results in the following theorem:

THEOREM 7.9

Hypothesis. *The roots λ_1 and λ_2 of the characteristic equation (7.70) associated with the system (7.66) are the conjugate complex numbers $a \pm bi$.*

Conclusion. *The system (7.66) has two real linearly independent solutions of the form*

$$\begin{aligned} x &= e^{at}(A_1 \cos bt - A_2 \sin bt), \\ y &= e^{at}(B_1 \cos bt - B_2 \sin bt), \end{aligned} \quad \text{and} \quad \begin{aligned} x &= e^{at}(A_2 \cos bt + A_1 \sin bt), \\ y &= e^{at}(B_2 \cos bt + B_1 \sin bt), \end{aligned}$$

where A_1, A_2, B_1, and B_2 are definite real constants. The general solution of the system (7.66) may thus be written

$$\begin{aligned} x &= e^{at}[c_1(A_1 \cos bt - A_2 \sin bt) + c_2(A_2 \cos bt + A_1 \sin bt)], \\ y &= e^{at}[c_1(B_1 \cos bt - B_2 \sin bt) + c_2(B_2 \cos bt + B_1 \sin bt)], \end{aligned}$$

where c_1 and c_2 are arbitrary constants.

▶ Example 7.18

$$\frac{dx}{dt} = 3x + 2y, \qquad \frac{dy}{dt} = -5x + y. \tag{7.91}$$

We assume a solution of the form (7.67):

$$x = Ae^{\lambda t}, \qquad y = Be^{\lambda t}. \tag{7.92}$$

Substituting (7.92) into (7.91) we obtain

$$A\lambda e^{\lambda t} = 3Ae^{\lambda t} + 2Be^{\lambda t},$$
$$B\lambda e^{\lambda t} = -5Ae^{\lambda t} + Be^{\lambda t},$$

and this leads at once to the algebraic system

$$(3 - \lambda)A + 2B = 0, \qquad -5A + (1 - \lambda)B = 0, \tag{7.93}$$

in the unknown λ. For nontrivial solutions of this system we must have

$$\begin{vmatrix} 3 - \lambda & 2 \\ -5 & 1 - \lambda \end{vmatrix} = 0.$$

Expanding this, we obtain the characteristic equation

$$\lambda^2 - 4\lambda + 13 = 0.$$

The roots of this equation are the conjugate complex numbers $2 \pm 3i$.

Setting $\lambda = 2 + 3i$ in (7.93), we obtain

$$(1 - 3i)A + 2B = 0,$$
$$-5A + (-1 - 3i)B = 0.$$

A simple nontrivial solution of this system is $A = 2$, $B = -1 + 3i$. Using these values we obtain the complex solution

$$x = 2e^{(2+3i)t},$$
$$y = (-1 + 3i)e^{(2+3i)t},$$

of the given system (7.91). Using Euler's formula this takes the form

$$x = e^{2t}[(2 \cos 3t) + i(2 \sin 3t)],$$
$$y = e^{2t}[(-\cos 3t - 3 \sin 3t) + i(3 \cos 3t - \sin 3t)].$$

Since both the real and imaginary parts of this solution of system (7.91) are themselves solutions of (7.91), we thus obtain the two real solutions

$$\begin{aligned} x &= 2e^{2t} \cos 3t, \\ y &= -e^{2t}(\cos 3t + 3 \sin 3t), \end{aligned} \tag{7.94}$$

and

$$\begin{aligned} x &= 2e^{2t} \sin 3t, \\ y &= e^{2t}(3 \cos 3t - \sin 3t). \end{aligned} \tag{7.95}$$

Finally, since the two solutions (7.94) and (7.95) are linearly independent we may write the general solution of the system (7.91) in the form

$$\begin{aligned} x &= 2e^{2t}(c_1 \cos 3t + c_2 \sin 3t), \\ y &= e^{2t}[c_1(-\cos 3t - 3 \sin 3t) + c_2(3 \cos 3t - \sin 3t)], \end{aligned}$$

where c_1 and c_2 are arbitrary constants.

Exercises

Find the general solution of each of the linear systems in Exercises 1 through 22.

1. $\dfrac{dx}{dt} = 5x - 2y,$ $\quad \dfrac{dy}{dt} = 4x - y.$

2. $\dfrac{dx}{dt} = 5x - y,$ $\quad \dfrac{dy}{dt} = 3x + y.$

3. $\dfrac{dx}{dt} = x + 2y,$ $\quad \dfrac{dy}{dt} = 3x + 2y.$

4. $\dfrac{dx}{dt} = 2x + 3y,$ $\quad \dfrac{dy}{dt} = -x - 2y.$

5. $\dfrac{dx}{dt} = 3x + y,$ $\quad \dfrac{dy}{dt} = 4x + 3y.$

6. $\dfrac{dx}{dt} = 6x - y,$ $\quad \dfrac{dy}{dt} = 3x + 2y.$

7. $\dfrac{dx}{dt} = 3x - 4y,$ $\quad \dfrac{dy}{dt} = 2x - 3y.$

8. $\dfrac{dx}{dt} = 2x - y,$ $\quad \dfrac{dy}{dt} = 9x + 2y.$

9. $\dfrac{dx}{dt} = x + 3y,$ $\quad \dfrac{dy}{dt} = 3x + y.$

10. $\dfrac{dx}{dt} = 3x + 2y,$ $\quad \dfrac{dy}{dt} = 6x - y.$

11. $\frac{dx}{dt} = 3x - y,$
$\frac{dy}{dt} = 4x - y.$

12. $\frac{dx}{dt} = 7x + 4y,$
$\frac{dy}{dt} = -x + 3y.$

13. $\frac{dx}{dt} = 5x + 4y,$
$\frac{dy}{dt} = -x + y.$

14. $\frac{dx}{dt} = x - 2y,$
$\frac{dy}{dt} = 2x - 3y.$

15. $\frac{dx}{dt} = x - 4y,$
$\frac{dy}{dt} = x + y.$

16. $\frac{dx}{dt} = 2x - 3y,$
$\frac{dy}{dt} = 3x + 2y.$

17. $\frac{dx}{dt} = x - 3y,$
$\frac{dy}{dt} = 3x + y.$

18. $\frac{dx}{dt} = 5x - 4y,$
$\frac{dy}{dt} = 2x + y.$

19. $\frac{dx}{dt} = 4x - 2y,$
$\frac{dy}{dt} = 5x + 2y.$

20. $\frac{dx}{dt} = x - 5y,$
$\frac{dy}{dt} = 2x - y.$

21. $\frac{dx}{dt} = 3x - 2y,$
$\frac{dy}{dt} = 2x + 3y.$

22. $\frac{dx}{dt} = 6x - 5y,$
$\frac{dy}{dt} = x + 2y.$

23. Consider the linear system

$$t\frac{dx}{dt} = a_1x + b_1y,$$

$$t\frac{dy}{dt} = a_2x + b_2y,$$

where a_1, b_1, a_2, and b_2 are real constants. Show that the transformation $t = e^w$ transforms this system into a linear system with constant coefficients.

24. Use the result of Exercise 23 to solve the system

$$t\frac{dx}{dt} = x + y,$$

$$t\frac{dy}{dt} = -3x + 5y.$$

25. Use the result of Exercise 23 to solve the system

$$t\frac{dx}{dt} = 2x + 3y,$$

$$t\frac{dy}{dt} = 2x + y.$$

26. Consider the linear system

$$\frac{dx}{dt} = a_1x + b_1y,$$

$$\frac{dy}{dt} = a_2x + b_2y,$$

where a_1, b_1, a_2, and b_2 are real constants. Show that the condition $a_2b_1 > 0$ is sufficient, but not necessary, for the system to have two real linearly independent solutions of the form

$$x = Ae^{\lambda t},$$

$$y = Be^{\lambda t}.$$

27. Suppose that the roots of the characteristic equation (7.70) of the system (7.66) are real and equal; and let λ denote their common value. Also assume that the system (7.66) *is* such that $a_1 = b_2 \neq 0$ and $a_2 = b_1 = 0$. Show that in this special subcase there exist two linearly independent solutions of the form (7.67).

28. Suppose that the roots of the characteristic equation (7.70) of the system (7.66) are real and equal; and let λ denote their common value. Also assume that the system (7.66) is *not* such that $a_1 = b_2 \neq 0$ and $a_2 = b_1 = 0$. Then show that there exists *no* nontrivial solution of the form

$$x = Ate^{\lambda t},$$

$$y = Bte^{\lambda t},$$

which is linearly independent of the "basic" solution of the form (7.67).

29. Referring to the conclusion of Theorem 7.8, show that $B_1/A_1 = B/A$ in the case under consideration.

7.5 Matrices and Vectors

A. *The Most Basic Concepts*

The study of matrices and vectors is a large and important subject, and an entire chapter the size of this one would be needed to present all of the most fundamental concepts and results. Therefore, after defining a matrix, we shall introduce only those very special concepts and results which we shall need and use in this book. For the reader familiar with matrices and vectors, this section will be a very simple review of a few select topics. On the other hand, for the reader unfamiliar with matrices and vectors, the detailed treatment here will provide just what is needed for an understanding of the rest of the chapter.

DEFINITIONS

A matrix *is defined to be a rectangular array*

$$\mathbf{A} = \begin{pmatrix} a_{11} & a_{12} & \cdots & a_{1n} \\ a_{21} & a_{22} & \cdots & a_{2n} \\ \vdots & \vdots & & \vdots \\ a_{m1} & a_{m2} & \cdots & a_{mn} \end{pmatrix}$$

of elements a_{ij} $(i = 1, 2, \ldots, m; j = 1, 2, \ldots, n)$, *arranged in m (horizontal)* rows *and n (vertical)* columns. *The matrix is denoted by the boldface letter* **A**, *as indicated; and the element in its ith row and jth column, by* a_{ij}, *as suggested. We write* $\mathbf{A} = (a_{ij})$, *and call* **A** *an* m × n matrix.

We shall be concerned with the two following special sizes of matrices.

1. A *square matrix* is a matrix for which the number of rows is the same as the number of columns. If the common number of rows and columns is *n*, we call the matrix an *n* × *n square matrix*. We write

$$\mathbf{A} = \begin{pmatrix} a_{11} & a_{12} & \cdots & a_{1n} \\ a_{21} & a_{22} & \cdots & a_{2n} \\ \vdots & \vdots & & \vdots \\ a_{n1} & a_{n2} & \cdots & a_{nn} \end{pmatrix}.$$

2. A *vector* (or *column vector*) is a matrix having just one column. If the vector has *n* rows (and, of course, one column), we call it an *n* × 1 *vector* (or *n* × 1 *column vector*). We write

$$\mathbf{x} = \begin{pmatrix} x_1 \\ x_2 \\ \vdots \\ x_n \end{pmatrix}.$$

The elements of a matrix (and hence, in particular, of a vector) may be real numbers, real functions, real function values, or simply "variables." We usually denote square matrices by boldface Roman or Greek capital letters and vectors by boldface Roman or Greek lowercase letters.

Let us give a few specific examples.

The matrix

$$\mathbf{A} = \begin{pmatrix} 1 & 3 & 6 & 1 \\ -2 & 0 & 4 & -5 \\ 7 & 5 & -3 & 2 \\ 4 & -1 & 3 & -6 \end{pmatrix}$$

is a 4 × 4 square matrix of real numbers; whereas $\mathbf{\Phi}$ defined by

$$\mathbf{\Phi}(t) = \begin{pmatrix} t^2 & t+1 & 5 \\ t & t^2 & 3t \\ 1 & 0 & 2t-1 \end{pmatrix}$$

is a 3×3 square matrix of real functions defined for all real t.

The vector

$$\mathbf{c} = \begin{pmatrix} 3 \\ 1 \\ 2 \\ 5 \\ -2 \end{pmatrix}$$

is a 5×1 column vector of real numbers; the vector ϕ defined by

$$\phi(t) = \begin{pmatrix} e^t \\ te^t \\ 2e^t + 1 \end{pmatrix}$$

is a 3×1 column vector of real functions defined for all real t; and the vector

$$\mathbf{x} = \begin{pmatrix} x_1 \\ x_2 \\ x_3 \\ x_4 \end{pmatrix}$$

is a 4×1 column vector in the four variables x_1, x_2, x_3, x_4.

The elements of a vector are usually called its *components*. Given an $n \times 1$ column vector, the element in the ith row, for each $i = 1, 2, \ldots, n$, is then called its *ith component*.

For example, the third component of the column vector **c** illustrated above is 2, and its fourth component is 5.

For any given positive integer n, the $n \times 1$ column vector with all components equal to zero is called the *zero vector* and is denoted by **0**. Thus if $n = 4$, we have the 4×1 zero vector

$$\mathbf{0} = \begin{pmatrix} 0 \\ 0 \\ 0 \\ 0 \end{pmatrix}.$$

Now consider an $n \times n$ square matrix $\mathbf{A} = (a_{ij})$. The *principal diagonal* of **A** is the diagonal of elements from the upper left corner to the lower right corner of **A**; and the *diagonal elements* of **A** are the set of elements which lie along this principal diagonal, that is, the elements $a_{11}, a_{22}, \ldots, a_{nn}$. Now, for any given positive integer n, the $n \times n$ square matrix in which all the diagonal elements are one and all the other elements are zero is called the *identity matrix* and is denoted by **I**. That is, $\mathbf{I} = (a_{ij})$, where $a_{ij} = 1$ for all $i = j$ and $a_{ij} = 0$ for $i \neq j$ ($i = 1, 2, \ldots, n; j = 1, 2, \ldots, n$). Thus if $n = 4$, the 4×4 identity matrix is

$$\mathbf{I} = \begin{pmatrix} 1 & 0 & 0 & 0 \\ 0 & 1 & 0 & 0 \\ 0 & 0 & 1 & 0 \\ 0 & 0 & 0 & 1 \end{pmatrix}.$$

DEFINITION

We say that two $m \times n$ *matrices* $\mathbf{A} = (a_{ij})$ *and* $\mathbf{B} = (b_{ij})$ *are* equal *if and only if each element of one is equal to the corresponding element of the other. That is,* $\mathbf{A}$ *and* $\mathbf{B}$ *are equal if and only if* $a_{ij} = b_{ij}$ *for* $i = 1, 2, \ldots, m$ *and* $j = 1, 2, \ldots, n$. *If* $\mathbf{A}$ *and* $\mathbf{B}$ *are equal, we write* $\mathbf{A} = \mathbf{B}$.

Thus, for example, the two 3×3 square matrices

$$\mathbf{A} = \begin{pmatrix} a_{11} & a_{12} & a_{13} \\ a_{21} & a_{22} & a_{23} \\ a_{31} & a_{32} & a_{33} \end{pmatrix} \quad \text{and} \quad \mathbf{B} = \begin{pmatrix} 6 & 5 & 7 \\ -1 & 0 & 8 \\ 0 & -2 & -4 \end{pmatrix}$$

are equal if and only if

$$a_{11} = 6, \quad a_{12} = 5, \quad a_{13} = 7, \quad a_{21} = -1, \quad a_{22} = 0,$$
$$a_{23} = 8, \quad a_{31} = 0, \quad a_{32} = -2, \quad \text{and} \quad a_{33} = -4.$$

We then write $\mathbf{A} = \mathbf{B}$.

Likewise, the vectors

$$\mathbf{x} = \begin{pmatrix} x_1 \\ x_2 \\ x_3 \\ x_4 \end{pmatrix} \quad \text{and} \quad \mathbf{c} = \begin{pmatrix} -3 \\ 7 \\ 2 \\ -6 \end{pmatrix}$$

are equal if and only if

$$x_1 = -3, \quad x_2 = 7, \quad x_3 = 2, \quad x_4 = -6.$$

We then write $\mathbf{x} = \mathbf{c}$.

DEFINITION Addition of Matrices

The sum of two $m \times n$ *matrices* $\mathbf{A} = (a_{ij})$ *and* $\mathbf{B} = (b_{ij})$ *is defined to be the* $m \times n$ *matrix* $\mathbf{C} = (c_{ij})$, *where* $c_{ij} = a_{ij} + b_{ij}$, *for* $i = 1, 2, \ldots, m$ *and* $j = 1, 2, \ldots, n$. *We write* $\mathbf{C} = \mathbf{A} + \mathbf{B}$.

We may describe the addition of matrices by saying that the sum of two $m \times n$ matrices is the $m \times n$ matrix obtained by adding element-by-element.

Thus, for example, the sum of the two 3×3 square matrices

$$\mathbf{A} = \begin{pmatrix} 1 & 4 & -5 \\ 6 & 2 & 0 \\ 9 & 8 & 3 \end{pmatrix} \quad \text{and} \quad \mathbf{B} = \begin{pmatrix} -5 & 4 & 0 \\ 1 & -1 & 3 \\ 6 & 2 & 7 \end{pmatrix}$$

is the 3×3 square matrix

$$\mathbf{C} = \begin{pmatrix} 1 - 5 & 4 + 4 & -5 + 0 \\ 6 + 1 & 2 - 1 & 0 + 3 \\ 9 + 6 & 8 + 2 & 3 + 7 \end{pmatrix} = \begin{pmatrix} -4 & 8 & -5 \\ 7 & 1 & 3 \\ 15 & 10 & 10 \end{pmatrix}.$$

We write $\mathbf{C} = \mathbf{A} + \mathbf{B}$.

Likewise, the sum of the vectors

$$\mathbf{x} = \begin{pmatrix} x_1 \\ x_2 \\ x_3 \\ x_4 \end{pmatrix} \quad \text{and} \quad \mathbf{y} = \begin{pmatrix} y_1 \\ y_2 \\ y_3 \\ y_4 \end{pmatrix} \quad \text{is} \quad \mathbf{x} + \mathbf{y} = \begin{pmatrix} x_1 + y_1 \\ x_2 + y_2 \\ x_3 + y_3 \\ x_4 + y_4 \end{pmatrix}.$$

DEFINITION Multiplication by a Number

The product of the $m \times n$ matrix $\mathbf{A} = (a_{ij})$ and the number c *is defined to be the* $m \times n$ *matrix* $\mathbf{B} = (b_{ij})$, *where* $b_{ij} = ca_{ij}$ *for* $i = 1, 2, \ldots, m$ *and* $j = 1, 2, \ldots, n$. *We write* $\mathbf{B} = c\mathbf{A}$.

We may describe the multiplication of an $m \times n$ matrix $\mathbf{A}$ by a number c by saying that the product so formed is the $m \times n$ matrix which results from multiplying each individual element of $\mathbf{A}$ by the number c.

Thus, for example, the product of the 3×3 square matrix

$$\mathbf{A} = \begin{pmatrix} 2 & -3 & 6 \\ -7 & 4 & -1 \\ 0 & 5 & 2 \end{pmatrix}$$

by the number 3 is the 3×3 square matrix

$$\mathbf{B} = \begin{pmatrix} 3 \cdot 2 & 3 \cdot (-3) & 3 \cdot 6 \\ 3 \cdot (-7) & 3 \cdot 4 & 3 \cdot (-1) \\ 3 \cdot 0 & 3 \cdot 5 & 3 \cdot 2 \end{pmatrix} = \begin{pmatrix} 6 & -9 & 18 \\ -21 & 12 & -3 \\ 0 & 15 & 6 \end{pmatrix}.$$

We write $\mathbf{B} = 3\mathbf{A}$.

Likewise, the product of the vector

$$\mathbf{x} = \begin{pmatrix} x_1 \\ x_2 \\ x_3 \\ x_4 \end{pmatrix}$$

by the number 5 is the vector

$$\mathbf{y} = \begin{pmatrix} 5x_1 \\ 5x_2 \\ 5x_3 \\ 5x_4 \end{pmatrix}.$$

We write $\mathbf{y} = 5\mathbf{x}$.

DEFINITION

Let $\mathbf{x}_1, \mathbf{x}_2, \ldots, \mathbf{x}_m$ *be* m $n \times 1$ *vectors, and let* $c_1, c_2, \ldots, c_m$ *be* m *numbers. Then an element of the form*

$$c_1\mathbf{x}_1 + c_2\mathbf{x}_2 + \cdots + c_m\mathbf{x}_m$$

is an $n \times 1$ *vector called a* linear combination *of the vectors* $\mathbf{x}_1, \mathbf{x}_2, \ldots, \mathbf{x}_m$.

For example, consider the four 3×1 vectors

$$\mathbf{x}_1 = \begin{pmatrix} 2 \\ -1 \\ 3 \end{pmatrix}, \quad \mathbf{x}_2 = \begin{pmatrix} 3 \\ 2 \\ 1 \end{pmatrix}, \quad \mathbf{x}_3 = \begin{pmatrix} 1 \\ -3 \\ -2 \end{pmatrix}, \quad \mathbf{x}_4 = \begin{pmatrix} 4 \\ 5 \\ 0 \end{pmatrix}$$

and the four real numbers 2, 4, 5, and -3. Then

$$2\mathbf{x}_1 + 4\mathbf{x}_2 + 5\mathbf{x}_3 - 3\mathbf{x}_4 = 2\begin{pmatrix}2\\-1\\3\end{pmatrix} + 4\begin{pmatrix}3\\2\\1\end{pmatrix} + 5\begin{pmatrix}1\\-3\\-2\end{pmatrix} - 3\begin{pmatrix}4\\5\\0\end{pmatrix}$$

$$= \begin{pmatrix}4\\-2\\6\end{pmatrix} + \begin{pmatrix}12\\8\\4\end{pmatrix} + \begin{pmatrix}5\\-15\\-10\end{pmatrix} + \begin{pmatrix}-12\\-15\\0\end{pmatrix}$$

$$= \begin{pmatrix}4+12+5-12\\-2+8-15-15\\6+4-10+0\end{pmatrix} = \begin{pmatrix}9\\-24\\0\end{pmatrix}$$

is a linear combination of $\mathbf{x}_1$, $\mathbf{x}_2$, $\mathbf{x}_3$, and $\mathbf{x}_4$.

DEFINITION

Let

$$\mathbf{A} = \begin{pmatrix}a_{11} & a_{12} & \cdots & a_{1n}\\ a_{21} & a_{22} & \cdots & a_{2n}\\ \vdots & \vdots & & \vdots\\ a_{n1} & a_{n2} & \cdots & a_{nn}\end{pmatrix} \quad \textit{and} \quad \mathbf{x} = \begin{pmatrix}x_1\\x_2\\ \vdots\\x_n\end{pmatrix}$$

be an $n \times n$ square matrix and an $n \times 1$ vector, respectively. Then the product $\mathbf{Ax}$ of the $n \times n$ matrix $\mathbf{A}$ by the $n \times 1$ vector $\mathbf{x}$ *is defined to be the $n \times 1$ vector*

$$\mathbf{Ax} = \begin{pmatrix}a_{11}x_1 + a_{12}x_2 + \cdots + a_{1n}x_n\\ a_{21}x_1 + a_{22}x_2 + \cdots + a_{2n}x_n\\ \vdots\\ a_{n1}x_1 + a_{n2}x_2 + \cdots + a_{nn}x_n\end{pmatrix}.$$

Note that $\mathbf{Ax}$ is a vector. If we denote it by $\mathbf{y}$ and write

$$\mathbf{y} = \mathbf{Ax},$$

where

$$\mathbf{y} = \begin{pmatrix}y_1\\y_2\\ \vdots\\y_n\end{pmatrix},$$

then we have

$$\begin{aligned}y_1 &= a_{11}x_1 + a_{12}x_2 + \cdots + a_{1n}x_n,\\ y_2 &= a_{21}x_1 + a_{22}x_2 + \cdots + a_{2n}x_n,\\ &\ \ \vdots\\ y_n &= a_{n1}x_1 + a_{n2}x_2 + \cdots + a_{nn}x_n;\end{aligned}$$

that is, in general, for each $i = 1, 2, \ldots, n$,

$$y_i = a_{i1}x_1 + a_{i2}x_2 + \cdots + a_{in}x_n = \sum_{j=1}^{n} a_{ij}x_j.$$

For example, if

$$\mathbf{A} = \begin{pmatrix} 2 & -4 & 7 \\ 5 & 3 & -8 \\ -3 & 6 & 1 \end{pmatrix} \quad \text{and} \quad \mathbf{x} = \begin{pmatrix} x_1 \\ x_2 \\ x_3 \end{pmatrix},$$

then

$$\mathbf{y} = \mathbf{A}\mathbf{x} = \begin{pmatrix} 2x_1 - 4x_2 + 7x_3 \\ 5x_1 + 3x_2 - 8x_3 \\ -3x_1 + 6x_2 + x_3 \end{pmatrix}.$$

Before introducing the next concept, we state and illustrate two useful results, leaving their proofs to the reader (see Exercise 7 at the end of this section).

Result A. *If* **A** *is an* $n \times n$ *square matrix and* **x** *and* **y** *are* $n \times 1$ *column vectors, then*

$$\mathbf{A}(\mathbf{x} + \mathbf{y}) = \mathbf{A}\mathbf{x} + \mathbf{A}\mathbf{y}.$$

Result B. *If* **A** *is an* $n \times n$ *square matrix,* **x** *is an* $n \times 1$ *column vector, and* c *is a number, then*

$$\mathbf{A}(c\mathbf{x}) = c(\mathbf{A}\mathbf{x}).$$

▶ Example 7.19. We illustrate Results A and B using the matrix

$$\mathbf{A} = \begin{pmatrix} 2 & 1 & 3 \\ -1 & 4 & 1 \\ 3 & -2 & 5 \end{pmatrix},$$

the vectors

$$\mathbf{x} = \begin{pmatrix} x_1 \\ x_2 \\ x_3 \end{pmatrix} \quad \text{and} \quad \mathbf{y} = \begin{pmatrix} y_1 \\ y_2 \\ y_3 \end{pmatrix},$$

and the number $c = 3$.

Illustrating Result A, we have

$$\mathbf{A}(\mathbf{x} + \mathbf{y}) = \begin{pmatrix} 2 & 1 & 3 \\ -1 & 4 & 1 \\ 3 & -2 & 5 \end{pmatrix} \begin{pmatrix} x_1 + y_1 \\ x_2 + y_2 \\ x_3 + y_3 \end{pmatrix} = \begin{pmatrix} 2(x_1 + y_1) + (x_2 + y_2) + 3(x_3 + y_3) \\ -(x_1 + y_1) + 4(x_2 + y_2) + (x_3 + y_3) \\ 3(x_1 + y_1) - 2(x_2 + y_2) + 5(x_3 + y_3) \end{pmatrix}$$

$$= \begin{pmatrix} 2x_1 + x_2 + 3x_3 \\ -x_1 + 4x_2 + x_3 \\ 3x_1 - 2x_2 + 5x_3 \end{pmatrix} + \begin{pmatrix} 2y_1 + y_2 + 3y_3 \\ -y_1 + 4y_2 + y_3 \\ 3y_1 - 2y_2 + 5y_3 \end{pmatrix}$$

$$= \begin{pmatrix} 2 & 1 & 3 \\ -1 & 4 & 1 \\ 3 & -2 & 5 \end{pmatrix} \begin{pmatrix} x_1 \\ x_2 \\ x_3 \end{pmatrix} + \begin{pmatrix} 2 & 1 & 3 \\ -1 & 4 & 1 \\ 3 & -2 & 5 \end{pmatrix} \begin{pmatrix} y_1 \\ y_2 \\ y_3 \end{pmatrix} = \mathbf{A}\mathbf{x} + \mathbf{A}\mathbf{y}.$$

Illustrating Result B, we have

$$\mathbf{A}(c\mathbf{x}) = \begin{pmatrix} 2 & 1 & 3 \\ -1 & 4 & 1 \\ 3 & -2 & 5 \end{pmatrix} \begin{pmatrix} 3x_1 \\ 3x_2 \\ 3x_3 \end{pmatrix} = \begin{pmatrix} 6x_1 + 3x_2 + 9x_3 \\ -3x_1 + 12x_2 + 3x_3 \\ 9x_1 - 6x_2 + 15x_3 \end{pmatrix}$$

$$= 3 \begin{pmatrix} 2x_1 + x_2 + 3x_3 \\ -x_1 + 4x_2 + x_3 \\ 3x_1 - 2x_2 + 5x_3 \end{pmatrix} = c(\mathbf{A}\mathbf{x}).$$

We have seen examples of vectors whose components are numbers and also of vectors whose components are real functions defined on an interval $[a, b]$. We now distinguish between these two types by means of the following definitions:

DEFINITIONS

1. A vector all of whose components are numbers is called a constant vector.

2. A vector all of whose components are real functions defined on an interval $[a, b]$ is called a vector function.

DEFINITION

Let ϕ defined by

$$\phi(t) = \begin{pmatrix} \phi_1(t) \\ \phi_2(t) \\ \vdots \\ \phi_n(t) \end{pmatrix}$$

be an $n \times 1$ vector function whose components $\phi_1(t), \phi_2(t), \ldots, \phi_n(t)$ are differentiable on an interval $[a, b]$. Then the derivative *of ϕ is the vector function defined by*

$$\frac{d\phi(t)}{dt} = \begin{pmatrix} \dfrac{d\phi_1(t)}{dt} \\ \dfrac{d\phi_2(t)}{dt} \\ \vdots \\ \dfrac{d\phi_n(t)}{dt} \end{pmatrix}$$

for all $t \in [a, b]$.

Thus the derivative of a given vector function all of whose components are differentiable is the vector function obtained from the given vector function by differentiating each component of the given vector function.

▶ Example 7.20. The derivative of the vector function ϕ defined for all t by

$$\phi(t) = \begin{pmatrix} 4t^3 \\ 2t^2 + 3t \\ 2e^{3t} \end{pmatrix}$$

is the vector function defined for all t by

$$\frac{d\phi(t)}{dt} = \begin{pmatrix} 12t^2 \\ 4t + 3 \\ 6e^{3t} \end{pmatrix}.$$

Exercises

1. In each case, find the sum $\mathbf{A} + \mathbf{B}$ of the given matrices.

   (a) $\mathbf{A} = \begin{pmatrix} 2 & 3 \\ 5 & 4 \end{pmatrix}$ and $\mathbf{B} = \begin{pmatrix} 6 & -1 \\ -7 & 2 \end{pmatrix}$.

(b) $\mathbf{A} = \begin{pmatrix} 2 & 1 & 3 \\ -1 & 0 & 5 \\ -4 & 3 & -2 \end{pmatrix}$ and $\mathbf{B} = \begin{pmatrix} 7 & -1 & 6 \\ 2 & 4 & -3 \\ 5 & -5 & 1 \end{pmatrix}$.

(c) $\mathbf{A} = \begin{pmatrix} -5 & 0 & 4 \\ -2 & -1 & -3 \\ 6 & 2 & 5 \end{pmatrix}$ and $\mathbf{B} = \begin{pmatrix} 7 & -2 & -3 \\ 6 & -3 & 1 \\ -2 & 1 & -3 \end{pmatrix}$.

2. In each case, find the product $c\mathbf{A}$ of the given matrix $\mathbf{A}$ and the number c.

(a) $\mathbf{A} = \begin{pmatrix} 1 & 2 \\ 7 & -3 \end{pmatrix}$, $c = 3$.

(b) $\mathbf{A} = \begin{pmatrix} 1 & -3 & 5 \\ 6 & -2 & 0 \\ -3 & 1 & 2 \end{pmatrix}$, $c = -4$.

(c) $\mathbf{A} = \begin{pmatrix} 5 & -1 & 2 \\ 4 & -3 & -2 \\ 0 & 3 & -6 \end{pmatrix}$, $c = -3$.

3. In each case, find the indicated linear combination of the given vectors

$$\mathbf{x}_1 = \begin{pmatrix} 3 \\ 2 \\ -1 \\ 4 \end{pmatrix}, \quad \mathbf{x}_2 = \begin{pmatrix} -1 \\ 3 \\ 5 \\ -2 \end{pmatrix}, \quad \mathbf{x}_3 = \begin{pmatrix} 2 \\ 4 \\ 0 \\ 6 \end{pmatrix}, \quad \text{and} \quad \mathbf{x}_4 = \begin{pmatrix} -1 \\ 2 \\ -3 \\ 5 \end{pmatrix}.$$

(a) $2\mathbf{x}_1 + 3\mathbf{x}_2 - \mathbf{x}_3$.

(b) $3\mathbf{x}_1 - 2\mathbf{x}_2 + 4\mathbf{x}_4$.

(c) $-\mathbf{x}_1 + 5\mathbf{x}_2 - 2\mathbf{x}_3 + 3\mathbf{x}_4$.

4. In each case, find the product $\mathbf{A}\mathbf{x}$ of the given matrix $\mathbf{A}$ by the given vector $\mathbf{x}$.

(a) $\mathbf{A} = \begin{pmatrix} 2 & 1 & -4 \\ 5 & -2 & 3 \\ 1 & -3 & 2 \end{pmatrix}$, $\mathbf{x} = \begin{pmatrix} x_1 \\ x_2 \\ x_3 \end{pmatrix}$.

(b) $\mathbf{A} = \begin{pmatrix} -3 & -5 & 7 \\ 0 & 4 & 1 \\ -2 & 1 & 3 \end{pmatrix}$, $\mathbf{x} = \begin{pmatrix} 2 \\ 3 \\ -2 \end{pmatrix}$.

(c) $\mathbf{A} = \begin{pmatrix} 1 & 0 & -3 \\ 2 & -5 & 4 \\ -3 & 1 & 2 \end{pmatrix}$, $\mathbf{x} = \begin{pmatrix} x_1 + x_2 \\ x_1 + 2x_2 \\ x_2 - x_3 \end{pmatrix}$.

5. Illustrate Results A and B of this subsection using the matrix

$$\mathbf{A} = \begin{pmatrix} 3 & -1 & 2 \\ 5 & 4 & -3 \\ -5 & 1 & 2 \end{pmatrix},$$

the vectors

$$\mathbf{x} = \begin{pmatrix} x_1 \\ x_2 \\ x_3 \end{pmatrix} \quad \text{and} \quad \mathbf{y} = \begin{pmatrix} y_1 \\ y_2 \\ y_3 \end{pmatrix},$$

and the number $c = 4$.

6. In each case, find the derivative of the vector function ϕ which is defined.

(a) $$\phi(t) = \begin{pmatrix} 5t^2 \\ -6t^3 + t^2 \\ 2t^2 - 5t \end{pmatrix}.$$

(b) $$\phi(t) = \begin{pmatrix} e^{3t} \\ (2t + 3)e^{3t} \\ t^2 e^{3t} \end{pmatrix}.$$

(c) $$\phi(t) = \begin{pmatrix} \sin 3t \\ \cos 3t \\ t \sin 3t \\ t \cos 3t \end{pmatrix}.$$

7. Prove Results A and B of the text (page 292).

B. Linear Independence and Dependence

Before proceeding, we state without proof the following two theorems from algebra.

THEOREM A

A system of n homogeneous linear algebraic equations in n unknowns has a nontrivial solution if and only if the determinant of coefficients of the system is equal to zero.

THEOREM B

A system of n linear algebraic equations in n unknowns has a unique solution if and only if the determinant of coefficients of the system is unequal to zero.

DEFINITION

A set of m constant vectors $\mathbf{v}_1, \mathbf{v}_2, \ldots, \mathbf{v}_m$ is linearly dependent *if there exists a set of m numbers $c_1, c_2, \ldots, c_m$, not all of which are zero, such that*

$$c_1\mathbf{v}_1 + c_2\mathbf{v}_2 + \cdots + c_m\mathbf{v}_m = \mathbf{0}.$$

▶ Example 7.21. The set of three constant vectors

$$\mathbf{v}_1 = \begin{pmatrix} 2 \\ 3 \\ 1 \end{pmatrix}, \quad \mathbf{v}_2 = \begin{pmatrix} 1 \\ -1 \\ 2 \end{pmatrix}, \quad \mathbf{v}_3 = \begin{pmatrix} 7 \\ 3 \\ 8 \end{pmatrix}$$

is linearly dependent, since there exists the set of three numbers 2, 3, and -1, none of which are zero, such that

$$2\mathbf{v}_1 + 3\mathbf{v}_2 + (-1)\mathbf{v}_3 = \mathbf{0}.$$

DEFINITION

A set of m constant vectors is linearly independent *if and only if the set is not linearly dependent. That is, a set of m constant vectors $\mathbf{v}_1, \mathbf{v}_2, \ldots, \mathbf{v}_m$ is linearly independent if the relation*

$$c_1\mathbf{v}_1 + c_2\mathbf{v}_2 + \cdots + c_m\mathbf{v}_m = \mathbf{0}$$

implies that

$$c_1 = c_2 = \cdots = c_m = 0.$$

▶ Example 7.22. The set of three constant vectors

$$\mathbf{v}_1 = \begin{pmatrix} 1 \\ 1 \\ 1 \end{pmatrix}, \quad \mathbf{v}_2 = \begin{pmatrix} -1 \\ 2 \\ 0 \end{pmatrix}, \quad \text{and} \quad \mathbf{v}_3 = \begin{pmatrix} 0 \\ 2 \\ 1 \end{pmatrix}$$

is linearly independent. For suppose we have

$$c_1\mathbf{v}_1 + c_2\mathbf{v}_2 + c_3\mathbf{v}_3 = \mathbf{0}; \tag{7.96}$$

that is,

$$c_1 \begin{pmatrix} 1 \\ 1 \\ 1 \end{pmatrix} + c_2 \begin{pmatrix} -1 \\ 2 \\ 0 \end{pmatrix} + c_3 \begin{pmatrix} 0 \\ 2 \\ 1 \end{pmatrix} = \begin{pmatrix} 0 \\ 0 \\ 0 \end{pmatrix}.$$

This is equivalent to the system

$$\begin{aligned} c_1 - c_2 &= 0, \\ c_1 + 2c_2 + 2c_3 &= 0, \\ c_1 + c_3 &= 0, \end{aligned} \tag{7.97}$$

of three homogeneous linear algebraic equations in the three unknowns c_1, c_2, c_3. The determinant of coefficients of this system is

$$\begin{vmatrix} 1 & -1 & 0 \\ 1 & 2 & 2 \\ 1 & 0 & 1 \end{vmatrix} = 1 \neq 0.$$

Thus by Theorem A, with $n = 3$, the system (7.97) has only the trivial solution $c_1 = c_2 = c_3 = 0$. Thus for the three given constant vectors, the relation (7.96) implies $c_1 = c_2 = c_3 = 0$; and so these three vectors are indeed linearly independent.

DEFINITION

The set of m vector functions $\boldsymbol{\phi}_1, \boldsymbol{\phi}_2, \ldots, \boldsymbol{\phi}_m$ *is* linearly dependent *on an interval* $a \le t \le b$ *if there exists a set of m numbers* $c_1, c_2, \ldots, c_m$, *not all zero, such that*

$$c_1\boldsymbol{\phi}_1(t) + c_2\boldsymbol{\phi}_2(t) + \cdots + c_m\boldsymbol{\phi}_m(t) = \mathbf{0}$$

for all $t \in [a, b]$.

▶ Example 7.23. Consider the set of three vector functions $\boldsymbol{\phi}_1$, $\boldsymbol{\phi}_2$, and $\boldsymbol{\phi}_3$, defined for all t by

$$\boldsymbol{\phi}_1(t) = \begin{pmatrix} e^{2t} \\ 2e^{2t} \\ 5e^{2t} \end{pmatrix}, \quad \boldsymbol{\phi}_2(t) = \begin{pmatrix} e^{2t} \\ 4e^{2t} \\ 11e^{2t} \end{pmatrix}, \quad \text{and} \quad \boldsymbol{\phi}_3(t) = \begin{pmatrix} e^{2t} \\ e^{2t} \\ 2e^{2t} \end{pmatrix},$$

respectively. This set of vector functions is linearly dependent on any interval $a \le t \le b$. To see this, note that

$$3 \begin{pmatrix} e^{2t} \\ 2e^{2t} \\ 5e^{2t} \end{pmatrix} + (-1) \begin{pmatrix} e^{2t} \\ 4e^{2t} \\ 11e^{2t} \end{pmatrix} + (-2) \begin{pmatrix} e^{2t} \\ e^{2t} \\ 2e^{2t} \end{pmatrix} = \begin{pmatrix} 0 \\ 0 \\ 0 \end{pmatrix};$$

and hence there exists the set of three numbers 3, -1, and -2, none of which are zero, such that

$$3\boldsymbol{\phi}_1(t) + (-1)\boldsymbol{\phi}_2(t) + (-2)\boldsymbol{\phi}_3(t) = \mathbf{0}$$

for all $t \in [a, b]$.

DEFINITION

A set of m vector functions is linearly independent *on an interval if and only if the set is not linearly dependent on that interval. That is, a set of m vector functions* $\boldsymbol{\phi}_1, \boldsymbol{\phi}_2, \ldots, \boldsymbol{\phi}_m$ *is linearly independent on an interval* $a \le t \le b$ *if the relation*

$$c_1\boldsymbol{\phi}_1(t) + c_2\boldsymbol{\phi}_2(t) + \cdots + c_m\boldsymbol{\phi}_m(t) = \mathbf{0}$$

for all $t \in [a, b]$ *implies that*

$$c_1 = c_2 = \cdots = c_m = 0.$$

▶ Example 7.24. Consider the set of two vector functions $\boldsymbol{\phi}_1$ and $\boldsymbol{\phi}_2$, defined for all t by

$$\boldsymbol{\phi}_1(t) = \begin{pmatrix} e^t \\ e^t \end{pmatrix} \quad \text{and} \quad \boldsymbol{\phi}_2(t) = \begin{pmatrix} e^{2t} \\ 2e^{2t} \end{pmatrix},$$

respectively. We shall show that $\boldsymbol{\phi}_1$ and $\boldsymbol{\phi}_2$ are linearly independent on any interval $a \le t \le b$. To do this, we assume the contrary; that is, we assume that $\boldsymbol{\phi}_1$ and $\boldsymbol{\phi}_2$ are linear *dependent* on $[a, b]$. Then there exist numbers c_1 and c_2, not both zero, such that

$$c_1\boldsymbol{\phi}_1(t) + c_2\boldsymbol{\phi}_2(t) = \mathbf{0},$$

for all $t \in [a, b]$. Then

$$c_1e^t + c_2e^{2t} = 0,$$
$$c_1e^t + 2c_2e^{2t} = 0;$$

and multiplying each equation through by e^{-t}, we have

$$c_1 + c_2e^t = 0,$$
$$c_1 + 2c_2e^t = 0,$$

for all $t \in [a, b]$. This implies that $c_1 + c_2e^t = c_1 + 2c_2e^t$ and hence $1 = 2$, which is an obvious contradiction. Thus the assumption that $\boldsymbol{\phi}_1$ and $\boldsymbol{\phi}_2$ are linearly dependent on $[a, b]$ is false, and so these two vector functions are linearly independent on that interval.

Note. If a set of m vector functions $\boldsymbol{\phi}_1, \boldsymbol{\phi}_2, \ldots, \boldsymbol{\phi}_m$ is linearly dependent on an interval $a \le t \le b$, then it readily follows that for each fixed $t_0 \in [a, b]$, the corresponding set of m constant vectors $\boldsymbol{\phi}_1(t_0), \boldsymbol{\phi}_2(t_0), \ldots, \boldsymbol{\phi}_m(t_0)$ is linearly dependent. However, the analogous statement for a set of m linearly independent vector functions is not valid. That is, if a set of m vector functions $\boldsymbol{\phi}_1, \boldsymbol{\phi}_2, \ldots, \boldsymbol{\phi}_m$ is linearly independent on an interval $a \le t \le b$, then it is *not* necessarily true that for each fixed $t_0 \in [a, b]$, the corresponding set of m constant vectors $\boldsymbol{\phi}_1(t_0), \boldsymbol{\phi}_2(t_0), \ldots, \boldsymbol{\phi}_m(t_0)$ is linearly independent. Indeed, the corresponding set of constant vectors $\boldsymbol{\phi}_1(t_0)$, $\boldsymbol{\phi}_2(t_0), \ldots, \boldsymbol{\phi}_m(t_0)$ may be linearly *dependent* for *each* $t_0 \in [a, b]$. See Exercise 6 at the end of this section.

Exercises

1. In each case, show that the given set of constant vectors is linearly dependent.

(a) $\mathbf{v}_1 = \begin{pmatrix} 3 \\ -1 \\ 2 \end{pmatrix}, \quad \mathbf{v}_2 = \begin{pmatrix} 13 \\ 5 \\ -4 \end{pmatrix}, \quad \mathbf{v}_3 = \begin{pmatrix} 2 \\ 4 \\ -5 \end{pmatrix}.$

(b) $\mathbf{v}_1 = \begin{pmatrix} 2 \\ 4 \\ 5 \end{pmatrix}, \quad \mathbf{v}_2 = \begin{pmatrix} 2 \\ 6 \\ 8 \end{pmatrix}, \quad \mathbf{v}_3 = \begin{pmatrix} 1 \\ -1 \\ -2 \end{pmatrix}.$

(c) $\mathbf{v}_1 = \begin{pmatrix} 1 \\ 2 \\ -3 \end{pmatrix}, \quad \mathbf{v}_2 = \begin{pmatrix} 2 \\ 1 \\ -3 \end{pmatrix}, \quad \mathbf{v}_3 = \begin{pmatrix} 7 \\ -1 \\ -6 \end{pmatrix}.$

2. In each case, show that the given set of constant vectors is linearly independent.

(a) $\mathbf{v}_1 = \begin{pmatrix} 2 \\ 1 \\ 0 \end{pmatrix}, \quad \mathbf{v}_2 = \begin{pmatrix} 1 \\ 0 \\ 3 \end{pmatrix}, \quad \mathbf{v}_3 = \begin{pmatrix} 0 \\ -1 \\ 1 \end{pmatrix}.$

(b) $\mathbf{v}_1 = \begin{pmatrix} 1 \\ 2 \\ 0 \end{pmatrix}, \quad \mathbf{v}_2 = \begin{pmatrix} -2 \\ 0 \\ 1 \end{pmatrix}, \quad \mathbf{v}_3 = \begin{pmatrix} 1 \\ -1 \\ 2 \end{pmatrix}.$

(c) $\mathbf{v}_1 = \begin{pmatrix} 1 \\ 2 \\ -1 \end{pmatrix}, \quad \mathbf{v}_2 = \begin{pmatrix} 1 \\ 1 \\ 1 \end{pmatrix}, \quad \mathbf{v}_3 = \begin{pmatrix} 1 \\ -1 \\ 3 \end{pmatrix}.$

3. In each case, show that the set of vector functions $\boldsymbol{\phi}_1, \boldsymbol{\phi}_2, \boldsymbol{\phi}_3$, defined for all t as indicated, is linearly dependent on any interval $a \leq t \leq b$.

(a) $\boldsymbol{\phi}_1(t) = \begin{pmatrix} 2e^{3t} \\ 3e^{3t} \\ -e^{3t} \end{pmatrix}, \quad \boldsymbol{\phi}_2(t) = \begin{pmatrix} 4e^{3t} \\ -5e^{3t} \\ 5e^{3t} \end{pmatrix}, \quad \boldsymbol{\phi}_3(t) = \begin{pmatrix} 5e^{3t} \\ 2e^{3t} \\ e^{3t} \end{pmatrix}.$

(b) $\boldsymbol{\phi}_1(t) = \begin{pmatrix} \sin t + \cos t \\ 2 \sin t \\ -\cos t \end{pmatrix}, \quad \boldsymbol{\phi}_2(t) = \begin{pmatrix} 2 \sin t \\ 3 \sin t - \cos t \\ -\sin t \end{pmatrix},$

$\boldsymbol{\phi}_3(t) = \begin{pmatrix} 4 \cos t \\ 2 \cos t \\ 2 \sin t - 4 \cos t \end{pmatrix}.$

4. In each case, show that the set of vector functions $\boldsymbol{\phi}_1$ and $\boldsymbol{\phi}_2$ defined for all t as indicated, is linearly independent on any interval $a \leq t \leq b$.

(a) $\boldsymbol{\phi}_1(t) = \begin{pmatrix} e^{t} \\ 2e^{t} \end{pmatrix}$ and $\boldsymbol{\phi}_2(t) = \begin{pmatrix} e^{3t} \\ 4e^{3t} \end{pmatrix}.$

(b) $\boldsymbol{\phi}_1(t) = \begin{pmatrix} 2e^{2t} \\ -e^{2t} \end{pmatrix}$ and $\boldsymbol{\phi}_2(t) = \begin{pmatrix} e^{-t} \\ 3e^{-t} \end{pmatrix}.$

5. Show that the set of two vector functions ϕ_1 and ϕ_2 defined for all t by

$$\phi_1(t) = \begin{pmatrix} t \\ 0 \end{pmatrix} \quad \text{and} \quad \phi_2(t) = \begin{pmatrix} t^2 \\ 0 \end{pmatrix},$$

respectively, is linearly independent on any interval $a \leq t \leq b$.

6. Consider the vector functions ϕ_1 and ϕ_2 defined by

$$\phi_1(t) = \begin{pmatrix} t \\ 1 \end{pmatrix} \quad \text{and} \quad \phi_2(t) = \begin{pmatrix} te^t \\ e^t \end{pmatrix},$$

respectively. Show that the constant vectors $\phi_1(t_0)$ and $\phi_2(t_0)$ are linearly dependent for each t_0 in the interval $0 \leq t \leq 1$, but that the vector functions ϕ_1 and ϕ_2 are linearly independent on $0 \leq t \leq 1$.

7. Let

$$\boldsymbol{\alpha}^{(i)} = \begin{pmatrix} \alpha_{1i} \\ \alpha_{2i} \\ \vdots \\ \alpha_{ni} \end{pmatrix} \qquad (i = 1, 2, \ldots, n),$$

be a set of n linearly independent vectors. Show that

$$\begin{vmatrix} \alpha_{11} & \alpha_{12} & \cdots & \alpha_{1n} \\ \alpha_{21} & \alpha_{22} & \cdots & \alpha_{2n} \\ \vdots & \vdots & & \vdots \\ \alpha_{n1} & \alpha_{n2} & \cdots & \alpha_{nn} \end{vmatrix} \neq 0.$$

C. *Characteristic Values and Characteristic Functions*

Let $\mathbf{A}$ be a given $n \times n$ square matrix of real numbers, and let S denote the set of all $n \times 1$ column vectors of numbers. Now consider the equation

$$\mathbf{A}\mathbf{x} = \lambda\mathbf{x} \tag{7.98}$$

in the unknown vector $\mathbf{x} \in S$, where λ is a number. Clearly the zero vector $\mathbf{0}$ is a solution of this equation for every number λ. We investigate the possibility of finding nonzero vectors $\mathbf{x} \in S$ which are solutions of (7.98) for some choice of the number λ. In other words, we seek numbers λ corresponding to which there exist nonzero vectors $\mathbf{x}$ which satisfy (7.98). These desired values of λ and the corresponding desired nonzero vectors are designated in the following definitions.

DEFINITIONS

A characteristic value (*or* eigenvalue) *of the matrix* $\mathbf{A}$ *is a number* λ *for which the equation* $\mathbf{A}\mathbf{x} = \lambda\mathbf{x}$ *has a nonzero vector solution* $\mathbf{x}$.

A characteristic vector (*or* eigenvector) *of* $\mathbf{A}$ *is a nonzero vector* $\mathbf{x}$ *such that* $\mathbf{A}\mathbf{x} = \lambda\mathbf{x}$ *for some number* λ.

We proceed to solve this problem. Suppose

$$\mathbf{A} = \begin{pmatrix} a_{11} & a_{12} & \cdots & a_{1n} \\ a_{21} & a_{22} & \cdots & a_{2n} \\ \vdots & \vdots & & \vdots \\ a_{n1} & a_{n2} & \cdots & a_{nn} \end{pmatrix}$$

is the given $n \times n$ square matrix of real numbers, and let

$$\mathbf{x} = \begin{pmatrix} x_1 \\ x_2 \\ \vdots \\ x_n \end{pmatrix}.$$

Then Equation (7.98) may be written

$$\begin{pmatrix} a_{11} & a_{12} & \cdots & a_{1n} \\ a_{21} & a_{22} & \cdots & a_{2n} \\ \vdots & \vdots & & \vdots \\ a_{n1} & a_{n2} & \cdots & a_{nn} \end{pmatrix} \begin{pmatrix} x_1 \\ x_2 \\ \vdots \\ x_n \end{pmatrix} = \lambda \begin{pmatrix} x_1 \\ x_1 \\ \vdots \\ x_n \end{pmatrix},$$

and hence, multiplying the indicated entities,

$$\begin{pmatrix} a_{11}x_1 + a_{12}x_2 + \cdots + a_{1n}x_n \\ a_{21}x_1 + a_{22}x_2 + \cdots + a_{2n}x_n \\ \vdots \\ a_{n1}x_1 + a_{n2}x_2 + \cdots + a_{nn}x_n \end{pmatrix} = \begin{pmatrix} \lambda x_1 \\ \lambda x_2 \\ \vdots \\ \lambda x_n \end{pmatrix}.$$

Equating corresponding components of these two equal vectors, we have

$$\begin{aligned} a_{11}x_1 + a_{12}x_2 + \cdots + a_{1n}x_n &= \lambda x_1, \\ a_{21}x_1 + a_{22}x_2 + \cdots + a_{2n}x_n &= \lambda x_2, \\ &\vdots \\ a_{n1}x_1 + a_{n2}x_2 + \cdots + a_{nn}x_n &= \lambda x_n; \end{aligned}$$

and rewriting this, we obtain

$$\begin{aligned} (a_{11} - \lambda)x_1 + a_{12}x_2 + \cdots + a_{1n}x_n &= 0, \\ a_{21}x_1 + (a_{22} - \lambda)x_2 + \cdots + a_{2n}x_n &= 0, \\ &\vdots \\ a_{n1}x_1 + a_{n2}x_2 + \cdots + (a_{nn} - \lambda)x_n &= 0. \end{aligned} \tag{7.99}$$

Thus we see that (7.98) holds if and only if (7.99) does. Now we are seeking nonzero vectors $\mathbf{x}$ which satisfy (7.98). Thus a nonzero vector $\mathbf{x}$ satisfies (7.98) if and only if its set of components $x_1, x_2, \ldots, x_n$ is a nontrivial solution of (7.99). By Theorem A of Section 7.5B, the system (7.99) has nontrivial solutions if and only if its determinant of coefficients is equal to zero, that is, if and only if

$$\begin{vmatrix} a_{11} - \lambda & a_{12} & \cdots & a_{1n} \\ a_{21} & a_{22} - \lambda & \cdots & a_{2n} \\ \vdots & \vdots & & \vdots \\ a_{n1} & a_{n2} & \cdots & a_{nn} - \lambda \end{vmatrix} = 0. \tag{7.100}$$

It is easy to see that (7.100) is a polynomial equation of the nth degree in the unknown λ. In matrix notation it is written

$$|\mathbf{A} - \lambda\mathbf{I}| = 0,$$

where $\mathbf{I}$ is the $n \times n$ identity matrix (see Section 7.5A). Thus Equation (7.98) has a nonzero vector solution $\mathbf{x}$ for a certain value of λ if and only if λ satisfies the nth-degree polynomial equation (7.100). That is, the number λ is a characteristic value of the matrix $\mathbf{A}$ if and only if it satisfies this polynomial equation. We now designate this equation and also state the alternative definition of characteristic value which we have thus obtained.

DEFINITION

Let $\mathbf{A} = (a_{ij})$ *be an* $n \times n$ *square matrix of real numbers. The* characteristic equation *of* $\mathbf{A}$ *is the nth degree polynomial equation*

$$\begin{vmatrix} a_{11} - \lambda & a_{12} & \cdots & a_{1n} \\ a_{21} & a_{22} - \lambda & \cdots & a_{2n} \\ \vdots & \vdots & & \vdots \\ a_{n1} & a_{n2} & \cdots & a_{nn} - \lambda \end{vmatrix} = 0 \tag{7.100}$$

in the unknown λ; *and the* characteristic values *of* $\mathbf{A}$ *are the roots of this equation.*

Since the characteristic equation (7.100) of $\mathbf{A}$ is a polynomial equation of the nth degree, it has n roots. These roots may be real or complex, but of course they may or may not all be distinct. If a certain repeated root occurs m times, where $1 < m \leq n$, then we say that that root has *multiplicity* m. If we count each nonrepeated root once and each repeated root according to its multiplicity, then we can say that the $n \times n$ matrix $\mathbf{A}$ has precisely n characteristic values, say $\lambda_1, \lambda_2, \ldots, \lambda_n$.

Corresponding to each characteristic value λ_k of $\mathbf{A}$ there is a characteristic vector $\mathbf{x}_k$ $(k = 1, 2, \ldots, n)$. Further, if $\mathbf{x}_k$ is a characteristic vector of $\mathbf{A}$ corresponding to characteristic value λ_k, then so is $c\mathbf{x}_k$ for any nonzero number c. We shall be concerned with the linear independence of the various characteristic vectors of $\mathbf{A}$. Concerning this, we state the following two results without proof.

Result C. *Suppose each of the n characteristic values* $\lambda_1, \lambda_2, \ldots, \lambda_n$ *of the* $n \times n$ *square matrix* $\mathbf{A}$ *is distinct (that is, nonrepeated); and let* $\mathbf{x}_1, \mathbf{x}_2, \ldots, \mathbf{x}_n$ *be a set of n respective corresponding characteristic vectors of* $\mathbf{A}$. *Then the set of these n characteristic vectors is linearly independent.*

Result D. *Suppose the* $n \times n$ *square matrix* $\mathbf{A}$ *has a characteristic value of multiplicity m, where* $1 < m \leq n$. *Then this repeated characteristic value having multiplicity m has p linearly independent characteristic vectors corresponding to it, where* $1 \leq p \leq m$.

Now suppose $\mathbf{A}$ has at least one characteristic value of multiplicity m, where $1 < m \leq n$; and further suppose that for this repeated characteristic value, the number p of Result D is strictly less than m; that is, p is such that $1 \leq p < m$. Then corresponding to this characteristic value of multiplicity m, there are less than m linearly independent characteristic vectors. It follows at once that the matrix $\mathbf{A}$ must then have *less than n* linearly independent characteristic vectors. Thus we are led to the following result:

Result E. *If the $n \times n$ matrix* **A** *has one or more repeated characteristic values, then there may exist less than n linearly independent characteristic vectors of* **A**.

Before giving an example of finding the characteristic values and corresponding characteristic vectors of a matrix, we introduce a very special class of matrices whose characteristic values and vectors have some interesting special properties. This is the class of so-called real *symmetric* matrices, which we shall define below. First, however, we give a preliminary definition.

DEFINITION

Let $\mathbf{A} = (a_{ij})$ be an $m \times n$ matrix. The transpose *of* **A** *is the $n \times m$ matrix $\mathbf{B} = (b_{ij})$, where $b_{ij} = a_{ji}$, $i = 1, 2, \ldots, n$, $j = 1, 2, \ldots, m$. That is, the transpose of an $m \times n$ matrix* **A** *is the $n \times m$ matrix obtained from* **A** *by interchanging the rows and columns of* **A**. *We denote the transpose of* **A** *by $\mathbf{A}^T$.*

For example, the transpose of the 3×3 square matrix

$$\mathbf{A} = \begin{pmatrix} 1 & 2 & 6 \\ 3 & 4 & 7 \\ 5 & 9 & 5 \end{pmatrix}$$

is the 3×3 square matrix

$$\mathbf{A}^T = \begin{pmatrix} 1 & 3 & 5 \\ 2 & 4 & 9 \\ 6 & 7 & 5 \end{pmatrix}.$$

Note, in particular, that the transpose of the $n \times 1$ *column* vector

$$\mathbf{x} = \begin{pmatrix} x_1 \\ x_2 \\ \vdots \\ x_n \end{pmatrix}$$

is the $1 \times n$ *row* vector

$$\mathbf{x}^T = (x_1, x_2, \ldots, x_n).$$

DEFINITION

A square matrix **A** *of real numbers is called a real* symmetric *matrix if $\mathbf{A}^T = \mathbf{A}$.*

For example, the 3×3 square matrix

$$\mathbf{A} = \begin{pmatrix} 2 & -1 & 4 \\ -1 & 0 & 3 \\ 4 & 3 & 1 \end{pmatrix}$$

is a real symmetric matrix, since $\mathbf{A}^T = \mathbf{A}$.

Concerning real symmetric matrices, we state without proof the following interesting results:

Result F. *All of the characteristic values of a real symmetric matrix are real numbers.*

Result G. *If* **A** *is an* $n \times n$ *real symmetric square matrix, then there exist* n *linearly independent characteristic vectors of* **A**, *whether the* n *characteristic values of* **A** *are all distinct or whether one or more of these characteristic values is repeated.*

▶ Example 7.25. Find the characteristic values and characteristic vectors of the matrix

$$\mathbf{A} = \begin{pmatrix} 7 & -1 & 6 \\ -10 & 4 & -12 \\ -2 & 1 & -1 \end{pmatrix}.$$

Solution. The characteristic equation of **A** is

$$\begin{vmatrix} 7-\lambda & -1 & 6 \\ -10 & 4-\lambda & -12 \\ -2 & 1 & -1-\lambda \end{vmatrix} = 0.$$

Evaluating the determinant in the left member, we find that this equation may be written in the form

$$\lambda^3 - 10\lambda^2 + 31\lambda - 30 = 0.$$

or

$$(\lambda - 2)(\lambda - 3)(\lambda - 5) = 0.$$

Thus the characteristic values of **A** are

$$\lambda = 2, \quad \lambda = 3, \quad \text{and} \quad \lambda = 5.$$

The characteristic vectors corresponding to $\lambda = 2$ are the nonzero vectors

$$\mathbf{x} = \begin{pmatrix} x_1 \\ x_2 \\ x_3 \end{pmatrix}$$

such that

$$\begin{pmatrix} 7 & -1 & 6 \\ -10 & 4 & -12 \\ -2 & 1 & -1 \end{pmatrix}\begin{pmatrix} x_1 \\ x_2 \\ x_3 \end{pmatrix} = 2\begin{pmatrix} x_1 \\ x_2 \\ x_3 \end{pmatrix}.$$

Thus, x_1, x_2, x_3 must be a nontrivial solution of the system

$$\begin{aligned} 7x_1 - x_2 + 6x_3 &= 2x_1, \\ -10x_1 + 4x_2 - 12x_3 &= 2x_2, \\ -2x_1 + x_2 - x_3 &= 2x_3; \end{aligned}$$

that is,

$$\begin{aligned} 5x_1 - x_2 + 6x_3 &= 0, \\ -10x_1 + 2x_2 - 12x_3 &= 0, \\ -2x_1 + x_2 - 3x_3 &= 0. \end{aligned}$$

Note that the second of these three equations is merely a constant multiple of the first. Thus we seek nonzero numbers x_1, x_2, x_3 which satisfy the first and third of these equations. Writing these two as equations in the unknowns x_2 and x_3, we have

$$-x_2 + 6x_3 = -5x_1,$$
$$x_2 - 3x_3 = 2x_1.$$

Solving for x_2 and x_3, we find

$$x_2 = -x_1 \quad \text{and} \quad x_3 = -x_1.$$

We see at once that $x_1 = k$, $x_2 = -k$, $x_3 = -k$ is a solution of this for every real k. Hence the characteristic vectors corresponding to the characteristic value $\lambda = 2$ are the vectors

$$\mathbf{x} = \begin{pmatrix} k \\ -k \\ -k \end{pmatrix},$$

where k is an arbitrary nonzero number. In particular, letting $k = 1$, we obtain the particular characteristic vector

$$\begin{pmatrix} 1 \\ -1 \\ -1 \end{pmatrix}$$

corresponding to the characteristic value $\lambda = 2$.

Proceeding in like manner, one can find the characteristic vectors corresponding to $\lambda = 3$ and those corresponding to $\lambda = 5$. We give only a few highlights of these computations and leave the details to the reader. We find that the components x_1, x_2, x_3 of the characteristic vectors corresponding to $\lambda = 3$ must be a nontrivial solution of the system

$$4x_1 - x_2 + 6x_3 = 0,$$
$$-10x_1 + x_2 - 12x_3 = 0,$$
$$-2x_1 + x_2 - 4x_3 = 0.$$

From these we find that

$$x_2 = -2x_1 \quad \text{and} \quad x_3 = -x_1,$$

and hence $x_1 = k$, $x_2 = -2k$, $x_3 = -k$ is a solution for every real k. Hence the characteristic vectors corresponding to the characteristic value $\lambda = 3$ are the vectors

$$\mathbf{x} = \begin{pmatrix} k \\ -2k \\ -k \end{pmatrix},$$

where k is an arbitrary nonzero number. In particular, letting $k = 1$, we obtain the particular characteristic vector

$$\begin{pmatrix} 1 \\ -2 \\ -1 \end{pmatrix}$$

corresponding to the characteristic value $\lambda = 3$.

Finally, we proceed to find the characteristic vectors corresponding to $\lambda = 5$. We find that the components x_1, x_2, x_3 of these vectors must be a nontrivial solution of the system

$$\begin{aligned} 2x_1 - x_2 + 6x_3 &= 0, \\ -10x_1 - x_2 - 12x_3 &= 0, \\ -2x_1 + x_2 - 6x_3 &= 0. \end{aligned}$$

From these we find that

$$x_2 = -2x_1 \quad \text{and} \quad 3x_3 = -2x_1.$$

We find that $x_1 = 3k$, $x_2 = -6k$, $x_3 = -2k$ satisfies this for every real k. Hence the characteristic vectors corresponding to the characteristic value $\lambda = 5$ are the vectors

$$\mathbf{x} = \begin{pmatrix} 3k \\ -6k \\ -2k \end{pmatrix},$$

where k is an arbitrary nonzero number. In particular, letting $k = 1$, we obtain the particular characteristic vector

$$\begin{pmatrix} 3 \\ -6 \\ -2 \end{pmatrix}$$

corresponding to the characteristic value $\lambda = 5$.

Exercises

In each of Exercises 1 through 6 find all the characteristic values and vectors of the matrix.

1. $\begin{pmatrix} 1 & 2 \\ 3 & 2 \end{pmatrix}$.

2. $\begin{pmatrix} 3 & 2 \\ 6 & -1 \end{pmatrix}$.

3. $\begin{pmatrix} 1 & 1 & -1 \\ 2 & 3 & -4 \\ 4 & 1 & -4 \end{pmatrix}$.

4. $\begin{pmatrix} 1 & -1 & -1 \\ 1 & 3 & 1 \\ -3 & -6 & 6 \end{pmatrix}$.

5. $\begin{pmatrix} 1 & -1 & -1 \\ 1 & 3 & 1 \\ -3 & 1 & -1 \end{pmatrix}$.

6. $\begin{pmatrix} 1 & 1 & 0 \\ 1 & 0 & 1 \\ 0 & 1 & 1 \end{pmatrix}$.

7.6 Basic Theory of Linear Systems in Normal Form: *n* Equations in *n* Unknown Functions

A. *Introduction*

We consider the normal form of linear system of n first-order differential equations in n unknown functions $x_1, x_2, \ldots, x_n$. As noted in Section 7.1A, this system is of the form

$$\begin{aligned}
\frac{dx_1}{dt} &= a_{11}(t)x_1 + a_{12}(t)x_2 + \cdots + a_{1n}(t)x_n + F_1(t),\\
\frac{dx_2}{dt} &= a_{21}(t)x_1 + a_{22}(t)x_2 + \cdots + a_{2n}(t)x_n + F_2(t),\\
&\ \ \vdots\\
\frac{dx_n}{dt} &= a_{n1}(t)x_1 + a_{n2}(t)x_2 + \cdots + a_{nn}(t)x_n + F_n(t).
\end{aligned} \tag{7.101}$$

We shall assume that all of the functions defined by $a_{ij}(t)$, $i = 1, 2, \ldots, n$, $j = 1, 2, \ldots, n$, and $F_i(t)$, $i = 1, 2, \ldots, n$, are continuous on a real interval $a \le t \le b$. If all $F_i(t) = 0$, $i = 1, 2, \ldots, n$, for all t, then the system (7.101) is called *homogeneous*. Otherwise, the system is called *nonhomogeneous.*

▶ Example 7.26. The system

$$\begin{aligned}
\frac{dx_1}{dt} &= 7x_1 - x_2 + 6x_3,\\
\frac{dx_2}{dt} &= -10x_1 + 4x_2 - 12x_3,\\
\frac{dx_3}{dt} &= -2x_1 + x_2 - x_3,
\end{aligned} \tag{7.102}$$

is a homogeneous linear system of the type (7.101) with $n = 3$ and having constant coefficients. The system

$$\begin{aligned}
\frac{dx_1}{dt} &= 7x_1 - x_2 + 6x_3 - 5t - 6,\\
\frac{dx_2}{dt} &= -10x_1 + 4x_2 - 12x_3 - 4t + 23,\\
\frac{dx_3}{dt} &= -2x_1 + x_2 - x_3 + 2,
\end{aligned} \tag{7.103}$$

is a nonhomogeneous linear system of the type (7.101) with $n = 3$, the nonhomogeneous terms being $-5t - 6$, $-4t + 23$, and 2, respectively.

We note that the system (7.101) can be written more compactly as

$$\frac{dx_i}{dt} = \sum_{j=1}^{n} a_{ij}(t)x_j + F_i(t) \qquad (i = 1, 2, \ldots, n).$$

We shall now proceed to express the system in an even more compact manner using vectors and matrices. We introduce the matrix **A** defined by

$$\mathbf{A}(t) = \begin{pmatrix} a_{11}(t) & a_{12}(t) & \cdots & a_{1n}(t) \\ a_{21}(t) & a_{22}(t) & \cdots & a_{2n}(t) \\ \vdots & \vdots & & \vdots \\ a_{n1}(t) & a_{n2}(t) & \cdots & a_{nn}(t) \end{pmatrix} \tag{7.104}$$

and the vectors **F** and **x** defined respectively by

$$\mathbf{F}(t) = \begin{pmatrix} F_1(t) \\ F_2(t) \\ \vdots \\ F_n(t) \end{pmatrix} \quad \text{and} \quad \mathbf{x} = \begin{pmatrix} x_1 \\ x_2 \\ \vdots \\ x_n \end{pmatrix}. \tag{7.105}$$

Then (1) by definition of the derivative of a vector, and (2) by multiplication of a matrix by a vector followed by addition of vectors, we have respectively

$$\frac{d\mathbf{x}}{dt} = \begin{pmatrix} \dfrac{dx_1}{dt} \\ \dfrac{dx_2}{dt} \\ \vdots \\ \dfrac{dx_n}{dt} \end{pmatrix}$$

and

$$\mathbf{A}(t)\mathbf{x} + \mathbf{F}(t) = \begin{pmatrix} a_{11}(t) & a_{12}(t) & \cdots & a_{1n}(t) \\ a_{21}(t) & a_{22}(t) & \cdots & a_{2n}(t) \\ \vdots & \vdots & & \vdots \\ a_{n1}(t) & a_{n2}(t) & \cdots & a_{nn}(t) \end{pmatrix} \begin{pmatrix} x_1 \\ x_2 \\ \vdots \\ x_n \end{pmatrix} + \begin{pmatrix} F_1(t) \\ F_2(t) \\ \vdots \\ F_n(t) \end{pmatrix}$$

$$= \begin{pmatrix} a_{11}(t)x_1 + a_{12}(t)x_2 + \cdots + a_{1n}(t)x_n + F_1(t) \\ a_{21}(t)x_1 + a_{22}(t)x_2 + \cdots + a_{2n}(t)x_n + F_2(t) \\ \vdots \\ a_{n1}(t)x_1 + a_{n2}(t)x_2 + \cdots + a_{nn}(t)x_n + F_n(t) \end{pmatrix}.$$

Comparing the components of $d\mathbf{x}/dt$ with the left members of (7.101) and the components of $\mathbf{A}(t)\mathbf{x} + \mathbf{F}(t)$ with the right members of (7.101), we see that system (7.101) can be expressed as the linear *vector* differential equation

$$\frac{d\mathbf{x}}{dt} = \mathbf{A}(t)\mathbf{x} + \mathbf{F}(t). \tag{7.106}$$

Conversely, if $\mathbf{A}(t)$ is given by (7.104) and $\mathbf{F}(t)$ and $\mathbf{x}$ are given by (7.105), then we see that the vector differential equation (7.106) can be expressed as the system (7.101). Thus the system (7.101) and the vector differential equation (7.106) both express the same relations and so are equivalent to one another. We refer to (7.106) as the *vector differential equation corresponding to the system* (7.101), and we shall sometimes call the system (7.101) the *scalar form of the vector differential equation* (7.106). Henceforth throughout this section, we shall usually write the system (7.101) as the corresponding vector differential equation (7.106).

▶ Example 7.27. The vector differential equation corresponding to the nonhomogeneous system (7.103) of Example 7.26 is

$$\frac{d\mathbf{x}}{dt} = \mathbf{A}(t)\mathbf{x} + \mathbf{F}(t),$$

where

$$\mathbf{A}(t) = \begin{pmatrix} 7 & -1 & 6 \\ -10 & 4 & -12 \\ -2 & 1 & -1 \end{pmatrix}, \quad \mathbf{x} = \begin{pmatrix} x_1 \\ x_2 \\ x_3 \end{pmatrix}, \quad \text{and} \quad \mathbf{F}(t) = \begin{pmatrix} -5t - 6 \\ -4t + 23 \\ 2 \end{pmatrix}.$$

Thus we can write this vector differential equation as

$$\frac{d\mathbf{x}}{dt} = \begin{pmatrix} 7 & -1 & 6 \\ -10 & 4 & -12 \\ -2 & 1 & -1 \end{pmatrix} \mathbf{x} + \begin{pmatrix} -5t - 6 \\ -4t + 23 \\ 2 \end{pmatrix},$$

where $\mathbf{x}$ is the vector with components x_1, x_2, x_3, as given above.

DEFINITION

By a solution *of the vector differential equation (7.106) we mean an* $n \times 1$ *column vector function*

$$\boldsymbol{\phi} = \begin{pmatrix} \phi_1 \\ \phi_2 \\ \vdots \\ \phi_n \end{pmatrix}, \tag{7.107}$$

whose components $\phi_1, \phi_2, \ldots, \phi_n$ *each have a continuous derivative on the real interval* $a \le t \le b$, *which is such that*

$$\frac{d\boldsymbol{\phi}(t)}{dt} = \mathbf{A}(t)\boldsymbol{\phi}(t) + \mathbf{F}(t) \tag{7.108}$$

for all t *such that* $a \le t \le b$. *In other words,* $\mathbf{x} = \boldsymbol{\phi}(t)$ *satisfies the vector differential equation (7.106) identically on* $a \le t \le b$. *That is, the components* $\phi_1, \phi_2, \ldots, \phi_n$ *of* $\boldsymbol{\phi}$ *are such that*

$$\begin{aligned} x_1 &= \phi_1(t), \\ x_2 &= \phi_2(t), \\ &\vdots \\ x_n &= \phi_n(t), \end{aligned} \tag{7.109}$$

simultaneously satisfy all n *equations of the scalar form (7.101) of the vector differential equation (7.106) for* $a \le t \le b$. *Hence we say that a* solution *of the system (7.101) is an ordered set of n real functions* $\phi_1, \phi_2, \ldots, \phi_n$, *each having continuous derivatives on* $a \le t \le b$, *such that*

$$\begin{aligned} x_1 &= \phi_1(t), \\ x_2 &= \phi_2(t), \\ &\vdots \\ x_n &= \phi_n(t), \end{aligned} \tag{7.109}$$

simultaneously satisfy all n equations of the system (7.101) for $a \le t \le b$.

▶ Example 7.28. The vector differential equation corresponding to the homogeneous linear system

$$\begin{aligned} \frac{dx_1}{dt} &= 7x_1 - x_2 + 6x_3, \\ \frac{dx_2}{dt} &= -10x_1 + 4x_2 - 12x_3, \\ \frac{dx_3}{dt} &= -2x_1 + x_2 - x_3, \end{aligned} \tag{7.102}$$

is

$$\frac{d\mathbf{x}}{dt} = \begin{pmatrix} 7 & -1 & 6 \\ -10 & 4 & -12 \\ -2 & 1 & -1 \end{pmatrix} \mathbf{x}, \quad \text{where } \mathbf{x} = \begin{pmatrix} x_1 \\ x_2 \\ x_3 \end{pmatrix}. \tag{7.110}$$

The column vector function $\boldsymbol{\phi}$ defined by

$$\boldsymbol{\phi}(t) = \begin{pmatrix} e^{3t} \\ -2e^{3t} \\ -e^{3t} \end{pmatrix}$$

is a solution of the vector differential equation (7.110) on every real interval $a \le t \le b$; for $\mathbf{x} = \boldsymbol{\phi}(t)$ satisfies (7.110) identically on $a \le t \le b$, that is,

$$\begin{pmatrix} 3e^{3t} \\ -6e^{3t} \\ -3e^{3t} \end{pmatrix} = \begin{pmatrix} 7 & -1 & 6 \\ -10 & 4 & -12 \\ -2 & 1 & -1 \end{pmatrix} \begin{pmatrix} e^{3t} \\ -2e^{3t} \\ -e^{3t} \end{pmatrix}.$$

Thus

$$\begin{aligned} x_1 &= e^{3t}, \\ x_2 &= -2e^{3t}, \\ x_3 &= -e^{3t}, \end{aligned} \tag{7.111}$$

simultaneously satisfy all three equations of the system (7.102) for $a \le t \le b$, and so we call (7.111) a solution of the system.

Theorem 7.10 is the basic existence and uniqueness theorem dealing with the vector differential equation (7.106). The statement and proof of this theorem, expressed in the scalar form, are outlined in Chapter 11 of the author's *Differential Equations*.

THEOREM 7.10
Consider the vector differential equation

$$\frac{d\mathbf{x}}{dt} = \mathbf{A}(t)\mathbf{x} + \mathbf{F}(t) \tag{7.106}$$

corresponding to the linear system (7.101) of n equations in n unknown functions. Let the components $a_{ij}(t)$, $i = 1, 2, \ldots, n$, $j = 1, 2, \ldots, n$, *of the matrix* $\mathbf{A}(t)$ *and the components* $F_i(t)$, $i = 1, 2, \ldots, n$, *of the vector* $\mathbf{F}(t)$ *all be continuous on the real interval* $a \le t \le b$. *Let* t_0 *be any point of the interval* $a \le t \le b$, *and let*

$$\mathbf{c} = \begin{pmatrix} c_1 \\ c_2 \\ \vdots \\ c_n \end{pmatrix}$$

be an $n \times 1$ *column vector of any n numbers* $c_1, c_2, \ldots, c_n$.
Then there exists a unique solution

$$\boldsymbol{\phi} = \begin{pmatrix} \phi_1 \\ \phi_2 \\ \vdots \\ \phi_n \end{pmatrix}$$

of the vector differential equation (7.106) such that

$$\boldsymbol{\phi}(t_0) = \mathbf{c}; \tag{7.112}$$

that is,

$$\begin{aligned} \phi_1(t_0) &= c_1, \\ \phi_2(t_0) &= c_2, \\ &\vdots \\ \phi_n(t_0) &= c_n, \end{aligned} \tag{7.113}$$

and this solution is defined on the entire interval $a \le t \le b$.

Interpreting this theorem in terms of the scalar form of the vector differential equation (7.106), that is, the system (7.101), we state the following: Under the stated continuity hypotheses on the functions a_{ij} and F_i, given any point t_0 in the interval $a \le t \le b$ and any n numbers $c_1, c_2, \ldots, c_n$, then there exists a unique solution

$$\begin{aligned} x_1 &= \phi_1(t), \\ x_2 &= \phi_2(t), \\ &\vdots \\ x_n &= \phi_n(t), \end{aligned}$$

such that

$$\begin{aligned} \phi_1(t_0) &= c_1, \\ \phi_2(t_0) &= c_2, \\ &\vdots \\ \phi_n(t_0) &= c_n, \end{aligned} \tag{7.113}$$

and this solution is defined for all t such that $a \le t \le b$.

B. Homogeneous Linear Systems

We now assume that all $F_i(t) = 0$, $i = 1, 2, \ldots, n$, for all t in the linear system (7.101) and consider the resulting *homogeneous* linear system

$$\begin{aligned}\frac{dx_1}{dt} &= a_{11}(t)x_1 + a_{12}(t)x_2 + \cdots + a_{1n}(t)x_n,\\ \frac{dx_2}{dt} &= a_{21}(t)x_1 + a_{22}(t)x_2 + \cdots + a_{2n}(t)x_n,\\ &\vdots\\ \frac{dx_n}{dt} &= a_{n1}(t)x_1 + a_{n2}(t)x_2 + \cdots + a_{nn}(t)x_n.\end{aligned} \tag{7.114}$$

The corresponding homogeneous vector equation is the equation of the form (7.106) for which $\mathbf{F}(t) = 0$ for all t and hence is

$$\frac{d\mathbf{x}}{dt} = \mathbf{A}(t)\mathbf{x}. \tag{7.115}$$

Throughout the remainder of Section 7.6 we shall always make the following assumption whenever we write or refer to the homogeneous vector differential equations (7.115): We shall assume that (7.115) is the vector differential equation corresponding to the homogeneous linear system (7.114) of n equations in n unknown functions and that the components $a_{ij}(t)$, $i = 1, 2, \ldots, n$, $j = 1, 2, \ldots, n$, of the $n \times n$ matrix $\mathbf{A}(t)$ are all continuous on the real interval $a \leq t \leq b$. Our first result concerning Equation (7.115) is an immediate consequence of Theorem 7.10.

COROLLARY TO THEOREM 7.10

Consider the homogeneous vector differential equation

$$\frac{d\mathbf{x}}{dt} = \mathbf{A}(t)\mathbf{x}. \tag{7.115}$$

Let t_0 be any point of $a \leq t \leq b$; and let

$$\boldsymbol{\phi} = \begin{pmatrix}\phi_1\\ \phi_2\\ \vdots\\ \phi_n\end{pmatrix}$$

be a solution of (7.115) such that $\boldsymbol{\phi}(t_0) = \mathbf{0}$, that is, such that

$$\phi_1(t_0) = \phi_2(t_0) = \cdots = \phi_n(t_0) = 0. \tag{7.116}$$

Then $\boldsymbol{\phi}(t) = \mathbf{0}$ for all t on $a \leq t \leq b$; that is,

$$\phi_1(t) = \phi_2(t) = \cdots = \phi_n(t) = 0$$

for all t on $a \leq t \leq b$.

Proof. Obviously $\boldsymbol{\phi}$ defined by $\boldsymbol{\phi}(t) = \mathbf{0}$ for *all* t on $a \leq t \leq b$ is *a* solution of the vector differential equation (7.115) which satisfies conditions (7.116). These conditions are of the form (7.113), where $c_1 = c_2 = \cdots = c_n = 0$; and by Theorem 7.10, there is a *unique* solution of the differential equation satisfying such a set of conditions. Thus $\boldsymbol{\phi}$ such that $\boldsymbol{\phi}(t) = \mathbf{0}$ for *all* t on $a \leq t \leq b$ is the *only* solution of (7.115) such that $\boldsymbol{\phi}(t_0) = \mathbf{0}$. *Q.E.D.*

THEOREM 7.11

A linear combination of m solutions of the homogeneous vector differential equation

$$\frac{d\mathbf{x}}{dt} = \mathbf{A}(t)\mathbf{x} \tag{7.115}$$

is also a solution of (7.115). That is, if the vector functions $\boldsymbol{\phi}_1, \boldsymbol{\phi}_2, \ldots, \boldsymbol{\phi}_m$ are m solutions of (7.115) and $c_1, c_2, \ldots, c_m$ are m numbers, then the vector function

$$\boldsymbol{\phi} = \sum_{k=1}^{m} c_k \boldsymbol{\phi}_k$$

is also a solution of (7.115).

Proof. We have

$$\frac{d}{dt}\left[\sum_{k=1}^{m} c_k \boldsymbol{\phi}_k(t)\right] = \sum_{k=1}^{m}\left[\frac{d}{dt} c_k \boldsymbol{\phi}_k(t)\right] = \sum_{k=1}^{m} c_k \left[\frac{d\boldsymbol{\phi}_k(t)}{dt}\right].$$

Now since each $\boldsymbol{\phi}_k$ is a solution of (7.115),

$$\frac{d\boldsymbol{\phi}_k(t)}{dt} = \mathbf{A}(t)\boldsymbol{\phi}_k(t) \quad \text{for } k = 1, 2, \ldots, m.$$

Thus we have

$$\frac{d}{dt}\left[\sum_{k=1}^{m} c_k \boldsymbol{\phi}_k(t)\right] = \sum_{k=1}^{m} c_k \mathbf{A}(t)\boldsymbol{\phi}_k(t).$$

We now use Results A and B of Section 7.5A. First applying Result B to each term in the right member above, and then applying Result A $(m - 1)$ times, we obtain

$$\sum_{k=1}^{m} c_k \mathbf{A}(t)\boldsymbol{\phi}_k(t) = \sum_{k=1}^{m} \mathbf{A}(t)[c_k \boldsymbol{\phi}_k(t)] = \mathbf{A}(t) \sum_{k=1}^{m} c_k \boldsymbol{\phi}_k(t).$$

Thus we have

$$\frac{d}{dt}\left[\sum_{k=1}^{m} c_k \boldsymbol{\phi}_k(t)\right] = \mathbf{A}(t)\left[\sum_{k=1}^{m} c_k \boldsymbol{\phi}_k(t)\right];$$

that is,

$$\frac{d\boldsymbol{\phi}(t)}{dt} = \mathbf{A}(t)\boldsymbol{\phi}(t),$$

for all t on $a \leq t \leq b$. Thus the linear combination

$$\boldsymbol{\phi} = \sum_{k=1}^{m} c_k \boldsymbol{\phi}_k$$

is a solution of (7.115). *Q.E.D.*

Before proceeding, the student should return to Section 7.5B and review the concepts of linear dependence and linear independence of vector functions.

In each of the next four theorems we shall be concerned with n vector functions, and we shall use the following common notation for the n vector functions of each of these theorems. We let $\boldsymbol{\phi}_1, \boldsymbol{\phi}_2, \ldots, \boldsymbol{\phi}_n$ be the n vector functions defined respectively by

$$\boldsymbol{\phi}_1(t) = \begin{pmatrix} \phi_{11}(t) \\ \phi_{21}(t) \\ \vdots \\ \phi_{n1}(t) \end{pmatrix}, \quad \boldsymbol{\phi}_2(t) = \begin{pmatrix} \phi_{12}(t) \\ \phi_{22}(t) \\ \vdots \\ \phi_{n2}(t) \end{pmatrix}, \ldots, \boldsymbol{\phi}_n(t) = \begin{pmatrix} \phi_{1n}(t) \\ \phi_{2n}(t) \\ \vdots \\ \phi_{nn}(t) \end{pmatrix}. \tag{7.117}$$

Carefully observe the notation scheme. For each vector, the first subscript of a component indicates the row of that component in the vector, whereas the second subscript indicates the vector of which the component is an element. For instance, ϕ_{35} would be the component occupying the third row of the vector $\boldsymbol{\phi}_5$.

DEFINITION

The $n \times n$ determinant

$$\begin{vmatrix} \phi_{11} & \phi_{12} & \cdots & \phi_{1n} \\ \phi_{21} & \phi_{22} & \cdots & \phi_{2n} \\ \vdots & \vdots & & \vdots \\ \phi_{n1} & \phi_{n2} & \cdots & \phi_{nn} \end{vmatrix} \tag{7.118}$$

is called the Wronskian *of the n vector functions $\boldsymbol{\phi}_1, \boldsymbol{\phi}_2, \ldots, \boldsymbol{\phi}_n$ defined by (7.117). We will denote it by $W(\boldsymbol{\phi}_1, \boldsymbol{\phi}_2, \ldots, \boldsymbol{\phi}_n)$ and its value at t by $W(\boldsymbol{\phi}_1, \boldsymbol{\phi}_2, \ldots, \boldsymbol{\phi}_n)(t)$.*

THEOREM 7.12

If the n vector functions $\boldsymbol{\phi}_1, \boldsymbol{\phi}_2, \ldots, \boldsymbol{\phi}_n$ defined by (7.117) are linearly dependent on $a \le t \le b$, then their Wronskian $W(\boldsymbol{\phi}_1, \boldsymbol{\phi}_2, \ldots, \boldsymbol{\phi}_n)(t)$ equals zero for all t on $a \le t \le b$.

Proof. We begin by employing the definition of linear dependence of vector functions on an interval: Since $\boldsymbol{\phi}_1, \boldsymbol{\phi}_2, \ldots, \boldsymbol{\phi}_n$ are linearly dependent on the interval $a \le t \le b$, there exist n numbers $c_1, c_2, \ldots, c_n$, not all zero, such that

$$c_1\boldsymbol{\phi}_1(t) + c_2\boldsymbol{\phi}_2(t) + \cdots + c_n\boldsymbol{\phi}_n(t) = \mathbf{0}$$

for all $t \in [a, b]$. Now using the definition (7.117) of $\boldsymbol{\phi}_1, \boldsymbol{\phi}_2, \ldots, \boldsymbol{\phi}_n$, and writing the preceding vector relation in the form of the n equivalent relations involving corresponding components, we have

$$\begin{aligned} c_1\phi_{11}(t) + c_2\phi_{12}(t) + \cdots + c_n\phi_{1n}(t) &= 0, \\ c_1\phi_{21}(t) + c_2\phi_{22}(t) + \cdots + c_n\phi_{2n}(t) &- 0, \\ &\vdots \\ c_1\phi_{n1}(t) + c_2\phi_{n2}(t) + \cdots + c_n\phi_{nn}(t) &= 0, \end{aligned}$$

for all $t \in [a, b]$. Thus, in particular, these must hold at an *arbitrary* point $t_0 \in [a, b]$.

Thus, letting $t = t_0$ in the preceding n relations, we obtain the homogeneous linear algebraic system

$$\begin{aligned}
\phi_{11}(t_0)c_1 + \phi_{12}(t_0)c_2 + \cdots + \phi_{1n}(t_0)c_n &= 0, \\
\phi_{21}(t_0)c_1 + \phi_{22}(t_0)c_2 + \cdots + \phi_{2n}(t_0)c_n &= 0, \\
&\vdots \\
\phi_{n1}(t_0)c_1 + \phi_{n2}(t_0)c_2 + \cdots + \phi_{nn}(t_0)c_n &= 0,
\end{aligned}$$

in the n unknowns $c_1, c_2, \ldots, c_n$. Since $c_1, c_2, \ldots, c_n$ are not all zero, the determinant of coefficients of the preceding system must be zero, by Theorem A of Section 7.5B. That is, we must have

$$\begin{vmatrix} \phi_{11}(t_0) & \phi_{12}(t_0) & \cdots & \phi_{1n}(t_0) \\ \phi_{21}(t_0) & \phi_{22}(t_0) & \cdots & \phi_{2n}(t_0) \\ \vdots & \vdots & & \vdots \\ \phi_{n1}(t_0) & \phi_{n2}(t_0) & \cdots & \phi_{nn}(t_0) \end{vmatrix} = 0.$$

But the left member of this is the Wronskian $W(\phi_1, \phi_2, \ldots, \phi_n)(t_0)$. Thus we have

$$W(\phi_1, \phi_2, \ldots, \phi_n)(t_0) = 0.$$

Since t_0 is an arbitrary point of $[a, b]$, we must have

$$W(\phi_1, \phi_2, \ldots, \phi_n)(t) = 0$$

for *all* t on $a \leq t \leq b$. *Q.E.D.*

Examine the proof of Theorem 7.12 and observe that it makes absolutely no use of the properties of solutions of differential equations. Thus it holds for arbitrary vector functions, whether they are solutions of a vector differential equation of the form (7.115) or not.

▶ **Example 7.29.** In Example 7.23 of Section 7.5B we saw that the three vector functions ϕ_1, ϕ_2, and ϕ_3 defined respectively by

$$\phi_1(t) = \begin{pmatrix} e^{2t} \\ 2e^{2t} \\ 5e^{2t} \end{pmatrix}, \quad \phi_2(t) = \begin{pmatrix} e^{2t} \\ 4e^{2t} \\ 11e^{2t} \end{pmatrix}, \quad \text{and} \quad \phi_3(t) = \begin{pmatrix} e^{2t} \\ e^{2t} \\ 2e^{2t} \end{pmatrix}$$

are linearly dependent on any interval $a \leq t \leq b$. Therefore, by Theorem 7.12, their Wronskian must equal zero for all t on $a \leq t \leq b$. Indeed, we find

$$W(\phi_1, \phi_2, \phi_3)(t) = \begin{vmatrix} e^{2t} & e^{2t} & e^{2t} \\ 2e^{2t} & 4e^{2t} & e^{2t} \\ 5e^{2t} & 11e^{2t} & 2e^{2t} \end{vmatrix} = 0 \quad \text{for all } t.$$

THEOREM 7.13

Let the vector functions $\phi_1, \phi_2, \ldots, \phi_n$ defined by (7.117) be n solutions of the homogeneous linear vector differential equation

$$\frac{d\mathbf{x}}{dt} = \mathbf{A}(t)\mathbf{x}. \tag{7.115}$$

If the Wronskian $W(\phi_1, \phi_2, \ldots, \phi_n)(t_0) = 0$ *at some* $t_0 \in [a, b]$, *then* $\phi_1, \phi_2, \ldots, \phi_n$ *are linearly dependent on* $a \le t \le b$.

Proof. Consider the linear algebraic system

$$\begin{aligned} c_1\phi_{11}(t_0) + c_2\phi_{12}(t_0) + \cdots + c_n\phi_{1n}(t_0) &= 0, \\ c_1\phi_{21}(t_0) + c_2\phi_{22}(t_0) + \cdots + c_n\phi_{2n}(t_0) &= 0, \\ &\vdots \\ c_1\phi_{n1}(t_0) + c_2\phi_{n2}(t_0) + \cdots + c_n\phi_{nn}(t_0) &= 0, \end{aligned} \tag{7.119}$$

in the n unknowns $c_1, c_2, \ldots, c_n$. Since the determinant of coefficients is

$$W(\phi_1, \phi_2, \ldots, \phi_n)(t_0) \quad \text{and} \quad W(\phi_1, \phi_2, \ldots, \phi_n)(t_0) = 0$$

by hypothesis, this system has a nontrivial solution by Theorem A of Section 7.5B. That is, there exist numbers $c_1, c_2, \ldots, c_n$, not all zero, which satisfy all n equations of system (7.119). These n equations are the n corresponding component relations equivalent to the one vector relation

$$c_1\boldsymbol{\phi}_1(t_0) + c_2\boldsymbol{\phi}_2(t_0) + \cdots + c_n\boldsymbol{\phi}_n(t_0) = \mathbf{0}. \tag{7.120}$$

Thus there exist numbers $c_1, c_2, \ldots, c_n$, not all zero, such that (7.120) holds.

Now consider the vector function $\boldsymbol{\phi}$ defined by

$$\boldsymbol{\phi}(t) = c_1\boldsymbol{\phi}_1(t) + c_2\boldsymbol{\phi}_2(t) + \cdots + c_n\boldsymbol{\phi}_n(t) \tag{7.121}$$

for all $t \in [a, b]$. Since $\boldsymbol{\phi}_1, \boldsymbol{\phi}_2, \ldots, \boldsymbol{\phi}_n$ are solutions of the differential equation (7.115), by Theorem 7.11, the linear combination $\boldsymbol{\phi}$ defined by (7.121) is also a solution of (7.115). Now from (7.120), we see that this solution $\boldsymbol{\phi}$ is such that $\boldsymbol{\phi}(t_0) = \mathbf{0}$. Thus by the corollary to Theorem 7.10, we must have $\boldsymbol{\phi}(t) = \mathbf{0}$ for *all* $t \in [a, b]$. That is, using the definition (7.121),

$$c_1\boldsymbol{\phi}_1(t) + c_2\boldsymbol{\phi}_2(t) + \cdots + c_n\boldsymbol{\phi}_n(t) = \mathbf{0}$$

for *all* $t \in [a, b]$, where $c_1, c_2, \ldots, c_n$ are not all zero. Thus, by definition, $\boldsymbol{\phi}_1, \boldsymbol{\phi}_2, \ldots, \boldsymbol{\phi}_n$ are linearly dependent on $a \le t \le b$. *Q.E.D.*

▶ Example 7.30. Consider the vector functions $\boldsymbol{\phi}_1, \boldsymbol{\phi}_2$, and $\boldsymbol{\phi}_3$ defined respectively by

$$\boldsymbol{\phi}_1(t) = \begin{pmatrix} e^{3t} \\ -2e^{3t} \\ -e^{3t} \end{pmatrix}, \quad \boldsymbol{\phi}_2(t) = \begin{pmatrix} 2e^{3t} \\ -4e^{3t} \\ -2e^{3t} \end{pmatrix}, \quad \text{and} \quad \boldsymbol{\phi}_3(t) = \begin{pmatrix} -3e^{3t} \\ 6e^{3t} \\ 3e^{3t} \end{pmatrix}.$$

It is easy to verify that $\boldsymbol{\phi}_1, \boldsymbol{\phi}_2$, and $\boldsymbol{\phi}_3$ are all solutions of the homogeneous linear vector differential equation

$$\frac{d\mathbf{x}}{dt} = \begin{pmatrix} 7 & -1 & 6 \\ -10 & 4 & -12 \\ -2 & 1 & -1 \end{pmatrix} \mathbf{x}, \quad \text{where} \quad \mathbf{x} = \begin{pmatrix} x_1 \\ x_2 \\ x_3 \end{pmatrix} \tag{7.110}$$

on every real interval $a \leq t \leq b$ (see Example 7.28). Thus, in particular, ϕ_1, ϕ_2, and ϕ_3 are solutions of (7.110) on every interval $[a, b]$ containing $t_0 = 0$. It is easy to see that

$$W(\phi_1, \phi_2, \phi_3)(0) = \begin{vmatrix} 1 & 2 & -3 \\ -2 & -4 & 6 \\ -1 & -2 & 3 \end{vmatrix} = 0.$$

Thus by Theorem 7.13, ϕ_1, ϕ_2, and ϕ_3 are linearly dependent on every $[a, b]$ containing 0. Indeed, note that

$$\phi_1(t) + \phi_2(t) + \phi_3(t) = \mathbf{0}$$

for all t on every interval $[a, b]$, and recall the definition of linear dependence.

Note. Theorem 7.13 is *not* true for vector functions $\phi_1, \phi_2, \ldots, \phi_n$ which are *not solutions* of a homogeneous linear vector differential equation (7.115). For example, consider the vector functions ϕ_1 and ϕ_2 defined respectively by

$$\phi_1(t) = \begin{pmatrix} t \\ 0 \end{pmatrix} \quad \text{and} \quad \phi_2(t) = \begin{pmatrix} t^2 \\ 0 \end{pmatrix}.$$

It can be shown that ϕ_1 and ϕ_2 are *not solutions* of any differential equation of the form (7.115) for which $n = 2$ (see Suggested Reading list at end of this chapter: Petrovski, *Ordinary Differential Equations*, Theorem, pages 110–111, for the method of doing so). Clearly

$$W(\phi_1, \phi_2)(t_0) = \begin{vmatrix} t_0 & t_0^2 \\ 0 & 0 \end{vmatrix} = 0$$

for *all* t_0 in *every* interval $a \leq t \leq b$. However, ϕ_1 and ϕ_2 are *not* linearly dependent. To show this, proceed as in Example 7.24; see Exercise 5 at the end of Section 7.5B.

THEOREM 7.14

Let the vector functions $\phi_1, \phi_2, \ldots, \phi_n$ defined by (7.117) be n solutions of the homogeneous linear vector differential equation

$$\frac{d\mathbf{x}}{dt} = \mathbf{A}(t)\mathbf{x} \tag{7.115}$$

on the real interval $[a, b]$. Then

$$\text{either} \quad W(\phi_1, \phi_2, \ldots, \phi_n)(t) = 0 \quad \text{for all } t \in [a, b],$$

$$\text{or} \quad W(\phi_1, \phi_2, \ldots, \phi_n)(t) = 0 \quad \text{for no } t \in [a, b].$$

Proof.

$$\text{Either} \quad W(\phi_1, \phi_2, \ldots, \phi_n)(t) = 0 \quad \text{for } some \ t \in [a, b]$$

$$\text{or} \quad W(\phi_1, \phi_2, \ldots, \phi_n)(t) = 0 \quad \text{for } no \ t \in [a, b].$$

If $W(\phi_1, \phi_2, \ldots, \phi_n)(t) = 0$ for *some* $t \in [a, b]$, then by Theorem 7.13, the solutions $\phi_1, \phi_2, \ldots, \phi_n$ are linearly dependent on $[a, b]$; and then by Theorem 7.12, $W(\phi_1, \phi_2, \ldots, \phi_n)(t) = 0$ for *all* $t \in [a, b]$. Thus the Wronksian of $\phi_1, \phi_2, \ldots, \phi_n$ *either* equals zero for *all* $t \in [a, b]$ *or* equals zero for *no* $t \in [a, b]$. *Q.E.D.*

THEOREM 7.15

Let the vector functions $\phi_1, \phi_2, \ldots, \phi_n$ defined by (7.117) be n solutions of the homogeneous linear vector differential equation

$$\frac{d\mathbf{x}}{dt} = \mathbf{A}(t)\mathbf{x} \tag{7.115}$$

on the real interval [a, b]. These n solutions $\phi_1, \phi_2, \ldots, \phi_n$ of (7.115) are linearly independent on [a, b] if and only if

$$W(\phi_1, \phi_2, \ldots, \phi_n)(t) \neq 0$$

for all $t \in [a, b]$.

Proof. By Theorems 7.12 and 7.13, the solutions $\phi_1, \phi_2, \ldots, \phi_n$ are linearly *dependent* on $[a, b]$ if and only if $W(\phi_1, \phi_2, \ldots, \phi_n)(t) = 0$ for *all* $t \in [a, b]$. Hence, $\phi_1, \phi_2, \ldots, \phi_n$ are linearly *independent* on $[a, b]$ if and only if $W(\phi_1, \phi_2, \ldots, \phi_n(t_0) \neq 0$ for *some* $t_0 \in [a, b]$. Then by Theorem 7.14, $W(\phi_1, \phi_2, \ldots, \phi_n)(t_0) \neq 0$ for *some* $t_0 \in [a, b]$ if and only if $W(\phi_1, \phi_2, \ldots, \phi_n)(t) \neq 0$ for *all* $t \in [a, b]$. *Q.E.D.*

▶ Example 7.31. Consider the vector functions ϕ_1, ϕ_2, and ϕ_3 defined respectively by

$$\phi_1(t) = \begin{pmatrix} e^{2t} \\ -e^{2t} \\ -e^{2t} \end{pmatrix}, \quad \phi_2(t) = \begin{pmatrix} e^{3t} \\ -2e^{3t} \\ -e^{3t} \end{pmatrix}, \quad \text{and} \quad \phi_3(t) = \begin{pmatrix} 3e^{5t} \\ -6e^{5t} \\ -2e^{5t} \end{pmatrix}. \tag{7.122}$$

It is easy to verify that ϕ_1, ϕ_2, and ϕ_3 are all solutions of the homogeneous linear vector differential equation

$$\frac{d\mathbf{x}}{dt} = \begin{pmatrix} 7 & -1 & 6 \\ -10 & 4 & -12 \\ -2 & 1 & -1 \end{pmatrix} \mathbf{x}, \quad \text{where} \quad \mathbf{x} = \begin{pmatrix} x_1 \\ x_2 \\ x_3 \end{pmatrix}, \tag{7.110}$$

on every real interval $a \leq t \leq b$. We evaluate

$$W(\phi_1, \phi_2, \phi_3)(t) = \begin{vmatrix} e^{2t} & e^{3t} & 3e^{5t} \\ -e^{2t} & -2e^{3t} & -6e^{5t} \\ -e^{2t} & -e^{3t} & -2e^{5t} \end{vmatrix} = -e^{10t} \neq 0$$

for all real t. Thus by Theorem 7.15, the solutions ϕ_1, ϕ_2, and ϕ_3 of (7.110) defined by (7.122) are linearly independent on every real interval $[a, b]$.

DEFINITIONS

Consider the homogeneous linear vector differential equation

$$\frac{d\mathbf{x}}{dt} = \mathbf{A}(t)\mathbf{x}, \tag{7.115}$$

where $\mathbf{x}$ *is an* $n \times 1$ *column vector.*

1. A set of n linearly independent solutions of (7.115) is called a fundamental set of solutions *of (7.115).*

2. *A matrix whose individual columns consist of a fundamental set of solutions of (7.115) is called a* fundamental matrix *of (7.115). That is, if the vector functions ϕ_1, $\phi_2, \ldots, \phi_n$ defined by (7.117) make up a fundamental set of solutions of (7.115), then the $n \times n$ square matrix*

$$\begin{pmatrix} \phi_{11}(t) & \phi_{12}(t) & \cdots & \phi_{1n}(t) \\ \phi_{21}(t) & \phi_{22}(t) & \cdots & \phi_{2n}(t) \\ \vdots & \vdots & & \vdots \\ \phi_{n1}(t) & \phi_{n2}(t) & \cdots & \phi_{nn}(t) \end{pmatrix}$$

is a fundamental matrix of (7.115).

▶ Example 7.32. In Example 7.31 we saw that the three vector functions ϕ_1, ϕ_2, and ϕ_3 defined respectively by

$$\phi_1(t) = \begin{pmatrix} e^{2t} \\ -e^{2t} \\ -e^{2t} \end{pmatrix}, \quad \phi_2(t) = \begin{pmatrix} e^{3t} \\ -2e^{3t} \\ -e^{3t} \end{pmatrix}, \quad \text{and} \quad \phi_3(t) = \begin{pmatrix} 3e^{5t} \\ -6e^{5t} \\ -2e^{5t} \end{pmatrix} \tag{7.122}$$

are linearly independent solutions of the differential equation

$$\frac{d\mathbf{x}}{dt} = \begin{pmatrix} 7 & -1 & 6 \\ -10 & 4 & -12 \\ -2 & 1 & -1 \end{pmatrix} \mathbf{x}, \qquad \text{where } \mathbf{x} = \begin{pmatrix} x_1 \\ x_2 \\ x_3 \end{pmatrix}, \tag{7.110}$$

on every real interval $[a, b]$. Thus these three solutions ϕ_1, ϕ_2, and ϕ_3 form a fundamental set of differential equation (7.110), and a fundamental matrix of the differential equation is

$$\begin{pmatrix} e^{2t} & e^{3t} & 3e^{5t} \\ -e^{2t} & -2e^{3t} & -6e^{5t} \\ -e^{2t} & -e^{3t} & -2e^{5t} \end{pmatrix}.$$

We know that the differential equation (7.110) of Examples 7.31 and 7.32 has the fundamental set of solutions ϕ_1, ϕ_2, ϕ_3 defined by (7.122). We now show that every vector differential equation (7.115) has fundamental sets of solutions.

THEOREM 7.16

There exist fundamental sets of solutions of the homogeneous linear vector differential equation

$$\frac{d\mathbf{x}}{dt} = \mathbf{A}(t)\mathbf{x}. \tag{7.115}$$

Proof. We begin by defining a special set of constant vectors $\mathbf{u}_1, \mathbf{u}_2, \ldots, \mathbf{u}_n$. We define

$$\mathbf{u}_1 = \begin{pmatrix} 1 \\ 0 \\ 0 \\ \vdots \\ 0 \\ 0 \end{pmatrix}, \mathbf{u}_2 = \begin{pmatrix} 0 \\ 1 \\ 0 \\ \vdots \\ 0 \\ 0 \end{pmatrix}, \ldots, \mathbf{u}_n = \begin{pmatrix} 0 \\ 0 \\ 0 \\ \vdots \\ 0 \\ 1 \end{pmatrix}.$$

That is, in general, for each $i = 1, 2, \ldots, n$, $\mathbf{u}_i$ has ith component one and all other components zero. Now let $\phi_1, \phi_2, \ldots, \phi_n$ be the n solutions of (7.115) which satisfy the conditions

$$\phi_i(t_0) = \mathbf{u}_i \qquad (i = 1, 2, \ldots, n),$$

that is,

$$\phi_1(t_0) = \mathbf{u}_1, \phi_2(t_0) = \mathbf{u}_2, \ldots, \phi_n(t_0) = \mathbf{u}_n,$$

where t_0 is an arbitrary (but fixed) point of $[a, b]$. Note that these solutions exist and are unique by Theorem 7.10. We now find

$$W(\phi_1, \phi_2, \ldots, \phi_n)(t_0) = W(\mathbf{u}_1, \mathbf{u}_2, \ldots, \mathbf{u}_n) = \begin{vmatrix} 1 & 0 & \cdots & 0 \\ 0 & 1 & \cdots & 0 \\ \vdots & \vdots & & \vdots \\ 0 & 0 & \cdots & 1 \end{vmatrix} = 1 \neq 0.$$

Then by Theorem 7.14, $W(\phi_1, \phi_2, \ldots, \phi_n)(t) \neq 0$ for all $t \in [a, b]$; and so by Theorem 7.15, solutions $\phi_1, \phi_2, \ldots, \phi_n$ are linearly independent on $[a, b]$. Thus, by definition, $\phi_1, \phi_2, \ldots, \phi_n$ form a fundamental set of differential equation (7.115). *Q.E.D.*

THEOREM 7.17

Let $\phi_1, \phi_2, \ldots, \phi_n$ defined by (7.117) be a fundamental set of solutions of the homogeneous linear vector differential equation

$$\frac{d\mathbf{x}}{dt} = \mathbf{A}(t)\mathbf{x} \tag{7.115}$$

and let ϕ be an arbitrary solution of (7.115) on the real interval $[a, b]$. Then ϕ can be represented as a suitable linear combination of $\phi_1, \phi_2, \ldots, \phi_n$; that is, there exist numbers $c_1, c_2, \ldots, c_n$ such that

$$\phi = c_1\phi_1 + c_2\phi_2 + \cdots + c_n\phi_n$$

on $[a, b]$.

Proof. Suppose $\phi(t_0) = \mathbf{u}_0$, where $t_0 \in [a, b]$ and

$$\mathbf{u}_0 = \begin{pmatrix} u_{10} \\ u_{20} \\ \vdots \\ u_{n0} \end{pmatrix}$$

is a constant vector. Consider the linear algebraic system

$$\begin{aligned} c_1\phi_{11}(t_0) + c_2\phi_{12}(t_0) + \cdots + c_n\phi_{1n}(t_0) &= u_{10}, \\ c_1\phi_{21}(t_0) + c_2\phi_{22}(t_0) + \cdots + c_n\phi_{2n}(t_0) &= u_{20}, \\ &\vdots \\ c_1\phi_{n1}(t_0) + c_2\phi_{n2}(t_0) + \cdots + c_n\phi_{nn}(t_0) &= u_{n0}, \end{aligned} \tag{7.123}$$

of n equations in the n unknowns $c_1, c_2, \ldots, c_n$. Since $\phi_1, \phi_2, \ldots, \phi_n$ is a fundamental set of solutions of (7.115) on $[a, b]$, we know that $\phi_1, \phi_2, \ldots, \phi_n$ are linearly independent solutions on $[a, b]$ and hence by Theorem 7.15, $W(\phi_1, \phi_2, \ldots, \phi_n)(t_0) \neq 0$. Now observe that $W(\phi_1, \phi_2, \ldots, \phi_n)(t_0)$ is the determinant of coefficients of system (7.123), and so this determinant of coefficients is unequal to zero. Thus by Theorem B of Section 7.5B, the system (7.123) has a unique solution for $c_1, c_2, \ldots, c_n$. That is, there exists a unique set of numbers $c_1, c_2, \ldots, c_n$ such that

$$c_1\phi_1(t_0) + c_2\phi_2(t_0) + \cdots + c_n\phi_n(t_0) = \mathbf{u}_0,$$

and hence such that

$$\phi(t_0) = \mathbf{u}_0 = \sum_{k=1}^{n} c_k\phi_k(t_0). \tag{7.124}$$

Now consider the vector function ψ defined by

$$\psi(t) = \sum_{k=1}^{n} c_k\phi_k(t).$$

By Theorem 7.11, the vector function ψ is also a solution of the vector differential equation (7.115). Now note that

$$\psi(t_0) = \sum_{k=1}^{n} c_k\phi_k(t_0).$$

Hence by (7.124), we obtain $\psi(t_0) = \phi(t_0)$. Thus by Theorem 7.10, we must have $\psi(t) = \phi(t)$ for *all* $t \in [a, b]$. That is,

$$\phi(t) = \sum_{k=1}^{n} c_k\phi_k(t)$$

for all $t \in [a, b]$. Thus ϕ is expressed as the linear combination

$$\phi = c_1\phi_1 + c_2\phi_2 + \cdots + c_n\phi_n$$

of $\phi_1, \phi_2, \ldots, \phi_n$, where $c_1, c_2, \ldots, c_n$ is the *unique* solution of system (7.123).

Q.E.D.

As a result of Theorem 7.17, we are led to make the following definition.

DEFINITION

Consider the homogeneous linear vector differential equation

$$\frac{d\mathbf{x}}{dt} = \mathbf{A}(t)\mathbf{x}, \tag{7.115}$$

where $\mathbf{x}$ *is an* $n \times 1$ *column vector. By a* general solution *of (7.115), we mean a solution of the form*

$$c_1\phi_1 + c_2\phi_2 + \cdots + c_n\phi_n,$$

where $c_1, c_2, \ldots, c_n$ *are n arbitrary numbers and* $\phi_1, \phi_2, \ldots, \phi_n$ *is a fundamental set of solutions of (7.115).*

▶ **Example 7.33.** Consider the differential equation

$$\frac{d\mathbf{x}}{dt} = \begin{pmatrix} 7 & -1 & 6 \\ -10 & 4 & -12 \\ -2 & 1 & -1 \end{pmatrix}\mathbf{x}, \quad \text{where } \mathbf{x} = \begin{pmatrix} x_1 \\ x_2 \\ x_3 \end{pmatrix}. \tag{7.110}$$

In Example 7.32 we saw that the three vector functions ϕ_1, ϕ_2, and ϕ_3 defined respectively by

$$\phi_1(t) = \begin{pmatrix} e^{2t} \\ -e^{2t} \\ -e^{2t} \end{pmatrix}, \quad \phi_2(t) = \begin{pmatrix} e^{3t} \\ -2e^{3t} \\ -e^{3t} \end{pmatrix}, \quad \text{and} \quad \phi_3(t) = \begin{pmatrix} 3e^{5t} \\ -6e^{5t} \\ -2e^{5t} \end{pmatrix} \tag{7.122}$$

form a fundamental set of differential equation (7.110). Thus by Theorem 7.17, if ϕ is an arbitrary solution of (7.110), then ϕ can be represented as a suitable linear combination of these three linearly independent solutions ϕ_1, ϕ_2, and ϕ_3 of (7.110). Further, if c_1, c_2, and c_3 are arbitrary numbers, we see from the definition that $c_1\phi_1 + c_2\phi_2 + c_3\phi_3$ is a general solution of (7.110). That is, a general solution of (7.110) is defined by

$$c_1 \begin{pmatrix} e^{2t} \\ -e^{2t} \\ -e^{2t} \end{pmatrix} + c_2 \begin{pmatrix} e^{3t} \\ -2e^{3t} \\ -e^{3t} \end{pmatrix} + c_3 \begin{pmatrix} 3e^{5t} \\ -6e^{5t} \\ -2e^{5t} \end{pmatrix}$$

and can be written as

$$\begin{aligned} x_1 &= c_1e^{2t} + c_2e^{3t} + 3c_3e^{5t}, \\ x_2 &= -c_1e^{2t} - 2c_2e^{3t} - 6c_3e^{5t}, \\ x_3 &= -c_1e^{2t} - c_2e^{3t} - 2c_3e^{5t}, \end{aligned}$$

where c_1, c_2, and c_3 are arbitrary numbers.

C. *Nonhomogeneous Linear Systems*

We return briefly to the nonhomogeneous linear vector differential equation

$$\frac{d\mathbf{x}}{dt} = \mathbf{A}(t)\mathbf{x} + \mathbf{F}(t), \tag{7.106}$$

where $\mathbf{A}(t)$ is given by (7.104) and $\mathbf{F}(t)$ and $\mathbf{x}$ are given by (7.105). We shall see the solutions of this nonhomogeneous equation are closely related to those of the corresponding homogeneous equation

$$\frac{d\mathbf{x}}{dt} = \mathbf{A}(t)\mathbf{x}. \tag{7.115}$$

THEOREM 7.18

Let ϕ_0 be any solution of the nonhomogeneous linear vector differential equation

$$\frac{d\mathbf{x}}{dt} = \mathbf{A}(t)\mathbf{x} + \mathbf{F}(t); \tag{7.106}$$

let $\phi_1, \phi_2, \ldots, \phi_n$ be a fundamental set of solutions of the corresponding homogeneous differential equation

$$\frac{d\mathbf{x}}{dt} = \mathbf{A}(t)\mathbf{x}; \tag{7.115}$$

and let $c_1, c_2, \ldots, c_n$ be n numbers.

Then: (1) the vector function

$$\phi_0 + \sum_{k=1}^{n} c_k\phi_k \tag{7.125}$$

is also a solution of the nonhomogeneous differential equation (7.106) for every choice of $c_1, c_2, \ldots, c_n$*; and*

(2) an arbitrary solution ϕ *of the nonhomogeneous differential equation (7.106) is of the form (7.125) for a suitable choice of* $c_1, c_2, \ldots, c_n$.

Proof. (1) We show that (7.125) satisfies (7.106) for all choices of $c_1, c_2, \ldots, c_n$. We have

$$\frac{d}{dt}\left[\phi_0(t) + \sum_{k=1}^{n} c_k\phi_k(t)\right] = \frac{d\phi_0(t)}{dt} + \frac{d}{dt}\left[\sum_{k=1}^{n} c_k\phi_k(t)\right].$$

Now since ϕ_0 satisfies (7.106), we have

$$\frac{d\phi_0(t)}{dt} = \mathbf{A}(t)\phi_0(t) + \mathbf{F}(t);$$

and since by Theorem 7.11 $\sum_{k=1}^{n} c_k\phi_k$ satisfies (7.115), we also have

$$\frac{d}{dt}\left[\sum_{k=1}^{n} c_k\phi_k(t)\right] = \mathbf{A}(t)\left[\sum_{k=1}^{n} c_k\phi_k(t)\right].$$

Thus

$$\begin{aligned}\frac{d}{dt}\left[\phi_0(t) + \sum_{k=1}^{n} c_k\phi_k(t)\right] &= \mathbf{A}(t)\phi_0(t) + \mathbf{F}(t) + \mathbf{A}(t)\left[\sum_{k=1}^{n} c_k\phi_k(t)\right] \\ &= \mathbf{A}(t)\left[\phi_0(t) + \sum_{k=1}^{n} c_k\phi_k(t)\right] + \mathbf{F}(t).\end{aligned}$$

That is,

$$\frac{d\psi(t)}{dt} = \mathbf{A}(t)\psi(t) + \mathbf{F}(t)$$

where

$$\psi = \phi_0 + \sum_{k=1}^{n} c_k\phi_k;$$

and so

$$\psi = \phi_0 + \sum_{k=1}^{n} c_k\phi_k$$

is a solution of (7.106) for every choice of $c_1, c_2, \ldots, c_n$.

(2) Now consider an arbitrary solution ϕ of (7.106), and evaluate the derivative of the difference $\phi - \phi_0$. We have

$$\frac{d}{dt}\left[\phi(t) - \phi_0(t)\right] = \frac{d\phi(t)}{dt} - \frac{d\phi_0(t)}{dt}.$$

Since both ϕ and ϕ_0 satisfy (7.106), we have respectively

$$\frac{d\phi(t)}{dt} = \mathbf{A}(t)\phi(t) + \mathbf{F}(t),$$

$$\frac{d\phi_0(t)}{dt} = \mathbf{A}(t)\phi_0(t) + \mathbf{F}(t).$$

Thus we obtain

$$\frac{d}{dt}[\phi(t) - \phi_0(t)] = [\mathbf{A}(t)\phi(t) + \mathbf{F}(t)] - [\mathbf{A}(t)\phi_0(t) + \mathbf{F}(t)],$$

which at once reduces to

$$\frac{d}{dt}[\phi(t) - \phi_0(t)] = \mathbf{A}(t)[\phi(t) - \phi_0(t)].$$

Thus $\phi - \phi_0$ satisfies the *homogeneous* differential equation (7.115). Hence by Theorem 7.17, there exist a suitable choice of numbers $c_1, c_2, \ldots, c_n$ such that

$$\phi - \phi_0 = \sum_{k=1}^{n} c_k\phi_k.$$

Thus the arbitrary solution ϕ of (7.106) is of the form

$$\phi = \phi_0 + \sum_{k=1}^{n} c_k\phi_k \tag{7.125}$$

for a suitable choice of $c_1, c_2, \ldots, c_n$. *Q.E.D.*

DEFINITION

Consider the nonhomogeneous linear vector differential equation (7.106) and the corresponding homogeneous linear vector differential equation (7.115). By a general solution *of (7.106), we mean a solution of the form*

$$c_1\phi_1 + c_2\phi_2 + \cdots + c_n\phi_n + \phi_0,$$

where $c_1, c_2, \ldots, c_n$ are n arbitrary numbers, $\phi_1, \phi_2, \ldots, \phi_n$ is a fundamental set of solutions of (7.115), and ϕ_0 is any solution of (7.106).

▶ Example 7.34. Consider the nonhomogeneous differential equation

$$\frac{d\mathbf{x}}{dt} = \begin{pmatrix} 7 & -1 & 6 \\ -10 & 4 & -12 \\ -2 & 1 & -1 \end{pmatrix}\mathbf{x} + \begin{pmatrix} -5t - 6 \\ -4t + 23 \\ 2 \end{pmatrix} \tag{7.103}$$

and the corresponding homogeneous differential equation

$$\frac{d\mathbf{x}}{dt} = \begin{pmatrix} 7 & -1 & 6 \\ -10 & 4 & -12 \\ -2 & 1 & -1 \end{pmatrix}\mathbf{x}, \quad \text{where } \mathbf{x} = \begin{pmatrix} x_1 \\ x_2 \\ x_3 \end{pmatrix}. \tag{7.102}$$

These were introduced in Example 7.26, where they were written out in component form; and (7.102) has been used in Example 7.33 and several other examples as well. In Example 7.33 we observed that ϕ_1, ϕ_2, ϕ_3 defined respectively by

$$\phi_1(t) = \begin{pmatrix} e^{2t} \\ -e^{2t} \\ -e^{2t} \end{pmatrix}, \quad \phi_2(t) = \begin{pmatrix} e^{3t} \\ -2e^{3t} \\ -e^{3t} \end{pmatrix}, \quad \text{and} \quad \phi_3(t) = \begin{pmatrix} 3e^{5t} \\ -6e^{5t} \\ -2e^{5t} \end{pmatrix} \tag{7.122}$$

form a fundamental set of the homogeneous differential equation (7.102) [or (7.110), as it is numbered there]. Now observe that the vector function ϕ_0 defined by

$$\phi_0(t) = \begin{pmatrix} 2t \\ 3t - 2 \\ -t + 1 \end{pmatrix}$$

is a solution of the nonhomogeneous differential equation (7.103). Thus a general solution of (7.103) is given by

$$\mathbf{x} = c_1\phi_1(t) + c_2\phi_2(t) + c_3\phi_3(t) + \phi_0(t),$$

that is,

$$\mathbf{x} = c_1 \begin{pmatrix} e^{2t} \\ -e^{2t} \\ -e^{2t} \end{pmatrix} + c_2 \begin{pmatrix} e^{3t} \\ -2e^{3t} \\ -e^{3t} \end{pmatrix} + c_3 \begin{pmatrix} 3e^{5t} \\ -6e^{5t} \\ -2e^{5t} \end{pmatrix} + \begin{pmatrix} 2t \\ 3t - 2 \\ -t + 1 \end{pmatrix},$$

where c_1, c_2, and c_3 are arbitrary numbers. Thus a general solution of (7.103) can be written as

$$\begin{aligned} x_1 &= c_1e^{2t} + c_2e^{3t} + 3c_3e^{5t} + 2t, \\ x_2 &= -c_1e^{2t} - 2c_2e^{3t} - 6c_3e^{5t} + 3t - 2, \\ x_3 &= -c_1e^{2t} - c_2e^{3t} - 2c_3e^{5t} - t + 1, \end{aligned}$$

where c_1, c_2, and c_3 are arbitrary numbers.

Exercises

In each of Exercises 1 through 6, determine whether or not the matrix in Column I is a fundamental matrix of the corresponding linear system in Column II.

	I	II
1.	$\begin{pmatrix} e^t & e^{2t} & e^{-3t} \\ e^t & 2e^{2t} & 7e^{-3t} \\ e^t & e^{2t} & 11e^{-3t} \end{pmatrix}$,	$\dfrac{d\mathbf{x}}{dt} = \begin{pmatrix} 1 & 1 & -1 \\ 2 & 3 & -4 \\ 4 & 1 & -4 \end{pmatrix} \mathbf{x}.$
2.	$\begin{pmatrix} e^{2t} & e^{3t} & e^{-2t} \\ 0 & -e^{3t} & -e^{-2t} \\ -e^{2t} & -e^{3t} & 4e^{-2t} \end{pmatrix}$,	$\dfrac{d\mathbf{x}}{dt} = \begin{pmatrix} 1 & -1 & -1 \\ 1 & 3 & 1 \\ -3 & 1 & -1 \end{pmatrix} \mathbf{x}.$
3.	$\begin{pmatrix} e^{4t} & 0 & 2e^{4t} \\ 2e^{4t} & 3e^t & 4e^{4t} \\ e^{4t} & e^t & 2e^{4t} \end{pmatrix}$,	$\dfrac{d\mathbf{x}}{dt} = \begin{pmatrix} 1 & -3 & 9 \\ 0 & -5 & 18 \\ 0 & -3 & 10 \end{pmatrix} \mathbf{x}.$

4. $\begin{pmatrix} e^t & 0 & e^{4t} \\ 0 & 3e^t & 2e^{4t} \\ 0 & e^t & e^{4t} \end{pmatrix}$, $\quad \dfrac{d\mathbf{x}}{dt} = \begin{pmatrix} 1 & -3 & 9 \\ 0 & -5 & 18 \\ 0 & -3 & 10 \end{pmatrix} \mathbf{x}.$

5. $\begin{pmatrix} e^t & e^{2t} & e^{2t} \\ e^t & -e^{2t} & 0 \\ 3e^t & 0 & e^{2t} \end{pmatrix}$, $\quad \dfrac{d\mathbf{x}}{dt} = \begin{pmatrix} 3 & 1 & -1 \\ 1 & 3 & -1 \\ 3 & 3 & -1 \end{pmatrix} \mathbf{x}.$

6. $\begin{pmatrix} e^t & te^t & e^{2t} \\ e^t & (t+1)e^t & 2e^{2t} \\ e^t & (t+2)e^t & 4e^{2t} \end{pmatrix}$, $\quad \dfrac{d\mathbf{x}}{dt} = \begin{pmatrix} 0 & 1 & 0 \\ 0 & 0 & 1 \\ 2 & -5 & 4 \end{pmatrix} \mathbf{x}.$

7.7 Homogeneous Linear Systems with Constant Coefficients: *n* Equations in *n* Unknown Functions

A. *Introduction*

We now consider the normal form of a homogeneous linear system of n first-order differential equations in n unknown functions $x_1, x_2, \ldots, x_n$, where all of the coefficients are constants. To be more specific, we shall discuss the case in which each coefficient is a real number. Hence the system to be considered is of the form

$$\begin{aligned} \frac{dx_1}{dt} &= a_{11}x_1 + a_{12}x_2 + \cdots + a_{1n}x_n, \\ \frac{dx_2}{dt} &= a_{21}x_1 + a_{22}x_2 + \cdots + a_{2n}x_n, \\ &\vdots \\ \frac{dx_n}{dt} &= a_{n1}x_1 + a_{n2}x_2 + \cdots + a_{nn}x_n, \end{aligned} \tag{7.126}$$

where all of the a_{ij}, $i = 1, 2, \ldots, n$, $j = 1, 2, \ldots, n$ are real numbers.

Introducing the $n \times n$ constant matrix of real numbers

$$\mathbf{A} = \begin{pmatrix} a_{11} & a_{12} & \cdots & a_{1n} \\ a_{21} & a_{22} & \cdots & a_{2n} \\ \vdots & \vdots & & \vdots \\ a_{n1} & a_{n2} & \cdots & a_{nn} \end{pmatrix} \tag{7.127}$$

and the vector

$$\mathbf{x} = \begin{pmatrix} x_1 \\ x_2 \\ \vdots \\ x_n \end{pmatrix}, \tag{7.128}$$

the system (7.126) can be expressed as the homogeneous linear vector differential equation

$$\frac{d\mathbf{x}}{dt} = \mathbf{A}\mathbf{x}. \tag{7.129}$$

The real constant matrix **A** which appears in (7.129) and is defined by (7.127) is called the *coefficient matrix* of (7.129).

We seek solutions of the system (7.126), that is, of the corresponding vector differential equation (7.129). We shall proceed by analogy with the presentation in Section 7.4A. Doing this, we seek nontrivial solutions of system (7.126) of the form

$$\begin{aligned} x_1 &= \alpha_1 e^{\lambda t}, \\ x_2 &= \alpha_2 e^{\lambda t}, \\ &\vdots \\ x_n &= \alpha_n e^{\lambda t}, \end{aligned} \tag{7.130}$$

where $\alpha_1, \alpha_2, \ldots, \alpha_n$, and λ are numbers. Letting

$$\boldsymbol{\alpha} = \begin{pmatrix} \alpha_1 \\ \alpha_2 \\ \vdots \\ \alpha_n \end{pmatrix} \tag{7.131}$$

and using (7.128), we see that the vector form of the desired solution (7.130) is

$$\mathbf{x} = \boldsymbol{\alpha} e^{\lambda t}.$$

Thus we seek solutions of the vector differential equation (7.129) which are of the form

$$\mathbf{x} = \boldsymbol{\alpha} e^{\lambda t}, \tag{7.132}$$

where $\boldsymbol{\alpha}$ is a constant vector and λ is a number.

Now substituting (7.132) into (7.129), we obtain

$$\lambda \boldsymbol{\alpha} e^{\lambda t} = \mathbf{A} \boldsymbol{\alpha} e^{\lambda t}$$

which reduces at once to

$$\mathbf{A}\boldsymbol{\alpha} = \lambda \boldsymbol{\alpha} \tag{7.133}$$

and hence to

$$(\mathbf{A} - \lambda \mathbf{I})\boldsymbol{\alpha} = \mathbf{0},$$

where $\mathbf{I}$ is the $n \times n$ identity matrix. Written out in terms of components, this is the system of n homogeneous linear algebraic equations

$$\begin{aligned} (a_{11} - \lambda)\alpha_1 + \quad a_{12}\alpha_2 + \cdots + \quad a_{1n}\alpha_n &= 0, \\ a_{21}\alpha_1 + (a_{22} - \lambda)\alpha_2 + \cdots + \quad a_{2n}\alpha_n &= 0, \\ &\vdots \\ a_{n1}\alpha_1 + \quad a_{n2}\alpha_2 + \cdots + (a_{nn} - \lambda)\alpha_n &= 0, \end{aligned} \tag{7.134}$$

in the n unknowns $\alpha_1, \alpha_2, \ldots, \alpha_n$. By Theorem A of Section 7.5B, this system has a nontrivial solution if and only if

$$\begin{vmatrix} a_{11} - \lambda & a_{12} & \cdots & a_{1n} \\ a_{21} & a_{22} - \lambda & \cdots & a_{2n} \\ \vdots & \vdots & & \vdots \\ a_{n1} & a_{n2} & \cdots & a_{nn} - \lambda \end{vmatrix} = 0; \tag{7.135}$$

that is, in matrix notation,

$$|\mathbf{A} - \lambda \mathbf{I}| = \mathbf{0}.$$

Now look back at Section 7.5C. Doing so, we recognize Equation (7.135) as the *characteristic equation* of the coefficient matrix $\mathbf{A} = (a_{ij})$ of the vector differential equation (7.129), we know that it is an nth-degree polynomial equation in λ, and we recall that its roots $\lambda_1, \lambda_2, \ldots, \lambda_n$ are the *characteristic values* of $\mathbf{A}$. Substituting each characteristic value λ_i $(i = 1, 2, \ldots, n)$, into system (7.134), we obtain the corresponding nontrivial solution

$$\alpha_1 = \alpha_{1i}, \alpha_2 = \alpha_{2i}, \ldots, \alpha_n = \alpha_{ni}$$

$(i = 1, 2, \ldots, n)$ of system (7.134). Since (7.134) is merely the component form of (7.133), we recognize that the vector defined by

$$\boldsymbol{\alpha}^{(i)} = \begin{pmatrix} \alpha_{1i} \\ \alpha_{2i} \\ \vdots \\ \alpha_{ni} \end{pmatrix} \qquad (i = 1, 2, \ldots, n) \tag{7.136}$$

is a *characteristic vector* corresponding to the characteristic value λ_i $(i = 1, 2, \ldots, n)$. Thus we see that if the vector differential equation

$$\frac{d\mathbf{x}}{dt} = \mathbf{A}\mathbf{x} \tag{7.129}$$

has a solution of the form

$$\mathbf{x} = \boldsymbol{\alpha} e^{\lambda t} \tag{7.132}$$

then the number λ must be a characteristic value λ_i of the coefficient matrix $\mathbf{A}$ and the vector $\boldsymbol{\alpha}$ must be a characteristic vector $\boldsymbol{\alpha}^{(i)}$ corresponding to this characteristic value λ_i.

B. *Case of n Distinct Characteristic Values*

Now suppose that each of the n characteristic values $\lambda_1, \lambda_2, \ldots, \lambda_n$ of the $n \times n$ square coefficient matrix $\mathbf{A}$ of the vector differential equation is *distinct* (that is, nonrepeated); and let $\boldsymbol{\alpha}^{(1)}, \boldsymbol{\alpha}^{(2)}, \ldots, \boldsymbol{\alpha}^{(n)}$ be a set of n respective corresponding characteristic vectors of $\mathbf{A}$. Then the n distinct vector functions $\mathbf{x}_1, \mathbf{x}_2, \ldots, \mathbf{x}_n$ defined respectively by

$$\mathbf{x}_1(t) = \boldsymbol{\alpha}^{(1)} e^{\lambda_1 t}, \mathbf{x}_2(t) = \boldsymbol{\alpha}^{(2)} e^{\lambda_2 t}, \ldots, \mathbf{x}_n(t) = \boldsymbol{\alpha}^{(n)} e^{\lambda_n t} \tag{7.137}$$

are solutions of the vector differential equation (7.129) on every real interval $[a, b]$. This is readily seen as follows: From (7.133), for each $i = 1, 2, \ldots, n$, we have

$$\lambda_i \boldsymbol{\alpha}^{(i)} = \mathbf{A}\boldsymbol{\alpha}^{(i)};$$

and using this and the definition (7.137) of $\mathbf{x}_i(t)$, we obtain

$$\frac{d\mathbf{x}_i(t)}{dt} = \lambda_i \boldsymbol{\alpha}^{(i)} e^{\lambda_i t} = \mathbf{A}\boldsymbol{\alpha}^{(i)} e^{\lambda_i t} = \mathbf{A}\mathbf{x}_i(t),$$

which clearly states that $\mathbf{x}_i(t)$ satisfies the vector differential equation

$$\frac{d\mathbf{x}}{dt} = \mathbf{A}\mathbf{x}, \tag{7.129}$$

on $[a, b]$.

Now consider the Wronskian of the n solutions $\mathbf{x}_1, \mathbf{x}_2, \ldots, \mathbf{x}_n$ defined by (7.137). We find

$$W(\mathbf{x}_1, \mathbf{x}_2, \ldots, \mathbf{x}_n)(t) = \begin{vmatrix} \alpha_{11}e^{\lambda_1 t} & \alpha_{12}e^{\lambda_2 t} & \cdots & \alpha_{1n}e^{\lambda_n t} \\ \alpha_{21}e^{\lambda_1 t} & \alpha_{22}e^{\lambda_2 t} & \cdots & \alpha_{2n}e^{\lambda_n t} \\ \vdots & \vdots & & \vdots \\ \alpha_{n1}e^{\lambda_1 t} & \alpha_{n2}e^{\lambda_2 t} & \cdots & \alpha_{nn}e^{\lambda_n t} \end{vmatrix}$$

$$= e^{(\lambda_1 + \lambda_2 + \cdots + \lambda_n)t} \begin{vmatrix} \alpha_{11} & \alpha_{12} & \cdots & \alpha_{1n} \\ \alpha_{21} & \alpha_{22} & \cdots & \alpha_{2n} \\ \vdots & \vdots & & \vdots \\ \alpha_{n1} & \alpha_{n2} & \cdots & \alpha_{nn} \end{vmatrix}.$$

By Result C of Section 7.5C, the n characteristic vectors $\boldsymbol{\alpha}^{(1)}, \boldsymbol{\alpha}^{(2)}, \ldots, \boldsymbol{\alpha}^{(n)}$ are linearly independent. Therefore (see Exercise 7 at end of Section 7.5B), it follows that

$$\begin{vmatrix} \alpha_{11} & \alpha_{12} & \cdots & \alpha_{1n} \\ \alpha_{21} & \alpha_{22} & \cdots & \alpha_{2n} \\ \vdots & \vdots & & \vdots \\ \alpha_{n1} & \alpha_{n2} & \cdots & \alpha_{nn} \end{vmatrix} \neq 0.$$

Further, it is clear that

$$e^{(\lambda_1 + \lambda_2 + \cdots + \lambda_n)t} \neq 0$$

for all t. Thus $W(\mathbf{x}_1, \mathbf{x}_2, \ldots, \mathbf{x}_n)(t) \neq 0$ for all t on $[a, b]$. Hence by Theorem 7.15, the solutions $\mathbf{x}_1, \mathbf{x}_2, \ldots, \mathbf{x}_n$ of vector differential equation (7.129) which are defined by (7.137) are linearly independent on $[a, b]$ and so form a fundamental set of solutions of (7.129) on $[a, b]$. Thus a general solution of (7.129) is given by

$$c_1\mathbf{x}_1 + c_2\mathbf{x}_2 + \cdots + c_n\mathbf{x}_n,$$

where $c_1, c_2, \ldots, c_n$ are n arbitrary numbers. We summarize the results obtained in the following theorem:

THEOREM 7.19

Consider the vector differential equation

$$\frac{d\mathbf{x}}{dt} = \mathbf{A}\mathbf{x} \tag{7.129}$$

where $\mathbf{A}$ *is an* $n \times n$ *real constant matrix. Suppose each of the n characteristic values* $\lambda_1, \lambda_2, \ldots, \lambda_n$ *of* $\mathbf{A}$ *is distinct; and let* $\boldsymbol{\alpha}^{(1)}, \boldsymbol{\alpha}^{(2)}, \ldots, \boldsymbol{\alpha}^{(n)}$ *be a set of n respective corresponding characteristic vectors of* $\mathbf{A}$*. Then on every real interval* $[a, b]$*, the n functions defined by*

$$\boldsymbol{\alpha}^{(1)}e^{\lambda_1 t}, \boldsymbol{\alpha}^{(2)}e^{\lambda_2 t}, \ldots, \boldsymbol{\alpha}^{(n)}e^{\lambda_n t}$$

form a linearly independent set (fundamental set) of solutions of (7.129); and

$$\mathbf{x} = c_1\boldsymbol{\alpha}^{(1)}e^{\lambda_1 t} + c_2\boldsymbol{\alpha}^{(2)}e^{\lambda_2 t} + \cdots + c_n\boldsymbol{\alpha}^{(n)}e^{\lambda_n t},$$

where $c_1, c_2, \ldots, c_n$ are n arbitrary numbers, is a general solution of (7.129) on $[a, b]$.

▶ **Example 7.35.** Consider the homogeneous linear system

$$\begin{aligned}\frac{dx_1}{dt} &= 7x_1 - x_2 + 6x_3,\\ \frac{dx_2}{dt} &= -10x_1 + 4x_2 - 12x_3,\\ \frac{dx_3}{dt} &= -2x_1 + x_2 - x_3,\end{aligned} \tag{7.138}$$

or in matrix form,

$$\frac{d\mathbf{x}}{dt} = \begin{pmatrix} 7 & -1 & 6 \\ -10 & 4 & -12 \\ -2 & 1 & -1 \end{pmatrix}\mathbf{x}, \quad \text{where } \mathbf{x} = \begin{pmatrix} x_1 \\ x_2 \\ x_3 \end{pmatrix}. \tag{7.139}$$

We assume a solution of the form

$$\mathbf{x} = \boldsymbol{\alpha}e^{\lambda t},$$

that is,

$$\begin{aligned} x_1 &= \alpha_1 e^{\lambda t},\\ x_2 &= \alpha_2 e^{\lambda t},\\ x_3 &= \alpha_3 e^{\lambda t}.\end{aligned} \tag{7.140}$$

Substituting (7.140) into (7.138) and dividing through by $e^{\lambda t} \neq 0$, we obtain

$$\begin{aligned}\alpha_1\lambda &= 7\alpha_1 - \alpha_2 + 6\alpha_3,\\ \alpha_2\lambda &= -10\alpha_1 + 4\alpha_2 - 12\alpha_3,\\ \alpha_3\lambda &= -2\alpha_1 + \alpha_2 - \alpha_3,\end{aligned}$$

or

$$\begin{aligned}(7-\lambda)\alpha_1 \quad - \alpha_2 \quad + 6\alpha_3 &= 0,\\ -10\alpha_1 + (4-\lambda)\alpha_2 \quad - 12\alpha_3 &= 0,\\ -2\alpha_1 \quad + \alpha_2 + (-1-\lambda)\alpha_3 &= 0.\end{aligned} \tag{7.141}$$

This homogeneous linear algebraic system in α_1, α_2, α_3 has a nontrivial solution if and only if the determinant of its coefficients equals zero, that is, if and only if

$$\begin{vmatrix} 7-\lambda & -1 & 6 \\ -10 & 4-\lambda & -12 \\ -2 & 1 & -1-\lambda \end{vmatrix} = 0. \tag{7.142}$$

Clearly this is the characteristic equation of the coefficient matrix

$$\mathbf{A} = \begin{pmatrix} 7 & -1 & 6 \\ -10 & 4 & -12 \\ -2 & 1 & -1 \end{pmatrix} \tag{7.143}$$

of the given system (7.138) [or (7.139)]. It is a cubic equation in λ; and its roots $\lambda_1, \lambda_2, \lambda_3$ are the characteristic values of the matrix $\mathbf{A}$ given by (7.143). Expanding the determinant involved, we see that the characteristic equation (7.142) of $\mathbf{A}$ may be written

$$\lambda^3 - 10\lambda^2 + 31\lambda - 30 = 0,$$

or in factored form,

$$(\lambda - 2)(\lambda - 3)(\lambda - 5) = 0.$$

Thus the roots of the characteristic equation (7.142) are

$$\lambda_1 = 2, \quad \lambda_2 = 3, \quad \text{and} \quad \lambda_3 = 5. \tag{7.144}$$

A characteristic vector corresponding to $\lambda_1 = 2$ is a nonzero vector

$$\begin{pmatrix} \alpha_1 \\ \alpha_2 \\ \alpha_3 \end{pmatrix} \tag{7.145}$$

whose components are a nontrivial solution $\alpha_1, \alpha_2, \alpha_3$ of the algebraic system (7.141) when $\lambda = 2$. Equivalently, it is a nonzero vector given by (7.145) such that

$$\begin{pmatrix} 7 & -1 & 6 \\ -10 & 4 & -12 \\ -2 & 1 & -1 \end{pmatrix}\begin{pmatrix} \alpha_1 \\ \alpha_2 \\ \alpha_3 \end{pmatrix} = 2\begin{pmatrix} \alpha_1 \\ \alpha_2 \\ \alpha_3 \end{pmatrix}.$$

Starting in either of these completely equivalent ways, we at once find that $\alpha_1, \alpha_2, \alpha_3$ must be a nontrivial solution of the system

$$\begin{aligned} 5\alpha_1 - \alpha_2 + 6\alpha_3 &= 0, \\ -10\alpha_1 + 2\alpha_2 - 12\alpha_3 &= 0, \\ -2\alpha_1 + \alpha_2 - 3\alpha_3 &= 0. \end{aligned}$$

We have already solved this algebra problem in Example 7.25 (except for the notational difference of having used x's there, where we are using α's here). Looking back at that example, we found that a characteristic vector corresponding to $\lambda_1 = 2$ is given by

$$\boldsymbol{\alpha}^{(1)} = \begin{pmatrix} 1 \\ -1 \\ -1 \end{pmatrix}.$$

Likewise, reference to Example 7.25 shows that characteristic vectors corresponding to $\lambda_2 = 3$ and $\lambda_3 = 5$ are

$$\boldsymbol{\alpha}^{(2)} = \begin{pmatrix} 1 \\ -2 \\ -1 \end{pmatrix} \quad \text{and} \quad \boldsymbol{\alpha}^{(3)} = \begin{pmatrix} 3 \\ -6 \\ -2 \end{pmatrix},$$

respectively. Thus a fundamental set of solutions of (7.138) [or (7.139)] is

$$\boldsymbol{\alpha}^{(1)}e^{\lambda_1 t}, \qquad \boldsymbol{\alpha}^{(2)}e^{\lambda_2 t}, \qquad \boldsymbol{\alpha}^{(3)}e^{\lambda_3 t},$$

that is,

$$\begin{pmatrix} 1 \\ -1 \\ -1 \end{pmatrix} e^{2t}, \quad \begin{pmatrix} 1 \\ -2 \\ -1 \end{pmatrix} e^{3t}, \quad \text{and} \quad \begin{pmatrix} 3 \\ -6 \\ -2 \end{pmatrix} e^{5t},$$

or, rewriting these slightly,

$$\begin{pmatrix} e^{2t} \\ -e^{2t} \\ -e^{2t} \end{pmatrix}, \quad \begin{pmatrix} e^{3t} \\ -2e^{3t} \\ -e^{3t} \end{pmatrix} \quad \text{and} \quad \begin{pmatrix} 3e^{5t} \\ -6e^{5t} \\ -2e^{5t} \end{pmatrix},$$

respectively. A general solution of the system may thus be expressed as

$$\begin{aligned} x_1 &= c_1 e^{2t} + c_2 e^{3t} + 3c_3 e^{5t}, \\ x_2 &= -c_1 e^{2t} - 2c_2 e^{3t} - 6c_3 e^{5t}, \\ x_3 &= -c_1 e^{2t} - c_2 e^{3t} - 2c_3 e^{5t}, \end{aligned}$$

where c_1, c_2, and c_3 are arbitrary numbers.

We return to the vector differential equation

$$\frac{d\mathbf{x}}{dt} = \mathbf{A}\mathbf{x}, \tag{7.129}$$

where $\mathbf{A}$ is an $n \times n$ real constant matrix and reconsider the result stated in Theorem 7.19. In that theorem we stated that if each of the n characteristic values $\lambda_1, \lambda_2, \ldots, \lambda_n$ of A is *distinct* and if $\boldsymbol{\alpha}^{(1)}, \boldsymbol{\alpha}^{(2)}, \ldots, \boldsymbol{\alpha}^{(n)}$ is a set of n respective corresponding characteristic vectors of $\mathbf{A}$, then the n functions defined by

$$\boldsymbol{\alpha}^{(1)}e^{\lambda_1 t}, \boldsymbol{\alpha}^{(2)}e^{\lambda_2 t}, \ldots, \boldsymbol{\alpha}^{(n)}e^{\lambda_n t}$$

form a fundamental set of solutions of (7.129). Note that although we assume that $\lambda_1, \lambda_2, \ldots, \lambda_n$ are *distinct*, we do *not* require that they be *real.* Thus *complex* characteristic values may be present. However, since A is a real matrix, any complex characteristic values must occur in conjugate pairs. Suppose $\lambda_1 = a + bi$ and $\lambda_2 = a - bi$ form such a pair. Then the corresponding solutions are

$$\boldsymbol{\alpha}^{(1)}e^{(a+bi)t} \quad \text{and} \quad \boldsymbol{\alpha}^{(2)}e^{(a-bi)t},$$

and these solutions are *complex* solutions. Thus if one or more conjugate complex pairs of characteristic values occur, the fundamental set defined by $\boldsymbol{\alpha}^{(i)}e^{\lambda_i t}$, $i = 1, 2, \ldots, n$, contains *complex* functions. However, in such a case, this fundamental set may be replaced by another fundamental set, all of whose members are *real* functions. This is accomplished exactly as explained in Section 7.4D and illustrated in Example 7.18.

C. Remarks on the Case of Repeated Characteristic Values

We again consider the vector differential equation

$$\frac{d\mathbf{x}}{dt} = \mathbf{A}\mathbf{x}, \tag{7.129}$$

where $\mathbf{A}$ is an $n \times n$ real constant matrix; but here we given a brief introduction to the case in which $\mathbf{A}$ has a repeated characteristic value. To be definite, we suppose that $\mathbf{A}$ has a real characteristic value λ_1 of multiplicity m, where $1 < m \leq n$, and that all the other characteristic values $\lambda_{m+1}, \lambda_{m+2}, \ldots, \lambda_n$ (if there are any) are distinct. By Result D of Section 7.5C, we know that the repeated characteristic value λ_1 of multiplicity m has p linearly independent characteristic vectors, where $1 \leq p \leq m$. Now consider two subcases: (1) $p = m$; and (2) $p < m$.

In subcase (1), $p = m$, there are m linearly independent characteristic vectors $\boldsymbol{\alpha}^{(1)}, \boldsymbol{\alpha}^{(2)}, \ldots, \boldsymbol{\alpha}^{(m)}$ corresponding to λ_1. Then the n functions defined by

$$\boldsymbol{\alpha}^{(1)}e^{\lambda_1 t}, \boldsymbol{\alpha}^{(2)}e^{\lambda_1 t}, \ldots, \boldsymbol{\alpha}^{(m)}e^{\lambda_1 t}, \boldsymbol{\alpha}^{(m+1)}e^{\lambda_{m+1} t}, \ldots, \boldsymbol{\alpha}^{(n)}e^{\lambda_n t}$$

form a linearly independent set of n solutions of differential equation (7.129); and a general solution of (7.129) is a linear combination of these n solutions having n arbitrary numbers as the "constants of combination."

▶ **Example 7.36.** Consider the homogeneous linear system

$$\begin{aligned} \frac{dx_1}{dt} &= 3x_1 + x_2 - x_3, \\ \frac{dx_2}{dt} &= x_1 + 3x_2 - x_3, \\ \frac{dx_3}{dt} &= 3x_1 + 3x_2 - x_3, \end{aligned} \tag{7.146}$$

or in matrix form,

$$\frac{d\mathbf{x}}{dt} = \begin{pmatrix} 3 & 1 & -1 \\ 1 & 3 & -1 \\ 3 & 3 & -1 \end{pmatrix} \mathbf{x}, \quad \text{where} \quad \mathbf{x} = \begin{pmatrix} x_1 \\ x_2 \\ x_3 \end{pmatrix}. \tag{7.147}$$

We assume a solution of the form

$$\mathbf{x} = \boldsymbol{\alpha} e^{\lambda t};$$

that is,

$$\begin{aligned} x_1 &= \alpha_1 e^{\lambda t}, \\ x_2 &= \alpha_2 e^{\lambda t}, \\ x_3 &= \alpha_3 e^{\lambda t}. \end{aligned} \tag{7.148}$$

Substituting (7.148) into (7.146) and dividing through by $e^{\lambda t} \neq 0$, we obtain

$$\begin{aligned} \alpha_1 \lambda &= 3\alpha_1 + \alpha_2 - \alpha_3, \\ \alpha_2 \lambda &= \alpha_1 + 3\alpha_2 - \alpha_3, \\ \alpha_3 \lambda &= 3\alpha_1 + 3\alpha_2 - \alpha_3, \end{aligned}$$

or

$$\begin{aligned}(3-\lambda)\alpha_1 + \alpha_2 - \alpha_3 &= 0,\\ \alpha_1 + (3-\lambda)\alpha_2 - \alpha_3 &= 0,\\ 3\alpha_1 + 3\alpha_2 + (-1-\lambda)\alpha_3 &= 0.\end{aligned} \tag{7.149}$$

This homogeneous linear algebraic system in $\alpha_1, \alpha_2, \alpha_3$ has a nontrivial solution if and only if the determinant of its coefficients equals zero, that is, if and only if

$$\begin{vmatrix} 3-\lambda & 1 & -1 \\ 1 & 3-\lambda & -1 \\ 3 & 3 & -1-\lambda \end{vmatrix} = 0. \tag{7.150}$$

Of course this is the characteristic equation of the coefficient matrix

$$\mathbf{A} = \begin{pmatrix} 3 & 1 & -1 \\ 1 & 3 & -1 \\ 3 & 3 & -1 \end{pmatrix} \tag{7.151}$$

of the given system (7.146) [or (7.147)]. It is a cubic equation in λ; and its roots λ_1, λ_2, λ_3 are the characteristic values of the matrix $\mathbf{A}$ given by (7.151). Expanding the determinant involved, we see that the characteristic equation (7.150) of $\mathbf{A}$ may be written

$$\lambda^3 - 5\lambda^2 + 8\lambda - 4 = 0,$$

or, in factored form,

$$(\lambda - 1)(\lambda - 2)(\lambda - 2) = 0.$$

Thus the roots of the characteristic equation (7.150) are

$$\lambda_1 = 1, \quad \lambda_2 = 2, \quad \text{and} \quad \lambda_3 = 2. \tag{7.152}$$

Note that the real number 1 is a *distinct* characteristic value of the coefficient matrix (7.151) of the given system (7.146); but the real number 2 is a *repeated* characteristic value of this coefficient matrix.

We first consider the distinct characteristic value $\lambda_1 = 1$. A characteristic vector corresponding to $\lambda_1 = 1$ is a nonzero vector

$$\begin{pmatrix} \alpha_1 \\ \alpha_2 \\ \alpha_3 \end{pmatrix} \tag{7.153}$$

whose components are a nontrivial solution $\alpha_1, \alpha_2, \alpha_3$ of the algebraic system (7.149) when $\lambda = 1$. Equivalently, it is a nonzero vector given by (7.153) such that

$$\begin{pmatrix} 3 & 1 & -1 \\ 1 & 3 & -1 \\ 3 & 3 & -1 \end{pmatrix}\begin{pmatrix} \alpha_1 \\ \alpha_2 \\ \alpha_3 \end{pmatrix} = 1\begin{pmatrix} \alpha_1 \\ \alpha_2 \\ \alpha_3 \end{pmatrix}.$$

Starting in either of these completely equivalent ways, we at once find that $\alpha_1, \alpha_2, \alpha_3$ must be nontrivial solutions of the system

$$\begin{aligned}2\alpha_1 + \alpha_2 - \alpha_3 &= 0,\\ \alpha_1 + 2\alpha_2 - \alpha_3 &= 0,\\ 3\alpha_1 + 3\alpha_2 - 2\alpha_3 &= 0.\end{aligned}$$

Note that $\alpha_1 = k, \alpha_2 = k, \alpha_3 = 3k$ is a solution of this system for every real k. Hence the characteristic vectors corresponding to the characteristic value $\lambda = 1$ are the vectors

$$\boldsymbol{\alpha} = \begin{pmatrix} k \\ k \\ 3k \end{pmatrix},$$

where k is an arbitrary nonzero number. In particular, letting $k = 1$, we obtain the particular characteristic vector

$$\boldsymbol{\alpha}^{(1)} = \begin{pmatrix} 1 \\ 1 \\ 3 \end{pmatrix}$$

corresponding to $\lambda = 1$. Thus the corresponding solution of the form

$$\boldsymbol{\alpha} e^{\lambda t}$$

of the given system (7.146) is

$$\boldsymbol{\alpha}^{(1)} e^{\lambda_1 t},$$

that is,

$$\begin{pmatrix} 1 \\ 1 \\ 3 \end{pmatrix} e^{t}. \tag{7.154}$$

We now turn to the repeated characteristic value $\lambda_2 = \lambda_3 = 2$. To be more specific, this characteristic value 2 has multiplicity $m = 2 < 3 = n$, where n of course denotes the common number of rows and columns of the coefficient matrix (7.151) of the given system (7.146). A characteristic vector corresponding to this double characteristic value $\lambda_2 = \lambda_3 = 2$ is a nonzero vector

$$\begin{pmatrix} \alpha_1 \\ \alpha_2 \\ \alpha_3 \end{pmatrix} \tag{7.155}$$

whose components are a nontrivial solution $\alpha_1, \alpha_2, \alpha_3$ of the algebraic system (7.149) when $\lambda = 2$. Equivalently, it is a nonzero vector given by (7.155) such that

$$\begin{pmatrix} 3 & 1 & -1 \\ 1 & 3 & -1 \\ 3 & 3 & -1 \end{pmatrix} \begin{pmatrix} \alpha_1 \\ \alpha_2 \\ \alpha_3 \end{pmatrix} = 2 \begin{pmatrix} \alpha_1 \\ \alpha_2 \\ \alpha_3 \end{pmatrix}.$$

Starting in either of these completely equivalent ways, we at once find that $\alpha_1, \alpha_2, \alpha_3$ must be nontrivial solutions of the system

$$\begin{aligned} \alpha_1 + \alpha_2 - \alpha_3 &= 0, \\ \alpha_1 + \alpha_2 - \alpha_3 &= 0, \\ 3\alpha_1 + 3\alpha_2 - 3\alpha_3 &= 0. \end{aligned}$$

Note that each of these three relations is equivalent to both of the other two, and so the only relationship among $\alpha_1, \alpha_2, \alpha_3$ is that given most simply by

$$\alpha_1 + \alpha_2 - \alpha_3 = 0. \tag{7.156}$$

Clearly there exist two linearly independent vectors of the form (7.155) whose components satisfy this relation (7.156). For example, if $\alpha_1 = 1$, $\alpha_2 = -1$, and $\alpha_3 = 0$, we obtain the vector

$$\boldsymbol{\alpha}^{(2)} = \begin{pmatrix} 1 \\ -1 \\ 0 \end{pmatrix};$$

and if $\alpha_1 = 1$, $\alpha_2 = 0$, and $\alpha_3 = 1$, we obtain the vector

$$\boldsymbol{\alpha}^{(3)} = \begin{pmatrix} 1 \\ 0 \\ 1 \end{pmatrix}.$$

First note that the components of each of these two vectors $\boldsymbol{\alpha}^{(2)}$ and $\boldsymbol{\alpha}^{(3)}$ *do satisfy* (7.156), and hence *each is a characteristic vector* corresponding to the double root $\lambda_2 = \lambda_3 = 2$. Next note that *these vectors* $\boldsymbol{\alpha}^{(2)}$ *and* $\boldsymbol{\alpha}^{(3)}$ *are indeed linearly independent* (merely use the definition of linear independence of a set of constant vectors). Thus the characteristic value $\lambda = 2$ of multiplicity $m = 2$ has the $p = 2$ linearly independent characteristic vectors

$$\boldsymbol{\alpha}^{(2)} = \begin{pmatrix} 1 \\ -1 \\ 0 \end{pmatrix} \quad \text{and} \quad \boldsymbol{\alpha}^{(3)} = \begin{pmatrix} 1 \\ 0 \\ 1 \end{pmatrix}$$

corresponding to it. Hence this is an illustration of subcase (1) of the discussion preceding this example. Thus, corresponding to the twofold characteristic value $\lambda = 2$, there are two linearly independent solutions of the form

$$\boldsymbol{\alpha}e^{\lambda t}$$

of the given system. These are

$$\boldsymbol{\alpha}^{(2)}e^{2t} \quad \text{and} \quad \boldsymbol{\alpha}^{(3)}e^{2t},$$

that is,

$$\begin{pmatrix} 1 \\ -1 \\ 0 \end{pmatrix} e^{2t} \quad \text{and} \quad \begin{pmatrix} 1 \\ 0 \\ 1 \end{pmatrix} e^{2t}, \tag{7.157}$$

respectively. Hence a fundamental set of solutions of the given system (7.146) [or (7.147)] consists of the three vectors defined by (7.154) and (7.157). These are

$$\begin{pmatrix} 1 \\ 1 \\ 3 \end{pmatrix} e^{t}, \quad \begin{pmatrix} 1 \\ -1 \\ 0 \end{pmatrix} e^{2t}, \quad \text{and} \quad \begin{pmatrix} 1 \\ 0 \\ 1 \end{pmatrix} e^{2t},$$

or rewriting these slightly,

$$\begin{pmatrix} e^{t} \\ e^{t} \\ 3e^{t} \end{pmatrix}, \quad \begin{pmatrix} e^{2t} \\ -e^{2t} \\ 0 \end{pmatrix}, \quad \text{and} \quad \begin{pmatrix} e^{2t} \\ 0 \\ e^{2t} \end{pmatrix},$$

respectively. A general solution of the system may thus be expressed as

$$x_1 = c_1 e^t + (c_2 + c_3)e^{2t},$$
$$x_2 = c_1 e^t - c_2 e^{2t},$$
$$x_3 = 3c_1 e^t + c_3 e^{2t},$$

where c_1, c_2, and c_3 are arbitrary numbers.

One type of vector differential equation (7.129) which *always* leads to subcase (1), $p = m$, in the case of a repeated characteristic value λ_1 is that in which the $n \times n$ coefficient matrix $\mathbf{A}$ of (7.129) is a real symmetric matrix. For then, by Result G of Section 7.5C, there always exist n linearly independent characteristic vectors of $\mathbf{A}$, regardless of whether the n characteristic values of $\mathbf{A}$ are all distinct or not.

We now turn to a very brief consideration of subcase (2), $p < m$. In this case, there are less than m linearly independent characteristic vectors $\boldsymbol{\alpha}^{(1)}$ corresponding to the characteristic value λ_1 of multiplicity m. Hence there are less than m linearly independent solutions of differential equation (7.129) of the form $\boldsymbol{\alpha}^{(1)}e^{\lambda_1 t}$ corresponding to λ_1; and so there is *not* a fundamental set of solutions of the form $\boldsymbol{\alpha}^{(k)}e^{\lambda_k t}$, where λ_k is a characteristic value of $\mathbf{A}$ and $\boldsymbol{\alpha}^{(k)}$ is a characteristic vector corresponding to λ_k. Clearly we must seek linearly independent solutions of another form.

To discover what other forms of solution to seek, we look back at the analogous situation in Section 7.4C. The results there suggest the following:

If λ_1 is a characteristic value of multiplicity $m = 2$ and $p = 1 < m$, then we seek linearly independent solutions of the form

$$\boldsymbol{\alpha}e^{\lambda_1 t} \quad \text{and} \quad \boldsymbol{\alpha}te^{\lambda_1 t} + \boldsymbol{\beta}e^{\lambda_1 t};$$

where $\boldsymbol{\alpha}$ is a characteristic vector corresponding to λ_1, that is, $\boldsymbol{\alpha}$ satisfies

$$(\mathbf{A} - \lambda_1\mathbf{I})\boldsymbol{\alpha} = \mathbf{0};$$

and $\boldsymbol{\beta}$ is a vector which satisfies the equation

$$(\mathbf{A} - \lambda_1\mathbf{I})\boldsymbol{\beta} = \boldsymbol{\alpha}.$$

If λ_1 is a characteristic value of multiplicity $m > 2$, and $p < m$, then the forms of the m linearly independent solutions corresponding to λ_1 depend upon whether $p = 1, 2, \ldots,$ or $m - 1$. However, we shall not consider such situations in this text.

Exercises

Find the general solution of each of the homogeneous linear systems in Exercises 1 through 12, where in each exercise

$$\mathbf{x} = \begin{pmatrix} x_1 \\ x_2 \\ x_3 \end{pmatrix}.$$

1. $\dfrac{d\mathbf{x}}{dt} = \begin{pmatrix} 1 & 1 & -1 \\ 2 & 3 & -4 \\ 4 & 1 & -4 \end{pmatrix}\mathbf{x}.$

2. $\dfrac{d\mathbf{x}}{dt} = \begin{pmatrix} 1 & -1 & -1 \\ 1 & 3 & 1 \\ -3 & -6 & 6 \end{pmatrix}\mathbf{x}.$

3. $\dfrac{d\mathbf{x}}{dt} = \begin{pmatrix} 1 & -1 & -1 \\ 1 & 3 & 1 \\ -3 & 1 & -1 \end{pmatrix} \mathbf{x}.$

4. $\dfrac{d\mathbf{x}}{dt} = \begin{pmatrix} 1 & -2 & 4 \\ -2 & 3 & 0 \\ 4 & 0 & 2 \end{pmatrix} \mathbf{x}.$

5. $\dfrac{d\mathbf{x}}{dt} = \begin{pmatrix} 1 & 2 & 2 \\ 2 & 0 & 3 \\ 2 & 3 & 0 \end{pmatrix} \mathbf{x}.$

6. $\dfrac{d\mathbf{x}}{dt} = \begin{pmatrix} 1 & 1 & 0 \\ 1 & 0 & 1 \\ 0 & 1 & 1 \end{pmatrix} \mathbf{x}.$

7. $\dfrac{d\mathbf{x}}{dt} = \begin{pmatrix} 1 & -2 & 0 \\ -2 & 3 & 0 \\ 0 & 0 & 2 \end{pmatrix} \mathbf{x}.$

8. $\dfrac{d\mathbf{x}}{dt} = \begin{pmatrix} \frac{7}{5} & \frac{2}{5} & -\frac{1}{5} \\ \frac{4}{5} & \frac{9}{5} & -\frac{2}{5} \\ \frac{2}{5} & \frac{2}{5} & \frac{4}{5} \end{pmatrix} \mathbf{x}.$

9. $\dfrac{d\mathbf{x}}{dt} = \begin{pmatrix} 1 & -3 & 9 \\ 0 & -5 & 18 \\ 0 & -3 & 10 \end{pmatrix} \mathbf{x}.$

10. $\dfrac{d\mathbf{x}}{dt} = \begin{pmatrix} 3 & \frac{2}{7} & -\frac{4}{7} \\ 0 & \frac{19}{7} & \frac{4}{7} \\ 0 & \frac{6}{7} & \frac{9}{7} \end{pmatrix} \mathbf{x}.$

11. $\dfrac{d\mathbf{x}}{dt} = \begin{pmatrix} 11 & 6 & 18 \\ 9 & 8 & 18 \\ -9 & -6 & -16 \end{pmatrix} \mathbf{x}.$

12. $\dfrac{d\mathbf{x}}{dt} = \begin{pmatrix} 1 & 9 & 9 \\ 0 & 19 & 18 \\ 0 & 9 & 10 \end{pmatrix} \mathbf{x}.$

Suggested Reading

I. Fundamental Theory and Methods

BOYCE, W., and R. DIPRIMA, *Elementary Differential Equations*, 2nd ed. (Wiley, New York, 1969).

BRAUER, F., and J. NOHEL, *Ordinary Differential Equations: A First Course* (Benjamin, New York, 1967).

FORD, L., *Differential Equations*, 2nd ed. (McGraw-Hill, New York, 1955).

GOLDBERG, J., and A. SCHWARTZ, *Systems of Ordinary Differential Equations: An Introduction* (Harper and Row, New York, 1972).

KAPLAN, W., *Ordinary Differential Equations* (Addison-Wesley, Reading, Mass., 1958).

KREIDER, D., R. KULLER, and D. OSTBERG, *Elementary Differential Equations* (Addison-Wesley, Reading, Mass., 1968).

RITGER, P., and N. ROSE, *Differential Equations with Applications* (McGraw-Hill, New York, 1968).

II. Advanced Theory and Methods

CODDINGTON, E., and N. LEVINSON, *Theory of Ordinary Differential Equations* (McGraw-Hill, New York, 1955).

COLE, R., *Theory of Ordinary Differential Equations* (Appleton-Century-Crofts, New York, 1968).

HUREWICZ, W. *Lectures on Ordinary Differential Equations* (MIT Press, Cambridge, Mass., 1958).

INCE, E., *Ordinary Differential Equations* (Dover Publications, New York, 1953).

PETROVSKI, V., *Ordinary Differential Equations* (Prentice-Hall, Englewood Cliffs, N.J., 1966).

REID, W. T., *Ordinary Differential Equations* (Wiley, New York, 1971).

III. Matrix Algebra

DAVIS, P., *The Mathematics of Matrices*, 2nd ed. (Xerox, Lexington, Mass., 1973).

EVES, H., *Elementary Matrix Algebra* (Allyn and Bacon, Boston, 1966).

HOFFMAN, K., and R. KUNZE, *Linear Algebra* (Prentice-Hall, Englewood Cliffs, N.J., 1961).

HOHN, F., *Elementary Matrix Algebra*, 2nd ed. (Macmillan, New York, 1966).

8 Approximate Methods of Solving First-Order Equations

In Chapter 2 we considered certain special types of first-order differential equations having closed-form solutions which can be obtained exactly. For a first-order differential equation which is not of one or another of these special types, it usually is not apparent how one should proceed in an attempt to obtain a solution exactly. Indeed, in most such cases the discovery of an exact closed-form solution in terms of elementary functions would be an unexpected luxury! Therefore one considers the possibilities of obtaining approximate solutions of first-order differential equations. In this chapter we shall introduce several approximate methods. In the study of each method in this chapter our primary concern will be to obtain familiarity with the procedure itself and to develop skill in applying it. In general we shall not be concerned here with theoretical justifications and extended discussions of accuracy and error. We shall leave such matters, important as they are, to more specialized and advanced treatises and instead shall concentrate on the formal details of the various procedures.

8.1 Graphical Methods

A. Line Elements and Direction Fields

In Chapter 1 we considered briefly the geometric significance of the first-order differential equation

$$\frac{dy}{dx} = f(x, y), \tag{8.1}$$

where f is a real function of x and y. The explicit solutions of (8.1) are certain real functions, and the graphs of these solution functions are curves in the xy plane called the *integral curves* of (8.1). At each point (x, y) at which $f(x, y)$ is defined, the differential equation (8.1) defines the slope $f(x, y)$ at the point (x, y) of the integral curve of (8.1) which passes through this point. Thus we may construct the tangent to an integral curve of (8.1) at a given point (x, y) without actually knowing the solution function of which this integral curve is the graph.

We proceed to do this. Through the point (x, y) we draw a short segment of the tangent to the integral curve of (8.1) which passes through this point. That is, through (x, y) we construct a short segment the slope of which is $f(x, y)$, as given by the differential equation (8.1). Such a segment is called a *line element* of the differential equation (8.1).

For example, let us consider the differential equation

$$\frac{dy}{dx} = 2x + y. \tag{8.2}$$

Here $f(x, y) = 2x + y$, and the slope of the integral curve of (8.2) which passes through the point (1, 2) has at this point the value

$$f(1, 2) = 4.$$

Thus through the point (1, 2) we construct a short segment of slope 4 or, in other words, of angle of inclination approximately 76° (see Figure 8.1). This short segment is the line element of the differential equation (8.2) at the point (1, 2). It is tangent to the integral curve of (8.2) which passes through this point.

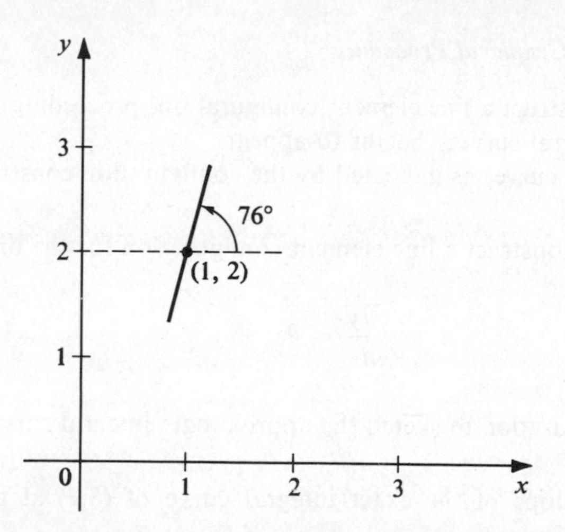

Line element of differential equation (8.2) at Point (1, 2).

FIGURE 8.1

Let us now return to the general equation (8.1). A line element of (8.1) can be constructed at every point (x, y) at which $f(x, y)$ in (8.1) is defined. Doing so for a selection of different points (x, y) leads to a configuration of selected line elements which indicates the directions of the integral curves at the various selected points. We shall refer to such a configuration as a *line element configuration*.

For each point (x, y) at which $f(x, y)$ is defined, the differential equation (8.1) thus defines a line segment with slope $f(x, y)$, or, in other words, a direction. Each such point, taken together with the corresponding direction so defined, constitutes the so-called *direction field* of the differential equation (8.1). We say that the differential equation (8.1) defines this direction field, and this direction field is represented graphically by a line element configuration. Clearly a more thorough and carefully constructed line element configuration gives a more accurate graphical representation of the direction field.

For a given differential equation of the form (8.1), let us assume that a "thorough and carefully constructed" line element configuration has been drawn. That is, we assume that line elements have been carefully constructed at a relatively large number of carefully chosen points. Then this resulting line element configuration will indicate the presence of a family of curves tangent to the various line elements constructed at the different points. This indicated family of curves is approximately the family of integral curves of the given differential equation. Actual smooth curves drawn tangent to the line elements as the configuration indicates will thus provide approximate graphs of the true integral curves.

Thus the construction of the line element configuration provides a procedure for approximately obtaining the solution of the differential equation in graphical form. We now summarize this basic graphical procedure and illustrate it with a simple example.

Summary of Basic Graphical Procedure

1. Carefully construct a line element configuration, proceeding until the family of "approximate integral curves" begins to appear.
2. Draw smooth curves as indicated by the configuration constructed in Step 1.

▶ Example 8.1. Construct a line element configuration for the differential equation

$$\frac{dy}{dx} = 2x + y, \tag{8.2}$$

and use this configuration to sketch the approximate integral curves.

Solution. The slope of the exact integral curve of (8.2) at any point (x, y) is given by

$$f(x, y) = 2x + y.$$

We evaluate this slope at a number of selected points and so determine the approximate inclination of the corresponding line element at each point selected. We then construct the line elements so determined. From the resulting configuration we sketch several of the approximate integral curves. A few typical inclinations are listed in Table 8.1 and the completed configuration with the approximate integral curves appears in Figure 8.2.

Comments. The basic graphical procedure outlined here is very general since it can be applied to any first-order differential equation of the form (8.1). However, the method has several obvious disadvantages. For one thing, although it provides the approximate graphs of the integral curves, it does not furnish analytic expressions for the solutions, either exactly or approximately. Furthermore, it is extremely tedious and time-consuming. Finally, the graphs obtained are only approximations to the graphs of the exact integral curves, and the accuracy of these approximate graphs is

TABLE 8.1

x	y	$\frac{dy}{dx}$ (*Slope*)	*Approximate inclination of line element*
$\frac{1}{2}$	$-\frac{1}{2}$	$\frac{1}{2}$	27°
$\frac{1}{2}$	0	1	45°
$\frac{1}{2}$	$\frac{1}{2}$	$\frac{3}{2}$	56°
$\frac{1}{2}$	1	2	63°
1	$-\frac{1}{2}$	$\frac{3}{2}$	56°
1	0	2	63°
1	$\frac{1}{2}$	$\frac{5}{2}$	68°
1	1	3	72°

uncertain. Of course, apparently better approximations can be obtained by constructing more complete and careful line element configurations, but this in turn increases the time and labor involved. We shall now consider a procedure by which the process may be speeded up considerably. This is the so-called *method of isoclines.*

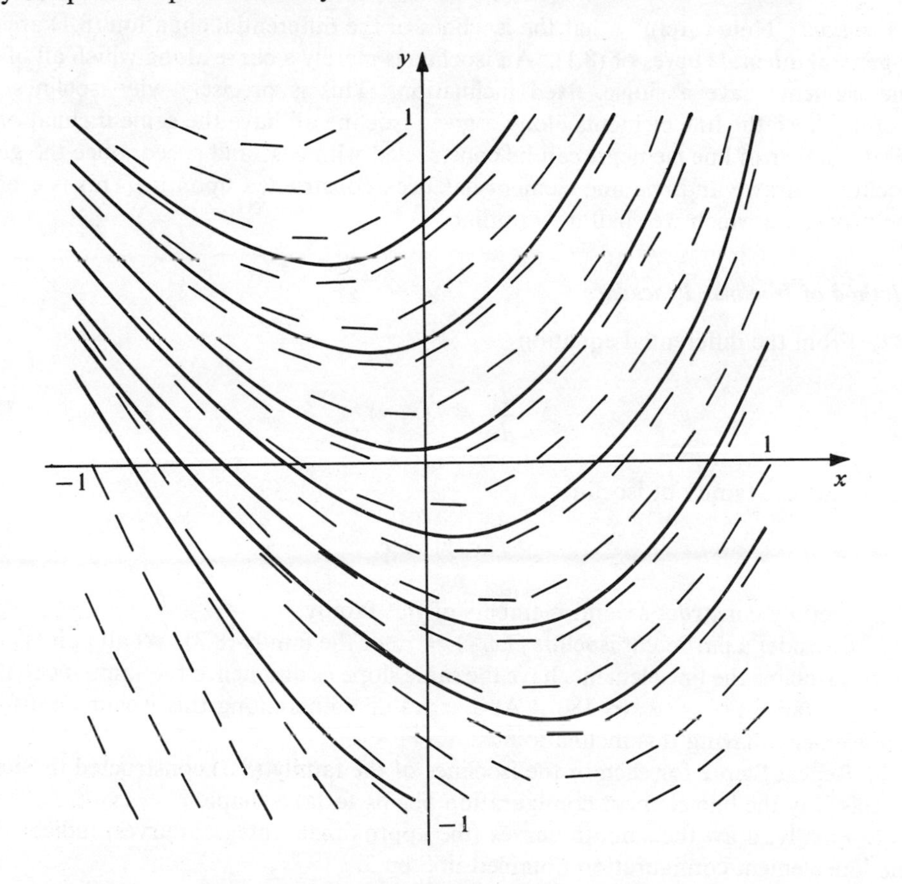

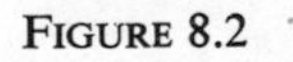
FIGURE 8.2

B. *The Method of Isoclines*

DEFINITION

Consider the differential equation

$$\frac{dy}{dx} = f(x, y). \tag{8.1}$$

A curve along which the slope $f(x, y)$ has a constant value c is called an isocline *of the differential equation (8.1). That is, the isoclines of (8.1) are the curves $f(x, y) = c$, for different values of the parameter c.*

For example, the isoclines of the differential equation

$$\frac{dy}{dx} = 2x + y \tag{8.2}$$

are the straight lines $2x + y = c$. These are of course the straight lines $y = -2x + c$ of slope -2 and y-intercept c.

Caution. Note carefully that the isoclines of the differential equation (8.1) are *not* in general integral curves of (8.1). An isocline is merely a curve along which all of the line elements have a single, fixed inclination. This is precisely why isoclines are useful. Since the line elements along a given isocline all have the same inclination, a great number of line elements can be constructed with ease and speed, once the given isocline is drawn and *one* line element has been constructed upon it. This is exactly the procedure which we shall now outline.

Method of Isoclines Procedure

1. From the differential equation

$$\frac{dy}{dx} = f(x, y) \tag{8.1}$$

determine the family of isoclines

$$f(x, y) = c, \tag{8.3}$$

and carefully construct several members of this family.

2. Consider a particular isocline $f(x, y) = c_0$ of the family (8.3). At all points (x, y) on this isocline the line elements have the same slope c_0 and hence the same inclination $\alpha_0 = \arctan c_0$, $0° \leq \alpha_0 < 180°$. At a series of points along this isocline construct line elements having this inclination α_0.

3. Repeat Step 2 for each of the isoclines of the family (8.3) constructed in Step 1. In this way the line element configuration begins to take shape.

4. Finally, draw the smooth curves (the approximate integral curves) indicated by the line element configuration obtained in Step 3.

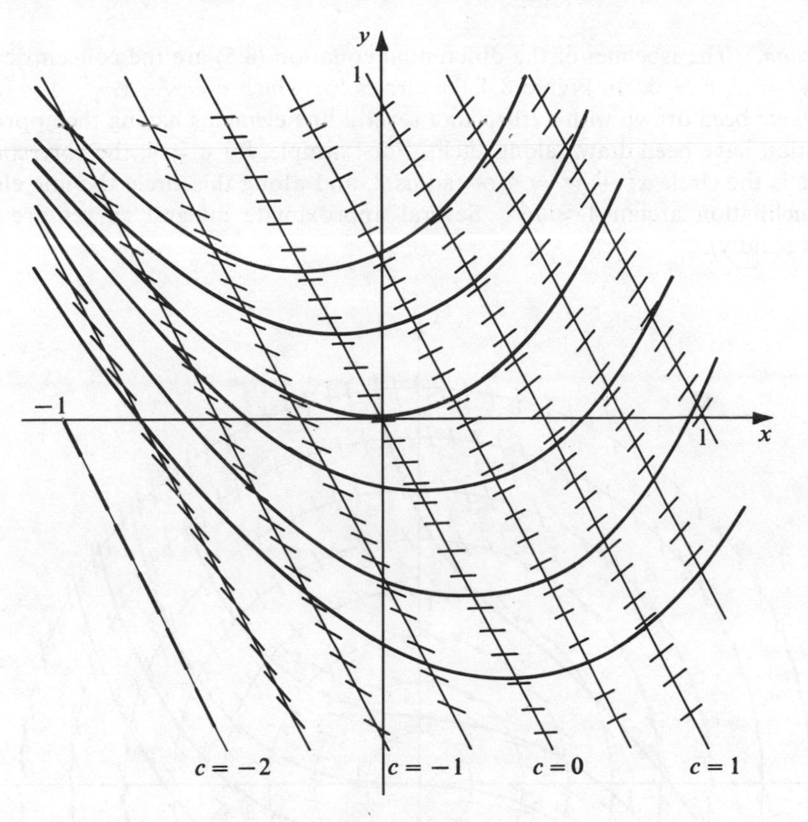

FIGURE 8.3

▶ Example 8.2. Employ the method of isoclines to sketch the approximate integral curves of

$$\frac{dy}{dx} = 2x + y. \tag{8.2}$$

Solution. We have already noted that the isoclines of the differential equation (8.2) are the straight lines $2x + y = c$ or

$$y = -2x + c. \tag{8.4}$$

In Figure 8.3 we construct these lines for $c = -2, -\frac{3}{2}, -1, -\frac{1}{2}, 0, \frac{1}{2}, 1, \frac{3}{2}$, and 2. On each of these we then construct a number of line elements having the appropriate inclination $\alpha = \arctan c$, $0° \leq \alpha < 180°$. For example, for $c = 1$, the corresponding isocline is $y = -2x + 1$, and on this line we construct line elements of inclination $\arctan 1 = 45°$. In the figure the isoclines are drawn with dashes and several of the approximate integral curves are shown (drawn solidly).

▶ Example 8.3. Employ the method of isoclines to sketch the approximate integral curves of

$$\frac{dy}{dx} = x^2 + y^2. \tag{8.5}$$

Solution. The isoclines of the differential equation (8.5) are the concentric circles $x^2 + y^2 = c$, $c > 0$. In Figure 8.4 the circles for which $c = \frac{1}{16}, \frac{1}{4}, \frac{9}{16}, 1, \frac{25}{16}, \frac{9}{4}, \frac{49}{16}$, and 4 have been drawn with dashes, and several line elements having the appropriate inclination have been drawn along each. For example, for $c = 4$, the corresponding isocline is the circle $x^2 + y^2 = 4$ of radius 2, and along this circle the line elements have inclination arctan $4 \approx 76°$. Several approximate integral curves are shown (drawn solidly).

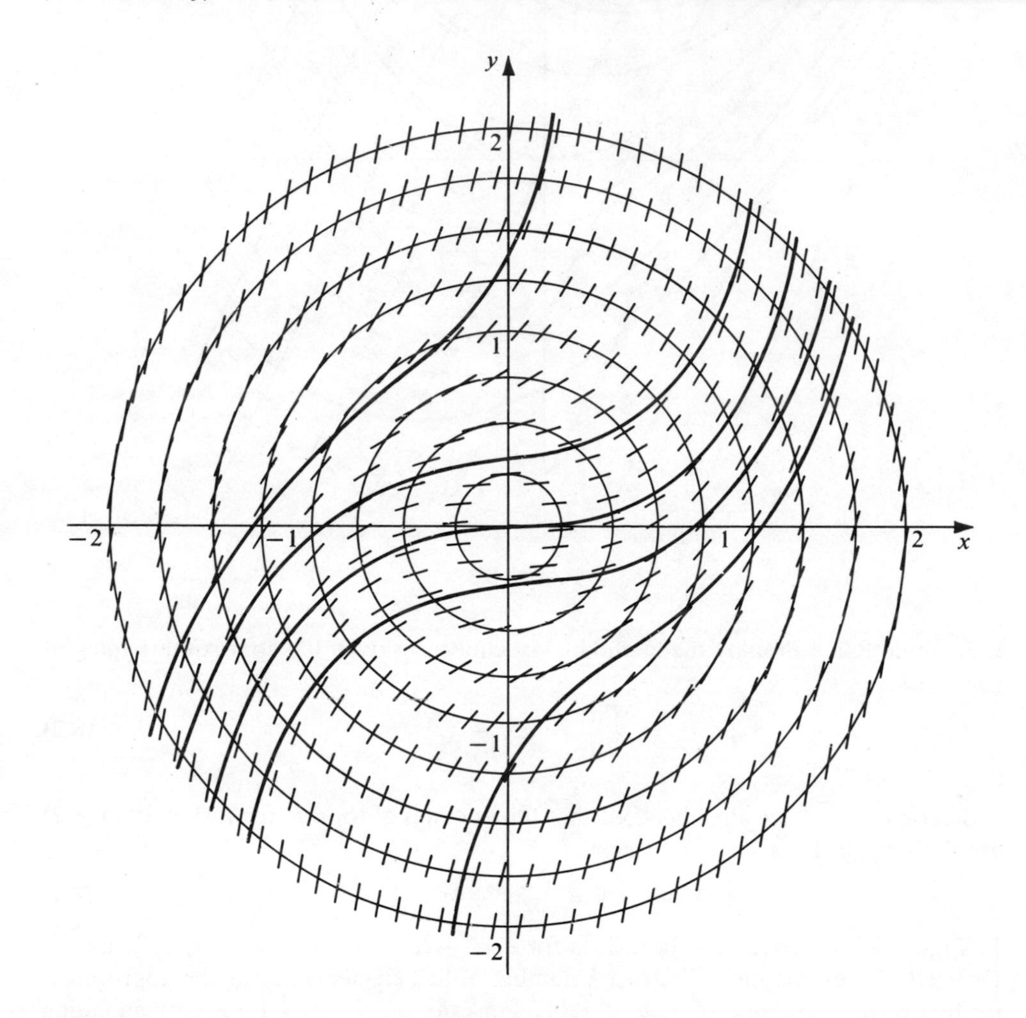

FIGURE 8.4

Exercises

Employ the method of isoclines to sketch the approximate integral curves of each of the differential equations in Exercises 1 through 12.

1. $\dfrac{dy}{dx} = 3x - y.$

2. $\dfrac{dy}{dx} = \dfrac{y}{x}.$

3. $\dfrac{dy}{dx} = \dfrac{y}{x^2}$.

4. $\dfrac{dy}{dx} = x^2 + 2y^2$.

5. $\dfrac{dy}{dx} = \dfrac{3x - y}{x + y}$.

6. $\dfrac{dy}{dx} = \sin x - y$.

7. $\dfrac{dy}{dx} = y^3 - x^2$.

8. $\dfrac{dy}{dx} = \dfrac{3x + 2y + x^2}{x + 2y}$.

9. $\dfrac{dy}{dx} = \dfrac{3x + y + x^3}{5x - y}$.

10. $\dfrac{dy}{dx} = \dfrac{\sin x + y}{x - y}$.

11. $\dfrac{dy}{dx} = \dfrac{(1 - x^2)y - x}{y}$.

12. $\dfrac{dy}{dx} = \dfrac{10(1 - x^2)y - x}{y}$.

8.2 Power Series Methods

A. *Introduction*

Although the graphical methods of the preceding section are very general, they suffer from several serious disadvantages. Not only are they tedious and subject to possible errors of construction, but they merely provide us with the approximate *graphs* of the solutions and do not furnish any analytic *expressions* for these solutions. Although we have now passed the naive state of searching for a closed-form solution in terms of elementary functions, we might still hope for solutions which can be represented as some type of infinite series. In particular, we shall seek solutions which are representable as power series.

We point out that not all first-order differential equations possess solutions which can be represented as power series, and it is beyond the scope of this book to consider conditions under which a first-order differential equation does possess such solutions. In order to explain the power series methods we shall *assume* that power series solutions actually do exist, realizing that this is an assumption which is not always justified.

Specifically, we consider the initial-value problem consisting of the differential equation

$$\frac{dy}{dx} = f(x, y) \tag{8.1}$$

and the initial condition

$$y(x_0) = y_0 \tag{8.6}$$

and *assume* that the differential equation (8.1) possesses a solution which is representable as a power series in powers of $(x - x_0)$. That is, we assume that the differential equation (8.1) has a solution of the form

$$y = c_0 + c_1(x - x_0) + c_2(x - x_0)^2 + \cdots = \sum_{n=0}^{\infty} c_n(x - x_0)^n \tag{8.7}$$

which is valid in some interval about the point x_0. We now consider methods of determining the coefficients $c_0, c_1, c_2, \ldots$ in (8.7) so that the series (8.7) actually does satisfy the differential equation (8.1).

B. The Taylor Series Method

We thus assume that the initial-value problem consisting of the differential equation (8.1) and the initial condition (8.6) has a solution of the form (8.7) which is valid in some interval about x_0. Then by Taylor's theorem, for each x in this interval the value $y(x)$ of this solution is given by

$$y(x) = y(x_0) + y'(x_0)(x - x_0) + \frac{y''(x_0)}{2!}(x - x_0)^2 + \cdots$$

$$= \sum_{n=0}^{\infty} \frac{y^{(n)}(x_0)}{n!}(x - x_0)^n. \quad (8.8)$$

From the initial condition (8.6), we have

$$y(x_0) = y_0,$$

and from the differential equation (8.1) itself,

$$y'(x_0) = f(x_0, y_0).$$

Substituting these values of $y(x_0)$ and $y'(x_0)$ into the series in (8.8), we obtain the first two coefficients of the desired series solution (8.7). Now differentiating the differential equation (8.1), we obtain

$$\frac{d^2y}{dx^2} = \frac{d}{dx}[f(x, y)] = f_x(x, y) + f_y(x, y)\frac{dy}{dx}$$

$$= f_x(x, y) + f_y(x, y)f(x, y), \quad (8.9)$$

where we use subscripts to denote partial differentiations. From this we obtain

$$y''(x_0) = f_x(x_0, y_0) + f_y(x_0, y_0)f(x_0, y_0).$$

Substituting this value of $y''(x_0)$ into (8.8), we obtain the third coefficient in the series solution (8.7). Proceeding in like manner, we differentiate (8.9) successively to obtain

$$\frac{d^3y}{dx^3}, \frac{d^4y}{dx^4}, \ldots, \frac{d^ny}{dx^n}, \ldots.$$

From these we obtain the values

$$y'''(x_0), y^{(iv)}(x_0), \ldots, y^{(n)}(x_0), \ldots.$$

Substituting these values into (8.8), we obtain the fourth and following coefficients in the series solution (8.7). Thus the coefficients in the series solution (8.7) are successively determined.

▶ Example 8.4. Use the Taylor series method to obtain a power series solution of the initial-value problem

$$\frac{dy}{dx} = x^2 + y^2, \quad (8.10)$$

$$y(0) = 1, \quad (8.11)$$

in powers of x.

Solution. Since we seek a solution in powers of x, we set $x_0 = 0$ in (8.7) and thus assume a solution of the form

$$y = c_0 + c_1 x + c_2 x^2 + \cdots = \sum_{n=0}^{\infty} c_n x^n.$$

By Taylor's theorem, we know that for each x in the interval where this solution is valid

$$y(x) = y(0) + y'(0)x + \frac{y''(0)}{2!}x^2 + \cdots = \sum_{n=0}^{\infty} \frac{y^{(n)}(0)}{n!}x^n. \qquad (8.12)$$

The initial condition (8.11) states that

$$y(0) = 1, \qquad (8.13)$$

and from the differential equation (8.10) we see that

$$y'(0) = 0^2 + 1^2 = 1. \qquad (8.14)$$

Differentiating (8.10) successively, we obtain

$$\frac{d^2y}{dx^2} = 2x + 2y\frac{dy}{dx}, \qquad (8.15)$$

$$\frac{d^3y}{dx^3} = 2 + 2y\frac{d^2y}{dx^2} + 2\left(\frac{dy}{dx}\right)^2, \qquad (8.16)$$

$$\frac{d^4y}{dx^4} = 2y\frac{d^3y}{dx^3} + 6\frac{dy}{dx}\frac{d^2y}{dx^2}. \qquad (8.17)$$

Substituting $x = 0$, $y = 1$, $\dfrac{dy}{dx} = 1$ into (8.15), we obtain

$$y''(0) = 2(0) + 2(1)(1) = 2. \qquad (8.18)$$

Substituting $y = 1$, $\dfrac{dy}{dx} = 1$, $\dfrac{d^2y}{dx^2} = 2$ into (8.16), we obtain

$$y'''(0) = 2 + 2(1)(2) + 2(1)^2 = 8. \qquad (8.19)$$

Finally, substituting $y = 1$, $\dfrac{dy}{dx} = 1$, $\dfrac{d^2y}{dx^2} = 2$, $\dfrac{d^3y}{dx^3} = 8$ into (8.17), we find that

$$y^{(iv)}(0) = (2)(1)(8) + (6)(1)(2) = 28. \qquad (8.20)$$

By successive differentiation of (8.17), we could proceed to determine

$$\frac{d^5y}{dx^5}, \frac{d^6y}{dx^6}, \ldots,$$

and hence obtain

$$y^{(v)}(0), y^{(vi)}(0), \ldots.$$

Now substituting the values given by (8.13), (8.14), (8.18), (8.19), and (8.20) into (8.12), we obtain the first five coefficients of the desired series solution. We thus find the solution

$$y = 1 + x + \frac{2}{2!}x^2 + \frac{8}{3!}x^3 + \frac{28}{4!}x^4 + \cdots$$

$$= 1 + x + x^2 + \frac{4}{3}x^3 + \frac{7}{6}x^4 + \cdots. \tag{8.21}$$

C. *The Method of Undetermined Coefficients*

We now consider an alternative method for obtaining the coefficients $c_0, c_1, c_2, \ldots$ in the assumed series solution

$$y = c_0 + c_1(x - x_0) + c_2(x - x_0)^2 + \cdots = \sum_{n=0}^{\infty} c_n(x - x_0)^n \tag{8.7}$$

of the problem consisting of the differential equation (8.1) with initial condition (8.6). We shall refer to this alternative method as the method of undetermined coefficients. In order to apply it we assume that $f(x, y)$ in the differential equation (8.1) is representable in the form

$$f(x, y) = a_{00} + a_{10}(x - x_0) + a_{01}(y - y_0) + a_{20}(x - x_0)^2 + a_{11}(x - x_0)(y - y_0) + a_{02}(y - y_0)^2 + \cdots. \tag{8.22}$$

The coefficients a_{ij} in (8.22) may be found by Taylor's formula for functions of two variables, although in many simple cases the use of this formula is unnecessary. Using the representation (8.22) for $f(x, y)$, the differential equation (8.1) takes the form

$$\frac{dy}{dx} = a_{00} + a_{10}(x - x_0) + a_{01}(y - y_0) + a_{20}(x - x_0)^2 + a_{11}(x - x_0)(y - y_0) + a_{02}(y - y_0)^2 + \cdots. \tag{8.23}$$

Now assuming that the series (8.7) converges in some interval $|x - x_0| < r$ $(r > 0)$ about x_0, we may differentiate (8.7) term by term and the resulting series will also converge on $|x - x_0| < r$ and represent $y'(x)$ there. Doing this we thus obtain

$$\frac{dy}{dx} = c_1 + 2c_2(x - x_0) + 3c_3(x - x_0)^2 + \cdots. \tag{8.24}$$

We note that in order for the series (8.7) to satisfy the initial condition (8.6) that $y = y_0$ at $x = x_0$, we must have $c_0 = y_0$ and hence

$$y - y_0 = c_1(x - x_0) + c_2(x - x_0)^2 + \cdots. \tag{8.25}$$

Now substituting (8.7) and (8.24) into the differential equation (8.23), and making use of (8.25), we find that

$$\begin{aligned} c_1 &+ 2c_2(x - x_0) + 3c_3(x - x_0)^2 + \cdots \\ &= a_{00} + a_{10}(x - x_0) + a_{01}[c_1(x - x_0) + c_2(x - x_0)^2 + \cdots] \\ &\quad + a_{20}(x - x_0)^2 + a_{11}(x - x_0)[c_1(x - x_0) + c_2(x - x_0)^2 + \cdots] \\ &\quad + a_{02}[c_1(x - x_0) + c_2(x - x_0)^2 + \cdots]^2 + \cdots. \end{aligned} \tag{8.26}$$

Performing the multiplications indicated in the right member of (8.26) and then combining like powers of $(x - x_0)$, we see that (8.26) takes the form

$$c_1 + 2c_2(x - x_0) + 3c_3(x - x_0)^2 + \cdots$$
$$= a_{00} + (a_{10} + a_{01}c_1)(x - x_0)$$
$$+ (a_{01}c_2 + a_{20} + a_{11}c_1 + a_{02}c_1^2)(x - x_0)^2 + \cdots. \tag{8.27}$$

In order that (8.27) be satisfied for all values of x in the interval $|x - x_0| < r$, the coefficients of like powers of $(x - x_0)$ on both sides of (8.27) must be equal. Equating these coefficients, we obtain

$$\begin{aligned} c_1 &= a_{00}, \\ 2c_2 &= a_{10} + a_{01}c_1, \\ 3c_3 &= a_{01}c_2 + a_{20} + a_{11}c_1 + a_{02}c_1^2, \\ &\vdots \end{aligned} \tag{8.28}$$

From the conditions (8.28) we determine successively the coefficients $c_1, c_2, c_3, \ldots$ of the series solution (8.7). From the first of conditions (8.28) we first obtain c_1 as the known coefficient a_{00}. Then from the second of conditions (8.28) we obtain c_2 in terms of the known coefficients a_{10} and a_{01} and the coefficient c_1 just determined. Thus we obtain $c_2 = \frac{1}{2}(a_{10} + a_{01}a_{00})$. In like manner, we proceed to determine c_3, $c_4, \ldots$. We observe that in general each coefficient c_n is thus given in terms of the known coefficients a_{ij} in the expansion (8.22) and the previously determined coefficients $c_1, c_2, \ldots, c_{n-1}$.

Finally, we substitute the coefficients $c_0, c_1, c_2, \ldots$ so determined into the series (8.7) and thereby obtain the desired solution.

▶ Example 8.5. Use the method of undetermined coefficients to obtain a power series solution of the initial-value problem

$$\frac{dy}{dx} = x^2 + y^2, \tag{8.10}$$

$$y(0) = 1, \tag{8.11}$$

in powers of x.

Solution. Since $x_0 = 0$, the assumed solution (8.7) is of the form

$$y = c_0 + c_1x + c_2x^2 + c_3x^3 + \cdots. \tag{8.29}$$

In order to satisfy the initial condition (8.11), we must have $c_0 = 1$ and hence the series (8.29) takes the form

$$y = 1 + c_1x + c_2x^2 + c_3x^3 + \cdots. \tag{8.30}$$

Differentiating (8.30), we obtain

$$\frac{dy}{dx} = c_1 + 2c_2x + 3c_3x^2 + 4c_4x^3 + \cdots. \tag{8.31}$$

For the differential equation (8.10) we have $f(x, y) = x^2 + y^2$. Since $x_0 = 0$ and $y_0 = 1$, we must expand $x^2 + y^2$ in the form

$$\sum_{i,j=0}^{\infty} a_{ij} x^i (y - 1)^j.$$

Since $y^2 = (y - 1)^2 + 2(y - 1) + 1$, the desired expansion is given by

$$x^2 + y^2 = 1 + 2(y - 1) + x^2 + (y - 1)^2.$$

Thus the differential equation (8.10) takes the form

$$\frac{dy}{dx} = 1 + 2(y - 1) + x^2 + (y - 1)^2. \tag{8.32}$$

Now substituting (8.30) and (8.31) into the differential equation (8.32), we obtain

$$\begin{aligned} &c_1 + 2c_2x + 3c_3x^2 + 4c_4x^3 + \cdots \\ &\quad = 1 + 2(c_1x + c_2x^2 + c_3x^3 + \cdots) + x^2 + (c_1x + c_2x^2 + \cdots)^2. \end{aligned} \tag{8.33}$$

Performing the indicated multiplications and collecting like powers of x in the right member of (8.33), we see that it takes the form

$$\begin{aligned} &c_1 + 2c_2x + 3c_3x^2 + 4c_4x^3 + \cdots \\ &\quad = 1 + 2c_1x + (2c_2 + 1 + c_1^2)x^2 + (2c_3 + 2c_1c_2)x^3 + \cdots. \end{aligned} \tag{8.34}$$

Equating the coefficients of the like powers of x in both members of (8.34), we obtain the conditions

$$\begin{aligned} c_1 &= 1, \\ 2c_2 &= 2c_1, \\ 3c_3 &= 2c_2 + 1 + c_1^2, \\ 4c_4 &= 2c_3 + 2c_1c_2, \\ &\vdots \end{aligned} \tag{8.35}$$

From the conditions (8.35), we obtain successively

$$\begin{aligned} c_1 &= 1, \\ c_2 &= c_1 = 1, \\ c_3 &= \tfrac{1}{3}(2c_2 + 1 + c_1^2) = \tfrac{4}{3}, \\ c_4 &= \tfrac{1}{4}(2c_3 + 2c_1c_2) = \tfrac{1}{4}(\tfrac{14}{3}) = \tfrac{7}{6}, \\ &\vdots \end{aligned} \tag{8.36}$$

Substituting the coefficients so determined in (8.36) into the series (8.30), we obtain the first five terms of the desired series solution. We thus find

$$y = 1 + x + x^2 + \tfrac{4}{3}x^3 + \tfrac{7}{6}x^4 + \cdots.$$

We note that this is of course the same series previously obtained by the Taylor series method and already given by (8.21).

Remark. We have made but little mention of the interval of convergence of the series involved in our discussion. We have merely assumed that a power series solution exists and converges on some interval $|x - x_0| < r$ $(r > 0)$ about the initial point x_0. In a practical problem the interval of convergence is of vital concern and should be determined, if possible. Another matter of great importance in a practical problem is the determination of the number of terms which have to be found in order to be certain of a sufficient degree of accuracy. We shall not consider these matters here. Our primary purpose has been merely to explain the details of the methods. We refer the reader to more advanced treatises for discussions of the important questions of convergence and accuracy.

Exercises

Obtain a power series solution in powers of x of each of the initial-value problems in Exercises 1 through 7 by (a) the Taylor series method and (b) the method of undetermined coefficients.

1. $\dfrac{dy}{dx} = x + y, \quad y(0) = 1.$
2. $\dfrac{dy}{dx} = x^2 + 2y^2, \quad y(0) = 4.$
3. $\dfrac{dy}{dx} = 1 + xy^2, \quad y(0) = 2.$
4. $\dfrac{dy}{dx} = x^3 + y^3, \quad y(0) = 3.$
5. $\dfrac{dy}{dx} = x + \sin y, \quad y(0) = 0.$
6. $\dfrac{dy}{dx} = 1 + x \sin y, \quad y(0) = 0.$
7. $\dfrac{dy}{dx} = e^x + x \cos y, \quad y(0) = 0.$

Obtain a power series solution in powers of $x - 1$ of each of the initial-value problems in Exercises 8 through 12 by (a) the Taylor series method and (b) the method of undetermined coefficients.

8. $\dfrac{dy}{dx} = x^2 + y^2, \quad y(1) = 4.$
9. $\dfrac{dy}{dx} = x^3 + y^2, \quad y(1) = 1.$
10. $\dfrac{dy}{dx} = x + y + y^2, \quad y(1) = 1.$
11. $\dfrac{dy}{dx} = x + \cos y, \quad y(1) = \pi.$
12. $\dfrac{dy}{dx} = x^2 + x \sin y, \quad y(1) = \dfrac{\pi}{2}.$

8.3 The Method of Successive Approximations

A. The Method

We again consider the initial-value problem consisting of the differential equation

$$\frac{dy}{dx} = f(x, y) \tag{8.1}$$

and the initial condition

$$y(x_0) = y_0. \tag{8.6}$$

We now outline the Picard method of successive approximations for finding a solution of this problem which is valid on some interval which includes the initial point x_0.

The first step of the Picard method is quite unlike anything which we have done before and at first glance it appears to be rather fruitless. For the first step actually consists of making a guess at the solution! That is, we choose a function ϕ_0 and call it a "zeroth approximation" to the actual solution. How do we make this guess? In other words, what function do we choose? Actually, many different choices could be made. The only thing that we know about the actual solution is that in order to satisfy the initial condition (8.6) it must assume the value y_0 at $x = x_0$. Therefore it would seem reasonable to choose for ϕ_0 a function which assumes this value y_0 at $x = x_0$. Although this requirement is not essential, it certainly seems as sensible as anything else. In particular, it is often convenient to choose for ϕ_0 the constant function which has the value y_0 for all x. While this choice is certainly not essential, it is often the simplest, most reasonable choice which quickly comes to mind.

In summary, then, the first step of the Picard method is to choose a function ϕ_0 which will serve as a zeroth approximation.

Having thus chosen a zeroth approximation ϕ_0, we now determine a first approximation ϕ_1 in the following manner. We determine $\phi_1(x)$ so that (1) it satisfies the differential equation obtained from (8.1) by replacing y in $f(x, y)$ by $\phi_0(x)$, and (2) it satisfies the initial condition (8.6). Thus ϕ_1 is determined such that

$$\frac{d}{dx}[\phi_1(x)] = f[x, \phi_0(x)] \tag{8.37}$$

and

$$\phi_1(x_0) = y_0. \tag{8.38}$$

We now assume that $f[x, \phi_0(x)]$ is continuous. Then ϕ_1 satisfies (8.37) and (8.38) if and only if

$$\phi_1(x) = y_0 + \int_{x_0}^{x} f[t, \phi_0(t)]\, dt. \tag{8.39}$$

From this equation the first approximation ϕ_1 is determined.

We now determine the second approximation ϕ_2 in a similar manner. The function ϕ_2 is determined such that

$$\frac{d}{dx}[\phi_2(x)] = f[x, \phi_1(x)] \tag{8.40}$$

and

$$\phi_2(x_0) = y_0. \tag{8.41}$$

Assuming that $f[x, \phi_1(x)]$ is continuous, then ϕ_2 satisfies (8.40) and (8.41) if and only if

$$\phi_2(x) = y_0 + \int_{x_0}^{x} f[t, \phi_1(t)]\, dt. \tag{8.42}$$

From this equation the second approximation ϕ_2 is determined.

We now proceed in like manner to determine a third approximation ϕ_3, a fourth approximation ϕ_4, and so on. The nth approximation ϕ_n is determined from

$$\phi_n(x) = y_0 + \int_{x_0}^{x} f[t, \phi_{n-1}(t)]\, dt, \tag{8.43}$$

where ϕ_{n-1} is the $(n - 1)$st approximation. We thus obtain a sequence of functions

$$\phi_0, \phi_1, \phi_2, \ldots, \phi_n, \ldots,$$

where ϕ_0 is chosen, ϕ_1 is determined from (8.39), ϕ_2 is determined from (8.42), ..., and in general ϕ_n is determined from (8.43) for $n \geq 1$.

Now just how does this sequence of functions relate to the actual solution of the initial-value problem under consideration? It can be proved, under certain general conditions and for x restricted to a sufficiently small interval about the initial point x_0, that (1) as $n \to \infty$ the sequence of functions ϕ_n defined by (8.43) for $n \geq 1$ approaches a limit function ϕ, and (2) this limit function ϕ satisfies both the differential equation (8.1) and the initial condition (8.6). That is, under suitable restrictions the function ϕ defined by

$$\phi = \lim_{n \to \infty} \phi_n$$

is the exact solution of the initial-value problem under consideration. Furthermore, the error in approximating the exact solution ϕ by the nth approximation ϕ_n will be arbitrarily small provided that n is sufficiently large and that x is sufficiently close to the initial point x_0.

B. An Example; Remarks on The Method

We illustrate the Picard method of successive approximations by applying it to the initial-value problem of Examples 8.4 and 8.5.

► **Example 8.6.** Use the method of successive approximations to find a sequence of functions which approaches the solution of the initial-value problem

$$\frac{dy}{dx} = x^2 + y^2, \tag{8.10}$$

$$y(0) = 1. \tag{8.11}$$

Solution. Our first step is to choose a function for the zeroth approximation ϕ_0. Since the initial value of y is 1, it would seem reasonable to choose for ϕ_0 the constant function which has the value 1 for all x. Thus, we let ϕ_0 be such that

$$\phi_0(x) = 1$$

for all x. The nth approximation ϕ_n for $n \geq 1$ is given by formula (8.43). Since $f(x, y) = x^2 + y^2$ in the differential equation (8.10), the formula (8.43) becomes in this case

$$\phi_n(x) = 1 + \int_0^x \{t^2 + [\phi_{n-1}(t)]^2\}\, dt, \qquad n \geq 1.$$

Using this formula for $n = 1, 2, 3, \ldots$, we obtain successively

$$\phi_1(x) = 1 + \int_0^x \{t^2 + [\phi_0(t)]^2\}\,dt = 1 + \int_0^x (t^2 + 1)\,dt = 1 + x + \frac{x^3}{3},$$

$$\phi_2(x) = 1 + \int_0^x \{t^2 + [\phi_1(t)]^2\}\,dt = 1 + \int_0^x \left[t^2 + \left(1 + t + \frac{t^3}{3}\right)^2\right] dt$$

$$= 1 + \int_0^x \left(1 + 2t + 2t^2 + \frac{2t^3}{3} + \frac{2t^4}{3} + \frac{t^6}{9}\right) dt$$

$$= 1 + x + x^2 + \frac{2x^3}{3} + \frac{x^4}{6} + \frac{2x^5}{15} + \frac{x^7}{63},$$

$$\phi_3(x) = 1 + \int_0^x \{t^2 + [\phi_2(t)]^2\}\,dt$$

$$= 1 + \int_0^x \left[t^2 + \left(1 + t + t^2 + \frac{2t^3}{3} + \frac{t^4}{6} + \frac{2t^5}{15} + \frac{t^7}{63}\right)^2\right] dt$$

$$= 1 + \int_0^x \left[1 + 2t + 4t^2 + \frac{10t^3}{3} + \frac{8t^4}{3} + \frac{29t^5}{15} + \frac{47t^6}{45} + \frac{164t^7}{315} + \frac{299t^8}{1260} + \frac{8t^9}{105} + \frac{184t^{10}}{4725} + \frac{t^{11}}{189} + \frac{4t^{12}}{945} + \frac{t^{14}}{3969}\right] dt$$

$$= 1 + x + x^2 + \frac{4x^3}{3} + \frac{5x^4}{6} + \frac{8x^5}{15} + \frac{29x^6}{90} + \frac{47x^7}{315} + \frac{41x^8}{630} + \frac{299x^9}{11{,}340} + \frac{4x^{10}}{525} + \frac{184x^{11}}{51{,}975} + \frac{x^{12}}{2268} + \frac{4x^{13}}{12{,}285} + \frac{x^{15}}{59{,}535}.$$

$$\vdots$$

We have chosen the zeroth approximation ϕ_0 and then found the next three approximations ϕ_1, ϕ_2, and ϕ_3 explicitly. We could proceed in like manner to find $\phi_4, \phi_5, \ldots$ explicitly and thus determine successively the members of the sequence $\{\phi_n\}$ which approaches the exact solution of the initial-value problem under consideration. However, we believe that the procedure is now clear, and for rather obvious reasons we shall not proceed further with this problem.

Remarks. The greatest disadvantage of the method of successive approximations is that it leads to tedious, involved, and sometimes impossible calculations. This is amply illustrated in Example 8.6. At best the calculations are usually very complicated, and in general it is impossible to carry through more than a few of the successive integrations exactly. Nevertheless, the method is of practical importance, for the first few approximations alone are sometimes quite accurate.

However, the principal use of the method of successive approximations is in proving existence theorems. Concerning this, we refer the reader to Chapter 10 of the author's *Differential Equations*.

Exercises

For each of the initial-value problems in Exercises 1 through 8 use the method of successive approximations to find the first three members ϕ_1, ϕ_2, ϕ_3 of a sequence of functions which approaches the exact solution of the problem.

1. $\dfrac{dy}{dx} = xy, \quad y(0) = 1.$
2. $\dfrac{dy}{dx} = x + y, \quad y(0) = 1.$
3. $\dfrac{dy}{dx} = x + y^2, \quad y(0) = 0.$
4. $\dfrac{dy}{dx} = 1 + xy^2, \quad y(0) = 0.$
5. $\dfrac{dy}{dx} = e^x + y^2, \quad y(0) = 0.$
6. $\dfrac{dy}{dx} = \sin x + y^2, \quad y(0) = 0.$
7. $\dfrac{dy}{dx} = 2x + y^3, \quad y(0) = 0.$
8. $\dfrac{dy}{dx} = 1 + 6xy^4, \quad y(0) = 0.$

8.4 Numerical Methods

A. *Introduction; A Problem for Illustration*

In this section we introduce certain basic numerical methods for approximating the solution of the initial-value problem consisting of the differential equation

$$\frac{dy}{dx} = f(x, y) \tag{8.1}$$

and the initial condition

$$y(x_0) = y_0. \tag{8.6}$$

Numerical methods employ the differential equation (8.1) and the condition (8.6) to obtain approximations to the values of the solution corresponding to various, selected values of x. To be more specific, let y denote the solution of the problem and let h denote a positive *increment* in x. The initial condition (8.6) tells us that $y = y_0$ at $x = x_0$. A numerical method will employ the differential equation (8.1) and the condition (8.6) to approximate successively the values of y at $x_1 = x_0 + h$, $x_2 = x_1 + h$, $x_3 = x_2 + h, \ldots$.

Let us denote these approximate values of y by $y_1, y_2, y_3, \ldots$, respectively. That is, we let y_n denote the approximate value of y at $x = x_n = x_0 + nh$, $n = 1, 2, 3, \ldots$. Now all that we know about y before starting is that $y = y_0$ at $x = x_0$. In order to get started we need a method which requires only the value y_n in order to obtain the next value y_{n+1}. For we can apply such a method with $n = 0$ and use the initial value y_0 to obtain the next value y_1. A method which uses only y_n to find y_{n+1}, and which therefore enables us to get started, is called a *starting method.* Once we have used a starting method to find y_1, we can repeat it with $n = 1$ to find y_2, with $n = 2$ to find y_3, and so forth. However, once we have several calculated values at our disposal, it is often convenient to change over to a method which uses both y_n and one or more *preceding* values $y_{n-1}, y_{n-2}, \ldots$ to find the next value y_{n+1}. Such a method, which enables us to continue once we have got sufficiently well started, is called a *continuing method.* Most of our attention in this text will be devoted to starting methods.

Our principal objective in this section is to present the actual details of certain basic numerical methods for solving first-order initial-value problems. In general, we shall not consider the theoretical justifications of these methods, nor shall we enter into detailed discussions of matters such as accuracy and error.

Before turning to the details of the numerical methods to be considered, we introduce a simple initial-value problem which we shall use for purposes of illustration throughout this section. We consider the problem

$$\frac{dy}{dx} = 2x + y, \tag{8.2}$$

$$y(0) = 1. \tag{8.44}$$

We have already employed the differential equation (8.2) to illustrate the graphical methods of Section 8.1. We note at once that it is a linear differential equation and hence it can be solved exactly. Using the methods of Section 2.3, we find at once that its general solution is

$$y = -2(x + 1) + ce^x, \tag{8.45}$$

where c is an arbitrary constant. Applying the initial condition (8.44) to (8.45), we find that the exact solution of the initial-value problem consisting of (8.2) and (8.44) is

$$y = -2(x + 1) + 3e^x. \tag{8.46}$$

We have chosen the problem consisting of (8.2) and (8.44) for illustrative purposes for two reasons. First, the differential equation (8.2) is so simple, that numerical methods may be applied to it without introducing involved computations which might obscure the main steps of the method to a beginner. Second, since the exact solution (8.46) of the problem has been found, we can compare approximate solution values obtained numerically with this exact solution and thereby gain some insight into the accuracy of numerical methods.

Of course in practice we would not solve a simple linear differential equation such as (8.2) by a numerical method. The methods of this section are actually designed for equations which can *not* be solved exactly and for equations which, although solvable exactly, have exact solutions so unwieldy that they are practically useless.

B. The Euler Method

The Euler method is very simple but not very practical. An understanding of it, however, paves the way for an understanding of the more practical (but also more complicated) methods which follow.

Let y denote the exact solution of the initial-value problem which consists of the differential equation

$$\frac{dy}{dx} = f(x, y) \tag{8.1}$$

and the initial condition

$$y(x_0) = y_0. \tag{8.6}$$

Let h denote a positive increment in x and let $x_1 = x_0 + h$. Then

$$\int_{x_0}^{x_1} f(x, y)\,dx = \int_{x_0}^{x_1} \frac{dy}{dx}\,dx = y(x_1) - y(x_0).$$

Since y_0 denotes the value $y(x_0)$ of the exact solution y at x_0, we have

$$y(x_1) = y_0 + \int_{x_0}^{x_1} f(x, y)\,dx. \tag{8.47}$$

If we assume that $f(x, y)$ varies slowly on the interval $x_0 \le x \le x_1$, then we can approximate $f(x, y)$ in (8.47) by its value $f(x_0, y_0)$ at the left endpoint x_0. Then

$$y(x_1) \approx y_0 + \int_{x_0}^{x_1} f(x_0, y_0)\,dx.$$

But

$$\int_{x_0}^{x_1} f(x_0, y_0)\,dx = f(x_0, y_0)(x_1 - x_0) + hf(x_0, y_0).$$

Thus

$$y(x_1) \approx y_0 + hf(x_0, y_0).$$

Thus we obtain the approximate value y_1 of y at $x_1 = x_0 + h$ by the formula

$$y_1 = y_0 + hf(x_0, y_0). \tag{8.48}$$

Having obtained y_1 by formula (8.48), we proceed in like manner to obtain y_2 by the formula $y_2 = y_1 + hf(x_1, y_1)$, y_3 by the formula $y_3 = y_2 + hf(x_2, y_2)$, and so forth. In general we find y_{n+1} in terms of y_n by the formula

$$y_{n+1} = y_n + hf(x_n, y_n). \tag{8.49}$$

Before illustrating the method, we give a useful geometric interpretation. The graph of the exact solution y is a curve C in the xy plane (see Figure 8.5). Let P denote the initial point (x_0, y_0) and let T be the tangent to C at P. Let Q be the point at which the line $x = x_1$ intersects C and let N be the point at which this line intersects T. Then the exact value of y at x_1 is represented by LQ. The approximate value y_1 is represented by LN, since $LN = LM + MN = y_0 + PM \tan\theta = y_0 + hf(x_0, y_0)$. The error in approximating the exact value of y at x_1 by y_1 is thus represented by NQ. The figure suggests that if h is sufficiently small, then this error NQ will also be small and hence that the approximation will be good.

It is clear that the Euler method is indeed simple, but it should also be apparent from our discussion why it is not very practical. If the increment h is *not* very small, then the errors in the approximations generally will not be small and thus the method will lead to quite inaccurate results. If the increment h *is* very small, then the computations will be more lengthy and so the method will involve tedious and time-consuming labor.

▶ Example 8.7. Apply the Euler method to the initial-value problem

$$\frac{dy}{dx} = 2x + y, \tag{8.2}$$

$$y(0) = 1. \tag{8.44}$$

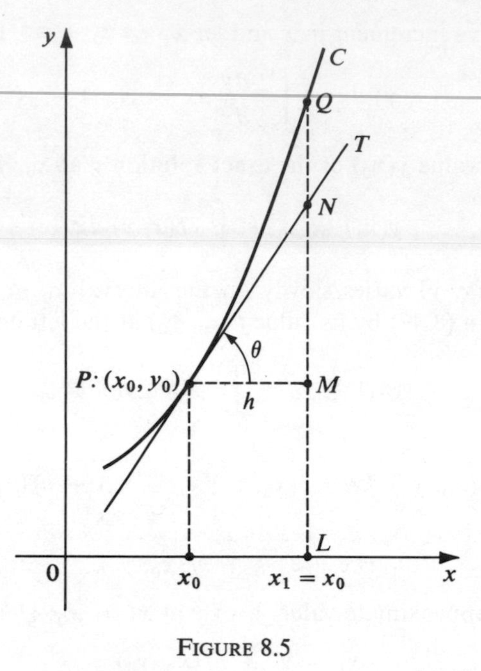

FIGURE 8.5

Employ the method to approximate the value of the solution y at $x = 0.2, 0.4, 0.6, 0.8$, and 1 using (1) $h = 0.2$, and (2) $h = 0.1$. Obtain results to three figures after the decimal point. Compare with the exact value determined from (8.46).

Solution. 1. We use formula (8.49) with $f(x, y) = 2x + y$ and $h = 0.2$. From the initial condition (8.44), we have $x_0 = 0$, $y_0 = 1$. We now proceed with the calculations.

(a) $x_1 = x_0 + h = 0.2, f(x_0, y_0) = f(0, 1) = 1.000,$
$y_1 = y_0 + hf(x_0, y_0) = 1.000 + 0.2(1.000) = 1.200.$

(b) $x_2 = x_1 + h = 0.4, f(x_1, y_1) = f(0.2, 1.200) = 1.600,$
$y_2 = y_1 + hf(x_1, y_1) = 1.200 + 0.2(1.600) = 1.520.$

(c) $x_3 = x_2 + h = 0.6, f(x_2, y_2) = f(0.4, 1.520) = 2.320,$
$y_3 = y_2 + hf(x_2, y_2) = 1.520 + 0.2(2.320) = 1.984.$

(d) $x_4 = x_3 + h = 0.8, f(x_3, y_3) = f(0.6, 1.984) = 3.184,$
$y_4 = y_3 + hf(x_3, y_3) = 1.984 + 0.2(3.184) = 2.621.$

(e) $x_5 = x_4 + h = 1.0, f(x_4, y_4) = f(0.8, 2.621) = 4.221,$
$y_5 = y_4 + hf(x_4, y_4) = 2.621 + 0.2(4.221) = 3.465.$

These results, corresponding to the various values of x_n, are collected in the second column of Table 8.2.

2. We now use formula (8.49) with $f(x, y) = 2x + y$ and $h = 0.1$. Of course we again have $x_0 = 0$, $y_0 = 1$. The calculations follow.

(a) $x_1 = x_0 + h = 0.1, f(x_0, y_0) = f(0, 1) = 1.000,$
$y_1 = y_0 + hf(x_0, y_0) = 1.000 + 0.1(1.000) = 1.100.$

(b) $x_2 = x_1 + h = 0.2, f(x_1, y_1) = f(0.1, 1.100) = 1.300,$
$y_2 = y_1 + hf(x_1, y_1) = 1.100 + 0.1(1.300) = 1.230.$

(c) $x_3 = x_2 + h = 0.3, f(x_2, y_2) = f(0.2, 1.230) = 1.630,$
$y_3 = y_2 + hf(x_2, y_2) = 1.230 + 0.1(1.630) = 1.393.$

(d) $x_4 = x_3 + h = 0.4, f(x_3, y_3) = f(0.3, 1.393) = 1.993,$
$y_4 = y_3 + hf(x_3, y_3) = 1.393 + 0.1(1.993) = 1.592.$

(e) $x_5 = x_4 + h = 0.5, f(x_4, y_4) = f(0.4, 1.592) = 2.392,$
$y_5 = y_4 + hf(x_4, y_4) = 1.592 + 0.1(2.392) = 1.831.$

(f) $x_6 = x_5 + h = 0.6, f(x_5, y_5) = f(0.5, 1.831) = 2.831,$
$y_6 = y_5 + hf(x_5, y_5) = 1.831 + 0.1(2.831) = 2.114.$

(g) $x_7 = x_6 + h = 0.7, f(x_6, y_6) = f(0.6, 2.114) = 3.314,$
$y_7 = y_6 + hf(x_6, y_6) = 2.114 + 0.1(3.314) = 2.445.$

(h) $x_8 = x_7 + h = 0.8, f(x_7, y_7) = f(0.7, 2.445) = 3.845,$
$y_8 = y_7 + hf(x_7, y_7) = 2.445 + 0.1(3.845) = 2.830.$

(i) $x_9 = x_8 + h = 0.9, f(x_8, y_8) = f(0.8, 2.830) = 4.430,$
$y_9 = y_8 + hf(x_8, y_8) = 2.830 + 0.1(4.430) = 3.273.$

(j) $x_{10} = x_9 + h = 1.0, f(x_9, y_9) = f(0.9, 3.273) = 5.073,$
$y_{10} = y_9 + hf(x_9, y_9) = 3.273 + 0.1(5.073) = 3.780.$

These results are collected in the third column of Table 8.2. The values of the exact solution y, computed from (8.46) to three figures after the decimal point, are listed in the fourth column of Table 8.2.

TABLE 8.2

x_n	y_n using $h = 0.2$	y_n using $h = 0.1$	y
0.0	1.000	1.000	1.000
0.1	—	1.100	1.116
0.2	1.200	1.230	1.264
0.3	—	1.393	1.450
0.4	1.520	1.592	1.675
0.5	—	1.831	1.946
0.6	1.984	2.114	2.266
0.7	—	2.445	2.641
0.8	2.621	2.830	3.076
0.9	—	3.273	3.579
1.0	3.465	3.780	4.155

From this table we compute the errors involved in both approximations at $x = 0.2$, 0.4, 0.6, 0.8, and 1.0. These errors are tabulated in Table 8.3.

TABLE 8.3

x_n	*Error using* $h = 0.2$	*Error using* $h = 0.1$
0.2	0.064	0.034
0.4	0.155	0.083
0.6	0.282	0.152
0.8	0.455	0.246
1.0	0.690	0.375

A study of these tables illustrates two important facts concerning the Euler method. First, for a fixed value of h, the error becomes greater and greater as we proceed over a larger and larger range away from the initial point. Second, for a fixed value of x_n, the error is smaller if the value of h is smaller.

Exercises

1. Consider the initial-value problem

$$\frac{dy}{dx} = x - 2y,$$

$$y(0) = 1.$$

(a) Apply the Euler method to approximate the values of the solution y at $x = 0.1, 0.2, 0.3$, and 0.4, using $h = 0.1$. Obtain results to three figures after the decimal point.

(b) Proceed as in part (a) using $h = 0.05$.

(c) Find the exact solution of the problem and determine its values at $x = 0.1$, 0.2, 0.3, and 0.4 (to three figures after the decimal point).

(d) Compare the results obtained in parts (a), (b), and (c). Tabulate errors as in Table 8.3.

2. Proceed as in Exercise 1 for the initial-value problem

$$\frac{dy}{dx} = x + y,$$

$$y(0) = 2.$$

C. The Modified Euler Method

In Section B we observed that the value $y(x_1)$ of the exact solution y of the initial-value problem

$$\frac{dy}{dx} = f(x, y), \tag{8.1}$$

$$y(x_0) = y_0, \tag{8.6}$$

at $x_1 = x_0 + h$ is given by

$$y(x_1) = y_0 + \int_{x_0}^{x_1} f(x, y)\, dx. \tag{8.47}$$

In the Euler method we approximated $f(x, y)$ in (8.47) by its value $f(x_0, y_0)$ at the left endpoint of the interval $x_0 \leq x \leq x_1$ and thereby obtained the approximation

$$y_1 = y_0 + hf(x_0, y_0) \tag{8.48}$$

for y at x_1. It seems reasonable that a more accurate value would be obtained if we were to approximate $f(x, y)$ by the *average* of its values at the left and right endpoints of $x_0 \leq x \leq x_1$, instead of simply by its value at the left endpoint x_0. This is essentially what is done in the modified Euler method, which we shall now explain.

In order to approximate $f(x, y)$ by the average of its values at x_0 and x_1, we need to know its value $f[x_1, y(x_1)]$ at x_1. However, we do not know the value $y(x_1)$ of y at x_1. We must find a first approximation $y_1^{(1)}$ for $y(x_1)$, and to do this we proceed just as we did at the start of the basic Euler method. That is, we take

$$y_1^{(1)} = y_0 + hf(x_0, y_0) \tag{8.50}$$

as the *first* approximation to the value of y at x_1. Then we approximate $f[x_1, y(x_1)]$ by $f(x_1, y_1^{(1)})$, using the value $y_1^{(1)}$ found by (8.50). From this we obtain

$$\frac{f(x_0, y_0) + f(x_1, y_1^{(1)})}{2}, \tag{8.51}$$

which is approximately the average of the values of $f(x, y)$ at the endpoints x_0, and x_1. We now replace $f(x, y)$ in (8.47) by (8.51) and thereby obtain

$$y_1^{(2)} = y_0 + \frac{f(x_0, y_0) + f(x_1, y_1^{(1)})}{2}\, h \tag{8.52}$$

as the *second* approximation to the value of y at x_1.

We now use the second approximation $y_1^{(2)}$ to obtain a second approximation $f(x_1, y_1^{(2)})$ for the value of $f(x, y)$ at x_1. From this we proceed to obtain

$$y_1^{(3)} = y_0 + \frac{f(x_0, y_0) + f(x_1, y_1^{(2)})}{2}\, h \tag{8.53}$$

as the *third* approximation to the value of y at x_1. Proceeding in this way we obtain a sequence of approximations

$$y_1^{(1)}, y_1^{(2)}, y_1^{(3)}, \ldots$$

to the value of the exact solution y at x_1. We proceed to compute the successive members of the sequence until we encounter two consecutive members which have the same value to the number of decimal places required. We take the common value of these two consecutive members as our approximation to the value of the solution y at x_1 and denote it by y_1.

Having finally approximated y at x_1 by y_1, we now move on and proceed to approximate y at $x_2 = x_1 + h$. We proceed in exactly the same way as we did in finding y_1. We find successively

$$\begin{aligned} y_2^{(1)} &= y_1 + hf(x_1, y_1), \\ y_2^{(2)} &= y_1 + \frac{f(x_1, y_1) + f(x_2, y_2^{(1)})}{2} h, \\ y_2^{(3)} &= y_1 + \frac{f(x_1, y_1) + f(x_2, y_2^{(2)})}{2} h, \\ &\vdots \end{aligned} \tag{8.54}$$

until two consecutive members of this sequence agree, thereby obtaining an approximation y_2 to the value of y at x_2.

Proceeding further in like manner one obtains an approximation y_3 to the value of y at x_3, and so forth.

▶ Example 8.8. Apply the modified Euler method to the initial-value problem

$$\frac{dy}{dx} = 2x + y, \tag{8.2}$$

$$y(0) = 1. \tag{8.44}$$

Employ the method to approximate the value of the solution y at $x = 0.2$ and $x = 0.4$ using $h = 0.2$. Obtain results to three figures after the decimal point. Compare with the results obtained using the basic Euler method with $h = 0.1$ and with the exact values (Example 8.7, Table 8.2).

Solution. Here $f(x, y) = 2x + y$, $x_0 = 0$, and $y_0 = 1$, and we are to use $h = 0.2$. We begin by approximating the value of y at $x_1 = x_0 + h = 0.2$. A first approximation $y_1^{(1)}$ to this value is found using formula (8.50). Since $f(x_0, y_0) = f(0, 1) = 1.000$, we have

$$y_1^{(1)} = y_0 + hf(x_0, y_0) = 1.000 + 0.2(1.000) = 1.200.$$

We now use (8.52) to find a second approximation $y_1^{(2)}$ to the desired value. Since $f(x_1, y_1^{(1)}) = f(0.2, 1.200) = 1.600$, we have

$$y_1^{(2)} = y_0 + \frac{f(x_0, y_0) + f(x_1, y_1^{(1)})}{2} h = 1.000 + \frac{1.000 + 1.600}{2}(0.2) = 1.260.$$

We next employ (8.53) to find a third approximation $y_1^{(3)}$. Since $f(x_1, y_1^{(2)}) = f(0.2, 1.260) = 1.660$, we find

$$y_1^{(3)} = y_0 + \frac{f(x_0, y_0) + f(x_1, y_1^{(2)})}{2} h = 1.000 + \frac{1.000 + 1.660}{2}(0.2) = 1.266.$$

Proceeding in like manner, we obtain fourth and fifth approximations $y_1^{(4)}$ and $y_1^{(5)}$, respectively. We find

$$y_1^{(4)} = y_0 + \frac{f(x_0, y_0) + f(x_1, y_1^{(3)})}{2} h = 1.000 + \frac{1.000 + 1.666}{2}(0.2) = 1.267$$

and

$$y_1^{(5)} = y_0 + \frac{f(x_0, y_0) + f(x_1, y_1^{(4)})}{2} h = 1.000 + \frac{1.000 + 1.667}{2}(0.2) = 1.267.$$

Since the approximations $y_1^{(4)}$ and $y_1^{(5)}$ are the same to the number of decimal places required, we take their common value as the approximation y_1 to the value of the solution y at $x_1 = 0.2$. That is, we take

$$y_1 = 1.267. \tag{8.55}$$

We now proceed to approximate the value of y at $x_2 = x_1 + h = 0.4$. For this purpose we employ the formulas (8.54), using $y_1 = 1.267$. We find successively

$$y_2^{(1)} = y_1 + hf(x_1, y_1) = 1.267 + 0.2(1.667) = 1.600,$$

$$y_2^{(2)} = y_1 + \frac{f(x_1, y_1) + f(x_2, y_2^{(1)})}{2} h = 1.267 + \frac{1.667 + 2.400}{2}(0.2) = 1.674,$$

$$y_2^{(3)} = y_1 + \frac{f(x_1, y_1) + f(x_2, y_2^{(2)})}{2} h = 1.267 + \frac{1.667 + 2.474}{2}(0.2) = 1.681,$$

$$y_2^{(4)} = y_1 + \frac{f(x_1, y_1) + f(x_2, y_2^{(3)})}{2} h = 1.267 + \frac{1.667 + 2.481}{2}(0.2) = 1.682,$$

$$y_2^{(5)} = y_1 + \frac{f(x_1, y_1) + f(x_2, y_2^{(4)})}{2} h = 1.267 + \frac{1.667 + 2.482}{2}(0.2) = 1.682.$$

Since the approximations $y_2^{(4)}$ and $y_2^{(5)}$ are both the same to the required number of decimal places, we take their common value as the approximation y_2 to the value of the solution y at $x_2 = 0.4$. That is, we take

$$y_2 = 1.682. \tag{8.56}$$

We compare the results (8.55) and (8.56) with those obtained using the basic Euler method with $h = 0.1$ and with the exact values. For this purpose the various results and the corresponding errors are listed in Table 8.4.

TABLE 8.4

		Using basic Euler method with $h = 0.1$		*Using modified Euler with* $h = 0.2$	
x_n	*Exact value of y (to three decimal places)*	*Approximation* y_n	*Error*	*Approximation* y_n	*Error*
0.2	1.264	1.230	0.034	1.267	0.003
0.4	1.675	1.592	0.083	1.682	0.007

The principal advantage of the modified Euler method over the basic Euler method is immediately apparent from a study of Table 8.4. The modified method is much more accurate. At $x = 0.4$ the error using the modified method with $h = 0.02$ is 0.007. The corresponding error 0.083 using the basic method with $h = 0.1$ is nearly twelve times as great, despite the fact that a smaller value of h was used in this case. Of course at each step the modified method involves more lengthy and complicated calculations than the basic method. We note, however, that the basic method would require many individual steps to give a result as accurate as that which the modified method can provide in a single step.

Exercises

1. Consider the initial-value problem

$$\frac{dy}{dx} = 3x + 2y,$$

$$y(0) = 1.$$

   (a) Apply the modified Euler method to approximate the values of the solution y at $x = 0.1, 0.2$, and 0.3 using $h = 0.1$. Obtain results to three figures after the decimal point.
   (b) Proceed as in part (a) using $h = 0.05$.
   (c) Find the exact solution of the problem and determine its values at $x = 0.1$, 0.2, and 0.3 (to three figures after the decimal point).
   (d) Compare the results obtained in parts (a), (b), and (c), and tabulate errors.

2. Proceed as in Exercise 1 for the initial-value problem

$$\frac{dy}{dx} = 2x - y,$$

$$y(0) = 3.$$

3. Consider the initial-value problem

$$\frac{dy}{dx} = x^2 + y^2,$$

$$y(0) = 1.$$

   (a) Apply the modified Euler method to approximate the values of the solution y at $x = 0.1, 0.2$, and 0.3, using $h = 0.1$. Obtain results to three figures after the decimal point.
   (b) Apply the Euler method to approximate the values of the solution y at $x = 0.1, 0.2$, and 0.3, using $h = 0.1$. Obtain results to three figures after the decimal point.
   (c) Compare the results obtained in parts (a) and (b) and tabulate errors.

D. *The Runge-Kutta Method*

We now consider the so-called Runge-Kutta method for approximating the values of the solution of the initial-value problem

$$\frac{dy}{dx} = f(x, y), \tag{8.1}$$

$$y(x_0) = y_0, \tag{8.6}$$

at $x_1 = x_0 + h$, $x_2 = x_1 + h$, and so forth. This method gives surprisingly accurate results without the need of using extremely small values of the interval h. We shall give no justification for the method but shall merely list the several formulas involved and explain how they are used.*

To approximate the value of the solution of the initial-value problem under consideration at $x_1 = x_0 + h$ by the Runge-Kutta method, we proceed in the following way. We calculate successively the numbers k_1, k_2, k_3, k_4, and K_0 defined by the formulas

$$\begin{aligned} k_1 &= hf(x_0, y_0), \\ k_2 &= hf\left(x_0 + \frac{h}{2}, y_0 + \frac{k_1}{2}\right), \\ k_3 &= hf\left(x_0 + \frac{h}{2}, y_0 + \frac{k_2}{2}\right), \\ k_4 &= hf(x_0 + h, y_0 + k_3), \end{aligned} \tag{8.57}$$

and

$$K_0 = \tfrac{1}{6}(k_1 + 2k_2 + 2k_3 + k_4).$$

Then we set

$$y_1 - y_0 + K_0 \tag{8.58}$$

and take this as the approximate value of the exact solution at $x_1 = x_0 + h$.

Having thus determined y_1, we proceed to approximate the value of the solution at $x_2 = x_1 + h$ in an exactly similar manner. Using $x_1 = x_0 + h$ and y_1 as determined by (8.58), we calculate successively the numbers k_1, k_2, k_3, k_4, and K_1 defined by

$$\begin{aligned} k_1 &= hf(x_1, y_1), \\ k_2 &= hf\left(x_1 + \frac{h}{2}, y_1 + \frac{k_1}{2}\right), \\ k_3 &= hf\left(x_1 + \frac{h}{2}, y_1 + \frac{k_2}{2}\right), \\ k_4 &= hf(x_1 + h, y_1 + k_3), \end{aligned} \tag{8.59}$$

and

$$K_1 = \tfrac{1}{6}(k_1 + 2k_2 + 2k_3 + k_4).$$

* See F. Hildebrand, *Introduction to Numerical Analysis* (McGraw-Hill, New York, 1956), Ch. 6, and H. Levy and E. Baggot, *Numerical Solutions of Differential Equations* (Dover, New York, 1950), Ch. 3.

Then we set

$$y_2 = y_1 + K_1 \tag{8.60}$$

and take this as the approximate value of the exact solution at $x_2 = x_1 + h$.

We proceed to approximate the value of the solution at $x_3 = x_2 + h$, $x_4 = x_3 + h$, and so forth, in an exactly similar manner. Letting y_n denote the approximate value obtained for the solution at $x_n = x_0 + nh$, we calculate successively k_1, k_2, k_3, k_4, and K_n defined by

$$k_1 = hf(x_n, y_n),$$

$$k_2 = hf\left(x_n + \frac{h}{2}, y_n + \frac{k_1}{2}\right),$$

$$k_3 = hf\left(x_n + \frac{h}{2}, y_n + \frac{k_2}{2}\right),$$

$$k_4 = hf(x_n + h, y_n + k_3),$$

and

$$K_n = \tfrac{1}{6}(k_1 + 2k_2 + 2k_3 + k_4).$$

Then we set

$$y_{n+1} = y_n + K_n$$

and take this as the approximate value of the exact solution at $x_{n+1} = x_n + h$.

▶ Example 8.9. Apply the Runge-Kutta method to the initial-value problem

$$\frac{dy}{dx} = 2x + y, \tag{8.2}$$

$$y(0) = 1. \tag{8.44}$$

Employ the method to approximate the value of the solution y at $x = 0.2$ and $x = 0.4$ using $h = 0.2$. Carry the intermediate calculations in each step to five figures after the decimal point, and round off the final results of each step to four such places. Compare with the exact value.

Solution. Here $f(x, y) = 2x + y$, $x_0 = 0$, $y_0 = 1$, and we are to use $h = 0.2$. Using these quantities we calculate successively k_1, k_2, k_3, k_4, and K_0 defined by (8.57). We first find

$$k_1 = hf(x_0, y_0) = 0.2f(0, 1) = 0.2(1) = 0.20000.$$

Then since

$$x_0 + \frac{h}{2} = 0 + \frac{1}{2}(0.2) = 0.1$$

and

$$y_0 + \frac{k_1}{2} = 1.00000 + \frac{1}{2}(0.20000) = 1.10000,$$

we find

$$k_2 = hf\left(x_0 + \frac{h}{2},\ y_0 + \frac{k_1}{2}\right) = 0.2f(0.1,\ 1.10000)$$

$$= 0.2(1.30000) = 0.26000.$$

Next, since

$$y_0 + \frac{k_2}{2} = 1.00000 + \frac{1}{2}(0.26000) = 1.13000,$$

we find

$$k_3 = hf\left(x_0 + \frac{h}{2},\ y_0 + \frac{k_2}{2}\right) = 0.2f(0.1,\ 1.13000)$$

$$= 0.2(1.33000) = 0.26600.$$

Since $x_0 + h = 0.2$ and $y_0 + k_3 = 1.00000 + 0.26600 = 1.26600$, we obtain

$$k_4 = hf(x_0 + h,\ y_0 + k_3) = 0.2f(0.2,\ 1.26600)$$
$$= 0.2(1.66600) = 0.33320.$$

Finally, we find

$$K_0 = \tfrac{1}{6}(k_1 + 2k_2 + 2k_3 + k_4) = \tfrac{1}{6}(0.20000 + 0.52000 + 0.53200 + 0.33320)$$
$$= 0.26420.$$

Then by (8.58) the approximate value of the solution at $x_1 = 0.2$ is

$$y_1 = 1 + 0.2642 = 1.2642. \qquad (8.61)$$

Now using y_1 as given by (8.61), we calculate successively k_1, k_2, k_3, k_4, and K_1 defined by (8.59). We first find

$$k_1 = hf(x_1, y_1) = 0.2f(0.2,\ 1.2642) = 0.2(1.6642) = 0.33284.$$

Then since

$$x_1 + \frac{h}{2} = 0.2 + \frac{1}{2}(0.2) = 0.3$$

and

$$y_1 + \frac{k_1}{2} = 1.26420 + \frac{1}{2}(0.33284) = 1.43062,$$

we find

$$k_2 = hf\left(x_1 + \frac{h}{2},\ y_1 + \frac{k_1}{2}\right) = 0.2f(0.3,\ 1.43062)$$

$$= 0.2(2.03062) = 0.40612.$$

Next, since

$$y_1 + \frac{k_2}{2} = 1.26420 + \frac{1}{2}(0.40612) = 1.46726,$$

we find

$$k_3 = hf\left(x_1 + \frac{h}{2},\ y_1 + \frac{k_2}{2}\right) = 0.2f(0.3,\ 1.46726)$$

$$= 0.2(2.06726) = 0.41345.$$

Since $x_1 + h = 0.4$ and $y_1 + k_3 = 1.26420 + 0.41345 = 1.67765$, we obtain

$$\begin{aligned} k_4 &= hf(x_1 + h, y_1 + k_3) = 0.2f(0.4, 1.67765) \\ &= 0.2(2.47765) = 0.49553. \end{aligned}$$

Finally, we find

$$\begin{aligned} K_1 &= \tfrac{1}{6}(k_1 + 2k_2 + 2k_3 + k_4) \\ &= \tfrac{1}{6}(0.33284 + 0.81224 + 0.82690 + 0.49553) = 0.41125. \end{aligned}$$

We round off K_1 according to the well-known rule of rounding off so as to leave the last digit retained an even one. Then by (8.60) we see that the approximate value of the solution at $x_2 = 0.4$ is

$$y_2 = 1.2642 + 0.4112 = 1.6754. \tag{8.62}$$

As before, the exact values are determined using (8.46). Rounded off to four places after the decimal point, the exact values at $x = 0.2$ and $x = 0.4$ are 1.2642 and 1.6754, respectively. The approximate value at $x = 0.2$ as given by (8.61) is therefore correct to four places after the decimal point, and the approximate value at $x = 0.4$ as given by (8.62) is likewise correct to four places!

The remarkable accuracy of the Runge-Kutta method in this problem is certainly apparent. In fact, if we employ the method to approximate the solution at $x = 0.4$ using $h = 0.4$ (that is, in only one step), we obtain the value 1.6752, which differs from the exact value 1.6754 by merely 0.0002.

Exercises

1. Consider the initial-value problem

$$\frac{dy}{dx} = 3x + 2y,$$

$$y(0) = 1.$$

   (a) Apply the Runge-Kutta method to approximate the values of the solution y at $x = 0.1, 0.2$, and 0.3, using $h = 0.1$. Carry the intermediate calculations in each step to five figures after the decimal point, and round off the final results of each step to four such places.

   (b) Find the exact solution of the problem and compare the results obtained in part (a) with the exact values.

2. Proceed as in Exercise 1 for the initial-value problem

$$\frac{dy}{dx} = 2x - y,$$

$$y(0) = 3.$$

3. Proceed as in part (a) of Exercise 1 for initial-value problem

$$\frac{dy}{dx} = x^2 + y^2,$$

$$y(0) = 1.$$

E. *The Milne Method*

The Euler method, the modified Euler method, and the Runge-Kutta method are all *starting* methods for the numerical solution of an initial-value problem. As we have already pointed out, a starting method is a method which can be used to start the solution. In contrast, a *continuing* method is one which cannot be used to start the solution but which can be used to continue it, once it is sufficiently well started. In this section we consider briefly a useful continuing method, the so-called Milne method. We shall give no justification for this method but shall merely list the formulas involved and indicate how they are employed.*

The Milne method can be used to approximate the value of the solution of the initial-value problem

$$\frac{dy}{dx} = f(x, y), \tag{8.1}$$

$$y(x_0) = y_0 \tag{8.6}$$

at $x_{n+1} = x_0 + (n + 1)h$, provided the values at the four previous points x_{n-3}, x_{n-2}, x_{n-1}, and x_n have been determined. We assume that these four previous values have been found and denote them by y_{n-3}, y_{n-2}, y_{n-1}, and y_n, respectively. Then we can use (8.1) to determine dy/dx at x_{n-2}, x_{n-1}, and x_n. That is, we can determine $y'_{n-2} = f(x_{n-2}, y_{n-2})$, $y'_{n-1} = f(x_{n-1}, y_{n-1})$, and $y'_n = f(x_n, y_n)$. These various numbers being determined, the Milne method proceeds as follows:

We first determine the number $y^{(1)}_{n+1}$ given by the formula

$$y^{(1)}_{n+1} = y_{n-3} + \frac{4h}{3}(2y'_n - y'_{n-1} + 2y'_{n-2}). \tag{8.63}$$

Having thus determined $y^{(1)}_{n+1}$, we next determine the number $y'^{(1)}_{n+1}$ given by

$$y'^{(1)}_{n+1} = f(x_{n+1}, y^{(1)}_{n+1}). \tag{8.64}$$

Finally, having determined $y'^{(1)}_{n+1}$, we proceed to determine the number $y^{(2)}_{n+1}$ given by

$$y^{(2)}_{n+1} = y_{n-1} + \frac{h}{3}(y'^{(1)}_{n+1} + 4y'_n + y'_{n-1}). \tag{8.65}$$

If the numbers $y^{(1)}_{n+1}$ determined from (8.63) and $y^{(2)}_{n+1}$ determined from (8.65) are the same to the number of decimal places required, then we take this common value to be the approximate value of the solution at x_{n+1} and denote it by y_{n+1}.

---

* See J. Scarborough, *Numerical Mathematical Analysis*, 6th ed. (Johns Hopkins, Baltimore, 1966).

If the numbers $y_{n+1}^{(1)}$ and $y_{n+1}^{(2)}$ so determined do not agree to the number of decimal places required and all of the calculations have been checked and appear to be correct, then we proceed in the following way. We calculate the number

$$E = \frac{y_{n+1}^{(2)} - y_{n+1}^{(1)}}{29},$$

which is the principal part of the error in the formula (8.65). If E is negligible with respect to the number of decimal places required, then we take the number $y_{n+1}^{(2)}$ given by (8.65) as the approximate value of the solution at x_{n+1} and denote it by y_{n+1}. On the other hand, if E is so large that it is not negligible with respect to the number of decimal places required, then the value of h employed is too large and a smaller value must be used.

We observe that once the values y_0, y_1, y_2, and y_3 have been determined, we can use the Milne formulas with $n = 3$ to determine y_4. Then when y_4 has been determined by the formulas, we can use them with $n = 4$ to determine y_5. Then proceeding in like manner, we can successively determine $y_6, y_7, \ldots$. But we must have y_0, y_1, y_2, and y_3 in order to start the Milne method. Of course y_0 is given exactly by the initial condition (8.6), and we can find y_1, y_2, and y_3 by one of the previously explained starting methods (for example, by the Runge-Kutta method).

▶ Example 8.10. Apply the Milne method to approximate the value at $x = 0.4$ of the solution y of the initial-value problem

$$\frac{dy}{dx} = 2x + y, \tag{8.2}$$

$$y(0) = 1, \tag{8.44}$$

assuming that the values at 0.1, 0.2, and 0.3 are 1.1155, 1.2642, and 1.4496, respectively.

Solution. We apply the formulas (8.63), (8.64), and (8.65) with $n = 3$. We set

$$\begin{aligned} x_0 &= 0, & y_0 &= 1.0000, \\ x_1 &= 0.1, & y_1 &= 1.1155, \\ x_2 &= 0.2, & y_2 &= 1.2642, \\ x_3 &= 0.3, & y_3 &= 1.4496, \end{aligned}$$

and $x_4 = 0.4$. Then using $f(x, y) = 2x + y$, we find

$$\begin{aligned} y_1' &= f(x_1, y_1) = f(0.1, 1.1155) = 1.3155, \\ y_2' &= f(x_2, y_2) = f(0.2, 1.2642) = 1.6642, \\ y_3' &= f(x_3, y_3) = f(0.3, 1.4496) = 2.0496. \end{aligned}$$

We now use (8.63) with $n = 3$ and $h = 0.1$ to determine $y_4^{(1)}$. We have

$$y_4^{(1)} = y_0 + \frac{4(0.1)}{3}(2y_3' - y_2' + 2y_1')$$

$$= 1.000 + \frac{0.4}{3}(4.0992 - 1.6642 + 2.6310) = 1.6755.$$

Having thus determined $y_4^{(1)}$, we use (8.64) with $n = 3$ to determine $y_4'^{(1)}$. We find

$$y_4'^{(1)} = f(x_4, y_4^{(1)}) = f(0.4, 1.6755) = 2.4755.$$

Finally, having determined $y_4'^{(1)}$, we use (8.65) with $n = 3$ and $h = 0.1$ to determine $y_4^{(2)}$. We obtain

$$y_4^{(2)} = y_2 + \frac{0.1}{3}(y_4'^{(1)} + 4y_3' + y_2')$$

$$= 1.2642 + \frac{0.1}{3}(2.4755 + 8.1984 + 1.6642)$$

$$= 1.6755.$$

Since the numbers $y_4^{(1)}$ and $y_4^{(2)}$ agree to four decimal places, we take their common value as the approximate value of the solution at $x_4 = 0.4$ and denote it by y_4. That is, we set

$$y_4 = 1.6755.$$

Using (8.46), the exact value at $x = 0.4$, rounded off to four places after the decimal point, is found to be 1.6754.

Exercises

1. Apply the Milne method to approximate the value at $x = 0.4$ of the solution y of the initial-value problem

$$\frac{dy}{dx} = 3x + 2y,$$

$$y(0) = 1,$$

assuming that the values at 0.1, 0.2, and 0.3 are 1.2375, 1.5607, and 1.9887, respectively.

2. Apply the Milne method to approximate the value at $x = 0.4$ of the solution y of the initial-value problem

$$\frac{dy}{dx} = 2x - y,$$

$$y(0) = 3,$$

assuming that the values at 0.1, 0.2, and 0.3 are 2.7242, 2.4937, and 2.3041, respectively.

Suggested Reading

I. Fundamental Texts

AGNEW, R., *Differential Equations*, 2nd ed. (McGraw-Hill, New York, 1960).
BOYCE, W., and R. DIPRIMA, *Elementary Differential Equations*, 2nd ed. (Wiley, New York, 1969).
BRAUER, F., and J. NOHEL, *Ordinary Differential Equations: A First Course* (Benjamin, New York, 1967).
RAINVILLE, E., and P. BEDIENT, *Elementary Differential Equations*, 4th ed. (MacMillan, New York, 1969).
RITGER, P., and N. ROSE, *Differential Equations with Applications* (McGraw-Hill, New York, 1968).

II. Specialized Texts

HILDEBRAND, F., *Introduction to Numerical Analysis* (McGraw-Hill, New York, 1956).
LEVY, H., and E. BAGGOTT, *Numerical Solutions of Differential Equations* (Dover, New York, 1950).
MILNE, W. E., *Numerical Solution of Differential Equations* (Dover, New York, 1970).
SCARBOROUGH, J., *Numerical Mathematical Analysis*, 6th ed. (Johns Hopkins, Baltimore, 1966).

The Laplace Transform

In this chapter we shall introduce a concept which is especially useful in the solution of initial-value problems. This concept is the so-called Laplace transform, which transforms a suitable function F of a real variable t into a related function f of a real variable s. When this transform is applied in connection with an initial-value problem involving a linear differential equation in an "unknown" function of t, it transforms the given initial-value problem into an algebraic problem involving the variable s. In Section 9.3 we shall indicate just how this transformation is accomplished and how the resulting algebraic problem is then employed to find the solution of the given initial-value problem. First, however, in Section 9.1 we shall introduce the Laplace transform itself and develop certain of its most basic and useful properties.

9.1 Definition, Existence, and Basic Properties of the Laplace Transform

A. Definition and Existence

DEFINITION

Let F be a real-valued function of the real variable t, defined for $t > 0$. Let s be a variable which we shall assume to be real, and consider the function f defined by

$$f(s) = \int_0^\infty e^{-st}F(t)\,dt, \tag{9.1}$$

for all values of s for which this integral exists. The function f defined by the integral (9.1) is called the Laplace transform *of the function F. We shall denote the Laplace transform f of F by $\mathscr{L}\{F\}$ and shall denote $f(s)$ by $\mathscr{L}\{F(t)\}$.*

In order to be certain that the integral (9.1) does exist for some range of values of s, we must impose suitable restrictions upon the function F under consideration. We shall do this shortly; however, first we shall directly determine the Laplace transforms of a few simple functions.

▶ **Example 9.1.** Consider the function F defined by

$$F(t) = 1, \qquad \text{for } t > 0.$$

Then

$$\mathscr{L}\{1\} = \int_0^\infty e^{-st}\cdot 1\,dt = \lim_{R\to\infty}\int_0^R e^{-st}\cdot 1\,dt = \lim_{R\to\infty}\left[\frac{-e^{-st}}{s}\right]_0^R$$

$$= \lim_{R\to\infty}\left[\frac{1}{s} - \frac{e^{-sR}}{s}\right] = \frac{1}{s}$$

for all $s > 0$. Thus we have

$$\mathscr{L}\{1\} = \frac{1}{s} \qquad (s > 0). \tag{9.2}$$

▶ **Example 9.2.** Consider the function F defined by

$$F(t) = t, \qquad \text{for } t > 0.$$

Then

$$\mathscr{L}\{t\} = \int_0^\infty e^{-st}\cdot t\,dt = \lim_{R\to\infty}\int_0^R e^{-st}\cdot t\,dt = \lim_{R\to\infty}\left[-\frac{e^{-st}}{s^2}(st+1)\right]_0^R$$

$$= \lim_{R\to\infty}\left[\frac{1}{s^2} - \frac{e^{-sR}}{s^2}(sR+1)\right] = \frac{1}{s^2}$$

for all $s > 0$. Thus

$$\mathscr{L}\{t\} = \frac{1}{s^2} \qquad (s > 0). \tag{9.3}$$

▶ **Example 9.3.** Consider the function f defined by

$$F(t) = e^{at}, \qquad \text{for } t > 0.$$

$$\mathscr{L}\{e^{at}\} = \int_0^\infty e^{-st}e^{at}\,dt = \lim_{R\to\infty}\int_0^R e^{(a-s)t}\,dt = \lim_{R\to\infty}\left[\frac{e^{(a-s)t}}{a-s}\right]_0^R$$

$$= \lim_{R\to\infty}\left[\frac{e^{(a-s)R}}{a-s} - \frac{1}{a-s}\right] = -\frac{1}{a-s} = \frac{1}{s-a} \qquad \text{for all } s > a.$$

Thus

$$\mathscr{L}\{e^{at}\} = \frac{1}{s-a} \qquad (s > a). \tag{9.4}$$

▶ **Example 9.4.** Consider the function F defined by

$$F(t) = \sin bt \qquad \text{for } t > 0.$$

$$\mathscr{L}\{\sin bt\} = \int_0^\infty e^{-st}\cdot \sin bt\,dt = \lim_{R\to\infty}\int_0^R e^{-st}\cdot \sin bt\,dt$$

$$= \lim_{R\to\infty}\left[-\frac{e^{-st}}{s^2+b^2}(s\sin bt + b\cos bt)\right]_0^R$$

$$= \lim_{R\to\infty}\left[\frac{b}{s^2+b^2} - \frac{e^{-sR}}{s^2+b^2}(s\sin bR + b\cos bR)\right]$$

$$= \frac{b}{s^2+b^2} \qquad \text{for all } s > 0.$$

Thus

$$\mathscr{L}\{\sin bt\} = \frac{b}{s^2 + b^2} \qquad (s > 0). \tag{9.5}$$

▶ Example 9.5. Consider the function F defined by

$$F(t) = \cos bt \qquad \text{for } t > 0.$$

$$\begin{aligned}
\mathscr{L}\{\cos bt\} &= \int_0^\infty e^{-st} \cdot \cos bt \, dt = \lim_{R\to\infty} \int_0^R e^{-st} \cos bt \, dt \\
&= \lim_{R\to\infty} \left[\frac{e^{-st}}{s^2 + b^2} (-s \cos bt + b \sin bt) \right]_0^R \\
&= \lim_{R\to\infty} \left[\frac{e^{-sR}}{s^2 + b^2} (-s \cos bR + b \sin bR) + \frac{s}{s^2 + b^2} \right] \\
&= \frac{s}{s^2 + b^2} \qquad \text{for all } s > 0.
\end{aligned}$$

Thus

$$\mathscr{L}\{\cos bt\} = \frac{s}{s^2 + b^2} \qquad (s > 0). \tag{9.6}$$

In each of the above examples we have seen directly that the integral (9.1) actually does exist for some range of values of s. We shall now determine a class of functions F for which this is always the case. To do so we first consider certain properties of functions.

DEFINITION

A function F is said to be piecewise continuous (*or* sectionally continuous) *on a finite interval $a \le t \le b$ if this interval can be divided into a finite number of subintervals such that (1) F is continuous in the interior of each of these subintervals, and (2) $F(t)$ approaches finite limits as t approaches either endpoint of each of the subintervals from its interior.*

Suppose F is piecewise continuous on $a \le t \le b$, and t_0, $a < t_0 < b$, is an endpoint of one of the subintervals of the above definition. Then the finite limit approached by $F(t)$ as t approaches t_0 from the left (that is, through smaller values of t) is called the *left-hand limit* of $F(t)$ as t approaches t_0, denoted by $\lim_{t\to t_0-} F(t)$ or by $F(t_0-)$. In like manner, the finite limit approached by $F(t)$ as t approaches t_0 from the right (through larger values) is called the *right-hand limit* of $F(t)$ as t approaches t_0, denoted by $\lim_{t\to t_0+} F(t)$ or $F(t_0+)$. We emphasize that at such a point t_0, both $F(t_0-)$ and $F(t_0+)$ are finite but they are not in general equal.

We point out that if F is continuous on $a \le t \le b$ it is necessarily piecewise continuous on this interval. Also, we note that if F is piecewise continuous on $a \le t \le b$, then F is integrable on $a \le t \le b$.

▶ Example 9.6. Consider the function F defined by

$$F(t) = \begin{cases} -1, & 0 < t < 2, \\ 1, & t > 2. \end{cases}$$

F is piecewise continuous on every finite interval $0 \le t \le b$, for every positive number b. At $t = 2$, we have

$$F(2-) = \lim_{t \to 2-} F(t) = -1,$$

$$F(2+) = \lim_{t \to 2+} F(t) = +1.$$

The graph of F is shown in Figure 9.1.

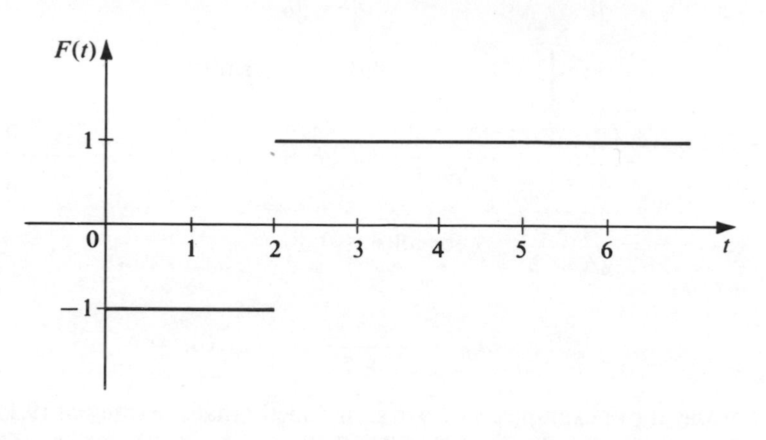

FIGURE 9.1

DEFINITION

A function F is said to be of exponential order *if there exists a constant α and positive constants t_0 and M such that*

$$e^{-\alpha t}|F(t)| < M \tag{9.7}$$

for all $t > t_0$ at which $F(t)$ is defined. More explicitly, if F is of exponential order corresponding to some definite constant α in (9.7), then we say that F is of exponential order $e^{\alpha t}$.

In other words, we say that F is of exponential order if a constant α exists such that the product $e^{-\alpha t}\,|F(t)|$ is bounded for all sufficiently large values of t. From (9.7) we have

$$|F(t)| < Me^{\alpha t} \tag{9.8}$$

for all $t > t_0$ at which $F(t)$ is defined. Thus if F is of exponential order and the values $F(t)$ of F become infinite as $t \to \infty$, these values cannot become infinite more rapidly than a multiple M of the corresponding values $e^{\alpha t}$ of some exponential function. We note that if F is of exponential order $e^{\alpha t}$, then F is also of exponential order $e^{\beta t}$ for any $\beta > \alpha$.

▶ Example 9.7. Every bounded function is of exponential order, with the constant $\alpha = 0$. Thus, for example, $\sin bt$ and $\cos bt$ are of exponential order.

▶ Example 9.8. The function F such that $F(t) = e^{at} \sin bt$ is of exponential order, with the constant $\alpha = a$. For we then have

$$e^{-\alpha t}|F(t)| = e^{-\alpha t}e^{at}|\sin bt| = |\sin bt|,$$

which is bounded for all t.

▶ Example 9.9. Consider the function F such that $F(t) = t^n$, where $n > 0$. Then $e^{-\alpha t}|F(t)|$ is $e^{-\alpha t}t^n$. For any $\alpha > 0$, $\lim_{t \to \infty} e^{-\alpha t}t^n = 0$. Thus there exists $M > 0$ and $t_0 > 0$ such that

$$e^{-\alpha t}|F(t)| = e^{-\alpha t}t^n < M$$

for $t > t_0$. Hence $F(t) = t^n$ is of exponential order, with the constant α equal to any positive number.

▶ Example 9.10. The function F such that $F(t) = e^{t^2}$ is *not* of exponential order, for in this case $e^{-\alpha t}|F(t)|$ is $e^{t^2 - \alpha t}$ and this becomes unbounded as $t \to \infty$, no matter what is the value of α.

We shall now proceed to obtain a theorem giving conditions on F which are sufficient for the integral (9.1) to exist. To obtain the desired result we shall need the following two theorems from advanced calculus, which we state without proof.

THEOREM A. Comparison Test for Improper Integrals

Hypothesis
1. Let g and G be real functions such that

$$0 \le g(t) \le G(t) \qquad on \quad a \le t < \infty.$$

2. Suppose $\int_a^\infty G(t)\,dt$ exists.
3. Suppose g is integrable on every finite closed subinterval of $a \le t < \infty$.

Conclusion. *Then $\int_a^\infty g(t)\,dt$ exists.*

THEOREM B

Hypothesis
1. Suppose the real function g is integrable on every finite closed subinterval of $a \le t < \infty$.
2. Suppose $\int_a^\infty |g(t)|\,dt$ exists.

Conclusion. *Then $\int_a^\infty g(t)\,dt$ exists.*

We now state and prove an existence theorem for Laplace transforms.

THEOREM 9.1

Hypothesis. *Let F be a real function which has the following properties:*
1. F is piecewise continuous in every finite closed interval $0 \le t \le b$ $(b > 0)$.

2. *F is of exponential order; that is, there exists α, $M > 0$, and $t_0 > 0$ such that*

$$e^{-\alpha t}|F(t)| < M \qquad \text{for} \quad t > t_0.$$

Conclusion. *The Laplace transform*

$$\int_0^\infty e^{-st}F(t)\,dt$$

of F exists for $s > \alpha$.

Proof. We have

$$\int_0^\infty e^{-st}F(t)\,dt = \int_0^{t_0} e^{-st}F(t)\,dt + \int_{t_0}^\infty e^{-st}F(t)\,dt.$$

By Hypothesis 1, the first integral of the right member exists. By Hypothesis 2,

$$e^{-st}|F(t)| < e^{-st}Me^{\alpha t} = Me^{-(s-\alpha)t}$$

for $t > t_0$. Also

$$\int_{t_0}^\infty Me^{-(s-\alpha)t}\,dt = \lim_{R\to\infty}\int_{t_0}^R Me^{-(s-\alpha)t}\,dt = \lim_{R\to\infty}\left[-\frac{Me^{-(s-\alpha)t}}{s-\alpha}\right]_{t_0}^R$$

$$= \lim_{R\to\infty}\left[\frac{M}{s-\alpha}\right][e^{-(s-\alpha)t_0} - e^{-(s-\alpha)R}] = \left[\frac{M}{s-\alpha}\right]e^{-(s-\alpha)t_0}$$

if $s > \alpha$.

Thus

$$\int_{t_0}^\infty Me^{-(s-\alpha)t}\,dt \qquad \text{exists for} \quad s > \alpha.$$

Finally, by Hypothesis 1, $e^{-st}|F(t)|$ is integrable on every finite closed subinterval of $t_0 \le t < \infty$. Thus, applying Theorem A with $g(t) = e^{-st}|F(t)|$ and $G(t) = Me^{-(s-\alpha)t}$, we see that

$$\int_{t_0}^\infty e^{-st}|F(t)|\,dt \qquad \text{exists if} \quad s > \alpha.$$

In other words,

$$\int_{t_0}^\infty |e^{-st}F(t)|\,dt \qquad \text{exists if} \quad s > \alpha,$$

and so by Theorem B

$$\int_{t_0}^\infty e^{-st}F(t)\,dt$$

also exists if $s > \alpha$. Thus the Laplace transform of F exists for $s > \alpha$. *Q.E.D.*

Let us look back at this proof for a moment. Actually we showed that if F satisfies the hypotheses stated, then

$$\int_{t_0}^\infty e^{-st}|F(t)|\,dt \qquad \text{exists if} \quad s > \alpha.$$

Further, Hypothesis 1 shows that

$$\int_0^{t_0} e^{-st}|F(t)|\,dt \qquad \text{exists.}$$

Thus

$$\int_0^{\infty} e^{-st}|F(t)|\,dt \qquad \text{exists if} \quad s > \alpha.$$

In other words, if F satisfies the hypotheses of Theorem 9.1, then not only does $\mathscr{L}\{F\}$ exist for $s > \alpha$, but also $\mathscr{L}\{|F|\}$ exists for $s > \alpha$. That is,

$$\int_0^{\infty} e^{-st}F(t)\,dt \qquad \text{is absolutely convergent for} \quad s > \alpha.$$

We point out that the conditions on F described in the hypothesis of Theorem 9.1 are not necessary for the existence of $\mathscr{L}\{F\}$. In other words, there exist functions F which do *not* satisfy the hypotheses of Theorem 9.1, but for which $\mathscr{L}\{F\}$ exists. For instance, suppose we replace Hypothesis 1 by the following less restrictive condition. Let us suppose that F is piecewise continuous in every finite closed interval $a \le t \le b$, where $a > 0$, and is such that $|t^nF(t)|$ remains bounded as $t \to 0^+$ for some number n, where $0 < n < 1$. Then, provided Hypothesis 2 remains satisfied, it can be shown that $\mathscr{L}\{F\}$ still exists. Thus for example, if $F(t) = t^{-1/3}$, $t > 0$, $\mathscr{L}\{F\}$ exists. For although F does *not* satisfy Hypothesis 1 of Theorem 9.1 [$F(t) \to \infty$ as $t \to 0^+$], it *does* satisfy the less restrictive requirement stated above (take $n = \frac{2}{3}$), and F is of exponential order.

B. *Basic Properties of the Laplace Transform*

THEOREM 9.2. The Linear Property

Let F_1 and F_2 be functions whose Laplace transforms exist, and let c_1 and c_2 be constants. Then

$$\mathscr{L}\{c_1F_1(t) + c_2F_2(t)\} = c_1\mathscr{L}\{F_1(t)\} + c_2\mathscr{L}\{F_2(t)\}. \tag{9.9}$$

Proof. This follows directly from the definition (9.1).

▶ Example 9.11. Use Theorem 9.2 to find $\mathscr{L}\{\sin^2 at\}$.

Since $\sin^2 at = (1 - \cos 2at)/2$, we have

$$\mathscr{L}\{\sin^2 at\} = \mathscr{L}\{\tfrac{1}{2} - \tfrac{1}{2}\cos 2at\}.$$

By Theorem 9.2,

$$\mathscr{L}\{\tfrac{1}{2} - \tfrac{1}{2}\cos 2at\} = \tfrac{1}{2}\mathscr{L}\{1\} - \tfrac{1}{2}\mathscr{L}\{\cos 2at\}.$$

By (9.2), $\mathscr{L}\{1\} = 1/s$, and by (9.6), $\mathscr{L}\{\cos 2at\} = s/(s^2 + 4a^2)$. Thus

$$\mathscr{L}\{\sin^2 at\} = \frac{1}{2}\cdot\frac{1}{s} - \frac{1}{2}\cdot\frac{s}{s^2 + 4a^2} = \frac{2a^2}{s(s^2 + 4a^2)}. \tag{9.10}$$

THEOREM 9.3

Hypothesis

1. Let F be a real function which is continuous for $t \geq 0$ *and of exponential order* $e^{\alpha t}$.

2. Let F' *(the derivative of F) be piecewise continuous in every finite closed interval* $0 \leq t \leq b$.

Conclusion. *Then* $\mathscr{L}\{F'\}$ *exists for* $s > \alpha$; *and*

$$\mathscr{L}\{F'(t)\} = s\mathscr{L}\{F(t)\} - F(0). \tag{9.11}$$

Proof. By definition of the Laplace transform,

$$\mathscr{L}\{F'(t)\} = \lim_{R\to\infty} \int_0^R e^{-st}F'(t)\,dt,$$

provided this limit exists. In any closed interval $0 \leq t \leq R$, $F'(t)$ has at most a finite number of discontinuities; denote these by $t_1, t_2, \ldots, t_n$, where

$$0 < t_1 < t_2 < \cdots < t_n \leq R.$$

Then we may write

$$\int_0^R e^{-st}F'(t)\,dt = \int_0^{t_1} e^{-st}F'(t)\,dt + \int_{t_1}^{t_2} e^{-st}F'(t)\,dt + \cdots + \int_{t_n}^{R} e^{-st}F'(t)\,dt.$$

Now the integrand of each of the integrals on the right is continuous. We may therefore integrate each by parts. Doing so, we obtain

$$\int_0^R e^{-st}F'(t)\,dt = \Big[e^{-st}F(t)\Big]_0^{t_1-} + s\int_0^{t_1} e^{-st}F(t)\,dt + \Big[e^{-st}F(t)\Big]_{t_1+}^{t_2-}$$

$$+ s\int_{t_1}^{t_2} e^{-st}F(t)\,dt + \cdots + \Big[e^{-st}F(t)\Big]_{t_n+}^{R-} + s\int_{t_n}^{R} e^{-st}F(t)\,dt.$$

By Hypothesis 1, F is continuous for $t \geq 0$. Thus

$$F(t_1-) = F(t_1+),\ F(t_2-) = F(t_2+), \ldots, F(t_n-) = F(t_n+).$$

Thus all of the integrated "pieces" add out, except for $e^{-st}F(t)|_{t=0}$ and $e^{-st}F(t)|_{t=R-}$, and there remains only

$$\int_0^R e^{-st}F'(t)\,dt = -F(0) + e^{-sR}F(R) + s\int_0^R e^{-st}F(t)\,dt.$$

But by Hypothesis 1 F is of exponential order $e^{\alpha t}$. Thus there exists $M > 0$ and $t_0 > 0$ such that $e^{-\alpha t}|F(t)| < M$ for $t > t_0$. Thus $|e^{-sR}F(R)| < Me^{-(s-\alpha)R}$ for $R > t_0$. Thus if $s > \alpha$,

$$\lim_{R\to\infty} e^{-sR}F(R) = 0.$$

Further,

$$\lim_{R\to\infty} s\int_0^R e^{-st}F(t)\,dt = s\mathscr{L}\{F(t)\}.$$

Thus, we have

$$\lim_{R\to\infty}\int_0^R e^{-st}F'(t)\,dt = -F(0) + s\mathscr{L}\{F(t)\},$$

and so $\mathscr{L}\{F'(t)\}$ exists for $s > \alpha$ and is given by (9.11). *Q.E.D.*

▶ Example 9.12. Consider the function defined by $F(t) = \sin^2 at$. This function satisfies the hypotheses of Theorem 9.3. Since $F'(t) = 2a \sin at \cos at$ and $F(0) = 0$, Equation (9.11) gives

$$\mathscr{L}\{2a \sin at \cos at\} = s\mathscr{L}\{\sin^2 at\}.$$

By (9.10),

$$\mathscr{L}\{\sin^2 at\} = \frac{2a^2}{s(s^2 + 4a^2)}.$$

Thus,

$$\mathscr{L}\{2a \sin at \cos at\} = \frac{2a^2}{s^2 + 4a^2}.$$

Since $2a \sin at \cos at = a \sin 2at$, we also have

$$\mathscr{L}\{\sin 2at\} = \frac{2a}{s^2 + 4a^2}.$$

Observe that this is the result (9.5), obtained in Example 9.4, with $b = 2a$.

We now generalize Theorem 9.3 and obtain the following result:

THEOREM 9.4

Hypothesis

1. Let F be a real function having a continuous $(n - 1)$st derivative $F^{(n-1)}$ (and hence $F, F', \ldots, F^{(n-2)}$ are also continuous) for $t \geq 0$; and assume that $F, F', \ldots, F^{(n-1)}$ are all of exponential order $e^{\alpha t}$.

2. Suppose $F^{(n)}$ is piecewise continuous in every finite closed interval $0 \leq t \leq b$.

Conclusion. *$\mathscr{L}\{F^{(n)}\}$ exists for $s > \alpha$ and*

$$\mathscr{L}\{F^{(n)}(t)\} = s^n\mathscr{L}\{F(t)\} - s^{n-1}F(0) - s^{n-2}F'(0) - s^{n-3}F''(0) - \cdots - F^{(n-1)}(0). \tag{9.12}$$

Outline of Proof. One first proceeds as in the proof of Theorem 9.3 to show that $\mathscr{L}\{F^{(n)}\}$ exists for $s > \alpha$ and is given by

$$\mathscr{L}\{F^{(n)}\} = s\mathscr{L}\{F^{(n-1)}\} - F^{(n-1)}(0).$$

One then completes the proof by mathematical induction.

▶ Example 9.13. We apply Theorem 9.4, with $n = 2$, to find $\mathscr{L}\{\sin bt\}$, which we have already found directly and given by (9.5). Clearly the function F defined by

$F(t) = \sin bt$ satisfies the hypotheses of the theorem with $\alpha = 0$. For $n = 2$, Equation (9.12) becomes

$$\mathscr{L}\{F''(t)\} = s^2\mathscr{L}\{F(t)\} - sF(0) - F'(0). \tag{9.13}$$

We have $F'(t) = b \cos bt$, $F''(t) = -b^2 \sin bt$, $F(0) = 0$, $F'(0) = b$. Substituting into Equation (9.13) we find

$$\mathscr{L}\{-b^2 \sin bt\} = s^2\mathscr{L}\{\sin bt\} - b,$$

and so

$$(s^2 + b^2)\mathscr{L}\{\sin bt\} = b.$$

Thus,

$$\mathscr{L}\{\sin bt\} = \frac{b}{s^2 + b^2} \qquad (s > 0),$$

which is the result (9.5), already found directly.

THEOREM 9.5. Translation Property

Hypothesis. *Suppose F is such that $\mathscr{L}\{F\}$ exists for $s > \alpha$.*

Conclusion. *For any constant a,*

$$\mathscr{L}\{e^{at}F(t)\} = f(s - a) \tag{9.14}$$

for $s > \alpha + a$, where $f(s)$ denotes $\mathscr{L}\{F(t)\}$.

Proof. $f(s) = \mathscr{L}\{F(t)\} = \int_0^\infty e^{-st}F(t)\,dt$.
Replacing s by $s - a$, we have

$$f(s - a) = \int_0^\infty e^{-(s-a)t}F(t)\,dt = \int_0^\infty e^{-st}[e^{at}F(t)]\,dt = \mathscr{L}\{e^{at}F(t)\}.$$

Q.E.D.

▶ Example 9.14. Find $\mathscr{L}\{e^{at}t\}$. We apply Theorem 9.5 with $F(t) = t$.

$$\mathscr{L}\{e^{at}t\} = f(s - a),$$

where $f(s) = \mathscr{L}\{F(t)\} = \mathscr{L}\{t\}$. By (9.3), $\mathscr{L}\{t\} = 1/s^2$ $(s > 0)$. That is, $f(s) = 1/s^2$ and so $f(s - a) = 1/(s - a)^2$. Thus

$$\mathscr{L}\{e^{at}t\} = \frac{1}{(s - a)^2} \qquad (s > a). \tag{9.15}$$

▶ Example 9.15. Find $\mathscr{L}\{e^{at} \sin bt\}$. We let $F(t) = \sin bt$. Then $\mathscr{L}\{e^{at} \sin bt\} = f(s - a)$, where

$$f(s) = \mathscr{L}\{\sin bt\} = \frac{b}{s^2 + b^2} \qquad (s > 0).$$

Thus

$$f(s-a) = \frac{b}{(s-a)^2 + b^2}$$

and so

$$\mathscr{L}\{e^{at} \sin bt\} = \frac{b}{(s-a)^2 + b^2} \qquad (s > a). \tag{9.16}$$

THEOREM 9.6

Hypothesis. *Suppose F is a function satisfying the hypotheses of Theorem 9.1 with Laplace transform f so that*

$$f(s) = \int_0^\infty e^{-st} F(t)\, dt;$$

and G is the function defined as follows:

$$G(t) = \begin{cases} 0, & 0 < t < a, \\ F(t-a), & t > a. \end{cases} \tag{9.17}$$

Conclusion

$$\mathscr{L}\{G(t)\} = e^{-as} f(s). \tag{9.18}$$

Proof

$$\mathscr{L}\{G(t)\} = \int_0^\infty e^{-st} G(t)\, dt = \int_0^a e^{-st} \cdot 0\, dt + \int_a^\infty e^{-st} F(t-a)\, dt$$

$$= \int_a^\infty e^{-st} F(t-a)\, dt.$$

Letting $t - a = \tau$, we obtain

$$\int_a^\infty e^{-st} F(t-a)\, dt = \int_0^\infty e^{-s(\tau+a)} F(\tau)\, d\tau = e^{-as} \int_0^\infty e^{-s\tau} F(\tau)\, d\tau = e^{-as} \mathscr{L}\{F(\tau)\}.$$

Thus

$$\mathscr{L}\{G(t)\} = e^{-as} f(s). \qquad \textit{Q.E.D.}$$

▶ Example 9.16. Find $\mathscr{L}\{G(t)\}$ if

$$G(t) = \begin{cases} 0, & 0 < t < \dfrac{\pi}{2}, \\ \sin t, & t > \dfrac{\pi}{2}. \end{cases}$$

Since $\sin t = \cos(t - \pi/2)$, we may write

$$G(t) = \begin{cases} 0, & 0 < t < \dfrac{\pi}{2}, \\ \cos\left(t - \dfrac{\pi}{2}\right), & t > \dfrac{\pi}{2}. \end{cases}$$

Thus Theorem 9.6 applies with $F(t) = \cos t$ and

$$\mathscr{L}\{G(t)\} = e^{-(\pi/2)s}f(s), \quad \text{where } f(s) = \mathscr{L}\{\cos t\} = \frac{s}{s^2 + 1}$$

(using (9.6) with $b = 1$). Therefore, we have

$$\mathscr{L}\{G(t)\} = \frac{se^{-(\pi/2)s}}{s^2 + 1}.$$

THEOREM 9.7

Hypothesis. *Suppose F is a function satisfying the hypotheses of Theorem 9.1, with Laplace transform f, where*

$$f(s) = \int_0^\infty e^{-st}F(t)\,dt. \tag{9.19}$$

Conclusion

$$\mathscr{L}\{t^nF(t)\} = (-1)^n \frac{d^n}{ds^n}[f(s)]. \tag{9.20}$$

Proof. Differentiate both sides of Equation (9.19) n times with respect to s. This differentiation is justified in the present case and yields

$$f'(s) = (-1)^1 \int_0^\infty e^{-st}tF(t)\,dt,$$

$$f''(s) = (-1)^2 \int_0^\infty e^{-st}t^2F(t)\,dt,$$

$$\vdots$$

$$f^{(n)}(s) = (-1)^n \int_0^\infty e^{-st}t^nF(t)\,dt,$$

from which the conclusion (9.20) is at once apparent. *Q.E.D.*

▶ Example 9.17. Find

$$\mathscr{L}\{t^2 \sin bt\}.$$

By Theorem 9.7,

$$\mathscr{L}\{t^2 \sin bt\} = (-1)^2 \frac{d^2}{ds^2}[f(s)],$$

where

$$f(s) = \mathscr{L}\{\sin bt\} = \frac{b}{s^2 + b^2}$$

(using (9.5)). From this,

$$\frac{d}{ds}[f(s)] = -\frac{2bs}{(s^2 + b^2)^2}$$

and

$$\frac{d^2}{ds^2}[f(s)] = \frac{6bs^2 - 2b^3}{(s^2 + b^2)^3}.$$

Thus,

$$\mathscr{L}\{t^2 \sin bt\} = \frac{6bs^2 - 2b^3}{(s^2 + b^2)^3}.$$

Exercises

1. Use the definition of the Laplace transform to find:
   (a) $\mathscr{L}\{t^2\}$,
   (b) $\mathscr{L}\{\sinh bt\}$.

2. Use Theorem 9.2 to find:
   (a) $\mathscr{L}\{\cos^2 at\}$,
   (b) $\mathscr{L}\{\sin at \sin bt\}$.

3. Use Theorem 9.2 to find $\mathscr{L}\{\sin^3 at\}$ and then employ Theorem 9.3 to obtain $\mathscr{L}\{\sin^2 at \cos at\}$.

4. Use Theorem 9.2 to find $\mathscr{L}\{\cos^3 at\}$ and then employ Theorem 9.3 to obtain $\mathscr{L}\{\cos^2 at \sin at\}$.

5. Given that $\mathscr{L}\{t^2\} = 2/s^3$, use Theorem 9.4 to find $\mathscr{L}\{t^4\}$.

6. Use Theorem 9.5 to find:
   (a) $\mathscr{L}\{e^{at}t^2\}$,
   (b) $\mathscr{L}\{e^{at} \sin^2 bt\}$.

7. Use Theorem 9.6 to find $\mathscr{L}\{G(t)\}$, where:

   (a) $G(t) = \begin{cases} 0, & 0 < t < \pi/2, \\ \cos t, & t > \pi/2. \end{cases}$

   (b) $G(t) = \begin{cases} 0, & 0 < t < 2, \\ t, & t > 2. \end{cases}$

8. Use Theorem 9.7 to find:
   (a) $\mathscr{L}\{t^2 \cos bt\}$,
   (b) $\mathscr{L}\{t^3 \sin bt\}$,
   (c) $\mathscr{L}\{t^4 e^{at}\}$.

9. Use the definition of the Laplace transform to find $\mathscr{L}\{F(t)\}$, where:

   (a) $F(t) = \begin{cases} 4, & 0 < t < 3, \\ 2, & t > 3. \end{cases}$

   (b) $F(t) = \begin{cases} t, & 0 < t < 2, \\ 3, & t > 2. \end{cases}$

   (c) $F(t) = \begin{cases} 0, & 0 < t < 1, \\ t, & 1 < t < 2, \\ 1, & t > 2. \end{cases}$

9.2 The Inverse Transform and the Convolution

A. The Inverse Transform

Thus far in this chapter we have been concerned with the following problem: Given a function F, defined for $t > 0$, to find its Laplace transform, which we denoted by $\mathscr{L}\{F\}$ or f. Now consider the inverse problem: Given a function f, to find a function F whose Laplace transform is the given f. We introduce the notation $\mathscr{L}^{-1}\{f\}$ to denote such a function F, denote $\mathscr{L}^{-1}\{f(s)\}$ by $F(t)$, and call such a function an *inverse transform* of f. That is,

$$F(t) = \mathscr{L}^{-1}\{f(s)\}$$

means that $F(t)$ is such that

$$\mathscr{L}\{F(t)\} = f(s).$$

Three questions arise at once:

1. Given a function f, does an inverse transform of f exist?
2. Assuming f does have an inverse transform, is this inverse transform unique?
3. How is an inverse transform found?

In answer to Question 1 we shall say "not necessarily," for there exist functions f which are not Laplace transforms of any function F. In order for f to be a transform it must possess certain continuity properties and also behave suitably as $s \to \infty$. To reassure the reader in a practical way we note that inverse transforms corresponding to numerous functions f have been determined and tabulated.

Now let us consider Question 2. Assuming that f is a function which *does have* an inverse transform, in what sense, if any, is this inverse transform unique? We answer this question in a manner which is adequate for our purposes by stating without proof the following theorem.

THEOREM 9.8

Hypothesis. *Let F and G be two functions which are continuous for $t \geq 0$ and which have the same Laplace transform f.*

Conclusion. *$F(t) = G(t)$ for all $t \geq 0$.*

Thus if it is known that a given function f has a *continuous* inverse transform F, then F is the *only* continuous inverse transform of f. Let us consider the following example.

▶ Example 9.18. By Equation (9.2), $\mathscr{L}\{1\} = 1/s$. Thus an inverse transform of the function f defined by $f(s) = 1/s$ is the *continuous* function F defined for all t by $F(t) = 1$. Thus by Theorem 9.8 there is no other *continuous* inverse transform of the function f such that $f(s) = 1/s$. However, discontinuous inverse transforms of this function f exist. For example, consider the function G defined as follows:

$$G(t) = \begin{cases} 1, & 0 < t < 3, \\ 2, & t = 3, \\ 1, & t > 3. \end{cases}$$

Then

$$\mathscr{L}\{G(t)\} = \int_0^\infty e^{-st}G(t)\,dt = \int_0^3 e^{-st}\,dt + \int_3^\infty e^{-st}\,dt$$

$$= \left[-\frac{e^{-st}}{s}\right]_0^3 + \lim_{R\to\infty}\left[-\frac{e^{-st}}{s}\right]_3^R = \frac{1}{s} \quad \text{if } s > 0.$$

Thus this discontinuous function G is also an inverse transform of f defined by $f(s) = 1/s$. However, we again emphasize that the only *continuous* inverse transform of f defined by $f(s) = 1/s$ is F defined for all t by $F(t) = 1$. Indeed we write

$$\mathscr{L}^{-1}\left\{\frac{1}{s}\right\} = 1,$$

with the understanding that F defined for all t by $F(t) = 1$ is the *unique continuous* inverse transform of f defined by $f(s) = 1/s$.

Finally, let us consider Question 3. Assuming a unique continuous inverse transform of f exists, how is it actually found? The direct determination of inverse transforms will not be considered in this book. Our primary means of finding the inverse transform of a given f will be to make use of a table of transforms. As already indicated, extensive tables of transforms have been prepared. A short table of this kind appears on page 392.

In using a table of transforms to find the inverse transform of a given f, certain preliminary manipulations often have to be performed in order to put the given $f(s)$ in a form to which the various entries in the table apply. Among the various techniques available, the method of partial fractions is often very useful. We shall illustrate its use in Example 9.20.

▶ Example 9.19. Using Table 9.1, find $\mathscr{L}^{-1}\left\{\dfrac{1}{s^2 + 6s + 13}\right\}$.

Solution. Looking in the $f(s)$ column of Table 9.1 we would first look for $f(s) = \dfrac{1}{as^2 + bs + c}$. However, we find no such $f(s)$; but we do find $f(s) = \dfrac{b}{(s+a)^2 + b^2}$ (number 11). We can put the given expression $\dfrac{1}{s^2 + 6s + 13}$ in this form as follows:

$$\frac{1}{s^2 + 6s + 13} = \frac{1}{(s+3)^2 + 4} = \frac{1}{2}\cdot\frac{2}{(s+3)^2 + 2^2}.$$

Thus, using number 11 of Table 9.1, we have

$$\mathscr{L}^{-1}\left\{\frac{1}{s^2 + 6s + 13}\right\} = \frac{1}{2}\mathscr{L}^{-1}\left\{\frac{2}{(s+3)^2 + 2^2}\right\} = \frac{1}{2}e^{-3t}\sin 2t.$$

▶ Example 9.20. Using Table 9.1, find $\mathscr{L}^{-1}\left\{\dfrac{1}{s(s^2 + 1)}\right\}$.

Solution. No entry of this form appears in the $f(s)$ column of Table 9.1. We employ the method of partial fractions. We have

$$\frac{1}{s(s^2+1)} = \frac{A}{s} + \frac{Bs+C}{s^2+1}$$

and hence

$$1 = (A+B)s^2 + Cs + A.$$

Thus

$$A + B = 0, \qquad C = 0, \quad \text{and} \quad A = 1.$$

From these equations, we have the partial fractions decomposition

$$\frac{1}{s(s^2+1)} = \frac{1}{s} - \frac{s}{s^2+1}.$$

Thus

$$\mathscr{L}^{-1}\left\{\frac{1}{s(s^2+1)}\right\} = \mathscr{L}^{-1}\left\{\frac{1}{s}\right\} - \mathscr{L}^{-1}\left\{\frac{s}{s^2+1}\right\}.$$

By number 1 of Table 9.1, $\mathscr{L}^{-1}\left\{\frac{1}{s}\right\} = 1$ and by number 4, $\mathscr{L}^{-1}\left\{\frac{s}{s^2+1}\right\} = \cos t$.

Thus

$$\mathscr{L}^{-1}\left\{\frac{1}{s(s^2+1)}\right\} = 1 - \cos t.$$

B. The Convolution

Another important procedure in connection with the use of tables of transforms is that furnished by the so-called convolution theorem which we shall state below. We first define the convolution of two functions F and G.

DEFINITION

Let F and G be two functions which are piecewise continuous on every finite closed interval $0 \le t \le b$ and of exponential order. The function denoted by $F * G$ and defined by

$$F(t) * G(t) = \int_0^t F(\tau)G(t-\tau)\,d\tau \tag{9.21}$$

is called the convolution *of the functions F and G.*

Let us change the variable of integration in (9.21) by means of the substitution $u = t - \tau$. We have

$$F(t) * G(t) = \int_0^t F(\tau)G(t-\tau)\,d\tau = -\int_t^0 F(t-u)G(u)\,du$$

$$= \int_0^t G(u)F(t-u)\,du = G(t) * F(t).$$

Thus we have shown that

$$F * G = G * F \tag{9.22}$$

Suppose that both F and G are piecewise continuous on every finite closed interval $0 \leq t \leq b$ and of exponential order e^{at}. Then it can be shown that $F * G$ is also piecewise continuous on every finite closed interval $0 \leq t \leq b$ and of exponential order $e^{(a+\epsilon)t}$, where ϵ is any positive number. Thus $\mathscr{L}\{F * G\}$ exists for s sufficiently large. More explicitly, it can be shown that $\mathscr{L}\{F * G\}$ exists for $s > a$.

We now prove the following important theorem concerning $\mathscr{L}\{F * G\}$.

THEOREM 9.9

Hypothesis. *Let the functions F and G be piecewise continuous on every finite closed interval $0 \leq t \leq b$ and of exponential order e^{at}.*

Conclusion

$$\mathscr{L}\{F * G\} = \mathscr{L}\{F\}\mathscr{L}\{G\} \tag{9.23}$$

for $s > a$.

Proof. By definition of the Laplace transform, $\mathscr{L}\{F * G\}$ is the function defined by

$$\int_0^\infty e^{-st}\left[\int_0^t F(\tau)G(t - \tau)\,d\tau\right] dt. \tag{9.24}$$

The integral (9.24) may be expressed as the iterated integral

$$\int_0^\infty \int_0^t e^{-st}F(\tau)G(t - \tau)\,d\tau\,dt. \tag{9.25}$$

Further, the iterated integral (9.25) is equal to the double integral

$$\iint_{R_1} e^{-st}F(\tau)G(t - \tau)\,d\tau\,dt, \tag{9.26}$$

where R_1 is the 45° wedge bounded by the lines $\tau = 0$ and $t = \tau$ (see Figure 9.2).

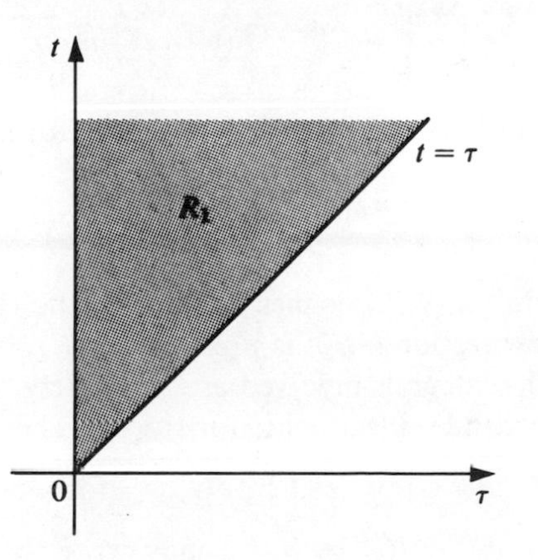

FIGURE 9.2

We now make the change of variable

$$\begin{aligned} u &= t - \tau, \\ v &= \tau, \end{aligned} \tag{9.27}$$

to transform the double integral (9.26). The change of variables (9.27) has Jacobian 1 and transforms the region R_1 in the τ, t plane into the first quadrant of the u, v plane. Thus the double integral (9.26) transforms into the double integral

$$\iint_{R_2} e^{-s(u+v)} F(v)G(u)\, du\, dv, \tag{9.28}$$

where R_2 is the quarter plane defined by $u > 0$, $v > 0$ (see Figure 9.3).

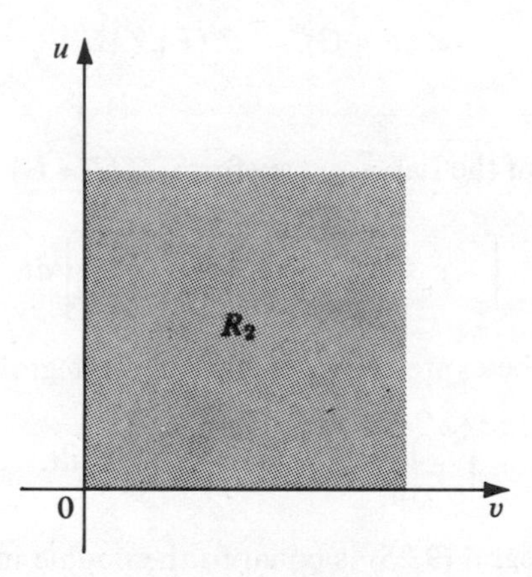

FIGURE 9.3

The double integral (9.28) is equal to the iterated integral

$$\int_0^\infty \int_0^\infty e^{-s(u+v)} F(v)G(u)\, du\, dv. \tag{9.29}$$

But the iterated integral (9.29) can be expressed in the form

$$\int_0^\infty e^{-sv} F(v)\, dv \int_0^\infty e^{-su} G(u)\, du. \tag{9.30}$$

But the left-hand integral in (9.30) defines $\mathscr{L}\{F\}$ and the right-hand integral defines $\mathscr{L}\{G\}$. Therefore the expression (9.30) is precisely $\mathscr{L}\{F\}\mathscr{L}\{G\}$.

We note that since the integrals involved are absolutely convergent for $s > a$, the operations performed are indeed legitimate for $s > a$. Therefore we have shown that

$$\mathscr{L}\{F * G\} = \mathscr{L}\{F\}\mathscr{L}\{G\} \qquad \text{for } s > a. \qquad \textit{Q.E.D.}$$

Denoting $\mathscr{L}\{F\}$ by f and $\mathscr{L}\{G\}$ by g, we may write the conclusion (9.23) in the form

$$\mathscr{L}\{F(t) * G(t)\} = f(s)g(s).$$

Hence, we have

$$\mathscr{L}^{-1}\{f(s)g(s)\} = F(t) * G(t) = \int_0^t F(\tau)G(t-\tau)\,d\tau, \tag{9.31}$$

and using (9.22), we also have

$$\mathscr{L}^{-1}\{f(s)g(s)\} = G(t) * F(t) = \int_0^t G(\tau)F(t-\tau)\,d\tau. \tag{9.32}$$

Suppose we are given a function h and are required to determine $\mathscr{L}^{-1}\{h(s)\}$. If we can express $h(s)$ as a product $f(s)g(s)$, where $\mathscr{L}^{-1}\{f(s)\} = F(t)$ and $\mathscr{L}^{-1}\{g(s)\} = G(t)$ are known, then we can apply either (9.31) or (9.32) to determine $\mathscr{L}^{-1}\{h(s)\}$.

▶ Example 9.21. Find $\mathscr{L}^{-1}\left\{\dfrac{1}{s(s^2+1)}\right\}$ using the convolution and Table 9.1.

Solution. We write $\dfrac{1}{s(s^2+1)}$ as the product $f(s)g(s)$, where $f(s) = \dfrac{1}{s}$ and $g(s) = \dfrac{1}{s^2+1}$. By Table 9.1, number 1, $F(t) = \mathscr{L}^{-1}\left\{\dfrac{1}{s}\right\} = 1$, and by number 3, $G(t) = \mathscr{L}^{-1}\left\{\dfrac{1}{s^2+1}\right\} = \sin t$. Thus by (9.31),

$$\mathscr{L}^{-1}\left\{\frac{1}{s(s^2+1)}\right\} = F(t) * G(t) = \int_0^t 1 \cdot \sin(t-\tau)\,d\tau,$$

and by (9.32),

$$\mathscr{L}^{-1}\left\{\frac{1}{s(s^2+1)}\right\} = G(t) * F(t) = \int_0^t \sin\tau \cdot 1\,d\tau.$$

The second of these two integrals is slightly more simple. Evaluating it, we have

$$\mathscr{L}^{-1}\left\{\frac{1}{s(s^2+1)}\right\} = 1 - \cos t.$$

Observe that we obtained this result in Example 9.20 by means of partial fractions.

Exercises

Use Table 9.1 to find $\mathscr{L}^{-1}\{f(s)\}$ for each of the functions f defined in Exercises 1 through 12:

1. $f(s) = \dfrac{2}{s^2+9}$.
2. $f(s) = \dfrac{3s}{s^2-4}$.
3. $f(s) = \dfrac{5}{(s-2)^4}$.
4. $f(s) = \dfrac{5s}{s^2+4s+4}$.
5. $f(s) = \dfrac{s+2}{s^2+4s+7}$.
6. $f(s) = \dfrac{s+10}{s^2+8s+20}$.

7. $f(s) = \dfrac{1}{s^3 + 4s^2 + 3s}$.

8. $f(s) = \dfrac{s + 1}{s^3 + 2s}$.

9. $f(s) = \dfrac{s + 3}{(s^2 + 4)^2}$.

10. $f(s) = \dfrac{s + 5}{s^4 + 3s^3 + 2s^2}$.

11. $f(s) = \dfrac{5}{(s + 2)^5}$.

12. $f(s) = \dfrac{2s + 7}{(s + 3)^4}$.

TABLE 9.1. LAPLACE TRANSFORMS

	$F(t)$	$f(s)$
1	1	$\dfrac{1}{s}$
2	e^{at}	$\dfrac{1}{s - a}$
3	$\sin bt$	$\dfrac{b}{s^2 + b^2}$
4	$\cos bt$	$\dfrac{s}{s^2 + b^2}$
5	$\sinh bt$	$\dfrac{b}{s^2 - b^2}$
6	$\cosh bt$	$\dfrac{s}{s^2 - b^2}$
7	$t^n \ (n = 1, 2, \ldots)$	$\dfrac{n!}{s^{n+1}}$
8	$t^n e^{at} \ (n = 1, 2, \ldots)$	$\dfrac{n!}{(s - a)^{n+1}}$
9	$t \sin bt$	$\dfrac{2bs}{(s^2 + b^2)^2}$
10	$t \cos bt$	$\dfrac{s^2 - b^2}{(s^2 + b^2)^2}$
11	$e^{-at} \sin bt$	$\dfrac{b}{(s + a)^2 + b^2}$
12	$e^{-at} \cos bt$	$\dfrac{s + a}{(s + a)^2 + b^2}$
13	$\dfrac{\sin bt - bt \cos bt}{2b^3}$	$\dfrac{1}{(s^2 + b^2)^2}$
14	$\dfrac{t \sin bt}{2b}$	$\dfrac{s}{(s^2 + b^2)^2}$

9.3 Laplace Transform Solution of Linear Differential Equations with Constant Coefficients

A. *The Method*

We now consider how the Laplace transform may be applied to solve the initial-value problem consisting of the nth-order linear differential equation with constant coefficients

$$a_0 \frac{d^n Y}{dt^n} + a_1 \frac{d^{n-1} Y}{dt^{n-1}} + \cdots + a_{n-1} \frac{dY}{dt} + a_n Y = B(t), \tag{9.33}$$

plus the initial conditions

$$Y(0) = c_0,\ Y'(0) = c_1, \ldots,\ Y^{(n-1)}(0) = c_{n-1}. \tag{9.34}$$

Theorem 4.1 (Chapter 4) assures us that this problem has a unique solution.

We now take the Laplace transform of both members of Equation (9.33). By Theorem 9.2, we have

$$a_0 \mathscr{L}\left\{\frac{d^n Y}{dt^n}\right\} + a_1 \mathscr{L}\left\{\frac{d^{n-1} Y}{dt^{n-1}}\right\} + \cdots + a_{n-1} \mathscr{L}\left\{\frac{dY}{dt}\right\} + a_n \mathscr{L}\{Y(t)\} = \mathscr{L}\{B(t)\}. \tag{9.35}$$

We now apply Theorem 9.4 to

$$\mathscr{L}\left\{\frac{d^n Y}{dt^n}\right\},\ \mathscr{L}\left\{\frac{d^{n-1} Y}{dt^{n-1}}\right\}, \ldots,\ \mathscr{L}\left\{\frac{dY}{dt}\right\}$$

in the left member of Equation (9.35). Using the initial conditions (9.34), we have

$$\begin{aligned}
\mathscr{L}\left\{\frac{d^n Y}{dt^n}\right\} &= s^n \mathscr{L}\{Y(t)\} - s^{n-1} Y(0) - s^{n-2} Y'(0) - \cdots - Y^{(n-1)}(0) \\
&= s^n \mathscr{L}\{Y(t)\} - c_0 s^{n-1} - c_1 s^{n-2} - \cdots - c_{n-1}, \\
\mathscr{L}\left\{\frac{d^{n-1} Y}{dt^{n-1}}\right\} &= s^{n-1} \mathscr{L}\{Y(t)\} - s^{n-2} Y(0) - s^{n-3} Y'(0) - \cdots - Y^{(n-2)}(0) \\
&= s^{n-1} \mathscr{L}\{Y(t)\} - c_0 s^{n-2} - c_1 s^{n-3} - \cdots - c_{n-2}, \\
&\ \vdots \\
\mathscr{L}\left\{\frac{dY}{dt}\right\} &= s \mathscr{L}\{Y(t)\} - Y(0) = s \mathscr{L}\{Y(t)\} - c_0.
\end{aligned}$$

Thus, letting $y(s)$ denote $\mathscr{L}\{Y(t)\}$ and $b(s)$ denote $\mathscr{L}\{B(t)\}$, Equation (9.35) becomes

$$\begin{aligned}
&[a_0 s^n + a_1 s^{n-1} + \cdots + a_{n-1} s + a_n] y(s) \\
&\qquad - c_0[a_0 s^{n-1} + a_1 s^{n-2} + \cdots + a_{n-1}] \\
&\qquad - c_1[a_0 s^{n-2} + a_1 s^{n-3} + \cdots + a_{n-2}] \\
&\qquad - \cdots - c_{n-2}[a_0 s + a_1] - c_{n-1} a_0 = b(s).
\end{aligned} \tag{9.36}$$

Since B is a known function of t, then b, assuming it exists and can be determined, is a known function of s. Thus Equation (9.36) is an algebraic equation in the "unknown" $y(s)$. We now solve the algebraic equation (9.36) to determine $y(s)$. Once $y(s)$ has been found, we then find the unique solution

$$Y(t) = \mathscr{L}^{-1}\{y(s)\}$$

of the given initial-value problem using the table of transforms.

We summarize this procedure as follows:

1. Take the Laplace transform of both sides of the differential equation (9.33), applying Theorem 9.4 and using the initial conditions (9.34) in the process, and equate the results to obtain the algebraic equation (9.36) in the "unknown" $y(s)$.
2. Solve the algebraic equation (9.36) thus obtained to determine $y(s)$.
3. Having found $y(s)$, employ the table of transforms to determine the solution $Y(t) = \mathscr{L}^{-1}\{y(s)\}$ of the given initial-value problem.

B. *Examples*

We shall now consider several detailed examples which will illustrate the procedure outlined above.

▶ **Example 9.22.** Solve the initial-value problem

$$\frac{dY}{dt} - 2Y = e^{5t}, \tag{9.37}$$

$$Y(0) = 3 \tag{9.38}$$

Step 1. Taking the Laplace transform of both sides of the differential equation (9.37), we have

$$\mathscr{L}\left\{\frac{dY}{dt}\right\} - 2\mathscr{L}\{Y(t)\} = \mathscr{L}\{e^{5t}\}. \tag{9.39}$$

Using Theorem 9.4 with $n = 1$ (or Theorem 9.3) and denoting $\mathscr{L}\{Y(t)\}$ by $y(s)$, we may express $\mathscr{L}\{dY/dt\}$ in terms of $y(s)$ and $Y(0)$ as follows:

$$\mathscr{L}\left\{\frac{dY}{dt}\right\} = sy(s) - Y(0).$$

Applying the initial condition (9.38), this becomes

$$\mathscr{L}\left\{\frac{dY}{dt}\right\} = sy(s) - 3.$$

Using this, the left member of Equation (9.39) becomes $sy(s) - 3 - 2y(s)$. From Table 9.1, number 2, $\mathscr{L}\{e^{5t}\} = 1/(s - 5)$. Thus Equation (9.39) reduces to the algebraic equation

$$[s - 2]y(s) - 3 = \frac{1}{s - 5} \tag{9.40}$$

in the unknown $y(s)$.

Step 2. We now solve Equation (9.40) for $y(s)$. We have

$$[s - 2]y(s) = \frac{3s - 14}{s - 5}$$

and so

$$y(s) = \frac{3s - 14}{(s - 2)(s - 5)}.$$

Step 3. We must now determine

$$\mathscr{L}^{-1}\left\{\frac{3s - 14}{(s - 2)(s - 5)}\right\}.$$

We employ partial fractions. We have

$$\frac{3s - 14}{(s - 2)(s - 5)} = \frac{A}{s - 2} + \frac{B}{s - 5},$$

and so $3s - 14 = A(s - 5) + B(s - 2)$. From this we find that

$$A = \tfrac{8}{3} \quad \text{and} \quad B = \tfrac{1}{3},$$

and so

$$\mathscr{L}^{-1}\left\{\frac{3s - 14}{(s - 2)(s - 5)}\right\} = \frac{8}{3}\mathscr{L}^{-1}\left\{\frac{1}{s - 2}\right\} + \frac{1}{3}\mathscr{L}^{-1}\left\{\frac{1}{s - 5}\right\}.$$

Using number 2 of Table 9.1,

$$\mathscr{L}^{-1}\left\{\frac{1}{s - 2}\right\} = e^{2t} \quad \text{and} \quad \mathscr{L}^{-1}\left\{\frac{1}{s - 5}\right\} = e^{5t}.$$

Thus the solution of the given initial-value problem is

$$Y = \tfrac{8}{3}e^{2t} + \tfrac{1}{3}e^{5t}.$$

▶ Example 9.23. Solve the initial-value problem

$$\frac{d^2Y}{dt^2} - 2\frac{dY}{dt} - 8Y = 0, \tag{9.41}$$

$$Y(0) = 3, \tag{9.42}$$

$$Y'(0) = 6, \tag{9.43}$$

Step 1. Taking the Laplace transform of both sides of the differential equation (9.41), we have

$$\mathscr{L}\left\{\frac{d^2Y}{dt^2}\right\} - 2\mathscr{L}\left\{\frac{dY}{dt}\right\} - 8\mathscr{L}\{Y(t)\} = \mathscr{L}\{0\}. \tag{9.44}$$

Since $\mathscr{L}\{0\} = 0$, the right member of Equation (9.44) is simply 0. Denote $\mathscr{L}\{Y(t)\}$

by $y(s)$. Then, applying Theorem 9.4, we have the following expressions for $\mathscr{L}\{d^2Y/dt^2\}$ and $\mathscr{L}\{dY/dt\}$ in terms of $y(s)$, $Y(0)$, and $Y'(0)$:

$$\mathscr{L}\left\{\frac{d^2Y}{dt^2}\right\} = s^2y(s) - sY(0) - Y'(0),$$

$$\mathscr{L}\left\{\frac{dY}{dt}\right\} = sy(s) - Y(0).$$

Applying the initial conditions (9.42) and (9.43) to these expressions, they become:

$$\mathscr{L}\left\{\frac{d^2Y}{dt^2}\right\} = s^2y(s) - 3s - 6,$$

$$\mathscr{L}\left\{\frac{dY}{dt}\right\} = sy(s) - 3.$$

Now, using these expressions, Equation (9.44) becomes

$$s^2y(s) - 3s - 6 - 2sy(s) + 6 - 8y(s) = 0$$

or

$$[s^2 - 2s - 8]y(s) - 3s = 0. \tag{9.45}$$

Step 2. We now solve Equation (9.45) for $y(s)$. We have at once

$$y(s) = \frac{3s}{(s-4)(s+2)}.$$

Step 3. We must now determine

$$\mathscr{L}^{-1}\left\{\frac{3s}{(s-4)(s+2)}\right\}.$$

We shall again employ partial fractions. From

$$\frac{3s}{(s-4)(s+2)} = \frac{A}{s-4} + \frac{B}{s+2}$$

we find that $A = 2$, $B = 1$. Thus

$$\mathscr{L}^{-1}\left\{\frac{3s}{(s-4)(s+2)}\right\} = 2\mathscr{L}^{-1}\left\{\frac{1}{s-4}\right\} + \mathscr{L}^{-1}\left\{\frac{1}{s+2}\right\}.$$

By Table 9.1, number 2, we find

$$\mathscr{L}^{-1}\left\{\frac{1}{s-4}\right\} = e^{4t} \quad \text{and} \quad \mathscr{L}^{-1}\left\{\frac{1}{s+2}\right\} = e^{-2t}.$$

Thus the solution of the given initial-value problem is

$$Y = 2e^{4t} + e^{-2t}.$$

▶ Example 9.24. Solve the initial-value problem

$$\frac{d^2Y}{dt^2} + Y = e^{-2t} \sin t, \tag{9.46}$$

$$Y(0) = 0, \tag{9.47}$$

$$Y'(0) = 0. \tag{9.48}$$

Step 1. Taking the Laplace transform of both sides of the differential equation (9.46), we have

$$\mathscr{L}\left\{\frac{d^2Y}{dt^2}\right\} + \mathscr{L}\{Y(t)\} = \mathscr{L}\{e^{-2t} \sin t\}. \tag{9.49}$$

Denoting $\mathscr{L}\{Y(t)\}$ by $y(s)$ and applying Theorem 9.4, we express $\mathscr{L}\{d^2Y/dt^2\}$ in terms of $y(s)$, $Y(0)$, and $Y'(0)$ as follows:

$$\mathscr{L}\left\{\frac{d^2Y}{dt^2}\right\} = s^2y(s) - sY(0) - Y'(0).$$

Applying the initial conditions (9.47) and (9.48) to this expression, it becomes simply

$$\mathscr{L}\left\{\frac{d^2Y}{dt^2}\right\} = s^2y(s);$$

and thus the left member of Equation (9.49) becomes $s^2y(s) + y(s)$. By number 11, Table 9.1, the right member of Equation (9.49) becomes

$$\frac{1}{(s + 2)^2 + 1}.$$

Thus Equation (9.49) reduces to the algebraic equation

$$(s^2 + 1)y(s) = \frac{1}{(s + 2)^2 + 1} \tag{9.50}$$

in the unknown $y(s)$.

Step 2. Solving Equation (9.50) for $y(s)$, we have

$$y(s) = \frac{1}{(s^2 + 1)[(s + 2)^2 + 1]}.$$

Step 3. We must now determine

$$\mathscr{L}^{-1}\left\{\frac{1}{(s^2 + 1)[(s + 2)^2 + 1]}\right\}.$$

We may use either partial fractions or the convolution. We shall illustrate both methods.

1. Use of Partial Fractions. We have

$$\frac{1}{(s^2 + 1)(s^2 + 4s + 5)} = \frac{As + B}{s^2 + 1} + \frac{Cs + D}{s^2 + 4s + 5}.$$

From this we find

$$1 = (As + B)(s^2 + 4s + 5) + (Cs + D)(s^2 + 1)$$
$$= (A + C)s^3 + (4A + B + D)s^2 + (5A + 4B + C)s + (5B + D).$$

Thus we obtain the equations

$$A + C = 0,$$
$$4A + B + D = 0,$$
$$5A + 4B + C = 0,$$
$$5B + D = 1.$$

From these equations we find that

$$A = -\tfrac{1}{8}, \qquad B = \tfrac{1}{8}, \qquad C = \tfrac{1}{8}, \qquad D = \tfrac{3}{8},$$

and so

$$\mathscr{L}^{-1}\left\{\frac{1}{(s^2 + 1)(s^2 + 4s + 5)}\right\} = -\frac{1}{8}\mathscr{L}^{-1}\left\{\frac{s}{s^2 + 1}\right\} + \frac{1}{8}\mathscr{L}^{-1}\left\{\frac{1}{s^2 + 1}\right\}$$
$$+ \frac{1}{8}\mathscr{L}^{-1}\left\{\frac{s}{s^2 + 4s + 5}\right\}$$
$$+ \frac{3}{8}\mathscr{L}^{-1}\left\{\frac{1}{s^2 + 4s + 5}\right\}. \qquad (9.51)$$

In order to determine

$$\frac{1}{8}\mathscr{L}^{-1}\left\{\frac{s}{s^2 + 4s + 5}\right\} + \frac{3}{8}\mathscr{L}^{-1}\left\{\frac{1}{s^2 + 4s + 5}\right\}, \qquad (9.52)$$

we write

$$\frac{s}{s^2 + 4s + 5} = \frac{s + 2}{(s + 2)^2 + 1} - \frac{2}{(s + 2)^2 + 1}.$$

Thus the expression (9.52) becomes

$$\frac{1}{8}\mathscr{L}^{-1}\left\{\frac{s + 2}{(s + 2)^2 + 1}\right\} + \frac{1}{8}\mathscr{L}^{-1}\left\{\frac{1}{(s + 2)^2 + 1}\right\},$$

and so (9.51) may be written

$$\mathscr{L}^{-1}\left\{\frac{1}{(s^2 + 1)(s^2 + 4s + 5)}\right\} = -\frac{1}{8}\mathscr{L}^{-1}\left\{\frac{s}{s^2 + 1}\right\} + \frac{1}{8}\mathscr{L}^{-1}\left\{\frac{1}{s^2 + 1}\right\}$$
$$+ \frac{1}{8}\mathscr{L}^{-1}\left\{\frac{s + 2}{(s + 2)^2 + 1}\right\}$$
$$+ \frac{1}{8}\mathscr{L}^{-1}\left\{\frac{1}{(s + 2)^2 + 1}\right\}.$$

Now using Table 9.1, numbers 4, 3, 12, and 11, respectively, we obtain the solution

$$Y(t) = -\tfrac{1}{8}\cos t + \tfrac{1}{8}\sin t + \tfrac{1}{8}e^{-2t}\cos t + \tfrac{1}{8}e^{-2t}\sin t$$

or

$$Y(t) = \frac{1}{8}(\sin t - \cos t) + \frac{e^{-2t}}{8}(\sin t + \cos t). \tag{9.53}$$

2. Use of the Convolution. We write $\dfrac{1}{(s^2+1)[(s+2)^2+1]}$ as the product $f(s)g(s)$, where $f(s) = \dfrac{1}{s^2+1}$ and $g(s) = \dfrac{1}{(s+2)^2+1}$. By Table 9.1, number 3, $F(t) = \mathscr{L}^{-1}\left\{\dfrac{1}{s^2+1}\right\} = \sin t$, and by number 11, $G(t) = \mathscr{L}^{-1}\left\{\dfrac{1}{(s+2)^2+1}\right\} = e^{-2t}\sin t$. Thus by Theorem 9.9 using (9.31) or (9.32), we have, respectively,

$$\mathscr{L}^{-1}\left\{\frac{1}{(s^2+1)[(s+2)^2+1]}\right\} = F(t) * G(t) = \int_0^t \sin\tau \cdot e^{-2(t-\tau)}\sin(t-\tau)\,d\tau$$

or

$$\mathscr{L}^{-1}\left\{\frac{1}{(s^2+1)[(s+2)^2+1]}\right\} = G(t) * F(t) = \int_0^t e^{-2\tau}\sin\tau\cdot\sin(t-\tau)\,d\tau.$$

The second of these integrals is slightly more simple; it reduces to

$$(\sin t)\int_0^t e^{-2\tau}\sin\tau\cos\tau\,d\tau - (\cos t)\int_0^t e^{-2\tau}\sin^2\tau\,d\tau.$$

Introducing double-angle formulas this becomes

$$\frac{\sin t}{2}\int_0^t e^{-2\tau}\sin 2\tau\,d\tau - \frac{\cos t}{2}\int_0^t e^{-2\tau}\,d\tau + \frac{\cos t}{2}\int_0^t e^{-2\tau}\cos 2\tau\,d\tau.$$

Carrying out the indicated integrations we find that this becomes

$$-\sin t\left[\frac{e^{-2\tau}}{8}(\sin 2\tau + \cos 2\tau)\right]_0^t + \frac{\cos t}{4}\left[e^{-2\tau}\right]_0^t + \cos t\left[\frac{e^{-2\tau}}{8}(\sin 2\tau - \cos 2\tau)\right]_0^t$$

$$= -\frac{e^{-2t}}{8}(\sin t\sin 2t + \sin t\cos 2t) + \frac{\sin t}{8} + \frac{e^{-2t}\cos t}{4} - \frac{\cos t}{4}$$

$$+ \frac{e^{-2t}}{8}(\cos t\sin 2t - \cos t\cos 2t) + \frac{\cos t}{8}.$$

Using double-angle formulas and simplifying, this reduces to

$$\frac{1}{8}(\sin t - \cos t) + \frac{e^{-2t}}{8}(\sin t + \cos t),$$

which is the solution (9.53) obtained above using partial fractions.

▶ **Example 9.25.** Solve the initial-value problem

$$\frac{d^3Y}{dt^3} + 4\frac{d^2Y}{dt^2} + 5\frac{dY}{dt} + 2Y = 10 \cos t, \tag{9.54}$$

$$Y(0) = 0, \tag{9.55}$$

$$Y'(0) = 0, \tag{9.56}$$

$$Y''(0) = 3. \tag{9.57}$$

Step 1. Taking the Laplace transform of both sides of the differential equation (9.54), we have

$$\mathscr{L}\left\{\frac{d^3Y}{dt^3}\right\} + 4\mathscr{L}\left\{\frac{d^2Y}{dt^2}\right\} + 5\mathscr{L}\left\{\frac{dY}{dt}\right\} + 2\mathscr{L}\{Y(t)\} = 10\mathscr{L}\{\cos t\}. \tag{9.58}$$

We denote $\mathscr{L}\{Y(t)\}$ by $y(s)$ and then apply Theorem 9.4 to express

$$\mathscr{L}\left\{\frac{d^3Y}{dt^3}\right\}, \quad \mathscr{L}\left\{\frac{d^2Y}{dt^2}\right\}, \quad \text{and} \quad \mathscr{L}\left\{\frac{dY}{dt}\right\}$$

in terms of $y(s)$, $Y(0)$, $Y'(0)$, and $Y''(0)$. We thus obtain

$$\mathscr{L}\left\{\frac{d^3Y}{dt^3}\right\} = s^3y(s) - s^2Y(0) - sY'(0) - Y''(0),$$

$$\mathscr{L}\left\{\frac{d^2Y}{dt^2}\right\} = s^2y(s) - sY(0) - Y'(0),$$

$$\mathscr{L}\left\{\frac{dY}{dt}\right\} = sy(s) - Y(0).$$

Applying the initial conditions (9.55), (9.56), and (9.57), these expressions become

$$\mathscr{L}\left\{\frac{d^3Y}{dt^3}\right\} = s^3y(s) - 3,$$

$$\mathscr{L}\left\{\frac{d^2Y}{dt^2}\right\} = s^2y(s),$$

$$\mathscr{L}\left\{\frac{dY}{dt}\right\} = sy(s).$$

Thus the left member of Equation (9.58) becomes

$$s^3y(s) - 3 + 4s^2y(s) + 5sy(s) + 2y(s)$$

or

$$[s^3 + 4s^2 + 5s + 2]y(s) - 3.$$

By number 4, Table 9.1,

$$10\mathscr{L}\{\cos t\} = \frac{10s}{s^2 + 1}.$$

Thus Equation (9.58) reduces to the algebraic equation

$$(s^3 + 4s^2 + 5s + 2)y(s) - 3 = \frac{10s}{s^2 + 1} \tag{9.59}$$

in the unknown $y(s)$.

Step 2. We now solve Equation (9.59) for $y(s)$. We have

$$(s^3 + 4s^2 + 5s + 2)y(s) = \frac{3s^2 + 10s + 3}{s^2 + 1}$$

or

$$y(s) = \frac{3s^2 + 10s + 3}{(s^2 + 1)(s^3 + 4s^2 + 5s + 2)}.$$

Step 3. We must now determine

$$\mathscr{L}^{-1}\left\{\frac{3s^2 + 10s + 3}{(s^2 + 1)(s^3 + 4s^2 + 5s + 2)}\right\}.$$

Let us not despair! We can again employ partial fractions to put the expression for $y(s)$ into a form where Table 9.1 can be used, but the work will be rather involved. We proceed by writing

$$\begin{aligned}\frac{3s^2 + 10s + 3}{(s^2 + 1)(s^3 + 4s^2 + 5s + 2)} &= \frac{3s^2 + 10s + 3}{(s^2 + 1)(s + 1)^2(s + 2)} \\ &= \frac{A}{s + 2} + \frac{B}{s + 1} + \frac{C}{(s + 1)^2} + \frac{Ds + E}{s^2 + 1}.\end{aligned} \tag{9.60}$$

From this we find

$$\begin{aligned}3s^2 + 10s + 3 &= A(s + 1)^2(s^2 + 1) + B(s + 2)(s + 1)(s^2 + 1) \\ &\quad + C(s + 2)(s^2 + 1) + (Ds + E)(s + 2)(s + 1)^2,\end{aligned} \tag{9.61}$$

or

$$\begin{aligned}3s^2 + 10s + 3 &= (A + B + D)s^4 + (2A + 3B + C + 4D + E)s^3 \\ &\quad + (2A + 3B + 2C + 5D + 4E)s^2 \\ &\quad + (2A + 3B + C + 2D + 5E)s + (A + 2B + 2C + 2E).\end{aligned}$$

From this we obtain the system of equations

$$\begin{aligned} A + B + D &= 0, \\ 2A + 3B + C + 4D + E &= 0, \\ 2A + 3B + 2C + 5D + 4E &= 3, \\ 2A + 3B + C + 2D + 5E &= 10, \\ A + 2B + 2C + 2E &= 3. \end{aligned} \tag{9.62}$$

Letting $s = -1$ in Equation (9.61), we find that $C = -2$; and letting $s = -2$ in this same equation results in $A = -1$. Using these values for A and C we find from the system (9.62) that

$$B = 2, \qquad D = -1, \quad \text{and} \quad E = 2.$$

Substituting these values thus found for A, B, C, D, and E into Equation (9.60), we see that

$$\begin{aligned}\mathscr{L}^{-1}\left\{\frac{3s^2 + 10s + 3}{(s^2 + 1)(s^3 + 4s^2 + 5s + 2)}\right\} &= -\mathscr{L}^{-1}\left\{\frac{1}{s + 2}\right\} + 2\mathscr{L}^{-1}\left\{\frac{1}{s + 1}\right\} \\ &\quad - 2\mathscr{L}^{-1}\left\{\frac{1}{(s + 1)^2}\right\} - \mathscr{L}^{-1}\left\{\frac{s}{s^2 + 1}\right\} \\ &\quad + 2\mathscr{L}^{-1}\left\{\frac{1}{s^2 + 1}\right\}.\end{aligned}$$

Using Table 9.1, numbers 2, 2, 8, 4, and 3, respectively, we obtain the solution

$$Y(t) = -e^{-2t} + 2e^{-t} - 2te^{-t} - \cos t + 2 \sin t.$$

▶ **Example 9.26.** Solve the initial-value problem

$$\frac{d^2Y}{dt^2} + 2\frac{dY}{dt} + 5Y = H(t), \tag{9.63}$$

where

$$H(t) = \begin{cases} 1, & 0 < t < \pi, \\ 0, & t > \pi, \end{cases} \tag{9.64}$$

$$Y(0) = 0, \tag{9.65}$$

$$Y'(0) = 0. \tag{9.66}$$

Step 1. We take the Laplace transform of both sides of the differential equation (9.63) to obtain

$$\mathscr{L}\left\{\frac{d^2Y}{dt^2}\right\} + 2\mathscr{L}\left\{\frac{dY}{dt}\right\} + 5\mathscr{L}\{Y(t)\} = \mathscr{L}\{H(t)\}. \tag{9.67}$$

Denoting $\mathscr{L}\{Y(t)\}$ by $y(s)$, using Theorem 9.4 as in the previous examples, and then applying the initial conditions (9.65) and (9.66), we see that the left member of Equation (9.67) becomes $[s^2 + 2s + 5]y(s)$. By the definition of the Laplace transform, from (9.64) we have

$$\mathscr{L}\{H(t)\} = \int_0^\infty e^{-st}H(t)\,dt = \int_0^\pi e^{-st}\,dt = \left[\frac{e^{-st}}{-s}\right]_0^\pi = \frac{1 - e^{-\pi s}}{s}.$$

Thus Equation (9.67) becomes

$$[s^2 + 2s + 5]y(s) = \frac{1 - e^{-\pi s}}{s}. \tag{9.68}$$

Step 2. We solve the algebraic equation (9.68) for $y(s)$ to obtain

$$y(s) = \frac{1 - e^{-\pi s}}{s(s^2 + 2s + 5)}.$$

Step 3. We must now determine

$$\mathscr{L}^{-1}\left\{\frac{1 - e^{-\pi s}}{s(s^2 + 2s + 5)}\right\}.$$

Let us write this as

$$\mathscr{L}^{-1}\left\{\frac{1}{s(s^2 + 2s + 5)}\right\} - \mathscr{L}^{-1}\left\{\frac{e^{-\pi s}}{s(s^2 + 2s + 5)}\right\},$$

and apply partial fractions to determine the first of these two inverse transforms. Writing

$$\frac{1}{s(s^2 + 2s + 5)} = \frac{A}{s} + \frac{Bs + C}{s^2 + 2s + 5},$$

we find at once that $A = \frac{1}{5}$, $B = -\frac{1}{5}$, $C = -\frac{2}{5}$. Thus

$$\begin{aligned}
\mathscr{L}^{-1}\left\{\frac{1}{s(s^2 + 2s + 5)}\right\} &= \frac{1}{5}\mathscr{L}^{-1}\left\{\frac{1}{s}\right\} - \frac{1}{5}\mathscr{L}^{-1}\left\{\frac{s + 2}{(s + 1)^2 + 4}\right\} \\
&= \frac{1}{5}\mathscr{L}^{-1}\left\{\frac{1}{s}\right\} - \frac{1}{5}\mathscr{L}^{-1}\left\{\frac{s + 1}{(s + 1)^2 + 4}\right\} \\
&\quad - \frac{1}{10}\mathscr{L}^{-1}\left\{\frac{2}{(s + 1)^2 + 4}\right\} \\
&= \frac{1}{5} - \frac{1}{5}e^{-t}\cos 2t - \frac{1}{10}e^{-t}\sin 2t,
\end{aligned}$$

using Table 9.1, numbers 1, 12, and 11, respectively. Letting

$$f(s) = \frac{1}{s(s^2 + 2s + 5)}$$

and

$$F(t) = \tfrac{1}{5} - \tfrac{1}{5}e^{-t}\cos 2t - \tfrac{1}{10}e^{-t}\sin 2t,$$

we thus have

$$\mathscr{L}^{-1}\{f(s)\} = F(t).$$

We now consider

$$\mathscr{L}^{-1}\left\{\frac{e^{-\pi s}}{s(s^2 + 2s + 5)}\right\} = \mathscr{L}^{-1}\{e^{-\pi s}f(s)\}.$$

By Theorem 9.6,

$$\mathscr{L}^{-1}\{e^{-\pi s}f(s)\} = G(t),$$

where

$$G(t) = \begin{cases} 0, & 0 < t < \pi, \\ F(t - \pi), & t > \pi. \end{cases}$$

Thus,

$$\mathscr{L}^{-1}\left\{\frac{e^{-\pi s}}{s(s^2+2s+5)}\right\}$$

$$= \begin{cases} 0, & 0 < t < \pi, \\ \frac{1}{5} - \frac{1}{5}e^{-(t-\pi)}\cos 2(t-\pi) - \frac{1}{10}e^{-(t-\pi)}\sin 2(t-\pi), & t > \pi \end{cases}$$

$$= \begin{cases} 0, & 0 < t < \pi, \\ \frac{1}{5} - \frac{1}{5}e^{-(t-\pi)}\cos 2t - \frac{1}{10}e^{-(t-\pi)}\sin 2t, & t > \pi. \end{cases}$$

Thus the solution is given by

$$Y(t) = \mathscr{L}^{-1}\left\{\frac{1-e^{\pi s}}{s(s^2+2s+5)}\right\} = F(t) - G(t)$$

$$= \begin{cases} \frac{1}{5} - \frac{1}{5}e^{-t}\cos 2t - \frac{1}{10}e^{-t}\sin 2t - 0, 0 < t < \pi, \\ \frac{1}{5} - \frac{1}{5}e^{-t}\cos 2t - \frac{1}{10}e^{-t}\sin 2t - \frac{1}{5} + \frac{1}{5}e^{-(t-\pi)}\cos 2t + \frac{1}{10}e^{-(t-\pi)}\sin 2t, \\ \qquad t > \pi, \end{cases}$$

or

$$Y(t) = \begin{cases} \frac{1}{5}\left[1 - e^{-t}\left(\cos 2t + \frac{1}{2}\sin 2t\right)\right], & 0 < t < \pi, \\ \frac{e^{-t}}{5}\left[(e^{\pi}-1)\cos 2t + \left(\frac{e^{\pi}-1}{2}\right)\sin 2t\right], & t > \pi. \end{cases}$$

Exercises

Use Laplace transforms to solve each of the initial-value problems in Exercises 1 through 12:

1. $\frac{dY}{dt} - Y = e^{3t}, \quad Y(0) = 2.$

2. $\frac{dY}{dt} + Y = 2\sin t, \quad Y(0) = -1.$

3. $\frac{d^2Y}{dt^2} - 5\frac{dY}{dt} + 6Y = 0,$

   $Y(0) = 1, \quad Y'(0) = 2.$

4. $\frac{d^2Y}{dt^2} + \frac{dY}{dt} - 12Y = 0,$

   $Y(0) = 4, \quad Y'(0) = -1.$

5. $\frac{d^2Y}{dt^2} - \frac{dY}{dt} - 2Y = 18e^{-t}\sin 3t,$

   $Y(0) = 0, \quad Y'(0) = 3.$

6. $\frac{d^2Y}{dt^2} + 2\frac{dY}{dt} + Y = te^{-2t},$

   $Y(0) = 1, \quad Y'(0) = 0.$

7. $\frac{d^3Y}{dt^3} - 5\frac{d^2Y}{dt^2} + 7\frac{dY}{dt} - 3Y = 20\sin t,$

   $Y(0) = 0, \quad Y'(0) = 0, \quad Y''(0) = -2.$

8. $$\frac{d^3Y}{dt^3} - 6\frac{d^2Y}{dt^2} + 11\frac{dY}{dt} - 6Y = te^{-4t},$$

$$Y(0) = 1, \quad Y'(0) = 0, \quad Y''(0) = -1.$$

9. $$\frac{d^2Y}{dt^2} - 3\frac{dY}{dt} + 2Y = H(t), \quad \text{where } H(t) = \begin{cases} 2, & 0 < t < 4, \\ 0, & t > 4, \end{cases}$$

$$Y(0) = 0, \quad Y'(0) = 0.$$

10. $$\frac{d^2Y}{dt^2} + 4\frac{dY}{dt} + 5Y = H(t), \quad \text{where } H(t) = \begin{cases} 1, & 0 < t < \frac{\pi}{2}, \\ 0, & t > \frac{\pi}{2}, \end{cases}$$

$$Y(0) = 0, \quad Y'(0) = 1.$$

11. $$\frac{d^2Y}{dt^2} + 5\frac{dY}{dt} + 6Y = H(t), \quad \text{where } H(t) = \begin{cases} 4, & 0 < t < 1, \\ 0, & t > 1, \end{cases}$$

$$Y(0) = 1, \quad Y'(0) = 0.$$

12. $$\frac{d^2Y}{dt^2} + 6\frac{dY}{dt} + 8Y = H(t), \quad \text{where } H(t) = \begin{cases} 3, & 0 < t < 2\pi, \\ 0, & t > 2\pi, \end{cases}$$

$$Y(0) = 1, \quad Y'(0) = -1.$$

9.4 Laplace Transform Solution of Linear Systems

A. The Method

We apply the Laplace transform method to find the solution of a first-order system

$$\begin{aligned} a_1\frac{dX}{dt} + a_2\frac{dY}{dt} + a_3X + a_4Y &= B_1(t), \\ b_1\frac{dX}{dt} + b_2\frac{dY}{dt} - b_3X + b_4Y &= B_2(t), \end{aligned} \tag{9.69}$$

where $a_1, a_2, a_3, a_4, b_1, b_2, b_3$, and b_4 are constants and B_1 and B_2 are known functions, which satisfies the initial conditions

$$X(0) = c_1 \quad \text{and} \quad Y(0) = c_2, \tag{9.70}$$

where c_1 and c_2 are constants.

The procedure is a straightforward extension of the method outlined in Section 9.3. Let $x(s)$ denote $\mathscr{L}\{X(t)\}$ and let $y(s)$ denote $\mathscr{L}\{Y(t)\}$. Then proceed as follows:

1. For each of the two equations of the system (9.69), take the Laplace transform of both sides of the equation, apply Theorem 9.3 and the initial conditions (9.70), and equate the results to obtain a linear algebraic equation in the two "unknowns" $x(s)$ and $y(s)$.

2. Solve the linear system of two algebraic equations in the two unknowns $x(s)$ and $y(s)$ thus obtained in Step 1 to explicitly determine $x(s)$ and $y(s)$.

3. Having found $x(s)$ and $y(s)$, employ the table of transforms to determine the solution $X(t) = \mathscr{L}^{-1}\{x(s)\}$ and $Y(t) = \mathscr{L}^{-1}\{y(s)\}$ of the given initial-value problem.

B. An Example

▶ **Example 9.27.** Use a Laplace transform to find the solution of the system

$$\frac{dX}{dt} - 6X + 3Y = 8e^t,$$
$$\frac{dY}{dt} - 2X - Y = 4e^t, \tag{9.71}$$

which satisfies the initial conditions

$$X(0) = -1,$$
$$Y(0) = 0. \tag{9.72}$$

Step 1. Taking the Laplace transform of both sides of each differential equation of system (9.71), we have

$$\mathscr{L}\left\{\frac{dX}{dt}\right\} - 6\mathscr{L}\{X(t)\} + 3\mathscr{L}\{Y(t)\} = \mathscr{L}\{8e^t\},$$
$$\mathscr{L}\left\{\frac{dY}{dt}\right\} - 2\mathscr{L}\{X(t)\} - \mathscr{L}\{Y(t)\} = \mathscr{L}\{4e^t\}. \tag{9.73}$$

Denote $\mathscr{L}\{X(t)\}$ by $x(s)$ and $\mathscr{L}\{Y(t)\}$ by $y(s)$. Then applying Theorem 9.3 and the initial conditions (9.72), we have

$$\mathscr{L}\left\{\frac{dX}{dt}\right\} = sx(s) - X(0) = sx(s) + 1,$$
$$\mathscr{L}\left\{\frac{dY}{dt}\right\} = sy(s) - Y(0) = sy(s). \tag{9.74}$$

Using Table 9.1, number 2, we find

$$\mathscr{L}\{8e^t\} = \frac{8}{s-1} \quad \text{and} \quad \mathscr{L}\{4e^t\} = \frac{4}{s-1}. \tag{9.75}$$

Thus, from (9.74) and (9.75), we see that Equations (9.73) become

$$sx(s) + 1 - 6x(s) + 3y(s) = \frac{8}{s-1},$$
$$sy(s) - 2x(s) - y(s) = \frac{4}{s-1},$$

which simplify to the form

$$(s - 6)x(s) + 3y(s) = \frac{8}{s-1} - 1,$$

$$-2x(s) + (s - 1)y(s) = \frac{4}{s-1},$$

or

$$\begin{aligned}(s - 6)x(s) + 3y(s) &= \frac{-s+9}{s-1}, \\ -2x(s) + (s - 1)y(s) &= \frac{4}{s-1}.\end{aligned} \tag{9.76}$$

Step 2. We solve the linear algebraic system of the two equations (9.76) in the two "unknowns" $x(s)$ and $y(s)$. We have

$$(s - 1)(s - 6)x(s) + 3(s - 1)y(s) = -s + 9,$$

$$-6x(s) + 3(s - 1)y(s) = \frac{12}{s-1}.$$

Subtracting we obtain

$$(s^2 - 7s + 12)x(s) = -s + 9 - \frac{12}{s-1},$$

from which we find

$$x(s) = \frac{-s^2 + 10s - 21}{(s-1)(s-3)(s-4)} = \frac{-s+7}{(s-1)(s-4)}.$$

In like manner, we find

$$y(s) = \frac{2s - 6}{(s-1)(s-3)(s-4)} = \frac{2}{(s-1)(s-4)}.$$

Step 3. We must now determine

$$X(t) = \mathcal{L}^{-1}\{x(s)\} = \mathcal{L}^{-1}\left\{\frac{-s+7}{(s-1)(s-4)}\right\}$$

and

$$Y(t) = \mathcal{L}^{-1}\{y(s)\} = \mathcal{L}^{-1}\left\{\frac{2}{(s-1)(s-4)}\right\}.$$

We first find $X(s)$. We use partial fractions and write

$$\frac{-s+7}{(s-1)(s-4)} = \frac{A}{s-1} + \frac{B}{s-4}.$$

From this we find

$$A = -2 \quad \text{and} \quad B = 1.$$

Thus

$$X(t) = -2\mathcal{L}^{-1}\left\{\frac{1}{s-1}\right\} + \mathcal{L}^{-1}\left\{\frac{1}{s-4}\right\},$$

and using Table 9.1, number 2, we obtain

$$X(t) = -2e^t + e^{4t}. \tag{9.77}$$

In like manner, we find $Y(s)$. Doing so, we obtain

$$Y(t) = -\tfrac{2}{3}e^t + \tfrac{2}{3}e^{4t}. \tag{9.78}$$

The pair defined by (9.77) and (9.78) constitute the solution of the given system (9.71) which satisfies the given initial conditions (9.72).

Exercises

In each of the following exercises, use the Laplace transform to find the solution of the given linear system which satisfies the given initial conditions.

1. $\dfrac{dX}{dt} + Y = 3e^{2t},$
   $\dfrac{dY}{dt} + X = 0,$
   $X(0) = 2, \quad Y(0) = 0.$

2. $\dfrac{dX}{dt} - 2Y = 0,$
   $\dfrac{dY}{dt} + X - 3Y = 2,$
   $X(0) = 3, \quad Y(0) = 0.$

3. $\dfrac{dX}{dt} + 2X - 4Y = 0,$
   $\dfrac{dY}{dt} - 2X = t,$
   $X(0) = 0, \quad Y(0) = 3,$

4. $\dfrac{dX}{dt} + X - 8Y = e^t,$
   $\dfrac{dY}{dt} - 2X + Y = 1,$
   $X(0) = 1, \quad Y(0) = 0.$

5. $\dfrac{dX}{dt} + X + Y = e^{-3t},$
   $\dfrac{dY}{dt} + X - 4Y = 1,$
   $X(0) = 1, \quad Y(0) = 2.$

6. $\dfrac{dX}{dt} - 4X + 2Y = 2t,$
   $\dfrac{dY}{dt} - 8X + 4Y = 1,$
   $X(0) = 3, \quad Y(0) = 5.$

7. $2\dfrac{dX}{dt} + \dfrac{dY}{dt} - X - Y = e^{-t},$
   $\dfrac{dX}{dt} + \dfrac{dY}{dt} + 2X + Y = e^t,$
   $X(0) = 2, \quad Y(0) = 1.$

8. $2\dfrac{dX}{dt} + \dfrac{dY}{dt} + X + 5Y = 4t,$
   $\dfrac{dX}{dt} + \dfrac{dY}{dt} + 2X + 2Y = 2,$
   $X(0) = 3, \quad Y(0) = -4.$

9. $\dfrac{d^2X}{dt^2} - 3\dfrac{dX}{dt} + \dfrac{dY}{dt} + 2X - Y = 0,$
   $\dfrac{dX}{dt} + \dfrac{dY}{dt} - 2X + Y = 0,$
   $X(0) = 0, \quad Y(0) = -1, \quad X'(0) = 0.$

10. $\dfrac{d^2X}{dt^2} - \dfrac{dY}{dt} = t + 1,$
    $\dfrac{dX}{dt} + \dfrac{dY}{dt} - 3X + Y = 2t - 1,$
    $X(0) = 0, \quad Y(0) = -\tfrac{11}{9}, \quad X'(0) = 0.$

Suggested Reading

AGNEW, R., *Differential Equations*, 2nd ed. (McGraw-Hill, New York, 1960).
BOYCE, W., and R. DIPRIMA, *Elementary Differential Equations*, 2nd ed. (Wiley, New York, 1969).
BRAUER, F., and J. NOHEL, *Ordinary Differential Equations: A First Course* (Benjamin, New York, 1967).
HOLL, D., C. MAPLE, and B. VINOGRADE, *Introduction to the Laplace Transform* (Appleton-Century-Crofts, New York, 1959).
KREIDER, D., R. KULLER, and D. OSTBERG, *Elementary Differential Equations* (Addison-Wesley, Reading, Mass., 1968).
RAINVILLE, E., and P. BEDIENT, *Elementary Differential Equations*, 4th ed. (Macmillan, New York, 1969).
RITGER, P., and N. ROSE, *Differential Equations with Applications* (McGraw-Hill, New York, 1968).
WIDDER, D. V., *Advanced Calculus*, 2nd ed. (Prentice-Hall, Englewood Cliffs, N.J., 1961).

Appendix

TABLE 1. Natural logarithms

x	$\ln x$	x	$\ln x$	x	$\ln x$	x	$\ln x$
		3.0	1.099	6.0	1.792	9.0	2.197
0.1	−2.303	3.1	1.131	6.1	1.808	9.1	2.208
0.2	−1.609	3.2	1.163	6.2	1.825	9.2	2.219
0.3	−1.204	3.3	1.194	6.3	1.841	9.3	2.230
0.4	−0.916	3.4	1.224	6.4	1.856	9.4	2.241
0.5	−0.693	3.5	1.253	6.5	1.872	9.5	2.251
0.6	−0.511	3.6	1.281	6.6	1.887	9.6	2.262
0.7	−0.357	3.7	1.308	6.7	1.902	9.7	2.272
0.8	−0.223	3.8	1.335	6.8	1.917	9.8	2.282
0.9	−0.105	3.9	1.361	6.9	1.932	9.9	2.293
1.0	0.000	4.0	1.386	7.0	1.946	10	2.303
1.1	0.095	4.1	1.411	7.1	1.960	20	2.996
1.2	0.182	4.2	1.435	7.2	1.974	30	3.401
1.3	0.262	4.3	1.459	7.3	1.988	40	3.689
1.4	0.336	4.4	1.482	7.4	2.001	50	3.912
1.5	0.405	4.5	1.504	7.5	2.015	60	4.094
1.6	0.470	4.6	1.526	7.6	2.028	70	4.248
1.7	0.531	4.7	1.548	7.7	2.041	80	4.382
1.8	0.588	4.8	1.569	7.8	2.054	90	4.500
1.9	0.642	4.9	1.589	7.9	2.067	100	4.605
2.0	0.693	5.0	1.609	8.0	2.079		
2.1	0.742	5.1	1.629	8.1	2.092		
2.2	0.788	5.2	1.649	8.2	2.105		
2.3	0.833	5.3	1.668	8.3	2.116		
2.4	0.875	5.4	1.686	8.4	2.128		
2.5	0.916	5.5	1.705	8.5	2.140		
2.6	0.956	5.6	1.723	8.6	2.152		
2.7	0.993	5.7	1.740	8.7	2.163		
2.8	1.030	5.8	1.758	8.8	2.175		
2.9	1.065	5.9	1.775	8.9	2.186		

TABLE 2. e^x and e^{-x}

x	e^x	e^{-x}	x	e^x	e^{-x}
0.0	1.000	1.0000	3.0	20.086	0.0498
0.1	1.105	0.9048	3.1	22.198	0.0450
0.2	1.221	0.8187	3.2	24.533	0.0408
0.3	1.350	0.7408	3.3	27.113	0.0369
0.4	1.492	0.6703	3.4	29.964	0.0334
0.5	1.649	0.6065	3.5	33.115	0.0302
0.6	1.822	0.5488	3.6	36.598	0.0273
0.7	2.014	0.4966	3.7	40.447	0.0247
0.8	2.226	0.4493	3.8	44.701	0.0224
0.9	2.460	0.4066	3.9	49.402	0.0202
1.0	2.718	0.3679	4.0	54.598	0.0183
1.1	3.004	0.3329	4.1	60.340	0.0166
1.2	3.320	0.3012	4.2	66.686	0.0150
1.3	3.669	0.2725	4.3	73.700	0.0136
1.4	4.055	0.2466	4.4	81.451	0.0123
1.5	4.482	0.2231	4.5	90.017	0.0111
1.6	4.953	0.2019	4.6	99.484	0.0101
1.7	5.474	0.1827	4.7	109.947	0.0091
1.8	6.050	0.1653	4.8	121.510	0.0082
1.9	6.686	0.1496	4.9	134.290	0.0074
2.0	7.389	0.1353	5	148.4	0.00674
2.1	8.166	0.1225	6	403.4	0.00248
2.2	9.025	0.1108	7	1096.6	0.00091
2.3	9.974	0.1003	8	2981.0	0.00034
2.4	11.023	0.0907	9	8103.1	0.00012
2.5	12.182	0.0821	10	22026.5	0.00005
2.6	13.464	0.0743			
2.7	14.880	0.0672			
2.8	16.445	0.0608			
2.9	18.174	0.0550			

Answers to Odd-Numbered Exercises

Section 1.1

1. Ordinary; first; linear.
3. Partial; second; linear.
5. Ordinary; fourth; nonlinear.
7. Ordinary; second; linear.
9. Ordinary; sixth; nonlinear.

Section 1.2

5. (a) 2, 3, −2. (b) −1, −2, 4.

Section 1.3

1. No; one of the supplementary conditions is not satisfied.
3. (a) $y = 3e^{4x} + 2e^{-3x}$. (b) $y = -2e^{-3x}$.
5. $y = 2x - 3x^2 + x^3$.
7. Yes.

Section 2.1

1. $3x^2 + 4xy + y^2 = c$.
3. $x + x^2y + 2y^2 = c$.
5. $3x^2y + 2y^2x - 5x - 6y = c$.
7. $y \tan x + \sec x + y^2 = c$.
9. $s^2 - s = ct$.
11. $x^2y - 3x + 2y^2 = 7$.
13. $y^2 \cos x - y \sin^2 x = 9$.
15. $-3y + 2x + y^2 = 2xy$.
17. (a) $A = \frac{3}{2}$; $2x^3 + 9x^2y + 12y^2 = c$.
    (b) $A = -2$; $2x^2 - 2y^2 - x = cxy^2$.
19. (a) $x^2y + c$. (b) $2x^{-1}y^{-3} - \frac{3}{2}x^2y^{-4} + c$.
21. (b) x^2. (c) $x^4 + x^3y^2 = c$.

Section 2.2

1. $(x^2 + 1)^2 y = c.$
3. $r^2 + s = c(1 - r^2 s).$
5. $r \sin^2 \theta = c.$
7. $(x + 1)^6(y^2 + 1) = c(x + 2)^4.$
9. $y^2 + xy = cx^3.$
11. $\sin \dfrac{y}{x} = cx.$
13. $(x^2 + y^2)^{3/2} = x^3 \ln cx^3.$
15. $x + 4 = (y + 2)^2 e^{-(y+1)}.$
17. $16(x + 3)(x + 2)^2 = 9(y^2 + 4)^2.$
19. $(2x + y)^2 = 12(y - x).$
23. (a) $x^3 - y^3 + 6xy^2 = c.$
    (b) $2x^3 + 3x^2 y + 3xy^2 = c.$

Section 2.3

1. $y = x^3 + cx^{-3}.$
3. $y = (x^3 + c)e^{-3x}.$
5. $x = 1 + ce^{1/t}.$
7. $3(x^2 + x)y = x^3 - 3x + c.$
9. $y = x^{-1}(1 + ce^{-x}).$
11. $r = (\theta + c) \cos \theta.$
13. $2(1 + \sin x)y = x + \sin x \cos x + c.$
15. $y = (1 + cx^{-1})^{-1}.$
17. $y = (2 + ce^{-8x^2})^{1/4}.$
19. $y = x^4 - 2x^2.$
21. $y = (e^x + 1)^2.$
23. $2r = \sin 2\theta + \sqrt{2} \cos \theta.$
25. $x^2 y^4 = x^4 + 15.$
27. $y = \begin{cases} 2(1 - e^{-x}), & 0 \le x < 1, \\ 2(e - 1)e^{-x}, & x \ge 1. \end{cases}$
29. $y = \begin{cases} e^{-x}(x + 1), & 0 \le x < 2, \\ 2e^{-x} + e^{-2}, & x \ge 2. \end{cases}$
31. (a) $y = \dfrac{ke^{-\lambda x}}{b - a\lambda} + ce^{-bx/a}$ if $\lambda \ne b/a;$
    $y = \dfrac{kxe^{-bx/a}}{a} + ce^{-bx/a}$ if $\lambda = b/a.$
35. (b) $y = \sin x - \cos x + \sin 2x - 2 \cos 2x + ce^{-x}.$
37. (a) $2x \sin y - x^2 = c.$
    (b) $y^2 + 2y + ce^{-x^2} - 1 = 0.$
39. $y = (x - 2 + ce^{-x})^{-1} + 1.$
41. $y = (2 + ce^{-2x^2})^{-1} + x.$

Section 2.4

1. $4x^5y + 4x^4y^2 + x^4 = c$.
3. $xy^2e^x + ye^x = c$.
5. $x^3y^4(xy + 1) = c$.
7. $5x^2 + 4xy + y^2 + 2x + 2y = c$.
9. $\ln [c(x^2 + y^2 - 2x + 2y + 2)] + 4 \arctan \left(\frac{y+1}{x-1}\right) = 0$.
11. $12x^2 + 16xy + 4y^2 + 4x + 8y - 89 = 0$.
13. $x + 2y - \ln |2x + 3y - 1| - 2 = 0$.
21. (a) $y = cx + c^2$. (b) $y = -x^2/4$.

Section 3.1

1. $x^2 + 3y^2 = k^2$.
3. $x^2 + y^2 - \ln y^2 = k$.
5. $x = y - 1 + ke^{-y}$.
7. $x^2y + y^3 = k$.
9. $y^2 = 2x - 1 + ke^{-2x}$.
11. $x^2 + y^2 = ky$.
13. $n = 3$.
15. $\ln (x^2 + y^2) + 2 \arctan (y/x) = c$.
17. $\ln |3x^2 + 3xy + 4y^2| - (2/\sqrt{39}) \arctan [(3x + 8y)/\sqrt{39}\, x] = c$.

Section 3.2

1. (a) $v = 8(1 - e^{-4t})$, $x = 2(4t - 1 + e^{-4t})$.
   (b) 8 ft/sec; 38 feet.
3. Rises 55.08 feet, or 61.08 feet above ground.
5. (a) 10.36 ft/sec. (b) 13.19 ft/sec.
7. $$v = \frac{(60{,}000\sqrt{5} + 12{,}500) + (60{,}000\sqrt{5} - 12{,}500)e^{0.256\sqrt{5}\,t}}{(1200 + 50\sqrt{5}) + (50\sqrt{5} - 1200)e^{0.256\sqrt{5}\,t}}.$$
9. $$v = \frac{\sqrt{5}\,(1 - e^{-8\sqrt{5}\,t})}{1 + e^{-8\sqrt{5}\,t}}.$$
11. 16.18 feet.
13. (a) 10.96 ft/sec. (b) 20.46 ft/sec.
15. 4.03 ft/sec.

Section 3.3

1. (a) 59.05%. (b) 2631 years.
3. (a) $50\sqrt{2}$ grams. (b) 12.5 grams. (c) 10.29 hours.
5. 40,833.
7. (a) \$1822.10. (b) approx 11.52 years.

9. (a) 112.31 lb. (b) 17.33 minutes.
11. (a) 318.53 lb. (b) 2.74 minutes.
13. (a) 0.072%. (b) 1.39 minutes.
15. (a) 63.33 °F. (b) 13.55 minutes.
17. 7.14 grams.

Section 4.1

7. (c) $y = 3e^{2x} - e^{3x}$.
9. (c) $y = 4x - x^2$.
15. $y = c_1e^{2x} + c_2(x + 1)$.
19. $y_p = -\frac{4}{3} - 2x + 3e^x$.

Section 4.2

1. $y = c_1e^{2x} + c_2e^{3x}$.
3. $y = c_1e^{x/2} + c_2e^{5x/2}$.
5. $y = c_1e^x + c_2e^{-x} + c_3e^{3x}$.
7. $y = (c_1 + c_2x)e^{4x}$.
9. $y = e^{2x}(c_1 \sin 3x + c_2 \cos 3x)$.
11. $y = c_1 \sin 3x + c_2 \cos 3x$.
13. $y = (c_1 + c_2x)e^x + c_3e^{3x}$.
15. $y = (c_1 + c_2x + c_3x^2)e^{2x}$.
17. $y = c_1e^x + c_2 \sin x + c_3 \cos x$.
19. $y = c_1 + c_2x + c_3x^2 + (c_4 + c_5x)e^x$.
21. $y = c_1e^{2x} + c_2e^{3x} + e^{-x}(c_3 \sin x + c_4 \cos x)$.
23. $$y = e^{\sqrt{2}\,x/2}\left(c_1 \sin \frac{\sqrt{2}\,x}{2} + c_2 \cos \frac{\sqrt{2}\,x}{2}\right) + e^{-\sqrt{2}\,x/2}\left(c_3 \sin \frac{\sqrt{2}\,x}{2} + c_4 \cos \frac{\sqrt{2}\,x}{2}\right).$$
25. $y = 2e^{4x} + e^{-3x}$.
27. $y = (13x + 3)e^{-2x}$.
29. $y = e^{2x} \sin 5x$.
31. $y = 3e^{-(1/3)x}[\sin \frac{2}{3}x + 2 \cos \frac{2}{3}x]$.
33. $y = e^x - 2e^{2x} + e^{3x}$.
35. $y = \frac{32}{9}e^{-x} - \frac{23}{9}e^{2x} + \frac{2}{3}xe^{2x}$.
39. $y = c_1 \sin x + c_2 \cos x + e^{-x}(c_3 \sin 2x + c_4 \cos 2x)$.

Section 4.3

1. $y = c_1e^x + c_2e^{2x} + 2x^2 + 6x + 7$.
3. $y = e^{-x}(c_1 \sin 2x + c_2 \cos 2x) + 2 \sin 2x - \cos 2x$.
5. $y = e^{-x}(c_1 \sin \sqrt{3}\,x + c_2 \cos \sqrt{3}\,x) + \dfrac{\sin 4x}{26} - \dfrac{3 \cos 4x}{52}$.

7. $y = c_1e^x + e^{-x}(c_2 \sin 2x + c_3 \cos 2x) - \dfrac{9 \sin 2x}{17} + \dfrac{2 \cos 2x}{17} - 2x^2$
$\dfrac{9x}{5} - \dfrac{82}{25}$.

9. $y = c_1e^{2x} + c_2e^{-3x} + 2xe^{2x} - 3e^{3x} + x + 2$.

11. $y = c_1e^{-x} + (c_2 + c_3x)e^{2x} + 2e^x - 2xe^{-x}$.

13. $y = c_1 + c_2 \sin x + c_3 \cos x + \dfrac{2x^3}{3} - 4x - 2x \sin x$.

15. $y = c_1e^x + c_2e^{2x} + c_3e^{3x} + \dfrac{x^2e^x}{4} + \dfrac{3xe^x}{4} + 4xe^{2x} + e^{4x}$.

17. $y = c_1 \sin x + c_2 \cos x - \frac{1}{4}x^2 \cos x + \frac{1}{4}x \sin x$.

19. $y = c_1 + c_2x + c_3e^x + c_4e^{-3x} - \dfrac{x^4}{2} - \dfrac{4x^3}{3} - \dfrac{19}{6}x^2 + 2x^2e^x - 9xe^x + \dfrac{e^{3x}}{27}$.

21. $y = -6e^x + 2e^{3x} + 3x^2 + 8x + 10$.

23. $y = e^{-2x} + 4e^{3x} - 2e^{2x} - xe^{3x}$.

25. $y = \frac{1}{5}[e^{2x}(\sin 3x + 2 \cos 3x) + \sin 3x + 3 \cos 3x]$.

27. $y = \dfrac{7e^{-x}}{20} - \dfrac{31e^{2x}}{40} + \dfrac{3xe^x}{4} + \dfrac{5e^x}{4} - \dfrac{\sin x}{10}$.

29. $y_p = Ax^3 + Bx^2 + Cx + D + Ee^{-2x}$.

31. $y_p = Ae^{-2x} + Bxe^{-2x} \sin x + Cxe^{-2x} \cos x$.

33. $y_p = Ax^2e^{-3x} \sin 2x + Bx^2e^{-3x} \cos 2x + Cxe^{-3x} \sin 2x$
$+ Dxe^{-3x} \cos 2x + Ex^2e^{-2x} \sin 3x + Fx^2e^{-2x} \cos 3x$
$+ Gxe^{-2x} \sin 3x + Hxe^{-2x} \cos 3x + Ie^{-2x} \sin 3x + Je^{-2x} \cos 3x$.

35. $y_p = Ax^4e^{2x} + Bx^3e^{2x} + Cx^2e^{3x} + Dxe^{3x} + Ee^{3x}$.

37. $y_p = Ax^3 \sin 2x + Bx^3 \cos 2x + Cx^2 \sin 2x + Dx^2 \cos 2x + Ex \sin 2x$
$+ Fx \cos 2x + Gx^5e^{2x} + Hx^4e^{2x} + Ix^3e^{2x} + Jx^2e^{2x} + Kxe^{2x}$.

39. $y_p = Ax^4 \sin x + Bx^4 \cos x + Cx^3 \sin x + Dx^3 \cos x + Ex^2 \sin x$
$+ Fx^2 \cos x$.

41. $y_p = A + Bx \sin 2x + Cx \cos 2x + Dxe^x + Exe^{-x}$
or $y_p = Ax \sin^2 x + Bx \cos^2 x + Cx \sin x \cos x + Dx \sinh x + Ex \cosh x$.

Section 4.4

1. $y = c_1 \sin x + c_2 \cos x + (\sin x)[\ln |\csc x - \cot x|]$.

3. $y = c_1 \sin x + c_2 \cos x + (\cos x)[\ln |\cos x|] + x \sin x$.

5. $y = c_1 \sin 2x + c_2 \cos 2x + \dfrac{\sin 2x}{4}[\ln |\sec 2x + \tan 2x|] - \dfrac{1}{4}$.

7. $y = e^{-2x}(c_1 \sin x + c_2 \cos x) + xe^{-2x} \sin x + (\ln |\cos x|)e^{-2x} \cos x$.

9. $y = c_1e^x + c_2xe^x - \dfrac{5x^3e^x}{36} + \dfrac{x^3e^x \ln x}{6}$.

11. $y = c_1e^x + c_2xe^x + \dfrac{e^x \sin^{-1} x}{4} + \dfrac{x^2e^x \sin^{-1} x}{2} + \dfrac{3xe^x\sqrt{1 - x^2}}{4}$.

13. $y = c_1e^{-x} + c_2e^{-2x} + (e^{-x} + e^{-2x})[\ln (1 + e^x)]$.

15. $y = c_1 \sin x + c_2 \cos x + (\sin x)[\ln(1 + \sin x)] - x \cos x - \dfrac{\cos^2 x}{1 + \sin x}.$

17. $y = c_1 x + c_2 x e^x - x^2 - x.$

19. $y = c_1 x + c_2(x + 1)^{-1} + x^2 - \dfrac{(2x^3 + 3x^2)(x + 1)^{-1}}{6}.$

Section 4.5

1. $y = c_1 x + c_2 x^3.$
3. $y = c_1 |x|^{1/2} + c_2 |x|^{3/2}.$
5. $y = c_1 \sin(\ln x^2) + c_2 \cos(\ln x^2).$
7. $y = c_1 x + c_2 x^2 + c_3 x^3.$
9. $y = (c_1 + c_2 \ln|x|)x^3 + c_3 x^{-2}.$
11. $y = c_1 x^2 + c_2 x^4 - 2x^3.$
13. $y = c_1 \sin(\ln x^2) + c_2 \cos(\ln x^2) + \dfrac{x \ln x^2}{5} - \dfrac{4x}{25}.$
15. $y = (c_1 + c_2 \ln|x|)x + c_3 x^2 + \dfrac{x^3}{4}.$
17. $y = -2x^2 + x^3.$
19. $y = \dfrac{1}{6}\left(\dfrac{x^{-2}}{2} + \dfrac{x^3}{3} - \ln x + \dfrac{1}{6}\right).$
21. $y = c_1(2x - 3) + c_2(2x - 3)^3.$

Section 4.6

1. (a) $f_1(x) = 2e^x - e^{2x}$, $f_2(x) = -e^x + e^{2x}$. (b) $5f_1(x) + 7f_2(x)$.

Section 5.2

1. $x = \dfrac{\cos 16t}{6}$; $\frac{1}{6}$ (ft), $\pi/8$ (sec), $8/\pi$ oscillations/sec.

3. (a) $x = \dfrac{\sin 8t}{4}$. (b) $\frac{1}{4}$ (ft), $\pi/4$ (sec), $4/\pi$ oscillations/sec.

   (c) $t = \dfrac{\pi}{48} + \dfrac{n\pi}{4}$ $(n = 0, 1, 2, \ldots)$. (d) $t = \dfrac{5\pi}{48} + \dfrac{n\pi}{4}$ $(n = 0, 1, 2, \ldots)$.

5. (a) $x = -\dfrac{\sin 10t}{5} + \dfrac{\cos 10t}{3}$. (b) $\dfrac{\sqrt{34}}{15}$ (ft); $\pi/5$ (sec); $5/\pi$ oscillations/sec.

   (c) 0.103 (sec); -3.888 ft/sec.

7. 18 lb.

Section 5.3

1. (a) $\dfrac{1}{4}\dfrac{d^2x}{dt^2} + 2\dfrac{dx}{dt} + 20x = 0$, $x(0) = \frac{1}{2}$, $x'(0) = 0$.

   (b) $x = e^{-4t}\left(\dfrac{\sin 8t}{4} + \dfrac{\cos 8t}{2}\right).$

(c) $x = \dfrac{\sqrt{5}}{4} e^{-4t} \cos(8t - \phi)$, where $\phi \approx 0.46$.

(d) $\pi/4$ (sec).

3. $x = (6t + \frac{3}{4})e^{-8t}$.

5. (a) $x = e^{-8t}\left(\dfrac{\sin 16t}{3} + \dfrac{2\cos 16t}{3}\right)$; $x = \dfrac{\sqrt{5}}{3} e^{-8t} \cos(16t - \phi)$, where $\phi \approx 0.46$.

(b) $\pi/8$ (sec); π. (c) 0.127.

7. (a) $x = (\frac{1}{4} + 2t)e^{-8t}$.

(b) $x = e^{-4t}\left(\dfrac{\sqrt{3}}{12} \sin 4\sqrt{3}\, t + \dfrac{1}{4} \cos 4\sqrt{3}\, t\right)$.

(c) $x = \left(\dfrac{3 + 2\sqrt{3}}{24}\right) e^{(-16+8\sqrt{3})t} + \left(\dfrac{3 - 2\sqrt{3}}{24}\right) e^{(-16-8\sqrt{3})t}$.

9. (a) 64. (b) $8\sqrt{3}$.

Section 5.4

1. $x = \dfrac{\cos 12t - \cos 20t}{4}$.

3. $x = -2te^{-8t} + \dfrac{\sin 8t}{4}$.

5. $x = e^{-8t}\left(\dfrac{\sqrt{2}}{2} \sin 4\sqrt{2}\, t + \cos 4\sqrt{2}\, t\right) + \sin 4t - \cos 4t$.

7. $x = e^{-2t}\left(-\dfrac{3 \sin 4t}{2} - 2\cos 4t\right) + \sin 2t + 2\cos 2t,\ 0 \le t \le \pi$;

$x = (e^{2\pi} - 1)e^{-2t}\left(\dfrac{3 \sin 4t}{2} + 2\cos 4t\right),\ t > \pi$.

9. (a) $x = \cos 7t - \cos 8t$.

Section 5.5

1. (a) $\dfrac{2\sqrt{2}}{\pi}$; $x = \dfrac{e^{-4t}(-\sqrt{3}\sin 4\sqrt{3}\,t - \cos 4\sqrt{3}\,t)}{18} + \dfrac{\sqrt{2}\sin 4\sqrt{2}\,t + \cos 4\sqrt{2}\,t}{18}$.

(b) 8; $x = \dfrac{t \sin 8t}{3}$.

3. (b) $2/\pi$; $3\sqrt{5}/4$. (c) $\sqrt{22}/2\pi$; $15\sqrt{23}/23$.

Section 5.6

1. $i = 4(1 - e^{-50t})$.

3. $q = \dfrac{1 - e^{-500t}}{50}$; $i = 10e^{-500t}$.

5. $i = e^{-80t}(-4.588 \sin 60t + 1.247 \cos 60t) - 1.247 \cos 200t + 1.331 \sin 200t$.

7. $q = e^{-Rt/2L}\left[\dfrac{Q_0\sqrt{c}\,R}{\sqrt{4L - R^2c}}\sin\left(\dfrac{\sqrt{4L - R^2c}}{2\sqrt{c}\,L}\,t\right) + Q_0\cos\left(\dfrac{\sqrt{4L - R^2c}}{2\sqrt{c}\,L}\,t\right)\right]$,

$i = -\dfrac{2Q_0}{\sqrt{4Lc - R^2c^2}}\,e^{-Rt/2L}\sin\left(\dfrac{\sqrt{4L - R^2c}}{2\sqrt{c}\,L}\,t\right)$.

Section 6.1

1. $y = c_0\left[1 + \sum_{n=1}^{\infty}\dfrac{(-1)^n x^{2n}}{[2\cdot 4\cdot 6\cdots(2n)]}\right] + c_1\left[x + \sum_{n=1}^{\infty}\dfrac{(-1)^n x^{2n+1}}{[3\cdot 5\cdot 7\cdots(2n+1)]}\right]$.

3. $y = c_0\left(1 - \dfrac{x^2}{2} - \dfrac{x^4}{24} + \cdots\right) + c_1\left(x - \dfrac{x^3}{3} - \dfrac{x^5}{30} + \cdots\right)$.

5. $y = c_0\left(1 - x^2 - \dfrac{x^3}{2} + \dfrac{x^4}{3} + \dfrac{11x^5}{40} + \cdots\right)$
$+ c_1\left(x - \dfrac{x^3}{2} - \dfrac{x^4}{4} + \dfrac{x^5}{8} + \cdots\right)$.

7. $y = c_0\left(1 - \dfrac{x^3}{6} + \dfrac{3x^5}{40} + \cdots\right) + c_1\left(x - \dfrac{x^3}{6} - \dfrac{x^4}{12} + \dfrac{3x^5}{40} + \cdots\right)$.

9. $y = c_0\left(1 + \dfrac{x^3}{6} + \dfrac{x^6}{18} + \cdots\right) + c_1\left(x + \dfrac{x^4}{6} + \dfrac{17x^7}{252} + \cdots\right)$.

11. $y = 1 + \sum_{n=1}^{\infty}\dfrac{x^{2n}}{[2\cdot 4\cdot 6\cdots(2n)]} = \sum_{n=0}^{\infty}\dfrac{x^{2n}}{2^n n!}$.

13. $y = 2 + 3x - \dfrac{7x^3}{6} - \dfrac{x^4}{2} + \dfrac{21x^5}{40} + \cdots$.

15. $y = c_0\left[1 - \dfrac{(x-1)^2}{2} + \dfrac{(x-1)^3}{2} - \dfrac{5(x-1)^4}{12} + \dfrac{(x-1)^5}{3} + \cdots\right]$
$+ c_1\left[(x-1) - \dfrac{(x-1)^2}{2} + \dfrac{(x-1)^3}{6} - \dfrac{(x-1)^5}{12} + \cdots\right]$.

17. $y = 2 + 4(x-1) - 4(x-1)^2 + \dfrac{4(x-1)^3}{3} - \dfrac{(x-1)^4}{3} + \dfrac{2(x-1)^5}{15} + \cdots$.

Section 6.2

1. $x = 0$ and $x = 3$ are regular singular points.

3. $x = 1$ is a regular singular point; $x = 0$ is an irregular singular point.

5. $y = C_1 x\left(1 - \dfrac{x^2}{14} + \dfrac{x^4}{616} - \cdots\right) + C_2 x^{-1/2}\left(1 - \dfrac{x^2}{2} + \dfrac{x^4}{40} - \cdots\right)$.

7. $y = C_1 x^{4/3}\left(1 - \dfrac{3x^2}{16} + \dfrac{9x^4}{896} - \cdots\right) + C_2 x^{2/3}\left(1 - \dfrac{3x^2}{8} + \dfrac{9x^4}{320} - \cdots\right)$.

9. $y = C_1 x^{1/3}\left(1 - \dfrac{3x^2}{16} + \dfrac{9x^4}{896} - \cdots\right) + C_2 x^{-1/3}\left(1 - \dfrac{3x^2}{8} + \dfrac{9x^4}{320} - \cdots\right)$.

11. $y = C_1\left(1 + x + \dfrac{3x^2}{10} + \cdots\right) + C_2 x^{1/3}\left(1 + \dfrac{7x}{12} + \dfrac{5x^2}{36} + \cdots\right)$.

13. $y = C_1 x^{1/2}\left(1 - \dfrac{x^2}{6} + \dfrac{x^4}{120} - \cdots\right) + C_2 x^{-1/2}\left(1 - \dfrac{x^2}{2} + \dfrac{x^4}{24} - \cdots\right)$.

15. $y = C_1\left[1 + \sum_{n=1}^{\infty} \frac{x^{2n}}{2^n n!}\right] + C_2 x^3\left[1 + \sum_{n=1}^{\infty} \frac{x^{2n}}{[5\cdot 7\cdot 9\cdots(2n+3)]}\right]$
$= C_1\left(1 + \frac{x^2}{2} + \frac{x^4}{8} + \cdots\right) + C_2 x^3\left(1 + \frac{x^2}{5} + \frac{x^4}{35} + \cdots\right).$

17. $y = C_1\left(1 - x + \frac{x^2}{2} - \frac{x^3}{6} + \cdots\right) + C_2 x^{-1}\left(1 + \frac{3x^2}{2} - \frac{x^3}{3} + \cdots\right)$
$= Ce^{-x} + C_2 x^{-1} e^x$, where $C = C_1 - C_2$.

19. $y = C_1 x\left[1 + 2\sum_{n=1}^{\infty} \frac{(-1)^n x^n}{n!\,(n+2)!}\right]$
$+ C_2\left[x^{-1}\left(-\frac{1}{2} - \frac{x}{2} + \frac{29x^2}{144} + \cdots\right) + \frac{1}{4} y_1(x) \ln|x|\right],$
where $y_1(x)$ denotes the solution of which C_1 is the coefficient.

21. $y = C_1 x^4\left(1 - \frac{x^2}{2} + \frac{x^4}{10} - \cdots\right)$
$+ C_2\left[x^{-2}\left(-\frac{1}{6} - \frac{x^2}{6} - \frac{x^4}{6} + \cdots\right) + \frac{2}{9} y_1(x) \ln|x|\right],$
where $y_1(x)$ denotes the solution of which C_1 is the coefficient.

23. $y = C_1\left[1 + \sum_{n=1}^{\infty} \frac{(-1)^n 2^n x^n}{(n!)^2}\right]$
$+ C_2\left[4x - 3x^2 + \frac{22x^3}{7} + \cdots + y_1(x) \ln|x|\right],$
where $y_1(x)$ denotes the solution of which C_1 is the coefficient.

25. $y = C_1 x\left[1 + \sum_{n=1}^{\infty} \frac{(-1)^n x^{2n}}{[2\cdot 4\cdot 6\cdots(2n)]^2}\right]$
$+ C_2\left[\frac{x^3}{4} - \frac{3x^5}{128} + \frac{11x^7}{13824} + \cdots + y_1(x) \ln|x|\right],$
where $y_1(x)$ denotes the solution of which C_1 is the coefficient.

Section 6.3

3. $y = \dfrac{c_1 \sin x + c_2 \cos x}{\sqrt{x}}.$

Section 7.1

1. $x = ce^{-2t},\ y = -\frac{2}{3}ce^{-2t} + \frac{1}{3}e^{4t} - \frac{1}{3}e^t.$

3. $x = ce^{-3t} + \frac{e^t}{4},\ y = -\frac{2ce^{-3t}}{3} + \frac{e^{3t}}{3} - \frac{e^t}{2}.$

5. $x = c_1 \sin t + c_2 \cos t,$
$y = -\left(\frac{3c_1 + c_2}{2}\right)\sin t + \left(\frac{c_1 - 3c_2}{2}\right)\cos t + \frac{e^t}{2} - \frac{e^{-t}}{2}.$

7. $x = c_1 e^{\sqrt{6}\,t} + c_2 e^{-\sqrt{6}\,t} - t + \frac{1}{6},$
$y = \frac{\sqrt{6}\,c_1 e^{\sqrt{6}\,t}}{6} - \frac{\sqrt{6}\,c_2 e^{-\sqrt{6}\,t}}{6} + \frac{t}{6} - \frac{1}{6} - \frac{e^{3t}}{3}.$

9. $x = c_1 e^t - \frac{\sin t}{2}$, $y = -\frac{c_1 e^t}{3} + \frac{\sin t}{2}$.
11. $x = c_1 e^{4t} + c_2 e^{-2t} - t + 1$, $y = -c_1 e^{4t} + c_2 e^{-2t} + t$.
13. $x = c_1 + c_2 e^{-2t} + 2t^2 + t$, $y = (1 - c_1) - 3c_2 e^{-2t} - t^2 - 3t$.
15. $x = c_1 e^t + c_2 e^{-2t} - te^t$;
$y = (\frac{1}{3} - c_1)e^t - \frac{1}{3}c_2 e^{-2t} + te^t$.
17. $x = c_1 e^{3t} + \frac{t}{3} - \frac{2}{9}$, $y = c_2 e^t - \frac{5c_1 e^{3t}}{2} - \frac{t}{3} - \frac{4}{9}$.
19. $x = c_1 + c_2 t + c_3 e^t + c_4 e^{-t} - \frac{t^2}{2}$,

$$y = (c_1 - c_2 + 1) + (c_2 + 1)t + c_4 e^{-t} - \frac{t^2}{2}.$$

21. $x = c_1 + c_2 e^t + c_3 e^{-3t} - \frac{t^2}{6} - \frac{14t}{9}$,

$$y = \left(3c_1 + \frac{17}{9}\right) + c_2 e^t - 3c_3 e^{-3t} - \frac{t^2}{2} - \frac{4t}{3}.$$

23. $\frac{dx_1}{dt} = x_2$,

$$\frac{dx_2}{dt} = -2x_1 + 3x_2 + t^2.$$

25. $\frac{dx_1}{dt} = x_2$,

$$\frac{dx_2}{dt} = x_3,$$

$$\frac{dx_3}{dt} = -2t^3 x_2 - tx_3 + 5t^4.$$

Section 7.2

1. $x_1 = 2\cos t - \cos 2t$, $x_2 = 4\cos t + \cos 2t$.
3. $i_1 = -\frac{10e^{-1000t}}{3} - \frac{5e^{-4000t}}{3} + 5$,

$$i_2 = -\frac{10e^{-1000t}}{3} + \frac{5e^{-4000t}}{6} + \frac{5}{2}.$$

Section 7.3

1. (c) $x = 2e^{5t} - e^{-t}$, $y = e^{5t} + e^{-t}$.

Section 7.4

1. $x = c_1 e^t + c_2 e^{3t}$, $y = 2c_1 e^t + c_2 e^{3t}$.
3. $x = 2c_1 e^{4t} + c_2 e^{-t}$, $y = 3c_1 e^{4t} - c_2 e^{-t}$.

5. $x = c_1e^t + c_2e^{5t}$, $y = -2c_1e^t + 2c_2e^{5t}$.
7. $x = 2c_1e^t + c_2e^{-t}$, $y = c_1e^t + c_2e^{-t}$.
9. $x = c_1e^{4t} + c_2e^{-2t}$, $y = c_1e^{4t} - c_2e^{-2t}$.
11. $x = c_1e^t + c_2te^t$, $y = 2c_1e^t + c_2(2t - 1)e^t$.
13. $x = -2c_1e^{3t} + c_2(2t + 1)e^{3t}$, $y = c_1e^{3t} - c_2te^{3t}$.
15. $x = 2e^t(-c_1 \sin 2t + c_2 \cos 2t)$, $y = e^t(c_1 \cos 2t + c_2 \sin 2t)$.
17. $x = e^t(c_1 \cos 3t + c_2 \sin 3t)$, $y = e^t(c_1 \sin 3t - c_2 \cos 3t)$.
19. $x = 2e^{3t}(c_1 \cos 3t + c_2 \sin 3t)$,
    $y = e^{3t}[c_1(\cos 3t + 3 \sin 3t) + c_2(\sin 3t - 3 \cos 3t)]$.
21. $x = e^{3t}(c_1 \cos 2t + c_2 \sin 2t)$, $y = e^{3t}(c_1 \sin 2t - c_2 \cos 2t)$.
25. $x = 3c_1t^4 + c_2t^{-1}$, $y = 2c_1t^4 - c_2t^{-1}$.

Section 7.5A

1. (b) $\begin{pmatrix} 9 & 0 & 9 \\ 1 & 4 & 2 \\ 1 & -2 & -1 \end{pmatrix}$.

2. (b) $\begin{pmatrix} -4 & 12 & -20 \\ -24 & 8 & 0 \\ 12 & -4 & -8 \end{pmatrix}$.

3. (b) $\begin{pmatrix} 7 \\ 8 \\ -25 \\ -36 \end{pmatrix}$.

4. (b) $\begin{pmatrix} -35 \\ 10 \\ -7 \end{pmatrix}$.

6. (b) $\begin{pmatrix} 3e^{3t} \\ (6t + 11)e^{3t} \\ (3t^2 + 2t)e^{3t} \end{pmatrix}$.

Section 7.5C

1. Characteristic values: -1 and 4;
   Respective corresponding characteristic vectors:
   $$\begin{pmatrix} k \\ -k \end{pmatrix} \quad \text{and} \quad \begin{pmatrix} 2k \\ 3k \end{pmatrix},$$
   where in each vector k is an arbitrary nonzero real number.
3. Characteristic values: 1, 2, and -3;
   Respective corresponding characteristic vectors:
   $$\begin{pmatrix} k \\ k \\ k \end{pmatrix}, \quad \begin{pmatrix} k \\ 2k \\ k \end{pmatrix}, \quad \text{and} \quad \begin{pmatrix} k \\ 7k \\ 11k \end{pmatrix},$$
   where in each vector k is an arbitrary nonzero real number.

5. Characteristic values: 2, 3, and -2;
Respective corresponding characteristic vectors:

$$\begin{pmatrix} k \\ 0 \\ -k \end{pmatrix}, \begin{pmatrix} k \\ -k \\ -k \end{pmatrix}, \text{ and } \begin{pmatrix} k \\ -k \\ 4k \end{pmatrix},$$

where in each vector k is an arbitrary nonzero real number.

Section 7.6

1. yes.
3. no.
5. yes.

Section 7.7

1. $x_1 = c_1e^t + c_2e^{2t} + c_3e^{-3t}$,
$x_2 = c_1e^t + 2c_2e^{2t} + 7c_3e^{-3t}$,
$x_3 = c_1e^t + c_2e^{2t} + 11c_3e^{-3t}$.
3. $x_1 = c_1e^{2t} + c_2e^{3t} + c_3e^{-2t}$,
$x_2 = -c_2e^{3t} - c_3e^{-2t}$,
$x_3 = -c_1e^{2t} - c_2e^{3t} + 4c_3e^{-2t}$.
5. $x_1 = c_1e^{5t} + 2c_2e^{-t}$,
$x_2 = c_1e^{5t} - c_2e^{-t} + c_3e^{-3t}$,
$x_3 = c_1e^{5t} - c_2e^{-t} - c_3e^{-3t}$.
7. $x_1 = -2c_1e^{(2+\sqrt{5})t} + 2c_2e^{(2-\sqrt{5})t}$,
$x_2 = (1 + \sqrt{5})c_1e^{(2+\sqrt{5})t} + (-1 + \sqrt{5})c_2e^{(2-\sqrt{5})t}$,
$x_3 = c_3e^{2t}$.
9. $x_1 = c_1e^{4t} + c_2e^t$,
$x_2 = 2c_1e^{4t} + 3c_2e^t + 3c_3e^t$,
$x_3 = c_1e^{4t} + c_2e^t + c_3e^t$.
11. $x_1 = c_1e^{-t} + 2c_2e^{2t}$,
$x_2 = c_1e^{-t} + 3c_3e^{2t}$,
$x_3 = -c_1e^{-t} - c_2e^{2t} - c_3e^{2t}$.

Section 8.2

1. $y = 1 + x + 2\sum_{n=2}^{\infty} \frac{x^n}{n!} = -x - 1 + 2e^x$.
3. $y = 2 + x + 2x^2 + \frac{4x^3}{3} + \frac{9x^4}{4} + \cdots$.
5. $y = \frac{x^2}{2} + \frac{x^3}{6} + \frac{x^4}{24} + \frac{x^5}{120} + \cdots$.
7. $y = x + x^2 + \frac{x^3}{6} - \frac{x^4}{12} + \cdots$.

9. $y = 1 + 2(x-1) + \dfrac{7(x-1)^2}{2} + \dfrac{14(x-1)^3}{3} + \dfrac{73(x-1)^4}{12} + \cdots.$

11. $y = \pi + \dfrac{(x-1)^2}{2} + \dfrac{(x-1)^5}{40} + \cdots.$

Section 8.3

1. $\phi_1(x) = 1 + \dfrac{x^2}{2}$, $\phi_2(x) = 1 + \dfrac{x^2}{2} + \dfrac{x^4}{8}$, $\phi_3(x) = 1 + \dfrac{x^2}{2} + \dfrac{x^4}{8} + \dfrac{x^6}{48}$.

3. $\phi_1(x) = \dfrac{x^2}{2}$, $\phi_2(x) = \dfrac{x^2}{2} + \dfrac{x^5}{20}$, $\phi_3(x) = \dfrac{x^2}{2} + \dfrac{x^5}{20} + \dfrac{x^8}{160} + \dfrac{x^{11}}{4400}$.

5. $\phi_1(x) = e^x - 1$, $\phi_2(x) = \dfrac{e^{2x}}{2} - e^x + x + \dfrac{1}{2}$,

$$\phi_3(x) = \frac{e^{4x}}{16} - \frac{e^{3x}}{3} + \frac{xe^{2x}}{2} + \frac{e^{2x}}{2} - 2xe^x + 2e^x + \frac{x^3}{3} + \frac{x^2}{2} + \frac{x}{4} - \frac{107}{48}.$$

7. $\phi_1(x) = x^2$, $\phi_2(x) = x^2 + \dfrac{x^7}{7}$, $\phi_3(x) = x^2 + \dfrac{x^7}{7} + \dfrac{x^{12}}{28} + \dfrac{3x^{17}}{833} + \dfrac{x^{22}}{7546}$.

Section 8.4B

1. (a) 0.800, 0.650, 0.540, 0.462. (b) 0.812, 0.670, 0.564, 0.488. (c) 0.823, 0.688, 0.586, 0.512.

Section 8.4C

1. (a) 1.239, 1.564, 1.995. (b) 1.238, 1.562, 1.991. (c) 1.237, 1.561, 1.989.
3. (a) 1.112, 1.255, 1.445. (b) 1.100, 1.222, 1.375.

Section 8.4D

1. (a) 1.2374, 1.5606, 1.9886. (b) 1.2374, 1.5606, 1.9887.
3. 1.1115, 1.2531, 1.4398.

Section 8.4E

1. 2.5447.

Section 9.1

1. (a) $\dfrac{2}{s^3}$. (b) $\dfrac{b}{s^2 - b^2}$.

3. $\mathscr{L}\{\sin^3 at\} = \dfrac{6a^3}{(s^2 + a^2)(s^2 + 9a^2)}$,

$\mathscr{L}\{\sin^2 at \cos at\} = \dfrac{2a^2 s}{(s^2 + a^2)(s^2 + 9a^2)}$.

5. $\dfrac{24}{s^5}$.

7. (a) $-\dfrac{e^{-\pi s/2}}{s^2+1}$. (b) $\dfrac{e^{-2s}(2s+1)}{s^2}$.

9. (a) $\dfrac{2}{s}(2 - e^{-3s})$ for $s > 0$.

(b) $\dfrac{1}{s^2} + \dfrac{e^{-2s}}{s} - \dfrac{e^{-2s}}{s^2}$ for $s > 0$.

(c) $\dfrac{(e^{-s} - e^{-2s})(s+1)}{s^2}$ for $s > 0$.

Section 9.2

1. $\dfrac{2 \sin 3t}{3}$.

3. $\dfrac{5t^3e^{2t}}{6}$.

5. $e^{-2t} \cos \sqrt{3}\, t$.

7. $\dfrac{1}{3} - \dfrac{e^{-t}}{2} + \dfrac{e^{-3t}}{6}$.

9. $\dfrac{4t \sin 2t + 3 \sin 2t - 6t \cos 2t}{16}$.

11. $\frac{5}{24}t^4e^{-2t}$.

Section 9.3

1. $Y = \dfrac{3e^t + e^{3t}}{2}$.

3. $Y = e^{2t}$.

5. $Y = 2e^{2t} - 3e^{-t} - e^{-t} \sin 3t + e^{-t} \cos 3t$.

7. $Y = (3 - 4t)e^t + \sin t - 3 \cos t$.

9. $Y = -2e^t + e^{2t} + 1,\ 0 < t < 4$;
$Y = 2(e^{-4} - 1)e^t + (1 - e^{-8})e^{2t},\ t > 4$.

11. $Y = \frac{2}{3} + e^{-2t} - \frac{2}{3}e^{-3t},\quad 0 < t < 1$,
$Y = \frac{4}{3} + (e^2 + 1)e^{-2t} - \frac{2}{3}(e^3 + 1)e^{-3t},\quad t > 1$.

Section 9.4

1. $X = -\dfrac{e^t}{2} + \dfrac{e^{-t}}{2} + 2e^{2t},\ Y = \dfrac{e^t}{2} + \dfrac{e^{-t}}{2} - e^{2t}$.

3. $X = \frac{13}{6}e^{2t} - \frac{49}{24}e^{-4t} - \frac{1}{2}t - \frac{1}{8}$,
$Y = \frac{13}{6}e^{2t} + \frac{49}{48}e^{-4t} - \frac{1}{4}t - \frac{3}{16}$.

5. $X = 3e^{-2t} - \frac{69}{40}e^{5t} - \frac{7}{8}e^{-3t} + \frac{3}{5}$,
$Y = \frac{1}{2}e^{-2t} + \frac{69}{40}e^{5t} - \frac{1}{8}e^{-3t} - \frac{1}{10}$.

7. $X = 8 \sin t + 2 \cos t$,
$$Y = -13 \sin t + \cos t + \frac{e^t}{2} - \frac{e^{-t}}{2}.$$

9. $X = 2e^t - e^{2t} - 1$,
$Y = e^t - 2$.

Index

B C D E F G H 7 9 8 7 6 5 4